AF698441

2005 IEEE INTERNATIONAL SOI CONFERENCE

PROCEEDINGS

IEEE

October 3 – 6, 2005

Hyatt Regency Waikiki
Honolulu, Hawaii

Sponsored by the IEEE Electron Devices Society

2005 IEEE INTERNATIONAL SOI CONFERENCE

Papers have been printed without editing as received from the authors.

All opinions expressed in the digest are those of the authors and are not binding on The Institute of Electrical and Electronics Engineers, Inc.

Publication of a paper in this digest is in no way intended to preclude publication of a fuller account of the paper elsewhere.

Copies of available volumes of this digest may be obtained from IEEE Service Center, 445 Hoes Lane, Piscataway, NJ 08855-1331, USA. Phone: (201) 981-0060 or (800) 678-IEEE, Telex: 833233, Fax: (201) 981-0027. Please refer to Catalog Numbers.

COPYRIGHT AND REPRINT PERMISSION: Abstracting is permitted with credit to the source. Libraries are permitted to photocopy beyond the limit of U.S. copyright law for private use of patrons those articles in this volume that carry a code at the bottom of the first page, provided the per-copy fee indicated in the code is paid through Copyright Clearance Center, 222 Rosewood Drive, Danvers, MA 01923. For other copying, reprint or republication permission, write to IEEE Copyrights Manager, IEEE Operations Center, 445 Hoes Lane, P.O. Box 1331, Piscataway, NJ 08855-1331. All rights reserved. Copyright ©2005 by the Institute of Electrical and Electronics Engineers, Inc.

IEEE Catalog Number....... 05CH37694

ISBN 0-7803-9212-4

Library of Congress .... N/A

ISSN 1078-621X

2005 IEEE INTERNATIONAL SOI CONFERENCE

OCTOBER, 2005
VOLUME 05CH37694

Hyatt Regency Waikiki
Honolulu, Hawaii

October 3 – 6, 2005

INTRODUCTION AND HISTORY

The IEEE International SOI Conference is a technical forum for the presentation and discussion of research results relating to semiconductor-on-insulator technology. The technical papers deal with a range of topics including materials preparation and characterization, device physics, modeling, circuit fabrication and applications. The Conference is sponsored by the Electron Devices Society of IEEE. Below is a brief history of the Conference. If you have additional information to complete the listing, please contact any committee member.

1975 IEEE SOS Technology Workshop
September 8-10, 1975
Waystation Hotel, South Lake Tahoe, CA
Conference Chairman: Frank Micheletti, Rockwell
Technical Program: Dave Dumin, Allied Chemical
Local Arrangements: Tom Toumbs
Treasurer: Robert Nielsen
42 papers submitted, 150 attendees

1976 IEEE SOS Technology Workshop
September 11-12, 1976
Stanford University Palo Alto, CA
Conference Chairman: Dave Dumin, Allied Chemical
Technical Program: Robert Nielson
Local Arrangements: R. Mueller
Treasurer: C.S. Kim
38 papers sumbmitted,136 attendees

1977 IEEE SOS Technology Workshop
September 28-30, 1977
Lodge at Vail, Vail, Colorado
Conference Chairman: Robert Stewart, RCA
Technical Program: D. Howard Phillips, Rockwell
Local Arrangements: Roger Stewart
Treasurer: R.M. Bergman
Banquet Speaker: Larry Lopp, Hewlett Packard
47 papers submitted, 150 attendees

1978 IEEE SOS Technology Workshop
October 4-6, 1978
Lodge at Vail, Vail, CO
Conference Chairman: D. Howard Phillips, Rockwell
Technical Program: Alfred Ipri, RCA
Local Arrangements: Roger Steward
Treasurer: Robert Nielsen
Banquet Speaker: Bobby Buchanan, Air Force Cambridge
27 paper submitted, 120 attendees

1979 IEEE SOS Technology Workshop
October 4-6, 1979
Carefree Inn, Carefree, Arizona
Conference Chairman: Alfred Ipri, RCA
Technical Program: R.J. Hollingsworth
Local Arrangements: C. Grenato
Treasurer: R.M. Bergman
33 papers submitted, 150 attendees

1980 IEEE SOS Technology Workshop
September 30-October 2, 1980
Stone Bridge Inn, Snowmass, CO
Conference Chairman: Pete Hudson
Technical Program: Steve Ramer
Treasurer: Sidney Marshall
27 papers submitted, 120 attendees

1981 IEEE SOS Technology Workshop
October 6-8, 1981
Sun River Resort, Sunriver, Oregon
Conference Chairman: Ranjeet Pancholy, Hughes
Technical Program: Mike Splinter
Local Arrangements: Quent Cassen
Treasurer: J.A. Mouten
Banquet Speaker: Peter Hudson, ERADCOM
37 papers submitted, 110 attendees

1982 IEEE SOS Technology Workshop*
October 5-7, 1982
Provincetown Inn, Provincetown, MA
Conference Chairman: Achilles G. Kokkas, RCA
Technical Program: Quent Cassen
Local Arrangements: Jim Ford
Treasurer: M. Czarnecki
Banquet Speaker: Frank Micheletti, Rockwell
38 papers submitted, 150 attendees
*First Workshop with SOI papers accepted

1983 IEEE SOS Technology Workshop
October 4-6, 1983
Snow King Resort, Jackson Hole, Wyoming
Conference Chairman: Quent Cassen, Aerospace
Technical Program: Jal K. Hakhur, Rockwell
Local Arrangements: Eliezer D. Richmont, NRL
Treasurer: Marian G. Peebles, Rockwell
Banquet Speaker: D. Howard Phillips, Rockwell
44 papers submitted, 140 attendees

1984 IEEE SOS Technology Workshop
October 2-4, 1984
Hilton Head Island & Beach Resort,
Hilton Head Island, SC
Conference Chairman: Jal K. Hakhu, Rockwell
Technical Program; Jim Ford, CDC
Local Arrangements: George Celler, AT&T Bells Labs
Treasurer: Jo Ann Landers, Rockwell
Banquet Speaker: Bob Conklin, REC Electronics
39 papers submitted, 140 attendees

1985 IEEE SOS/SOI Technology Conference
October 1-3, 1985
Prospector Square Resort, Park City UT
Conference Chairman: Jim Ford, CDC
Technical Program: George Celler, AT&T
Local Arrangements: Anil Gupta, Hughes
Treasurer: Jack Mee, Rockwell
Banquet Speaker: Jim Mendel, Stanford University
42 papers submitted, 140 attendees

1986 IEEE SOS/SOI Technology Conference
October 3-5, 1986
South Seas Plantation, Captiva Island, FL
Conference Chairman: George Celler, AT&T
Technical Program: Anil Gupta, Hughes
Local Arrangements: Marie Burham, Motorola
Treasurer: Brooke Jones, Rockwell
Banquet Speaker: William Gregory, Sandia National Labs
70 papers submitted-160 attendees

1987 IEEE SOS/SOI Technology Conference**
October 6-8, 1987
Tamarron Resort, Durango, CO
Conference Chair: Anil Gupta, Hughes
Technical Program: Marie Burham, Motorola
Local Arrangements: Jack Mee, Rockwell
Treasurer: Gracie Davis, Northrup Electronics
Banquet Speaker: Al Tasch
University of Texas at Austin
89 papers submitted 190 attendees
** The first time abstracts were published

1988 IEEE SOS/SOI Technology Conference
October 3-5, 1988
Sea Palms Resort, St. Simons Island, GA
Conference Chair: Marie Burnham, Motorola
Technical Program: Jean-Pierre Colinge, Hewlett-Packard
Local Arrangements: Gracie Davis, DNA
Treasurer: Michael Duffy, D. Sarnoff Research
Rump and Poster: Dimitris Ioannou, University of Maryland
Banquet Speaker: David Webb, University of North Dakota
101 papers submitted, 210 attendees

1989 IEEE SOI Technology Conference
October 3-5, 1989
High Sierra Hotel, State, S. Lake Tahoe, NV
Conference Chairman: Jean-Pierre Colinge, Hewlett Packard
Technical Program: Gracie Davis, DNA
Local Arrangements: Michael Duffy, D. Sarnoff Research
Treasurer: Jerry Brandewie, Rockwell
Rump and Poster: Dimitris Ioannou, University of Maryland
Banquet Speaker: John Moll, Hewlett-Packard
108 papers submitted, 250 attendees

1990 IEEE/SOI Technology Conference
October 2-4, 1990
Marriott's Casa Marina Resort, Key West, FL
Conference Chairman: Gracie Davis, DNA
Technical Program: Michael Duffy, D. Sarnoff Research
Local Arrangements: Jerry Brandewie, SEMATECH/Rockwell
Treasurer: John Schott, USAF RADC/ESR
Rump and Poster: Wade Krull, Harris Semiconductor
Banquet Speaker: Yuji Okuto, NEC Corp. Japan
86 papers submitted, 225 attendees
Best Paper Award: "Dual-MOSFET structure for suppression of kink in SOI MOSFETs at room and liquid helium temperatures", M.-H. Gao, J.-P. Colinge, L. Aauwers, S.H. Wu, and Claeys

1991 IEEE International SOI Conference
October 1-3, 1991
Hyatt Regency Beaver Creek Resort, Vail CO
General Chair: Michael Duffy, D. Sarnoff Research
Technical Program: Jerry Brandewie, SEMATECH/ Rockwell
Local Arrangements: John Schott, USA Rome Lab
Treasurer: Witek Maszara, Allied-Signal
Rump and Poster: Peter Roitman, NIST
Banquet Speaker: Jeff Miller, JPL
97 papers submitted, 180 attendees
Best Paper Award: "Correlation of the leakage and charge pumping in SIMOX gated diodes", H. Seghir, S. Cristoloveanu, R. Jerisian, J. Qualid, A. Auberton-Herve

1992 IEE International SOI Conference
October 6-8, 1992
Marriott at Sawgrass Resort, Ponte Vedra Beach, FL
General Chair: Jerry Brandewie, SEMATCH/Rockwell
Technical Program: John Schott, USAF Rome Lab.
Local Arrangements: Witek Maszara, Allied-Signal
Treasurer: Jay Schrankler, Honeywell
Rump and Poster: Don Mayer, Aerospace Corp.
Banquet Speaker: Will Stackhouse, JPL
94 papers submitted, 188 attendees
Best Paper Award: "Performance limitations of current deep submicron fully depleted SOI MOSFETs", J.G. Fossum, S. Krishnan, and P.-C. Yeh

1993 IEEE International SOI Conference
October 1-3. 1993
The Autry Resort, Palm Springs, CA
General Chair: John Schott, , USAF Rome Lab
Technical Program: Witek Maszara, Allied-Signal
Local Arrangements: Jay Schrankler, Honeywell
Treasurer: Don Mayer, The Aerospace Corp.
Rump and Poster: Scott Tyson, Mission Research Corp.
Banquet Speaker: R. Ervin Taylor, University of CA
75 papers submitted, 177 attendees
Best Paper Award: " A Unified Model of Threshold Voltage, Subthreshold Slope and Interface Coupling in Thin Film SOI MOSFETs", A.M. Ionescu, S. Cristoloveanu, A. Chovet, A. Hassein-Bey, A. Rusu

1994 IEEE SOI International Conference
October 6-8, 1994
White Elephant Resort, Nantucket Island, MA
General Chair: Witek Maszara, Allied Signal
Technical Program: Jay Schrankler, Honeywell
Local Arrangements: Don Mayer, The Aerospace Corp.
Treasurer: Jerry G. Fossum, University of Florida
Rump and Poster: Lisa Allen, Ibis Technology Corp.
Banquet Speaker: George Hazarigg, Science Foundation
120 Papers submitted, 203 attendees
Best Paper Award: "The Applications of Silicon-on-Insulator (SOI) Technology for the Fabrications of Fully Scanned Active Matrix Flat Panel Displays", A.C. Ipri, G. Dolny, F.L. Hsueh, R.G. Stewart, D. Jose, M. Spitzer, D-P. Vu, M. Batty, R. Khormaei, S. Thayer, T. Keyser, G. Becker, M. Tilton, R. Rhoades

1995 IEEE International SOI Conference***
October 2-5, 1995
Westward Look Resort, Tucson, AZ
General Chair: Jay Schrankler, Honeywell
Technical Program: Don Mayer, The Aerospace Corp.
Local Arrangements: Jerry G. Fossum, University of FL.
Treasurer: Lisa Allen, Ibis Technology Corp.
Rump and Poster: Ted W. Houston, Texas Instruments
Short Course Chair: James E. Chung, MIT
Banquet Speaker: Dr. Young-Soo Kim, Samsung
147 papers submitted, 242 attendees
Best Paper Award: "Comparison of Plasma-Induced Charging Damage in Bulk and SOI MOSFETs"M.J. Sherony, A.H. Chen, K.R. Mistry, D.A. Antoniadis, and B.S. Doyle
*** The first Short Course presented

1996 IEEE SOI International Conference
September 30-October 3, 1996
Sanibel Harbour Resort & Spa, Sanibel Island, FL
Conference Chair: Don Mayer, The Aerospace Corp.
Technical Program: Jerry G. Fossum, University of Florida
Local Arrangements: Lisa Allen, Ibis Technology Corp.
Treasurer: Akira Yoshino, NEC, Japan
Rump and Poster: Stephen Krause, Arizona State University
Short Course Chair: Jean Pierre Colinge
Banquet Speaker: Dr. Leon Alkalai, JPL
Best Paper Award: "Mechanistic Studies of Hydrophilic Wafer Bonding and Si Exfoliation for SOI Fabrication", M.K. Weldon, V. Marsico, Y.J. Chabal, S.B. Christman, E.E. Chaban, D.C. Jacobson, J.B.Sapjeta, A. Pinczuk, B.S. Dennis, A.P. Mills, C.A. Goodwin, D. –M. Hsieh

1997 IEEE SOI International Conference
October 6-9, 1997
Tenaya Lodge At Yosemite, Fish Camp, California
Conference Chair: Jerry Fossum, University of FL
Technical Program: Lisa P. Allen, Ibis Technology Corp.
Treasurer: Ted Houston, Texas Instruments
Rump and Poster: Subramanian S. Iyer, IBM Corp.
Short Course Chair: Jason Woo, UCLA
Banquet Speaker: Dr. Leon Alkalai, JPL
123 papers submitted, 240 attendees
Best Paper Award: "Mechanism of Silicon Exfoliation by Hydrogen Implantation and He, Li and Si Co-implantation", M.K. Weldon, V. Marsico, Y.J. Chabal, M. Collot, Y. Caudano, S.B. Christman, E.E. Chaban, D.C. Jacobson W.L. Brown, J.B. Sapjeta, C.A. Goodwin, D.-M. Hsieh, Bell Laboratories, Lucent Technologies, A. Agarwal, V.C. Venezia, and T.E. Haynes, Oak Ridge National Laboratory, and W.B. Jackson, Xerox Palo Alto Research Center, USA

1998 IEEE International SOI Conference
October 5-8, 1998
Indian River Plantation, Stuart, FL
General Chair: Lisa P. Allen, Ibis Technology Corp.
Technical Program: Ted Houston, Texas Instruments
Local Arrangements: Jason Woo, UCLA
Treasurer/Registration: Harold Hovel, IBM Corporation
Rump and Poster: John McKitterick, Allied-Signal
Short Course Chair: Dimitris Ioannou, Geo. Mason Univ.
Banquet Speaker: Dr. Jean Pierre Colinge
173 papers submitted, 240 attendees

1999 IEEE International SOI Conference
October 4-7, 1999
Doubletree Hotel Sonoma County, Rohnert Park, California
General Chair: Ted Houston, Texas Instruments
Technical Program: Jason Woo, UCLA
Local Arrangements: Harold Hovel, IBM Corporation
Treasurer/Registration: Dimitris Ioannou, Geo. Mason Univ.
Rump and Poster: Brian Doyle, Intel Corp.
Short Course Chair: Mike Liu, Honeywell
Publicity Chair: John Conley, Dynamics Research Corp.
Banquet Speaker: Jean Pierre Colinge, UC Davis
98 papers submitted, 235 attendees
Best Paper Award: "Ultimately Thin SOI MOSFETs: Special Characteristics and Mechanisms
T. Ernst, D. Munteanu, S. Cristoloveanu, T. Ouisse, N. Hefyene, S. Horiguchi, Y. Ono, Y. Takahashi & , K. Murase

2000 IEEE SOI International Conference
October 2-5, 2000
Sheraton Colonial Hotel & Golf Club, Wakefield, MA
General Chair: Jason Woo, UCLA
Technical Program: Harry Hovel, IBM Corporation
Local Arrangements: Dimitris Ioannou, Geo. Mason Univ.
Treasurer/Registration: Mike Liu, Honeywell
Rump and Poster: Mike Mendicino, Motorola
Short Course Chair: Brian Doyle, Intel Corp.
Publicity Chair: John Conley, JPL
103 papers submitted, 258 attendees
Best Paper Award: "A High-Performance Body-Charge-Modulated SOI Sense Amplifier", J.B. Kuang, D.H. Allen, C.T. Chuang

2001 IEEE SOI International Conference
October 1-4, 2001
Sheraton Tamarron Resort, Durango, CO
General Chair: Harry Hovel, IBM Corporation
Technical Program: Dimitris Ioannou, Geo. Mason Univ.
Local Arrangements: Mike Liu, Honeywell
Treasurer/Registration: Mike Mendicino, Motorola
Rump and Poster: John Conley, JPL
Short Course Chair: James Burns, MIT/Lincoln Lab
Publicity: Christophe Tretz, Advanced Micro Devices
111 papers submitted, 138 attendees
Best Paper Award: "Characterization of Fully Depleted SOI Transistors after Removal of the Silicon Substrate", J. Burns, K. Warner, P. Gouker

2002 IEEE SOI International Conference
October 7 - 10, 2002
Williamsburg Lodge, Williamsburg, VA
General Chair: Dimitris Ioannou, George Mason Univ.
Technical Program: Mike Liu, Honeywell
Local Arrangements: Mike Mendicino, Motorola
Treasurer/Registration: James Burns, MIT/Lincoln Lab
Rump and Poster: Olivier Faynot, CEA-LETI
Short Course & Publicity Chair: Christophe Tretz, AMD
123 papers submitted, 197 attendees
Best Paper Award: " Partial SOI/SON formation by He+ Implantation & Annealing" by A. Ogura

2003 IEEE SOI International Conference
September 29 - October 2, 2003
Newport Beach Marriott, Newport Beach, CA
General Chair: Mike Liu, Honeywell
Technical Program: Mike Mendicino, Motorola
Local Arrangements: James Burns, MIT/Lincoln Lab
Treasurer/Registration: Christophe Tretz, IBM
Rump and Poster: Pierre Fazan, EPFL
Short Course Chair Toshiro Hiramoto, University of Tokyo
Publicity Chair: Rajiv Joshi, IBM, Research Division
118 papers submitted, 181 attendees
Best Paper Award: "A New Block Refresh Concept for SOI Floating Body Memories" ,P. Fazan, S. Okhonin, M. Nagoga

2004 IEEE SOI International Conference
October 4 - 7, 2004
Francis Marion Hotel, Charleston, SC
General Chair: Mike Mendicino, Freescale Semiconductor
Technical Program: James Burns, MIT/Lincoln Lab
Local Arrangements: Christophe Tretz, IB M
Treasurer/Registration: Toshiro Hiramoto, University of Tokyo
Rump and Poster: Mario Pelella, AMD
Short Course Chair: Pierre Fazan, EPFL
Publicity Chair: Christophe Tretz, IBM
118 papers submitted, 181 attendees
Best Paper Award: "Experimental Gate Misalignment Analysis on Double Gate SOI MOSFETs" *J. Widiez, F. Daugé, M. Vinet, T. Poiroux, B. Previtali, M. Mouis, S. Deleonibus*

EXECUTIVE COMMITTEE

General Chair
James A. Burns
MIT Lincoln Laboratory
Advanced Silicon Technology
244 Wood St., Group 88, L-302
Lexington, MA 02173
(781)981-2654
jab@ll.mit.edu

Technical Program Chair
Christophe Tretz
IBM
5600 Cottle Rd., MS 49EA/010
San Jose, CA 95193
(408)888-3852
ctretz@us.ibm.com

Local Arrangements Chair
Toshiro Hiramoto
Institute of Industrial Science
University of Tokyo
4-6-1 Komaba, Meguro-ku
Tokyo 153-8505, Japan
Tel/Fax: +81-3-5452-6263
hiramoto@nano.iis.u-tokyo.ac.jp

Treasurer and Registration Chair
Pierre C. Fazan
EPFL
LEG ELB-ECUBLENS CH-1015
Lausanne, Switzerland
41 21 693 4604
Pierre.fazan@epfl.ch

SENIOR COMMITTEE

Rump and Poster Chair
Hector Sanchez
Freescale Semiconductor
7700 W. Parmer Ln. PL30
Austin, TX 78729-8084
(512)996-6244
hector.sanchez@freescale.com

Short Course Chair
Mario M. Pelella
AMD
One AMD Place
Sunnyvale, CA 94088
(408)749-4321
mario.pelella@amd.com

Conference Development
Theodore W. Houston, TI
ted-houston@ti.com

ADVISORY BOARD

Mike Liu, Honeywell, SSEC
Dimitris Ioannou, George Mason University
Michael Mendicino, Freescale Semiconductor

TECHNICAL COMMITTEE

Richard Brown, University of Utah
Samuel Fung, Taiwan Semiconductor Mfg.
Keith Jenkins, IBM
James Kuo, National Taiwan University
Harry Liu, Honeywell SSEC
Carlos Mazure, SOITEC
Kiyoshi Mitani, Shin-Etsu Handotai
Atsushi Ogura, Meiji University
Les Palkuti, DTRA/TDNR
Stephen Parke, Boise State University
Thierry Poiroux, CEA-LETI
Vyshnavi Suntharalingam, MIT Lincoln Lab
Wade Xiong, Texas Instruments
Rene Zingg, Phillips Semiconductors

2005 IEEE INTERNATIONAL SOI CONFERENCE

Hyatt Regency Waikiki, Honolulu, Hawaii ❖ 3 – 6 October, 2005

Dear Colleague,

On behalf of the Conference Committee, I am pleased to welcome you to the 2005 IEEE International Silicon on Insulator (SOI) Conference, the leading forum for the interaction of scientists and engineers interested in all aspects of SOI technology. Providing a comprehensive review of the current state of the technology, the Conference will comprise a Technical Program, the principal activity; a Short Course; and an evening Panel Discussion. In addition, the Conference experience will be enhanced by its location and informal atmosphere as well as the social events planned for attendees and their guests.

This year's Short Course is "SOI: An Enabling Technology for Low-Power Electronics" to be delivered by four specialists covering the following important applications: high-performance mobile microprocessors, low-power communication devices, ultra-low-power sensors and photonics, and 3D-SOI opportunities. The Short Course, organized by Dr. Mario Pelella of AMD, will be presented Monday, October 3rd.

The Technical Program Committee, chaired by Dr. Christopher Tretz of IBM, has selected a truly exceptional set of papers that substantially advance the field. The new information will broaden the perspectives of active specialists as well as provide the necessary background for newcomers to SOI technology. The Plenary Session, which kicks off the Technical Program on Tuesday, October 4th, will feature three invited speakers discussing SOI wafer developments, low-power circuit design, and SOI-based optoelectronic applications.

The topic of the Panel Discussion on Wednesday evening, October 5th, will be "Mixed-Signal Analog/Digital Circuit and System Design Issues in Scaled Technologies beyond 65 nm." Organized by Hector Sanchez of Freescale Semiconductor, the panel will comprise distinguished experts discussing their views on this important topic. We look forward to a lively discussion as the audience will be encouraged to challenge panel members with alternate opinions.

Our social program will include a welcome reception the evening of October 3rd; a banquet the evening of October 4th, with a presentation by Dr. Kenneth Chambers on the Pan_STARRS astronomical project at Mauna Kea Observatories; and a cookout the evening of October 5th. The cookout has been a favorite in the past and certainly will be special in our Hawaii setting. Wednesday afternoon is reserved for rest, relaxation, and enjoying Hawaii.

I would like to acknowledge the hard work and dedication of the executive and technical committees, as well as the conference management team, for putting together this excellent program. Even more important, let me express my appreciation to authors who submitted abstracts, because it is their work that is the core of the conference and that makes it a major contributor to the progress of SOI technology.

Sincerely,

James A. Burns
General Chair

c/o BACM, 578 Washington Blvd., #350, Marina del Rey, CA 90292
TEL: 310/305-7885 FAX: 310/305-1038 EMAIL: bobbi@bacminc.com
www.soiconference.org

PLENARY SESSION — Session 1

DEVICE CHARACTERIZATION — Session 2

CIRCUIT TECHNIQUES AND APPLICATIONS — Session 3

POSTERS — Session 4

NOTE: Poster numbers were chosen to associate the posters by technical content with the oral sessions. For example, P3.1, means the poster, P, is the first poster similar in technical content to the papers presented in session 3; P3.2 is the second poster with similar content to session 3, etc.

SOI MATERIALS TECHNOLOGY AND CHARACTERIZATION Session 5

MULTIPLE GATE DEVICES — Session 6

DEVICE PROCESS — Session 7

RADIATIONS AND OPTICS — Session 8

MEMORY DESIGN TECHNIQUES — Session 9

LATE NEWS Session 10

2005 IEEE INTERNATIONAL SOI CONFERENCE

Session #1

PLENARY SESSION

CHAIR

Christophe Tretz

IBM – San Jose Design Center

PLENARY SESSION Session 1

Status of Device Mobility Enhancement through Strained Silicon Engineering

Carlos Mazure and Ian Cayrefourcq
Soitec S.A., Parc Technologiques des Fontaines, 38190 Bernin, France
Email: carlos.mazure@soitec.fr

Strained silicon engineering has become a key innovation to enhance device on-current. It has allowed the IC industry to keep on the scaling path and assure a performance gain from one technology node to the next. Uniaxial strained silicon has already made its way into 90nm IC manufacturing. Biaxial wafer level strain may be the next step for boosting transistor performance. At the substrate level, strained silicon on insulator is a major innovation that offers higher carrier mobility, combining the advantages of SOI with those of strained silicon. This review focuses on the development of strained silicon, its impact on device properties and its scalability beyond 65nm design rules.

Introduction

Strained silicon engineering of the channel region has emerged as a powerfull tool to enhance electron and hole mobilities, and consequently the transistor on-currents (Ion). It has allowed the IC industry to keep on the scaling path and assure a performance gain from one technology node to the next. Strained silicon (sSi) is also the answer to help contain the IC power consumption issue by reducing the supply voltage V_{DD} and leakage currents for the same level of Ion.

The impact of mechanical stress on electron and hole mobilities have been thouroughly studied for the past 50 years [1, 2]. Since the early 90's uniaxial and biaxial strain-inducing techniques have received a considerable amount of attention from academia and the IC industry.

Both uniaxial and biaxial tensile strain induce an electron mobility enhancement since the physical mechanism is similar along the channel L_G. The biaxial tensile strain lifts the conduction band degeneracy between the in-plane and the perpendicular valleys, and as a consequence, it reduces the intervalley scattering, increases subband splitting and enhances carrier mobility by reducing the in-plane electron effective mass [3]. For holes the situation is more complex. Uniaxial compressive strain induces a valence band warpage along the channel resulting in a redistribution of carriers in the channel plane [4]. The mobility enhancement for biaxial tensile strain results from the splitting of light and heavy hole bands [5].

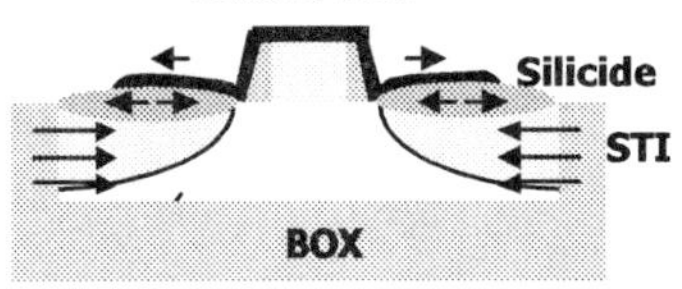

Figure 1: Illustration of process induced stressors.

Uniaxial Strain

Process induced stress has been known for many years as reported in the literature [6-12]. Source/Drain (S/D) and gate silicidation [9, 11], trench isolation (STI) [8, 12], and cap layer deposition [9, 10], all induce a certain level of stress which can be beneficial or detrimental to the transistor performance. Fig. 1 illustrates the stressor effects in a cross-section and fig. 2. shows a transistor layout illustrating the STI compressive stress.

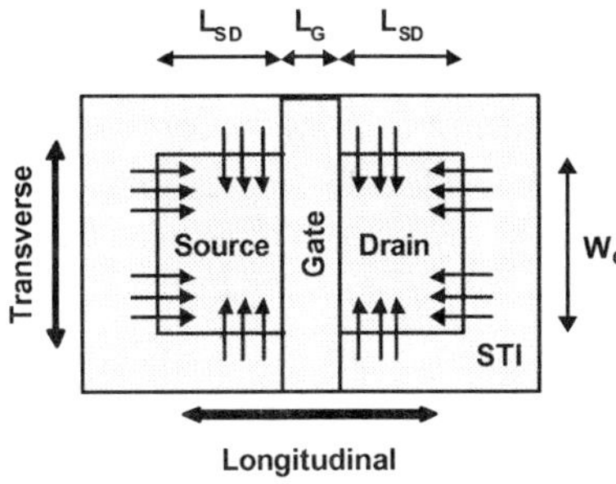

Figure 2: Transistor layout. L_G and W_G are the gate length and width, respectively. L_{SD} is the S/D length. The STI module induces a compressive stress in the channel region.

The transistor layout [7, 8, 11, 12] and gate to S/D misalignment [6] also have a significant

0-7803-9212-4/05/$20.00 ©2005 IEEE

effect on the resulting channel stress. The process induced stress effects are more pronounced for sub-μm geometries. A reduction of L_{SD} leads to a reduction of electron mobility [6, 12]. A reduction of transistor width W_G amplifies the compressive stress induced by STI in the channel, which leads to a hole mobility increase [12]. The challenge is to optimize all process modules, i.e. their uniaxial stressor components, to maximize beneficial effects while minimizing the negative contributions. The rule of thumb is that what enhances the NMOS mobility is detrimental to the PMOS, and vice versa, which translates into a narrow process window for simultaneous optimization of n- and p-channels. Fig. 3 illustrates the STI effects as a function of L_{SD} for different W_G.

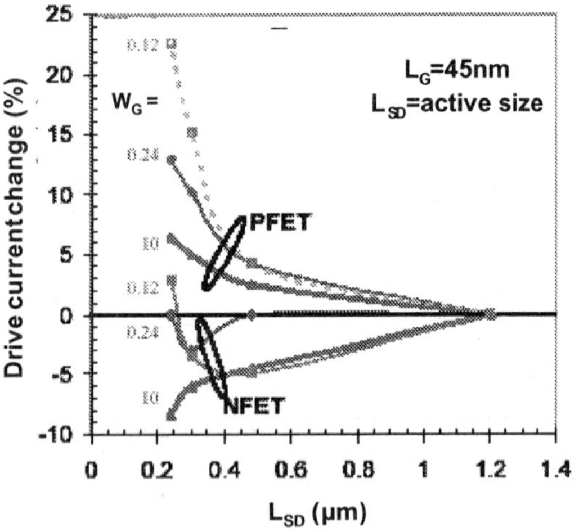

Figure 3: STI mechanical stress effect for n- and p-devices as a function of L_{SD} for W_G in μm. L_G = 45nm. By V. Chan et al. [12].

The development of advanced uniaxial stressors have added complexity to the CMOS process but have allowed for separate NMOS and PMOS Ion improvement. Examples of powerful uniaxial stressors are the embedded SiGe S/D for PMOS [13], dual stress liners (DSL) with compressive stress for PMOS and tensile for nMOS [14], and the stress memorization technique (SMT) which combines a polysilicon gate amorphisation with a strained nitride capping to modulate the tensile stress transferred via the gate to the channel [15]. The drawback of the uniaxial techniques is that these are device geometry and layout dependent [16], and the thickness of the liner stressor layers is limited by the shrinking of the gate pitch.

Biaxial Strain

The fundamental difference between the uniaxial and the biaxial sSi is that the former is introduced during CMOS processing while the later is built into the substrate [22-27]. As a consequence, biaxially sSi wafers offer tensile strain that is device layout and gate pitch independent, allowing IC design a high degree of freedom. The principle for wafer-level strain engineering is to generate a relaxed $Si_{1-x}Ge_x$ epitaxial template on a Si substrate [17, 18] followed by a Si epitaxial step (fig. 4). The epitaxial Si will retain the in-plane lattice constant of the template, resulting in a biaxial sSi film if the thickness is below a certain critical value.

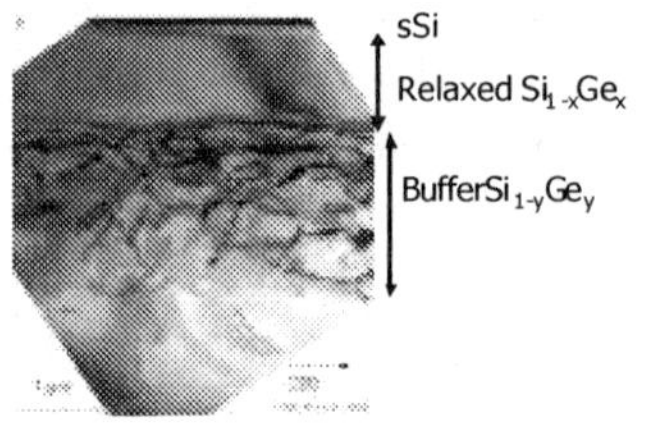

Figure 4: TEM cross-section of a virtual $Si_{1-x}Ge_x$ substrate.

Strained Si on virtual SiGe substrates enhance the device and circuit performance as expected but also result in a high off-state leakage level due to the presence of misfit and threading dislocations [19, 20]. For strained Si on insulator this is no longer an issue given that the top layer, i.e. the best crystal quality region, of the virtual substrate is transferred onto a handle wafer. Two types of substrates can be easily generated by layer transfer [21]: sSi directly on insulator (sSOI) [22-25] and sSi on relaxed SiGe on insulator (SGOI) [22, 27]. The strain amount and the resulting mobility and current drive increase depend on the Ge content of the SiGe template [26]. An alternative technique to fabricate SGOI is the condensation technique, which transforms an SOI wafer into a SGOI substrate. It has been thouroughly investigated [26] and will not be discussed here.

The effect of the strain on a (110) SiGe surface orientation depends on the crystallographic direction of the current flow in the plane. The maximum mobility is found for <100> and <110> directions, for electrons and holes, respectively. As for (100) surfaces, a strong dependence on gate electric field and the percentage of Ge is found [28].

For a sSi film grown on a fully relaxed $Si_{1-x}Ge_x$ (x=0.2) template, a biaxial stress level of 1.5 GPa is achieved [22, 25]. This leads to a NMOS mobility enhancement of 80%, and a 40% increase of current drive. If the Ge concentration x is increased to x>0.30, the same level of mobility enhancement is also achieved for PMOS [26, 29]. The fabrication of sSOI and SGOI substrates is similar [22] and has many steps in common with the Smart Cut SOI process flow [30] which maximizes the industrial synergy with high volume SOI manufacturing.

The thermal robustness of sSOI is shown in fig. 5. The mean strain of a 56nm sSi layer after a 2 hour anneal at different temperatures remains unchanged within the error bar. The full width half maximum (FWHM) of the Raman signal did not exhibit a broadening even at 1100°C. sSOI was annealed with an oxide cap of 10nm. Without an oxide cap, a broadening and a small shift of the Raman signal can be observed. From these annealing results it is clear that no strain relaxation for unpatterned sSOI is observed up to 1100°C if proper surface passivation is applied [25]. In contrast, for SGOI the thermal budget window is lower due to Ge up-diffusion from the underlying SiGe layer.

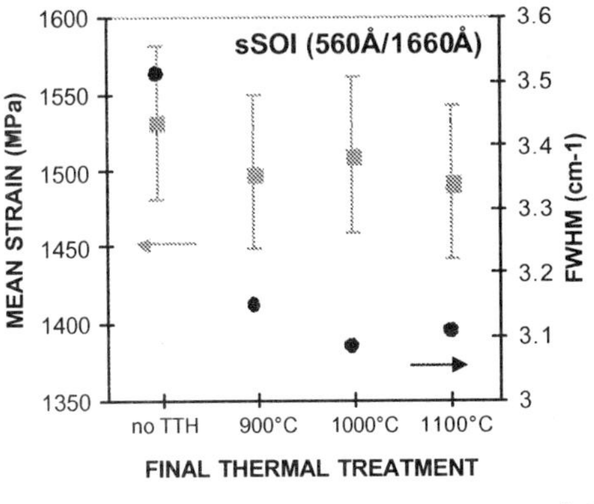

Figure 5: UV Raman stress measurement of sSOI as a function of annealing temperature for 2h. Strained Si= 56nm; buried oxide= 166nm. Laser: 363.8nm. Courtesy of M. Kennard.

NMOS fabricated on SGOI, L_G=45nm, show a mobility enhancement of 67% and a Ion increase of 18% with no indication of strain relaxation [27]. Transistor and gate oxide reliability studies give comparable results to SOI control samples (fig. 6) pointing out no detrimental effects due to the underlying SiGe film and minimal Ge up diffusion during CMOS processing [27], which constitutes one of the major concerns for the utilization of SGOI substrates.

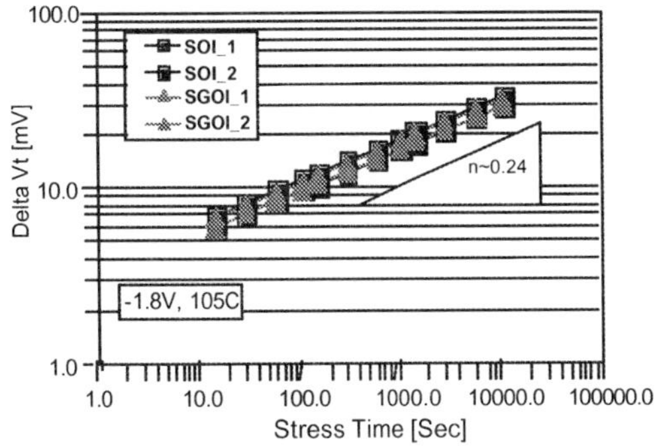

Figure 6: SGOI and SOI NMOS NBTI measurements. Slope and Vt shift are equivalent. Stress conditions: 3h, -1.8V, 105°C. By M. Sadaka et al. [27].

sSOI offers the benefits of sSi without the processing limitations imposed by a Ge-containing substrate, which explains the increased interest of the IC industry in this substrate. Typical sSi film thicknesses range from 10-20nm for fully depleted (FD) designs and up to 70nm for full compatibility with existing partially depleted (PD) architectures. Fig. 7 shows UV Raman stress measurements of sSOI for different sSi film thicknesses. No significant effect of the sSi thickness is observed on neither the strain Si peak, i.e. the stress level, nor a broadening of the FWHM Raman signal.

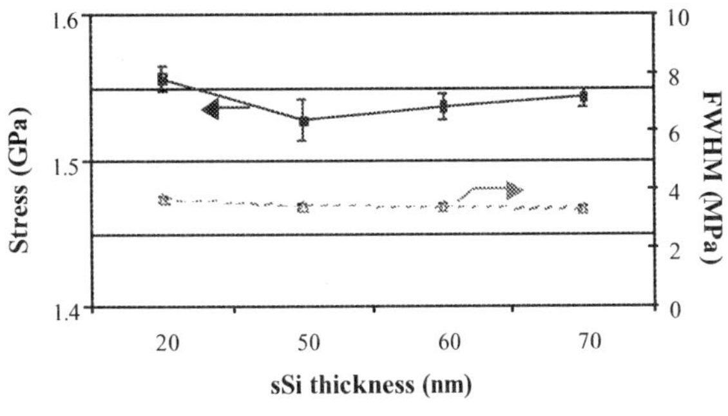

Figure 7: UV Raman stress measurement of sSOI as a function of sSi thickness. Laser: 363.8nm. Courtesy of M. Kennard.

The scalability of sSOI has been thoroughly investigated for ultrathin body sSOI with the fabrication of short channel devices [23, 24]. More recently, the impact of 40 to 50nm thick sSOI on PD MOSFETs has been reported [31], showing a 30% reduction in gate oxide leakage and a 60% improvement of the SRAM write margins.

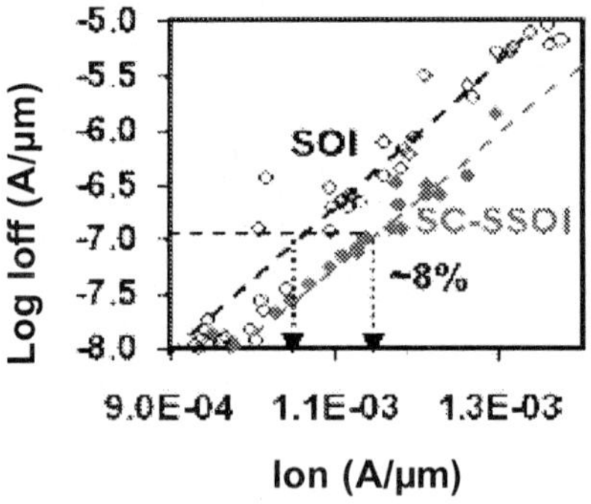

Figure 8: Current enhancement for 40nm PD sSOI nMOSFETs. By A. Thean et al. [31]. SC-sSOI stands for super critical sSOI.

Fig. 8 shows Ioff vs. Ion plots for 40nm PD sSOI devices. While for long channels a Ion enhancement of 50% is obtained, Ion increase drops to 25% for 150nm channels and to 8% for 40nm long devices (fig. 8). This is the result of the series resistances that dominate in the short channel range [31]. It is an issue that affects equally standard as well as strained devices. An improvement in S/D series resistances will proportionally benefit more strained than unstrained NMOS.

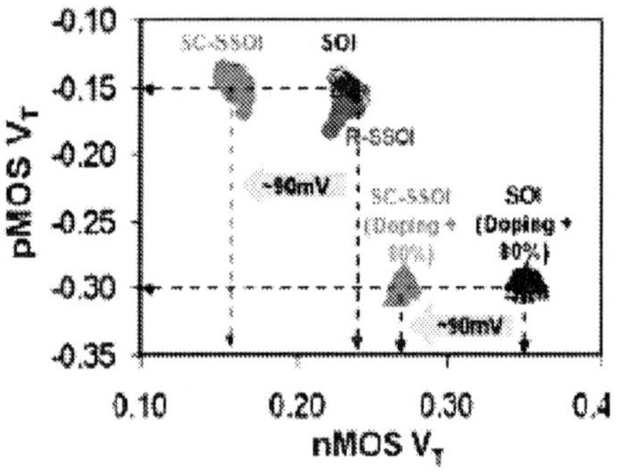

Figure 9: Low and high V_T for NMOS and PMOS devices for SC-sSOI and their processed matched control SOI. L_G=1µm, W_G=3µm. For intentionally relaxed sSOI (R-sSOI) nV_T shift disappears. By A. Thean et al. [31].

Nevertheless, the analysis [25] of published NMOS Ion results for different mobility enhancing techniques shows that biaxial tensile strain, in particular sSOI, is the most effective technique. In contrast, for PMOS the best results are obtained with compressive strain. A combination of sSOI with a uniaxial stressors would result in one very efficient solution.

An important finding is that NMOS V_T (nV_T) modulation by strain is insensitive to channel doping [31] as shown in fig. 9. If sSOI is intentionally relaxed, nV_T shift that is consistent with the smaller bandgap of the sSi films, disappears. As a result, the strain in the channel region can be monitored electrically through a threshold voltage V_T shift. It is found that minimal geometry devices (L_G=40nm, W_G=120nm) do remain highly strained.

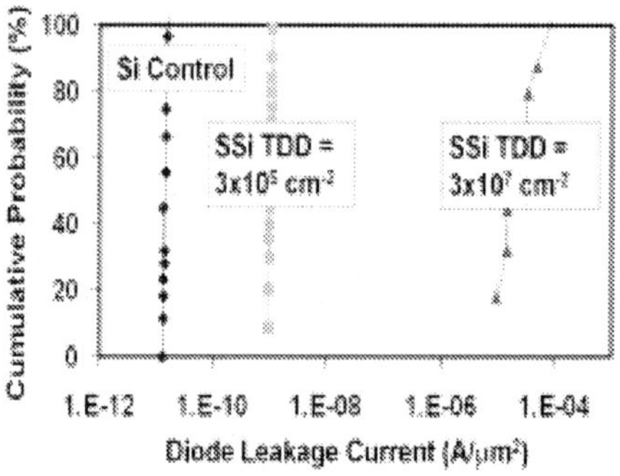

Figure 10: Diode leakage versus TDD by Q.-H. Xie et al. [32]. Courtesy of B.-Y. Nguyen, Freescale.

The threading dislocations density (TDD) of sSi remains the major focus the wafer making. TDD has been improved to $<10^4$ cm^{-2} , which according to fig. 10 would correspond to a diode leakage of 10^{-10}A/µm^2. Thus the sSOI quality is aproaching the requirements of IC manufacturing.

The combination of epitaxy and layer transfer opens up a multitude of possibilities to engineer the band structure and take the mobility enhancement beyond the present sSi levels [33-35] and design the solutions for the future. Fig. 11 shows an example of a dual channel structure where the band structure is engineered by epitaxy. For a sSi/strained SiGe dual channel a hole mobility increase of 2-3 can be obtained, even at high carrier densities as compared to SOI or SGOI substrates [33]. Significantly higher hole mobilities can be obtained if strained Ge is used instead of strained $Si_{1-x}Ge_x$ [34]. An attractive approach is to use a sSOI wafer as a template for strained $Si_{1-x}Ge_x$ instead of a relaxed SiGe template. The great advantage of this approach is that the NMOS is fabricated in a Ge free sSOI structure while the strained SiGe is kept only for PMOS [35].

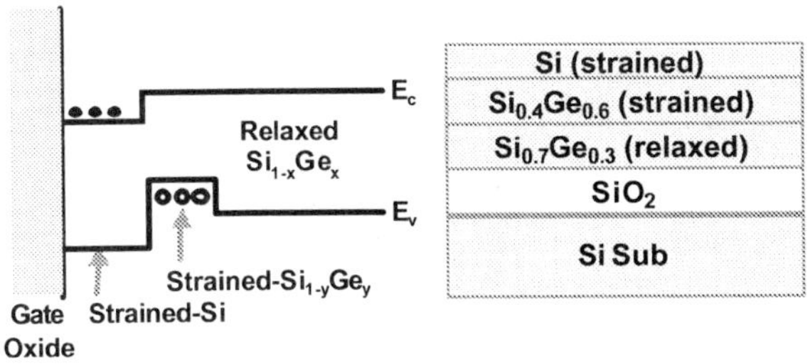

Figure 11: Schematic band diagram and the corresponding layer structure. Holes will tend to be confined in the strained SiGe buried channel while electrons will be confined to the surface sSi channel. By J. Jung et al. [33].

Conclusion

A review of uniaxial and biaxial strain techniques has been given, and in particular the sSOI potential for sub-65nm technology generations has been highligthed. Strained silicon on insulator (sSOI) is a key innovation at the substrate level that offers enhanced carrier mobility and increased on-current, while keeping leakage currents at a low level due to the buried oxide. It combines the advantages of SOI with those of strained silicon. sSOI is also an attractive platform for the PMOS dual channel enginnering.

Uniaxial strained silicon has already made its way into 90nm IC manufacturing. Biaxial wafer-level strain will be the next step for boosting transistor performance.

Acknowledgements

The work reviewed here is the result of the effort of many teams at different companies, institutes and universities. Special thanks in particular to G. Celler, B. Ghyselen, M. Kennard from SOITEC, N. Kernevez from LETI/CEA, G. Fitzgerald from MIT, B.Y. Nguyen from Freescale, M. Ieong from IBM, and D. Greenlaw from AMD.

References

[1] T. Ando, A. Fowler and F. Stern, Rev. Mod. Phys. **54**, 437 (1982). and references therein.

[2] C.S. Smith, Phys. Rev., **94**, 42 (1954).

[3] J. Welser, J.L. Hoyt, S. Tagaki, and J.F. Gibbons, IEDM Tech. Dig., 373 (1994).

[4] M.D. Giles, M. Armstrong, C. Auth, S.M. Cea, T. Ghani, T. Hoffmann, R. Kotlyar, P. Matagne, K. Mistry, R. Nagisetty, B.Obradovic, R. Shaheed, L. Shifren, M. Stettler, S. Tyagi, X. Wang, C. Weber, K. Zawadzki, Proc. Symp. VLSI Tech, 118 (2004).

[5] M. Fischetti, Z. Ren, P.M. Solomon, M. Yang and K. Rim, J. Appl. Phys., **94**, 1079 (2003).

[6] G. Scott, J. Lutze, M. Rubin, F. Nouri, and M. Manley, IEDM Tech. Dig., 827 (1999).

[7] H. Sayama, Y. Nishida, H. Oda, T. Oishi, S. Shimizu, T. Kunikiyo, K. Sonoda, Y. Inoue, and M. Inuishi, IEDM Tech. Dig., 657 (1999).

[8] T. Sanuki, A. Oishi, Y. Morimasa, S. Aota, T. Kinoshita, R. Hasumi, Y. Takegawa, K. Isobe, H. Yoshimura, M. Iwai, K. Sunouchi, and T. Noguchi, IEDM Tech. Dig., 65 (2003).

[9] C. Ge, C. Lin, C. Ko, C. Huang, Y. Huang, B. Chan, B. Perng, C. Sheu, P. Tsai, L. Yao, C. Wu, T. Lee, C. Chen, C. Wang, S. Lin, Y. Yeo, and C. Hu, IEDM Tech. Dig., 73 (2003).

[10] T. Komoda, A. Oishi, T. Sanuki, K. Kasai, H. Yoshimura, K. Ohno, M. Iwai, M. Saito, F. Matsuoka, N. Nagashima, and T. Noguchi, IEDM Tech. Dig., 217 (2004).

[11] A. Steegen, M. Stucchi, A. Lauwers, and K. Maex, IEDM Tech. Dig., 497 (1999).

[12] V. Chan, R. Rengarajan, N. Rovedo, W. Jin, T. Hook, P. Nguyen, J. Chen, E. Nowak, X. Chen, D. Lea, A. Chakravarti, V. Ku, S. Yang, A. Steegen, C. Baiocco, P. Shafer, H. Ng, S. Huang, and C. Wann, IEDM Tech. Dig., 77 (2003).

[13] T. Ghani, M. Armstrong, C. Auth, M. Bost, P. Charvat , G. Glass, T. Hoffmann, K. Johnson, C. Kenyon, J. Klaus, B. McIntyre, K. Mistry, A; Murthy, J. Sandford, M. Silberstein, S. Sivakumar, P. Smith, K. Zawadzki, S. Thompson, and M. Bohr, IEDM Tech. Dig., 978 (2003).

[14] H. S. Yang, R. Malik, S. Narasimha, Y. Li, R. Divakaruni, P. Agnello, S. Allen, A. Antreasyan, J. C. Arnold, K. Bandy, M. Belyansky, A. Bonnoit, G. Bronner, V. Chan, X. Chen, Z. Chen, D. Chidambarrao, A. Chou, W. Clark, S. W. Crowder, B. Engel, H. Harifuchi, S. F. Huang, R. Jagannathan, F. F. Jamin, Y. Kohyama, H. Kuroda, C. W. Lai, H. K. Lee, W. Lee, E. H. Lim, W. Lai, A. Mallikarjunan, K. Matsumoto, A. McKnight, J. Nayak, H. Y. Ng, S. Panda, R. Rengarajan, M. Steigerwalt, S. Subbanna,

K. Subramanian, J. Sudijono, G. Sudo, S. Sun, B. Tessier, Y. Toyoshima, P. Tran, R. Wise, R. Wong, I. Y. Yang, C. H. Wann, L. T. Su, M. Horstmann, T. Feudel, A. Wei, K. Frohberg, G. Burbach, M. Gerhardt, M. Lenski, R. Stephan, K. Wieczorek, M. Schaller, H. Salz, J. Hohage, H. Ruelke, J. Klais, P. Huebler, S. Luning, R. van Bentum, G. Grasshoff, C. Schwan, E. Ehrichs, S. Goad, J. Buller, S. Krishnan, D. Greenlaw, M. Raab, and N. Kepler, IEDM Tech. Dig., 1075 (2004).

[15] C.H. Chen, T.L. Lee, T.H. Hou, C.L. Chen, J.W. Hsu, K.L. Cheng, Y.H. Chiu, H.J. Tao, Y. Jin, C.H. Diaz, S.C. Chen, and M.S. Liang, Proc. Symp. VLSI Tech, 56 (2004).

[16] G. Eneman, P. Verheyen, R. Rooyackers, F. Nouri, L. Washington, R. Degareve, B. Kaczer, V. Moroz, A. DeKeersgieter, R. Schreutelkamp, M. Kawaguchi, Y. Kim, A. Samoilov, L. Smith, P.P. Absil, K. DeMeyer, M. Jurczak, S. Biesemans, Proc. Symp. VLSI Tech, 22 (2004).

[17] F.K. LeGoues, B.S. Meyerson and J.F. Morar, Phys. Rev. Lett., **66**, 2903 (1991).

[18] E.A. Fitzgerald, Y.-H. Xie, M.L. Green, D. Brasen, A.R. Kortan, J. Michel, Y.-J. Mii, and B.E. Weir, Appl. Phys. Lett., **59**, 811 (1991).

[19] H. C. Wang, Y. Wang, S. Chen, C. Ge, S. M. Ting, J. Kung, R. Hwang, H. Chiu, L. C. Sheu, P. Tsai, L. Yao, S. Chen, H. Tao, Y. Yeo, W. Lee, and C. Hu, IEDM Tech. Dig., 61 (2003).

[20] J.G. Fiorenza, G. Braithwaite, C. Leitz, M.T. Currie, Z.Y. Cheng, V.K. Yang, T. Langdo, J. Carlin, M. Sommerville, A. Lochtefeld, H. Badawi, M.T. Bulsara, Proc. of IRPS, 493 (2004).

[21] B. Ghyselen , J.-M. Hartmann, T. Ernst, C. Aulnette, B. Osternaud, Y. Bogumilowicz , A. Abbadie, P. Besson, O. Rayssac, A. Tiberj, N. Daval, I. Cayrefourq, F. Fournel, H. Moriceau, C. Di Nardo, F. Andrieu, V. Paillard, M. Cabié, L. Vincent, E. Snoeck, F. Cristiano, A. Rocher, A. Ponchet, A. Claverie, P. Boucaud , M.-N. Semeria, D. Bensahel, N. Kernevez and C. Mazure, Solid State Elect., **48**, 1285 (2004).

[22] B. Ghyselen , C. Aulnette, B. Osternaud, T. Akatsu, C. Mazure, C. Lagahe-Blanchard, S. Pocas, J.-M. Hartmann, P. Leduc, T. Ernst, H. Moriceau, Y. Campidelli, O. Kermarrec, P. Besson, Y. Morand, M. Rivoire, D. Bensahel and V. Paillard, ICSI3 Proc., 173 (2003).

[23] T.A. Langdo, A. Lochtefeld, M.T. Currie, R. Hammond, V.K. Yang, J.A. Carlin, C.J. Vineis, G. Braithwaite, H. Badawi, M.T. Bulsara, and E.A. Fitzgerald et al., IEEE Int. SOI Conf., 211 (2002).

[24] K. Rim, K. Chan, L. Shi, D. Boyd, J.Ott, N. Klymko, F. Cardone, L. Tai, S. Koester, M. Cobb, D. Canaperi, B. To, E. Duch, I. Babich, R. Carruthers, P. Saunders, G. Walker, Y. Zhang, M. Steen, M. Ieong., IEDM Tech. Dig., 49 (2003).

[25] I. Cayrefourcq, M. Kennard, F. Metral, C. Mazure, A. Thean, M. Sadaka, T. White, and B.Y. Nguyen, Silicon-On-Insulator Technology and Devices XII, ECS Proc., vol **2005-03**, 191 (2005).

[26] S. Tagaki, T. Mizuno, T. Tezuka, N. Sugiyama, T. Numata, K. Usuda, Y. Moriyama, S. Nakaharai, J. Koga, A. Tanabe, N. Hirashita, T. Maeda, EDM Tech. Dig., 57 (2003).

[27] M. Sadaka, A. Thean, A. Barr, D. Tekleab, S. Kalpat, T. White, T. Nguyen, R. Mora, P. Beckage, D. Jawarani, S. Zollner, M. Kottke, R. Liu, M. Canonico, Q.-H. Xie, X.-D. Wang, S. Parsons, D. Eades, M. Zavala, B.Y. Nguyen, C. Mazure, and J. Mogab, IEEE Int. SOI Conf, 209 (2004).

[28] T. Mizuno, N. Sugiyama, T. Tezuka, Y. Moriyama, S. Nakaharai, T. Maeda and S. Tagaki, IEDM Tech. Dig., 809 (2003).

[29] I. Aberg, O.O. Olubuyide, C. Ni Chléirigh, I. Lauer, D. A. Antoniadis, J. Li, R. Hull, J. L. Hoyt, Proc. Symp. VLSI Tech, 52 (2004).

[30] A.J. Auberton-Hervé and C. Maleville, IEEE Int. SOI Conf., 1 (2002).

[31] A. Thean, T. White, M. Sadaka, L. McCormick, M. Ramon, R. Mora, P. Beckage, M. Canonico, X.-D. Wang, S. Zollner, S. Murphy, V. Van Der Pas, M. Zavala, R. Roble, O. Zia, L-G. Kang, V. Kolagunta, N. Cave, J. Cheek, M. Mendocino, B.Y. Nguyen, M. Orlowski, S. Venkatesan, J. Mogab, C.H. Chang, Y.H.Chiu, H.C. Tuan, Y.C. See, M.S. Lian, Y.C. Sun, I. Cayrefourcq, F. Metral, M. Kennard, C. Mazure, Proc. Symp. VLSI Tech, 134 (2005).

[32] Q.-H. Xie, P. Feies, M. Kottke, X.-D. Wang, M. Canonico, D. Theodore, T. White, M. Sadaka, V. Vartanian, A. Thean, B.Y. Nguyen, A. Barr, S. Thomas, R. Liu; MRS Symp. E on Semiconductor Defect Engineering, E4.31, March (2005).C.

[33] J. Jung, M.L. Lee, S. Yu, E.A. Fitzgerald and D. Antoniadis, IEEE EDL, **24**, 460 (2003) .

[34] H. Shang, J. O. Chu, X. Wang, P.M. Mooney, K. Lee, J. Ott, K. Rim, K. Chan, K. W. Guarini, and M. Leong, Proc. Symp. VLSI Tech, 204 (2004).

[35] F. Andrieu, T. Ernst, O. Faynot, Y. Bogumilowicz, J.-M. Hartmann, J. Eymery, D. Lafond,Y.-M. Levaillant, C. Dupré, R. Powers, F. Fournel, C. Fenouillet-Beranger, A. Vandooren, B. Ghyselen, C. Mazure, N. Kernevez, G. Ghibaudo and S. Deleonibus, to be published in IEEE Int. SOI Conf (2005).

CMOS Photonics™ -- SOI Learns a New Trick

Cary Gunn, PhD
Luxtera, Inc., Carlsbad, CA
Email: cary@luxtera.com, Tel: (760) 448-3520 x116

Abstract

Luxtera has used Freescale Semiconductor's production 0.13µm SOI process to implement optical communications capability for high bandwidth LAN, SAN, shelf-to-shelf and chip-to-chip communications. These optical transceiver cores operate at 10Gbps and offer superior reach, power consumption, latency, die area, and scalability compared to emerging standards for electrical interconnect. They are monolithically fabricated alongside SOI CMOS circuitry in the same die. Thus, for the first time in history, high speed optical communications directly between silicon die can be accomplished at a price/performance point superior to traditional electrical interconnect.

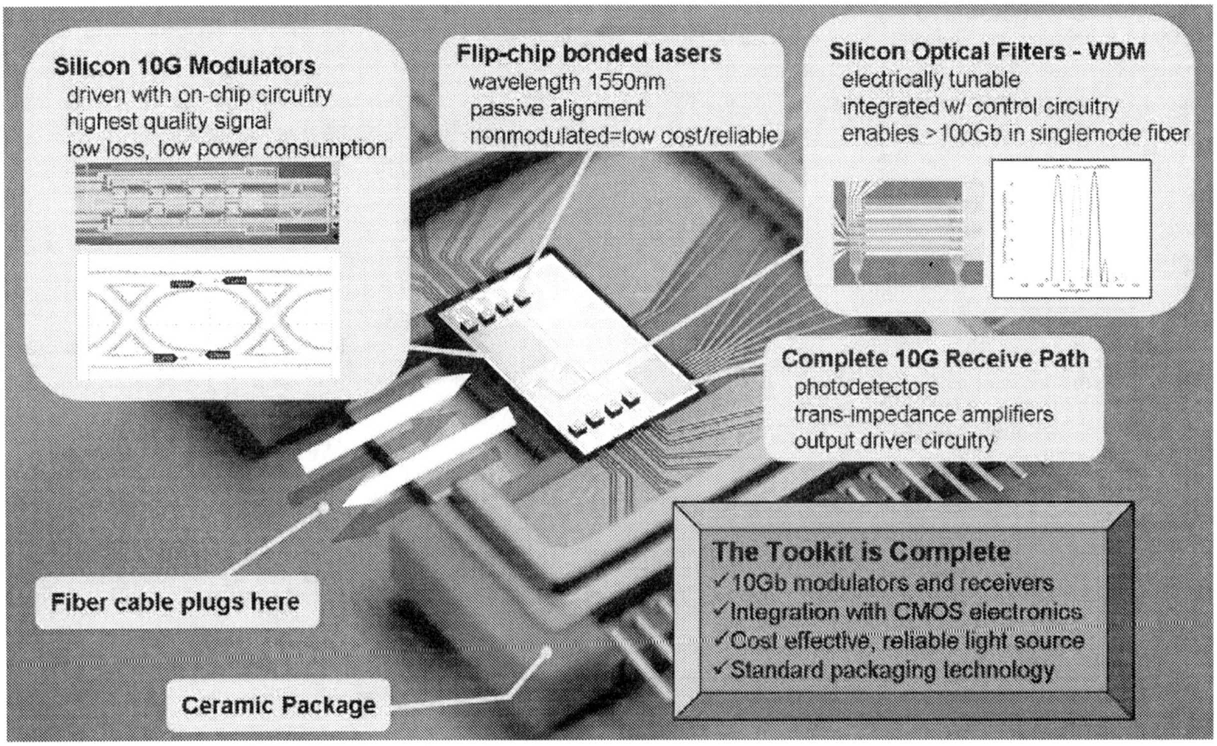

Figure 1. Overview of CMOS Photonics technology.

Motivation for Optical Interconnect

Electrical interconnect speed has continued to increase for many generations, despite the exponentially increasing technical barriers at each generation. As the signal bandwidth increases, physical effects such as the skin effect and impedance mismatches become more pronounced. This means that attenuation, signal reflections, crosstalk and dispersion have been increasingly difficult to combat in each new generation. This effect is particularly pronounced in the LAN/Ethernet market, where bandwidth has traditionally scaled by decades.

As Ethernet moved from 100Mb to 1Gb, the design of the transceivers became substantially more difficult, while the physical medium over which the signal traveled remained the same. Cat5e cabling, which contains 4 twisted pairs (TPs), has the bandwidth and dispersion properties to transmit 1Gb of data (250Mb/TP) for 300m using a reasonably powered analog front end and digital circuitry of moderate complexity. However, to increase the bandwidth/TP by another order of magnitude has required sacrifices on key performance metrics. The attenuation and dispersion of the cable have several key impacts on the

0-7803-9212-4/05/$20.00 ©2005 IEEE

system. First, a significant TP cable upgrade is required in order to get the bandwidth to transmit 10Gb data. There is a consensus in the standards body that Cat5 cabling will not suffice for the next generation, and either Cat6 or Cat7 cables must be used. Second, due to the attenuation, the amount of power required to drive the cable requires significant power from the analog front end. Such analog circuitry power consumption does not scale and will not improve with the next generation of CMOS. Third, the crosstalk and dispersion of the TPs, in combination with the attenuation, limit the projected reach of the standard to 55m, 1/2 of the traditional LAN reach.

In order to compensate for the crosstalk, reflections, and dispersion, a significant amount of digital processing is required in the transceiver. This adds significant power consumption and latency to the system, and requires a significant amount of silicon real estate. The IEEE standards body continues to slip the approval date while numerous companies attempt to find a workable solution.

The bottom line is that the adoption of 10GE has been delayed due to numerous new technical difficulties, and when implemented, it will have significant performance deficiencies and high costs. A comparison of these technologies is made in Table 1 below.

Table 1. Comparison of 1GE to 10GE. Electrical interconnects are increasingly difficult.

	1GE	10GE (projected)
Reach	100-220m	55m
Cable	Cat5e	Cat6/7
Power	<1W	7-15W
Latency	<100ns	1-3μs
Cost	\$3-5	\$200 -> \$50?

The same saga is being played out with all forms of electrical interconnect. Although at a completely different bandwidth*reach environment, backplane technologies are hitting the wall and find scaling increasingly difficult over the required distances. Chip to chip communications has long been a bottleneck in computer systems, and the move to multicore processors is likely to exacerbate this problem.

Across the board, electrical interconnect is consuming more power, requiring more expensive physical media, and costs more per port. No one at the companies battling these issues has a concept to increase the speed of their technology by another order of magnitude using electrical interconnects.

Legacy Optical Systems

Optical engineers have looked at the continued complexity required by electrical interconnect, and at each generation they have hoped that their technology will finally be a viable replacement for electrical systems. This has happened at the longer reaches. For data rates over 1Gb, and distances over 100m, optical interconnects dominate and have displaced legacy copper and microwave technologies.

Today, 10G optical ports with 10-1000km of reach are readily available. Using wavelength division multiplexing (WDM), optical systems are able to transmit many channels of data down a single fiber. Systems that transmit 40 wavelengths of light, each at 10Gb, are commercially available and implemented at various points in the worldwide telecommunications backbone.

The problem with traditional optical systems is that they are fundamentally expensive. Today, a typical high performance 10Gb transceiver costs ~\$500, and there are many good reasons for this:

1) **Too many standards**. Today there are more than 30 optical standards for 10G, each customized for a particular reach and fiber type. The highest performing standards can reach 80km unamplified, and more with optical amplifiers. The lower performing standards are limited to ~100m of reach over multi-mode optical fiber. There is not enough volume in each standard to drive costs down.
2) **No wafer-scale testability of many key components**. Although most components are processed using waferscale technology, most lasers and optical multiplexing technologies cannot be tested until the wafer has been diced and/or they are singulated and partially packaged. This has a profound impact on the cost of test, yield, and the cost of the final system.
3) **Manual assembly of many discrete components.** Today's optical transmitters and receivers are manually assembled and individually tested. This is required due to the custom equipment used to perform traditional optical alignment of lenses, lasers, photodetectors and fibers.
4) **Lack of coherent system design methodology.** Today, optical systems are cobbled together from many individual components, constructed by different companies in different material systems and using different technologies. System designers are forced to implement tight specifications in each of their component vendors in order to

control the variation in performance that happens independently of each other and often interact. This keeps the component costs very high. On the other hand, electrical transceiver designers control the entire transmitter and receiver design and are able to make global optimizations that bring system costs down.

Thus it is no surprise that optical communications has been known to be relevant only to the high performance, high cost applications that have not been addressable with electrical communications systems. However, this need not be the case. If optical transceivers can be made at the appropriate cost structure, they can offer significant performance advantages.

Why SOI is Important

The fundamental building block of an optical system is the optical waveguide. This is equivalent to the wire for electronics. Waveguides are used to route light from one portion of a die to another, and to connect individual components to each other. A waveguide is made of two portions, a core of relatively high index material and a cladding of relatively low index material. An optical fiber is an example of such a waveguide optimized for transmission over long distances. For a waveguide integrated into a small die, low transmission loss and a tight bending radius are essential. A waveguide constructed from a core of silicon, and a cladding of silicon dioxide is an ideal waveguide for integrated optics, having acceptably low loss and a tight bending radius. It also turns out that a typical SOI film can be patterned into effective waveguides at the same time that transistor bodies are etched. Figure 2 shows an oblique view of SOI waveguides constructed at Freescale, including a device known as a ring resonator. The dielectric films on top of the silicon have been removed in order to capture a clear image. These waveguides are 200nm thick and ~500nm wide. The radius of the ring is 2μm, demonstrating a tight bending radius. The typical loss from a Luxtera SOI waveguide today is 0.18 dB/mm.

A "Holy Grail" Technology

Integration of optical communications functionality into a CMOS platform is viewed as a "holy-grail" because it can deliver the superior performance of optics at the same (or better) cost structure of electronics. However, it is not enough that the technology simply works. In order to keep the costs low, the optical technology requires the following attributes:

1) Operation at the emerging industry standard data rate of 10Gb per channel.
2) Truly CMOS compatible processing. It is not good enough simply use silicon – the optical components have to coexist with standard, high-performance transistors, the processing steps and temperatures, and all other materials present in a CMOS wafer.
3) Wafer-scale testing of optical components is required to only package known good die.
4) Minimal use of exotic materials traditionally used in optical systems.
5) Automated assembly and optical alignment.
6) Use of conventional electronic packaging materials and processes.
7) A path for scalability to future generations using the same physical media and baseline technology.
8) Continued improvement in die area and power consumption of optical devices to allow monolithic integration of many channels.

Luxtera has demonstrated a technology that not only has superior technical performance, but also has ALL of the above attributes, allowing communications systems to be built with superior performance, and at a cost structure similar to electronics. Furthermore, these attributes allow scaling from 10G to 40G, 100G, etc.

CMOS Photonics™ Technology

There are several basic building blocks that must be available to system designers before they can build optical transceiver cores on a CMOS die. First, it is necessary to efficiently couple the light on/off the die, then the light must be modulated, and also detected. For the modulation and detection, optimized circuitry is required to drive the modulator and to amplify the received photocurrent. Thus, the basic building blocks are: fiber coupling interfaces, optical modulators with integrated drive circuitry, optical detectors with integrated TIAs, and a light source. A brief discussion of key building blocks will ensue. The performance of each and every one of these components is sufficient for realization of practical systems in the near future. CMOS Photonics is ready to be used in mainstream applications.

Fiber Coupling: The Holographic Lens

A fundamental issue facing optical technology in the past has been the interface between the optical fiber and the die. This is exacerbated in the case of SOI waveguides since the silicon waveguide is extremely small, with a core only 0.1μm² in cross sectional area. By comparison an optical fiber has a core with a cross sectional area 85μm², 3 orders of magnitude larger. Thus, coupling light from an optical fiber into a silicon waveguide is like pouring water from a firehose into a straw – most of the light will spill around the edges and be lost. Thus, it is important to be able to focus the light from a fiber down to the spot size of the silicon waveguide. Traditional approaches to this challenge require lenses, which are costly and require difficult and accurate alignment. Furthermore, it's required to route the waveguide to the edge of the die, which must then be diced without chipping, polished to optical smoothness, and coated with an anti-reflection coating. This is an expensive process and is fundamentally incompatible with wafer scale testing, since singulation is require to first expose the edge of the waveguide. Luxtera has addressed all of these problems with the holographic lens.

The holographic lens is a lithographically defined structure etched into the silicon. This structure is located on the surface of the die and is designed to accept light from normal incidence. Thus, a fiber can simply be brought near to the surface of the die, and the light is turned 90° and simultaneously focused into the waveguide. This is shown in Figure 2. The light travels down the fiber, which is normal to the surface of the die, and illuminates the fine structure of the holographic lens. The core of the fiber is represented by the dashed white lines. The lens itself is constructed from the gate polysilicon already present in the process, deposited and etched over the transistor silicon, which is used for the waveguides. The light then travels down the silicon waveguides, where it can interact with a component such as a ring resonator or a modulator. The holographic lens has evolved to be a very efficient coupling device, as shown in Figure 3.

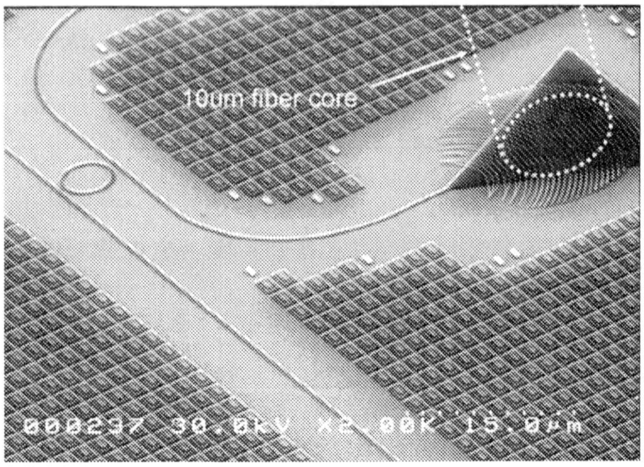

Figure 2. Oblique view of a holographic lens connected to silicon waveguides. The optical fiber illuminates the HL from a position normal to the surface. The core of the fiber is shown by the dashed lines.

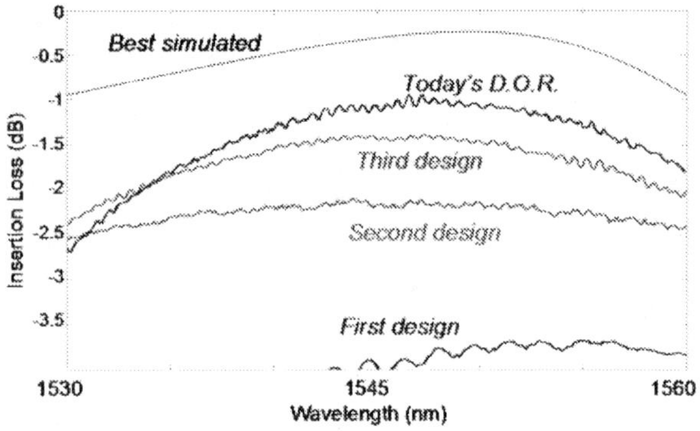

Figure 3. Evolution of Holographic Lens performance.

Today's design of record (DOR) holographic lens has insertion losses of ~1.5dB over a substantial portion of the typical optical communications band of 1530-1560nm. This lens is targeted for operation at 1545nm. As noted on Figure 3, the best simulated performance is 0.3dB coupling loss, which will bring the coupling loss on par with a conventional fiber connector used in patch panels. This performance is completely unprecedented in the field of silicon photonics, and is critical for development of practical parts for a number of reasons. First, it's quite obvious that low coupling loss is important in the performance of optical systems. Second, the holographic lens allows wafer scale test, both after the wafers are complete and during processing, thanks to the non-contact nature of the optical test. Luxtera's setup for wafer scale test is shown in Figure 4, where a fiber probe containing 16 fibers is brought near the surface of the wafer simultaneously with RF and DC electrical probes. Third, this allows flip-chip attachment of lasers, photodetectors, and fiber connectors directly to the top of the die using commercially available flip-chip equipment with practical alignment accuracies.

Wafer probe capability is particularly important to keep the cost of products down, and is taken for granted in the electronics industry. However, this is the first time complex optical functionality including modulators, optical multiplexers and waveguides have been testable prior to singulation. In this case, both optical and electrical testing are completed simultaneously in order to validate the entire system. This testing technique is compatible with use of probe cards in high volume testers. Luxtera believes this marks a windfall in the cost and complexity of optical systems, and it is a key reason why CMOS Photonics establishes a radically new cost structure for optical interconnect.

Figure 4. Wafer scale probe of a 200mm wafer using RF probes (top right), DC probes (sides), and a fiber probe (bottom left). Only a small portion of the wafer is seen.

10Gb Optical Modulator

One key piece of technology developed at Luxtera is an optical modulator capable of imprinting 10Gb data on light. The device Luxtera uses to accomplish this function is known as a Mach-Zehnder Modulator (MZM). In this device, light is split down two nearly identical paths and added back together after some distance. When light traveling down each path arrives at the addition point in phase with each other, it is not attenuated, but if the light in the two arms is phase shifted with respect to each other, the light destructively interferes, effectively extinguishing the light. Traditionally, when the light is on, it is considered a "1;" when extinguished, a "0." Thus development of a high efficiency splitter, combiner, and phase modulator are required to construct a good MZM. Such a device is shown in Figure 5.

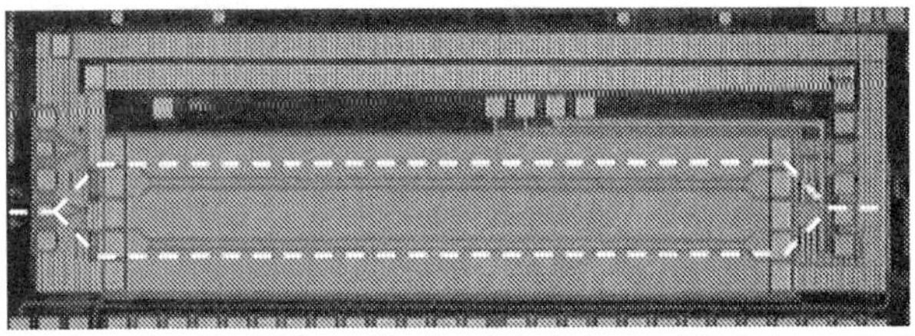

Figure 5. CMOS optical modulator with integrated driver. The dashed lines represent the optical waveguides, which are underneath the metalization.

Since the optical waveguides are buried down at the silicon layer, they are hard to see in this figure and are illustrated by the dashed white lines. The driver input pads are evident on the left and the power supply rails are shown across the top. Along the length of the MZM are transmission lines for carrying the signal input to the phase modulator, which are distributed down the length of the modulator. The pads mid-center is used to bias the modulator in quadrature, which is the optimum point for modulation. The modulator driver circuitry is on the left of the figure between the input and the transmission lines, and integrated terminations are present on the right hand side at the end of the transmission lines.

Typical results for a modulator are shown in Figure 6. Both the 10Gb eye diagram and frequency response are shown. Typical extinction ratio is 5dB, and the rise-fall time is clearly sufficient for error free 10Gb transmission.

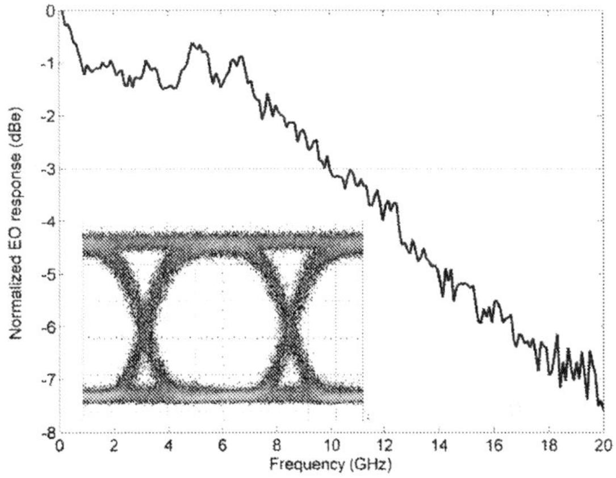

Figure 6. Silicon optical modulator 10Gb eye and frequency response.

Lasers – the optical power supply

Of course, light must first be present before it can be modulated. While there are a number of scientists working on squeezing light from silicon, Luxtera does not believe this work is practical in the near term, and as such has chosen to use a traditional continuous wave (CW) laser. Operation of a laser in CW mode simplifies laser design, and improves the yield of the laser. While a laser can be directly modulated by modulation of the drive current, this is detrimental to the lifetime and reliability of the device, and drives more stringent burn-in and testing requirements, all of which increase the cost of the laser. Luxtera has found that a CW laser, paired with a silicon modulator is the most cost-effective approach. However, in order for this to work well the laser must be flip-chipped on top of the silicon die using commercially available flip-chip technology. An oblique picture of lasers flip-chipped on top of a Luxtera die is shown in Figure 1. Note that the laser is quite small, only 0.3 mm^2. Figure 7 shows the top down view under higher magnification, including the silicon waveguides used to guide the light to the modulator. Such alignment is well within the capabilities of today's flip-chip equipment and can be done in very high volume at a very low cost.

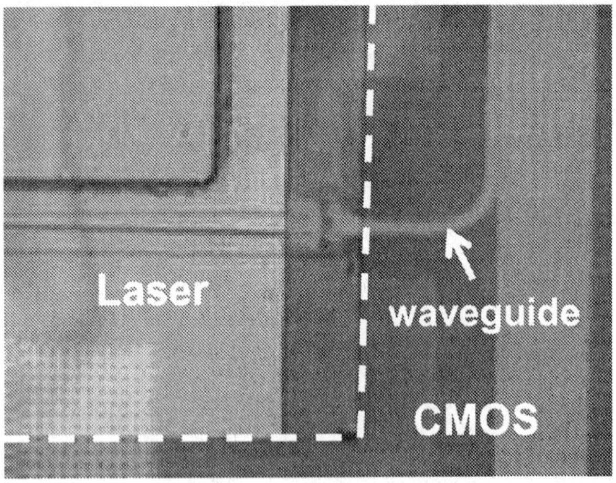

Figure 7. Top view of a flip-chipped laser on top of a CMOS die. The laser die is outlined by the dashed white lines. The light is coupled from the laser into the silicon waveguide on the CMOS die.

RF Circuitry

Of course, one of the key advantages of integration into CMOS is colocation of the RF driver and receiver circuitry with the optical components. The receiver transimpedence amplifier (TIA) and limiting amplifier (LA), which convert the photodector output current into a measurable voltage is a key building block of any optical system. This key circuit, among many others, has been demonstrated to work well, alongside the other CMOS Photonics components. Results from a typical TIA/LA circuit are shown in Figure 8. Note that the eye is open, and that the rise/fall time of the amplifier does not distort the signal. There are substantial advantages to monolithic design of photodetector circuits due the absence of the parasitics of the bondwires and packages between the detector and the initial amplifier.

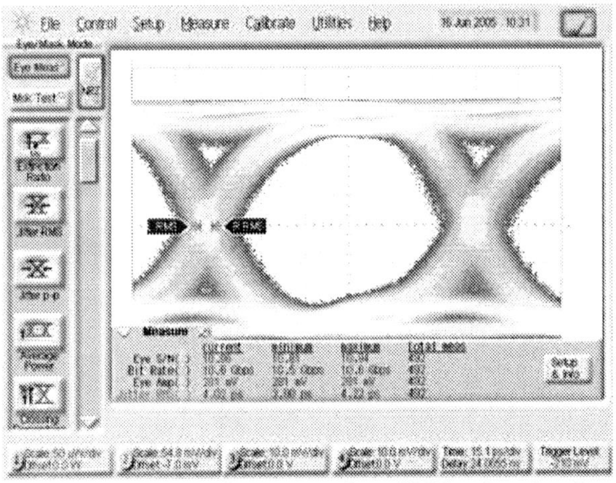

Figure 8. Eye diagram of a 10Gb signal after amplification by the CMOS transimpedance amplifier and limiting amplifier.

CMOS Photonics scales beyond 1Tb

An optical fiber, when deployed over distances required for LAN/SAN applications has the potential to carry 10Tbps. This bandwidth can be addressed with CMOS photoncs through use of optical multiplexing technology, as well as continued shrinkage in the dimensions of the core modulator and detector components. These elements will be discussed below.

Optical Multiplexers and Demultiplexers

A brief physics refresher: light belongs to the fundamental class of particles known as bosons, and electrons are fermions. The basic difference is that fermions strongly interact while bosons interact very weakly. This means that optical systems have an advantage over electrical systems in that many different signals can be sent down the same optical fiber with very little crosstalk. This is typically done in commercial optical systems today using a technology known as wavelength division multiplexing (WDM). Many different wavelengths of light are used, and each one is modulated independently. They are multiplexed and transmitted down a single fiber and demultiplexed at the receiver. While Luxtera is developing a number of optical multiplexing technologies, the most promising from a data density perspective is the ring filter, as has already been shown in Figure 2.

Data imprinted on many different wavelengths of light is incident down the fiber, and is coupled into the waveguide. Light that is resonant with the ring will be "dropped" to the opposing waveguide, after which it is sent to a receiver off the bottom of the picture. All other wavelengths will continue down the original waveguide to exit off the top of the picture, which is considered the "thru" port. Thus, a ring can function as an optical notch filter. An example of this is shown in Figure 9. Note that the loss of the light on the dropped wavelength is ~0.5dB, and the impact of the filter on the other wavelengths is negligible.

Use of a series of rings allows development of a WDM system using many different wavelengths. If each wavelength were modulated at 10Gb, a 100Gb system would be constructed using 10 rings, and a 400Gb system from 40. Since the dimensions of the rings are so small, many rings can be used in a small amount of silicon real estate.

Ring Modulators

While the MZ Modulator works quite well at 10Gb, it does require a substantial amount of real estate, approximately 2mm^2. While this is roughly equivalent in area to an electronic PHY at the same bandwidth, it

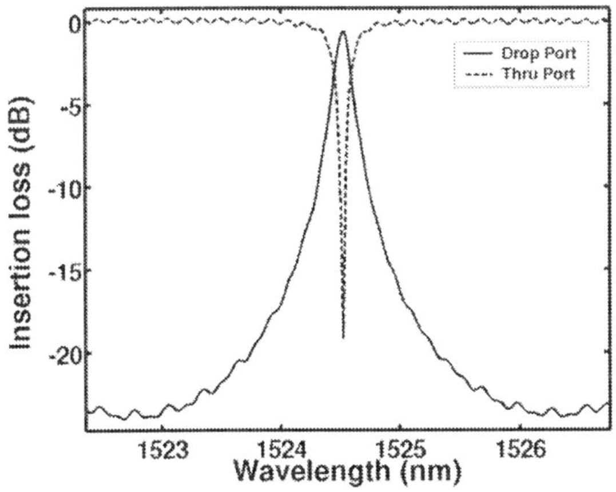

Figure 9. A ring resonator used as a notch filter.

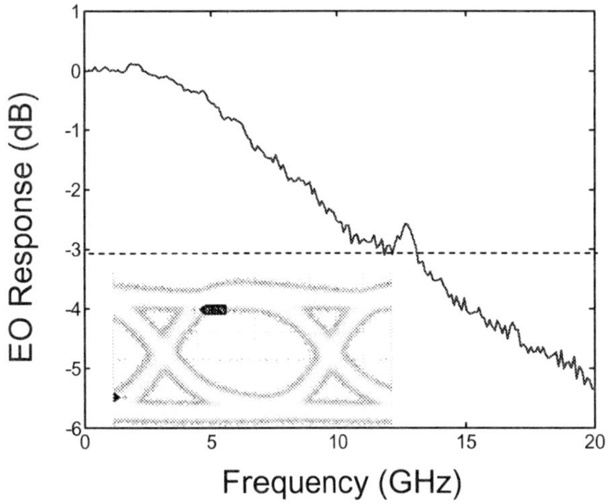

Figure 11. A typical ring modulator 10Gb eye diagram and frequency response.

would still be advantageous to further reduce the size of the device in order to scale to higher data densities. Again, a ring is a useful device. When employed as a rapidly tunable notch filter, light can be modulated quite rapidly. A picture of such a ring modulator is shown in Figure 10. In this case, the second waveguide is removed, and the modulation occurs on the "thru" waveguide. The ring radius is only 30μm. Figure 11 shows the measured results of the ring modulator at 10Gb, showing similar results to the MZ modulator, but in only a fraction of the area. The "spike" in the data at 13GHz is an artifact of the measurement equipment.

Conclusions

The technology required to implement CMOS Photonics has been proven, and product development is currently underway. All technology required for 10Gb operation has been demonstrated, in addition to technology required to scale to 100Gb and 1Tb. A single 10Gb channel today integrates 10s of optical components into a single die alongside circuitry of modest gate count, ~100k per transceiver. This is just the beginning of LSI integration of optical components, and 100Gb systems will have many 100s of optical components contained in a single die alongside millions of transistors. This technology is fundamentally enabled by SOI, and will continue to scale with future generations, with foreseeable integration into 90nm and 65nm process nodes. For the first time in history, the high bandwidth of optical interconnects can be achieved with the cost structure of CMOS.

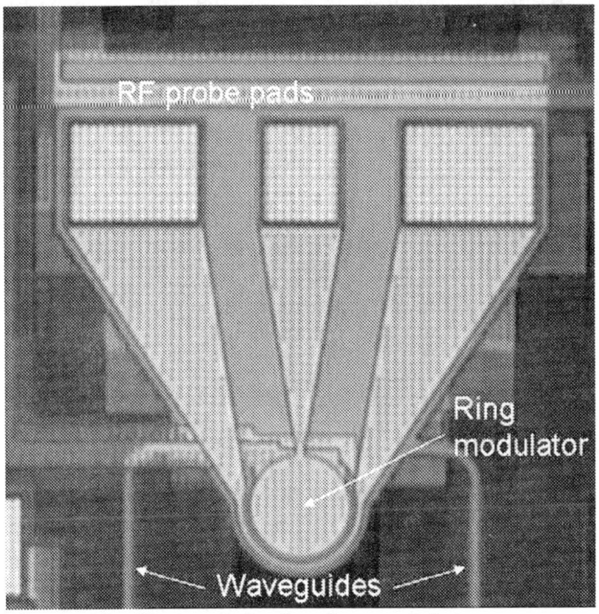

Figure 10. Photograph of a ring modulator.

Selected References

[1] R A Soref and B R Bennett. "Electrooptical Effects in Silicon" IEEE J of Quantum Electron, vol QE-23, no 1, Jan 1987, pp 123-9.

[2] Q Xu, B Schmidt, S Pradhan and M Lipson. "Micrometre-scale silicon electro-optic modulator." Nature Vol. 435, 19 May 2005, pp 325-7.

[3] J Lie, J Michel, W Giziewicz, D Pan, K Wada, D D Cannon, S Jongthammanurak, D T Danielson and L C Kimmerling, "High-performance, tensile-strained Ge p-i-n photodetectors on a Si platform." Appl Phys Lett. 87, 2005.

[4] L Liao, D Samara-Rubio, M Morse, A Liu, D Hodge, D Rubin, U D Keil, and T Franck. "High speed silicon Mach-Zehnder modulator." Opt. Express, vol 13, no. 8, 18 Apr 2005, pp 3129-3135.

[5] D Taillaert, W Bogaerts, P Bienstman, T F Krauss, P Van Daele, I Moerman, S Verstuyft, D De Mesel, and R Baets. IEEE J of Quantum Electron, vol 38, no 7, 7 Jul 2002. pp 949-955.

Using SOI to Achieve Low-Power Consumption in Digital

J.L. Pelloie
SOISIC – Miniparc Polytec, 60 rue des Berges, 38000 Grenoble, France
Email: jean-luc.pelloie@soisic.com, Tel: +33 4 38 78 47 00

Introduction

Power consumption has become a major concern in all electronic applications. Power dissipation is a heat source in high-speed processors used for PC or networking applications. Evacuating the generated heat has a cost at the system level (fans, limited number of processors on board) which may be saved by reducing the power consumption at circuit level. Power consumption is also critical in portable systems (cell phone, PDA, laptop) to maintain the battery lifetime and get the desired autonomy. Low-power meaning may be different depending on the applications and it is important to dissociate between those which require speed performance or not. Power optimization does not rely on one single technique but results from the combination of several techniques, implemented at different hierarchy levels during the design phase of the circuit. Most of the techniques used for bulk CMOS circuits remain valid for SOI circuits. Differences between bulk and SOI reside in the electrical behavior of the transistor, then all the high-level design techniques applied to reduce the power consumption (pipeline or parallel architecture, algorithm optimization) still need to be used. At circuit level, the power consumption results from the contribution of all the blocks: logic part (full custom or standard cells based), clock trees, memories (SRAM, DRAM), and I/Os. A power analysis must be run to identify the most consuming parts of the circuit. At system level, not only the circuit contribution is accounted for but also all the peripherals (screen, disk) to get a global estimation of the power budget.

Power consumption review

The power consumption of a digital circuit is generally divided in two components: static (or leakage) and dynamic: $P_{tot} = P_{stat} + P_{dyn}$. P_{stat} is dissipated during the quiescent periods, when the circuit is not active. This may correspond to long inactivity periods (stand-by mode) or short time periods between transitions. P_{dyn} is dissipated during the transitions (any transistor switching from off to on or on to off state). For advanced processes (90nm and below), the static current I_{stat} has different components depending on the state of the transistor.

When the transistor is off:

- I_{off}: subthreshold current which depends exponentially on the threshold voltage V_t
- I_{goff}: leakage between drain-source and gate due to band-to-band tunneling through the gate oxide in the drain-source/gate overlap region
- gate induced drain leakage (GIDL)

When the transistor is on:

- I_{gon}: leakage between the gate and the channel of the transistor plus drain-source/gate overlap region due to band-to-band tunneling through the gate oxide

A reverse junction leakage between drain-source and well or between wells must be added for bulk, which does not exist for SOI thanks to the dielectric isolation provided by the buried oxide.

P_{dyn} has two components: one is the power dissipated in the output capacitive load, the other is the short-circuit power dissipation in the cells. The capacitive output load C_L is the sum of several capacitances: cell output node C_{out} (junction and gate overlap capacitance), interconnect wires connected to the output (C_{wires}), input capacitances of all driven cells (C_{in}).

Reducing the power consumption is a global effort to reduce both P_{stat} and P_{dyn}. Depending on the applications, either P_{stat} or P_{dyn} may be the dominant component (for example P_{stat} for a cell phone which is in stand-by mode most of the time).

Static power

Circuits use many different types of cells such as inverter, Nand, Nor, more complex combinatorial cells (AOI, OAI, multiplexers) and sequential cells (latch, flip-flop). The static power results from the contribution of all the cells and for every cell, it depends on the logic state of the different inputs. As an example, Table 1 summarizes the static current expressions for the

0-7803-9212-4/05/$20.00 ©2005 IEEE

different combinations of the three inputs (A, B, C) of a Nand3 gate (Fig. 1).

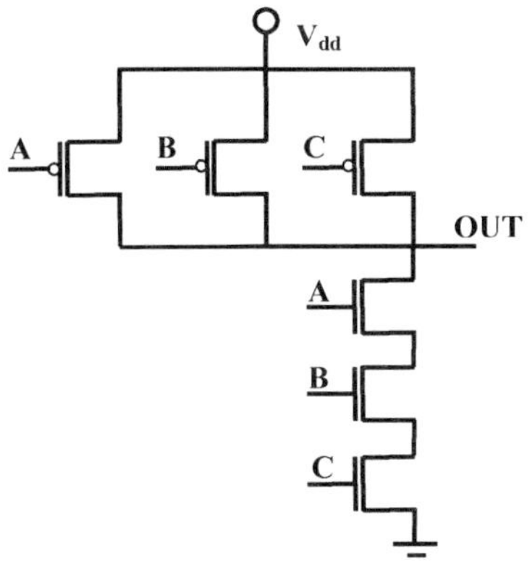

Fig. 1 – Nand3 schematics

A	B	C	I_{stat}
0	0	0	$3I_{gonP} + I_{offN3}$
0	0	1	$I_{gonN} + 2I_{gonP} + I_{offN2}$
0	1	0	$2I_{gonP} + I_{offN2}$
0	1	1	$2I_{gonN} + I_{gonP} + I_{offN}$
1	0	0	$2I_{gonP} + I_{offN2}$
1	0	1	$I_{gonN} + I_{gonP} + I_{offN}$
1	1	0	$I_{gonP} + I_{offN}$
1	1	1	$3(I_{gonN} + I_{offP})$

Table 1 – Nand3 static current expressions

N and P refer to NMOS and PMOS. $I_{offN}(I_{offP})$ is the NMOS (PMOS) subthreshold current. I_{offN2} and I_{offN3} are the subthreshold currents when two or three NMOS are in series. $I_{gonN}(I_{gonP})$ is the NMOS(PMOS) gate leakage. The I_{goff} component has been neglected. Reducing the static power requires to reduce both I_{off} and I_{gon} for advanced processes. I_{gon} may be reduced by increasing the gate oxide thickness T_{ox} or decreasing the power-supply voltage V_{dd}. Using a high-k dielectric material would be a convenient solution but it has still not been implemented in an advanced manufacturable process today. I_{gon} is the same for bulk and SOI if the same gate oxide and V_{dd} are used. I_{off} may be reduced by increasing V_t, either by increasing the doping level in the channel region or increasing T_{ox} which also reduces I_{gon}. Advanced SOI CMOS processes currently implement partially-depleted (PD) transistors. In this case I_{off} variation with V_{dd} is different between SOI and bulk (Fig. 2). In the bulk case, I_{off} depends on V_{dd} due to the DIBL effect (V_t lowering with V_{dd} increase). In the SOI case, on top of the DIBL effect, I_{off} is further amplified by the floating-body effect as the body is self-biased due to impact ionization and drain-body reverse junction currents (majority carriers are injected from the drain into the body).

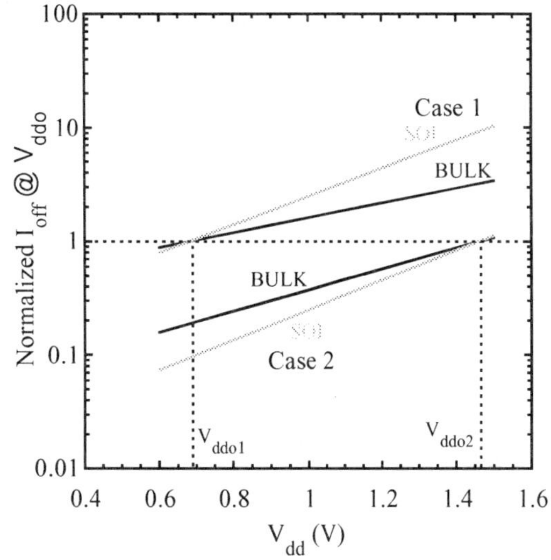

Fig. 2 – Bulk and SOI $I_{off}(V_{dd})$ (90nm process)

I_{off} may be the same for bulk and SOI at a given V_{ddo} value, depending on the targeted V_t (all processes offer several V_t options). Decreasing V_{dd} to reduce I_{off} is more efficient for SOI than bulk due to the reduction of the floating-body amplification. The drain-source voltage in cells using stacked transistors (Nand, Nor, etc..) is naturally reduced when several transistors in series are in off-state. For SOI cells, the I_{off} reduction is more important than for bulk. When comparing SOI to bulk P_{stat}, it is then necessary to compare all the cells used for circuit design. The comparison cannot be limited to a transistor comparison and the best way to achieve it is to run a static power analysis at circuit level.

I_{off} exhibits a strong dependence with temperature due to V_t variation. For a given application, I_{off} should be considered in the worst case conditions: highest V_{dd}, highest temperature, fastest process corner when accounting for process variations (generally corresponding to smallest gate length and thinnest gate oxide values). I_{off} variation with temperature is different between bulk and SOI. For PD SOI transistors, I_{off} depends on the body self-bias which mainly results from a balance between impact ionization and recombination at the junctions. Recombination increases when temperature increases and body self-bias consequently decreases. If bulk and SOI I_{off} are identical at a given temperature T_o at same V_{dd}, SOI I_{off} will be lower above T_o and higher below T_o.

It is important to note that reducing V_{dd} not only reduces I_{stat} but also directly reduces the static power as $P_{stat} = V_{dd} * I_{stat}$. P_{stat} can be reduced by lowering V_{dd} as far as the speed required by the application is maintained. This is easier to achieve with SOI as the transistor has intrinsically a higher drivability than bulk during transitions between logic states. To clarify this point, let us consider the same transistor architecture used for bulk and SOI (same channel doping and gate oxide thickness). At same V_{dd}, SOI will be faster due to SOI advantages that will be described in the next section and I_{off} will be higher due to body self-bias. As SOI is faster, V_{dd} may be decreased to $V_{ddSOI}<V_{dd}$ to match the speed achieved with bulk (given maximum frequency). Then I_{stat} has to be compared at V_{dd} for bulk and V_{ddSOI} for SOI. An immediate gain is the decrease of I_{gon} compared with bulk. If bulk and SOI I_{off} are matched, SOI I_{stat} is lower than bulk as I_{gon} is reduced. Fig.3 shows the measured $I_{stat}(V_{dd})$ variation at different temperatures for an SOI Nand3 gate. I_{stat} variation is always well fitted by $I_{stat} = I_{stato}\exp(kV_{dd})$ where I_{stato} and k depends on temperature, thus confirming the exponential dependence of I_{stat} on V_{dd}. Table 2 shows measured P_{stat} and speed reduction for an SOI Nand3 cell at 125°C when V_{dd} is lowered.

V_{dd} lowering	P_{stat} (%)	Speed (%)
1.1V → 1V	31	12
1.1V → 0.9V	53	29

Table 2 – 90nm SOI Nand3

However if a very low P_{stat} is targeted, high-Vt transistors with relatively thick gate oxide to minimize the gate leakage, so called low-leakage transistors, need to be used. The counterpart is that speed performance is strongly reduced and low V_{dd} operation is more difficult to achieve. Cells built with low-leakage or standard transistors will be mixed in different circuit blocks to achieve both low P_{stat} and high speed where it is required (critical paths).

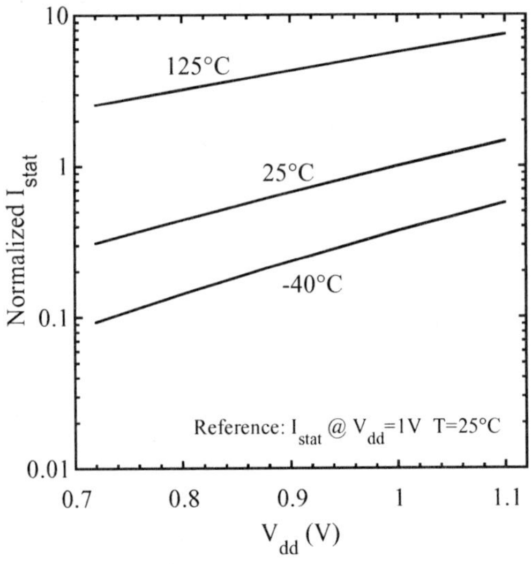

Fig. 3 – Nand3 $I_{stat}(V_{dd})$ (90nm SOI process)

Substrate or well reverse bias, increasing V_t, is used in bulk to reduce P_{stat} in stand-by mode. This technique requires to bring an extra power-supply source (negative for NMOS). This is only a partial solution for advanced processes as $V_t(V_{bs})$ variation is less pronounced and the gate leakage is not suppressed.

Dynamic power

P_{dyn} for a circuit running at clock frequency F is calculated from $P_{dyn} = (C_L V_{dd}^2 + E_{int})\alpha F$ where α is the activity factor (<1) as all the cells do not switch during one clock period. E_{int} is the internal energy dissipated during rise and fall output transitions due to the short-circuit current I_{sc}. When running dynamic power analysis at circuit level, C_L is determined from C_{in} and extracted C_{wires} once the circuit has been routed. The energy dissipated in C_{out} is generally included in E_{int}. E_{int} also depends on V_{dd}, corresponding to:

$$E_{int} = V_{dd}\left(\int_{rise} I_{sc}dt + \int_{fall} I_{sc}dt\right)$$

noting that I_{sc} also depends on V_{dd}. Total P_{dyn} results from the contribution of all the circuit blocks and all the cells in every block. Reducing P_{dyn} consists in reducing the different components. C_{wires} has been reduced by introducing low-k dielectrics in the interconnections but increases again as the interconnection pitch is reduced when going to more advanced processes. C_{wires} may be considered to be identical for bulk and SOI. C_{out} is smaller for SOI as the junction capacitance is reduced, however the resulting P_{dyn} reduction will not be significant in circuit areas where $C_{out}<<C_{wires}$.

The higher drivability of the SOI cells using PD transistors during switches results from the combination of two effects: capacitive coupling between the terminals and the body (Fig. 4) yielding a dynamic V_t lowering (during transitions) and no reverse substrate bias effect when transistors are stacked (Nand, Nor, etc..). Also the substrate and well ties used in bulk are not needed for SOI, the corresponding saved area is used to increase the width of the transistors, further enhancing the drivability of the cells.

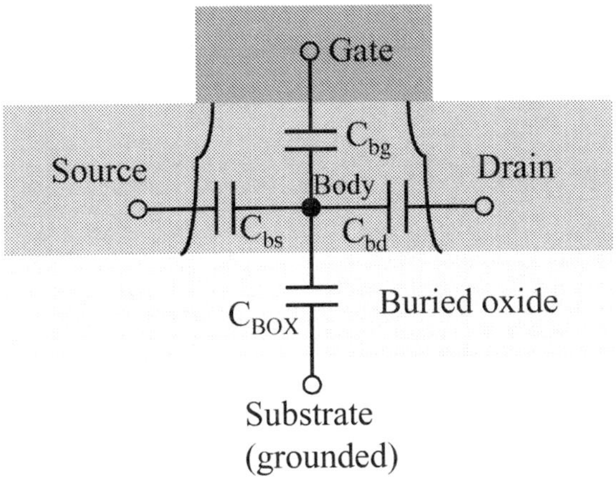

Fig. 4 – Capacitive network of PD SOI transistor

At same V_{dd}, area and load, an SOI cell is faster than bulk while P_{stat} is higher as explained in the previous section. P_{dyn} may be reduced in two different ways when using SOI. The first way is to reduce V_{dd} and achieve the same speed than bulk, reaching the speed requirement for a given application. Measured $P_{dyn}(V_{dd})$ variation for basic cells (Fig. 5) is well fitted by $P_{dyn} = P_oV_{dd}^2$. Scaling V_{dd} from 1V down to 0.8V reduces P_{dyn} by 36% as expected from the V_{dd}^2 dependence. For the same V_{dd} reduction, a 44% P_{dyn} reduction is measured on a 4 bits arithmetic logic unit (ALU), indicating that P_{dyn} may exhibit more than a quadratic dependence on V_{dd} at circuit level. The second way is to keep the same V_{dd}, still achieving the speed required by the application. Using bulk and SOI standard cells libraries, the tool synthesizing the circuit will select in the library the cells adapted to the load to reach the timing constraint. As SOI cells are faster, smaller cells will be used if the library offers a good drive strength granularity. This will result in a smaller circuit area and reduced P_{dyn} and P_{stat}.

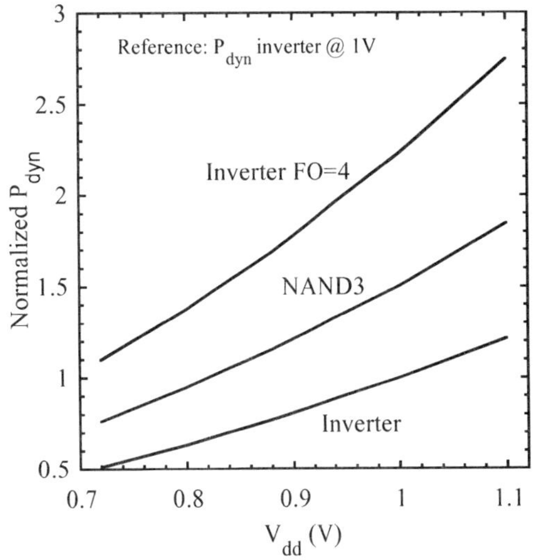

Fig. 5 – Pdyn(Vdd) for different cells (90nm SOI)

Summary

SOI offers an efficient static and dynamic power reduction through V_{dd} reduction without sacrifying speed performance. This V_{dd} reduction may be combined or not with other circuit techniques: MTCMOS where high-V_t transistors are connected in series between the core logic and ground/V_{dd}, gated clock cells where the clock is automatically diabled and not propagated. A low-power SOI circuit may be designed by gathering several blocks using different V_{dd} values, depending on the speed requirement for each block. In such a case level-shifters are used to adapt the V_{dd} difference between the blocks (V_{dd1} output to V_{dd2} input and V_{dd2} output to V_{dd1} input). This requires the implementation of several power-supply sources at system level. Implementing these techniques adds more design complexity and EDA tools must be able to support them. Making a comparison between bulk and SOI is not an easy task and it cannot be done only at transistor or cell level. The best way is to compare the results (speed, P_{stat}, P_{dyn}, area) obtained when a circuit is synthesized, placed and routed using bulk or SOI libraries, including the memories.

2005 IEEE INTERNATIONAL
SOI CONFERENCE

Session #2
DEVICE CHARACTERIZATION

CO-CHAIRS

Rene Zingg

Philips Semiconductors

Keith Jenkins

IBM/T.J. Watson Research Center

DEVICE CHARACTERIZATION Session 2

Evaluation of ESD characteristics for 65 nm SOI technology

S. Mitra, C. Putnam, R. Gauthier, R. Halbach, C. Seguin, A. Salman*
IBM Microelectronics Semiconductor Research and Development Center, Essex Junction, VT 05452
* AMD, One AMD place, Sunnyvale, CA 94088

Introduction

With aggressive scaling and continuous drive for higher performance requirements, Electrostatic Discharge (ESD) is becoming a major reliability challenge for advanced integrated circuits. Products must be designed with proper ESD protection circuits to provide adequate robustness and as the limits of the device capability are reached, factors like device reliability due to ESD sensitivity become more critical. In this paper, ESD characteristics of I/O elements offered in 65nm SOI technology are evaluated. Industry standard ESD protection levels (2kV HBM, 200V MM and 500V CDM) can be achieved with appropriate I/O circuit implementation.

Technology Features

The 65nm SOI technology is designed for both SRAM and logic applications and has the following features [1]:

- Twin well CMOS technology of Vdd up to 1.0V on partially depleted SOI
- Two different gate oxide thicknesses of 1.05nm (SG) and 2.35nm (DG)
- Thick oxide (DG) decoupling capacitors
- Low-k BEOL dielectric material

The suite of ESD elements investigated for this study includes

1. Traditional ESD devices
   a) P+/Nbody polysilicon bounded ESD diodes and diode strings
   b) RC-triggered SG and DG NFET power clamps (with 3 inverters driving the big FET)
2. Alternative ESD devices
   SBLK diode: alternative P+/Nwell diode formed by using silicide blocking (SBLK) instead of polysilicon [2]
3. Other I/O structures
   Silicided NFETs and PFETs (3 and 4 terminal floating body and H-body configurations)
4. Resistors: Buried Resistor (BR) and OP P^+ Poly Resistor
5. BEOL structures: Wires, contacts and vias

ESD Background

Three ESD models described by JEDEC and the ESD Association are Human Body Model (HBM), Machine Model (MM) and Charged Device Model (CDM).

An HBM discharge can occur between any two pins on a chip, and, as its name implies, is based on the interaction with a charged human. Products commonly target 2kV HBM robustness, which is a decaying exponential discharge with a peak current of 1.33A and time constant of 150ns. MM is similar to HBM but its waveform is a bi-directional damped oscillation with a peak current of 3.8A for 200V.

A CDM event occurs when a single pin of a triboelectrically charged module is grounded. This produces a pulse on the order of 1 to 2ns. CDM protection is generally required at 500V, typically corresponding to 5 to 10A of current.

Measurements Performed

Testing was performed at wafer level using an automated Transmission Line Pulse (TLP) tester with three different ESD pulses (100ns, 30ns and 1ns) and a wafer level HBM setup. Also, DC characteristics are tested to ensure adequate normal operating behavior prior to TLP testing. Under TLP, a square wave is delivered to the Device Under Test (DUT) and measurements are taken of the voltage across and the current through the DUT [3,4,5]. A 100ns TLP current of 2A corresponds roughly to the energy of an HBM event of 3kV and a Very Fast TLP (VFTLP) setup providing a 1 to 2ns pulse can be used to approximate a CDM ESD event. The HBM tester is designed with a series combination of a 100pF capacitor and a 1.5kΩ resistor, as is dictated by the JEDEC and ESD Association specifications.

Discussion of Results

The P+/Nbody diode performance was evaluated under TLP and DC testing. Fig. 1 shows TLP I-V characteristics for a 55μm finger width P+/Nbody diode at three representative perimeters under three different pulse widths (100ns, 30ns and 1ns). It is observed that the diode current handling capabilities are inversely proportional to pulse width, as expected. The diode failure region and the failure current (I_{fail}) for 100ns TLP are verified by simulating a 110mm perimeter diode and comparing the result with the measured data. It is observed in Fig. 2 that the measured data and the simulated result match well. The internal device temperature (T_{max}) rises with the higher pulse amplitude and T_{max} reaches ~1100 °C, corresponding approximately to the device failure temperature. The diode on-resistance (R_{on}) and failure current (I_{t2}) for 100ns TLP are plotted vs. perimeter in Fig. 3, where it is seen that R_{on} (measured from 0.2A to 1.0A) is inversely proportional to diode perimeter and I_{fail} is directly proportional to diode perimeter. The data also reveals that R_{on} in larger perimeter diodes drops less with perimeter as the wiring resistance begins to dominant the overall resistance [6].

Figs. 4 and 5 show the TLP I-V characteristics of RC-triggered silicided SG and DG NFET power clamp for two different widths. It is observed that for the same RC time constant (R=100kΩ and C=10pf), the wider power clamp (W=9000μm) is providing better ESD protection as expected -- 2A under 100ns TLP. HBM tests also showed

0-7803-9212-4/05/$20.00 ©2005 IEEE

it exceeding 4kV. Both power clamps turn on at ~0.5V. The higher R_{on} (~1Ω) observed in SG and DG power clamps, is a result of the lack of higher level metal strapping on the tested wafers. With the higher level metal strapping present, R_{on} would be ~0.5Ω.

Fig. 6 shows TLP I-V characteristics for a 55μm finger width SBLK diode at three representative perimeters under three different pulses widths (100ns, 30ns and 1ns). It is observed that diodes with perimeter ≥ 330μm can handle 2A under 100ns TLP. The scaling of I_{fail} and R_{on} across the three perimeters is shown in Fig. 7, which reveals higher R_{on} compared to the P+/Nbody diodes, but follows the same trend with perimeter as seen in Fig. 3. The SBLK diode design is still being optimized. To realize an optimal device design, R_{on} and I_{fail} are plotted against various anode/cathode spaces in Fig. 8, which shows that smaller anode/cathode space offers a lower R_{on} and higher I_{fail}. Further analysis on the measured and simulated anode capacitance of P+/Nbody poly-bounded and SBLK-bounded diode, shows that the SBLK diode has lower capacitance (0.3fF/μm) which is due to the fact that the gate capacitance increases the poly-bounded diode loading capacitance (0.35fF/μm). The simulation result matches well the measurements except the SBLK diode has some offset which could be due to the non-optimal process control of the SBLK mask compared to that of the poly mask which again also explains the higher resistance of the SBLK diode.

The TLP I-V characteristics for 100ns and 30ns pulse widths for silicided SG and DG NFETs are reported in Figs. 9 and 10, respectively. The subsequent variation of the trigger voltage (V_{t1}) with NFET design lengths are plotted in Fig. 11, which shows a drop of V_{t1} at the nominal L_{design} for 100ns TLP testing. Further investigations showed that V_{t1} at nominal L_{design} does not change significantly with pulse width variations. For SG and DG oxide PFETs, V_{t1} is ~1V higher than the corresponding NFET version at the same design length. SG and DG silicided NFETs are also characterized for the gate-oxide breakdown voltage by applying 100ns TLP at the gate while other terminals are all tied to ground. Table 1 presents a summary of the ESD parameters derived from the SG and DG silicided NFETs.

Other I/O structures that may see high current during an ESD event are resistors and BEOL wiring. For this technology, the resistors available are the BR and OP P+ Polysilicon resistor. These resistors need to be sized accordingly to handle ESD events. Fig. 12 shows the 100ns TLP characteristics of BR of various widths. It is seen that the I_{fail} is scaling with the width of the resistor.

Various BEOL structures comprising of metal levels, contacts and vias are also characterized to determine requirements to achieve industry standard HBM and CDM robustness using the TLP and HBM measurements. Fig. 13 shows ESD robustness of the M1 level by plotting I_{fail} and V_{fail} as extracted from the HBM measurement, vs M1 width for various M1 lengths. This clearly shows that a minimum M1 width of ~12.5μm is needed to handle of 3kV HBM. The HBM robustness of all the BEOL structures, in the form of normalized V_{fail} as a function of the width/length for the Mx (x=1, 2, 3) wires or as a function of the total number of CAs or vias, are presented in Table 2.

Conclusion

The ESD characteristics of I/O elements in 65nm SOI technology are thoroughly evaluated. With an appropriate design implementation using these discrete elements, industry standard ESD robustness can be achieved.

References

[1] E. Leobandung, et al, Symposium on VLSI Technology and Circuits, 2005.
[2] C. Putnam, et al, IEEE Intl. SOI Conference Proceedings, 2004.
[3] T. Maloney and N. Khurana, EOS/ESD Symposium, 1985.
[4] E. Grund and R. Gauthier, EOS/ESD Symposium, 2003.
[5] E. Grund and R. Gauthier, EOS/ESD Symposium, 2004.
[6] K. Chatty, et al, IEEE Intl. Rel. Physics Symposium, 2005.

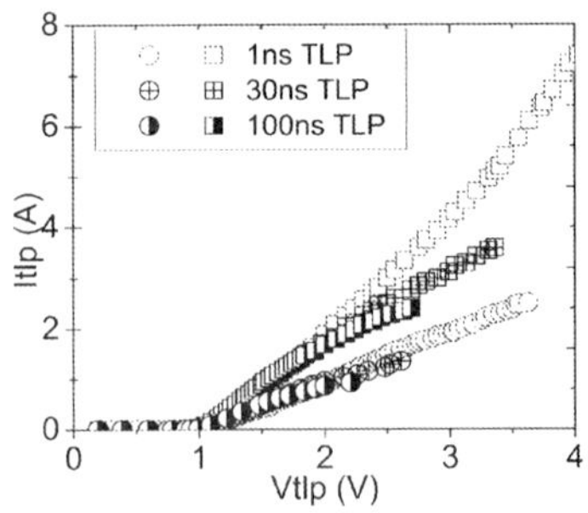

Figure 1: 100ns/30ns/1ns TLP I-V characteristics for 55μm finger width P+/Nbody diode for 110μm (○) and 330μm (□) perimeters.

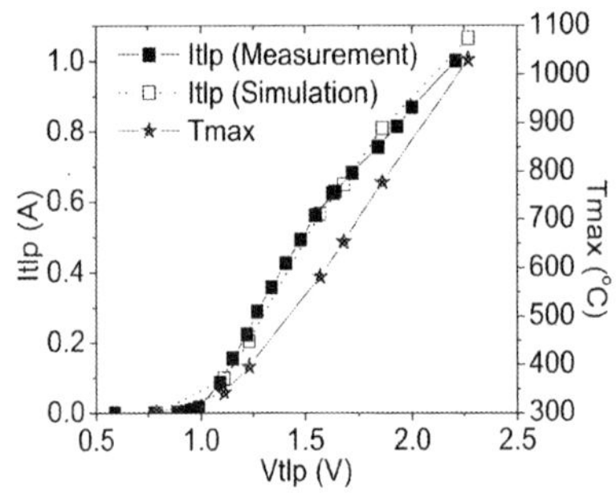

Figure 2: Simulation and hardware correlation of the 100ns TLP I-V characteristics for P+/Nbody diode of perimeter=110μm, showing that the diode goes in failure regime as the internal temperature reaches ~1100 °C

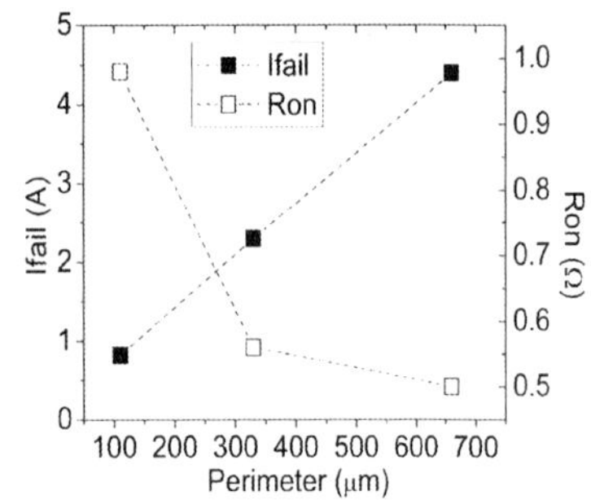

Figure 3: Variation of Failure Current (I_{fail}) and on-resistance (R_{on}) with P+/Nbody diode perimeter based on 100ns TLP I-V characteristics.

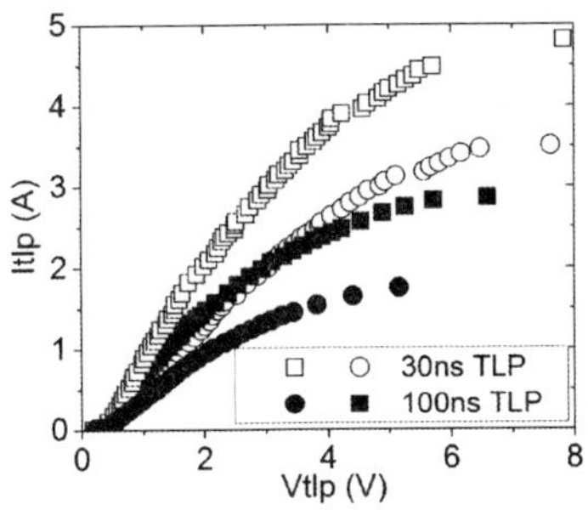

Figure 4: 100ns/30ns TLP I-V characteristics for thin-oxide RC-triggered NFET power clamp with widths 3300μm (○) and 9000μm (□).

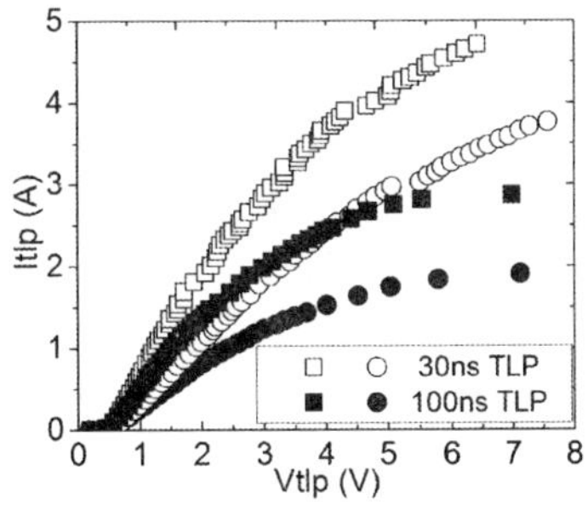

Figure 5: 100ns/30ns TLP I-V characteristics for thick-oxide RC-triggered NFET power clamp with widths 3300μm (○) and 9000μm (□)

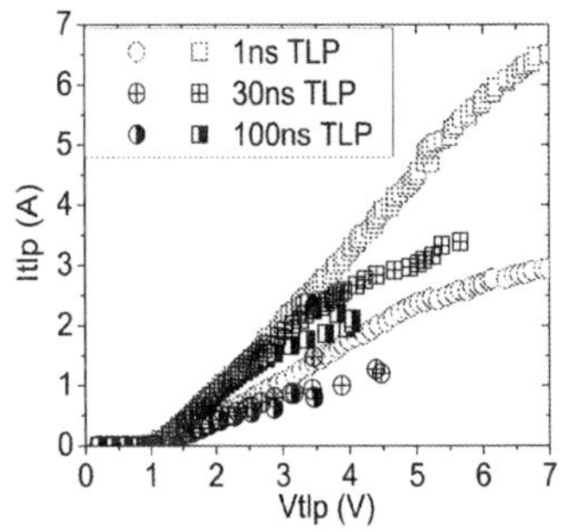

Figure 6: 100ns/30ns/1ns TLP I-V characteristics for 55μm finger width SBLK diode for 110μm (○) and 330μm (□) perimeters.

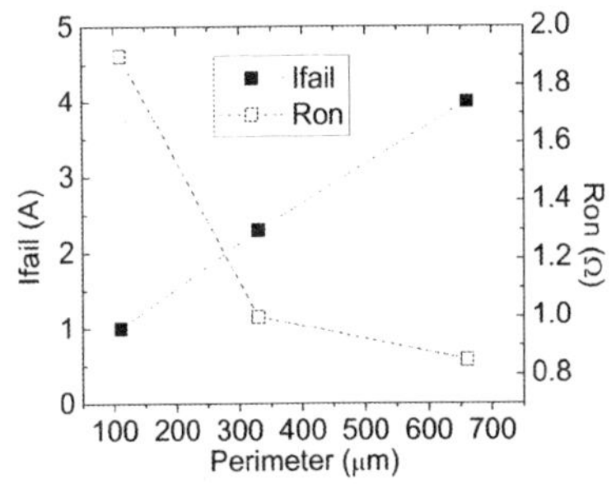

Figure 7: Variation of I_{fail} and R_{on} with cathode to anode spacing in 330μm perimeter diode based on 100ns TLP I-V characteristics.

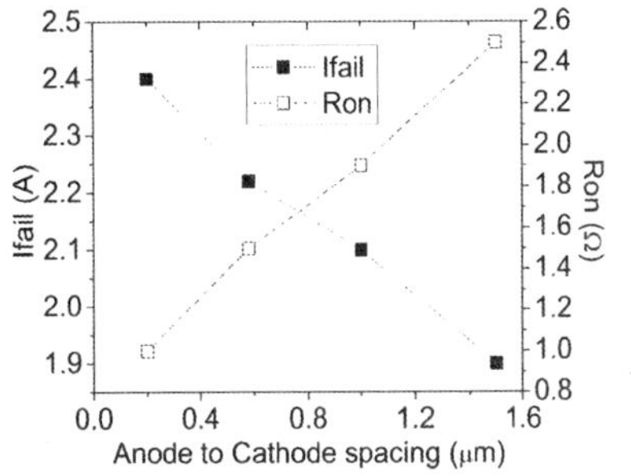

Figure 8: Variation of I_{fail} and R_{on} with cathode to anode spacing in 330μm perimeter diode based on 100ns TLP I-V characteristics.

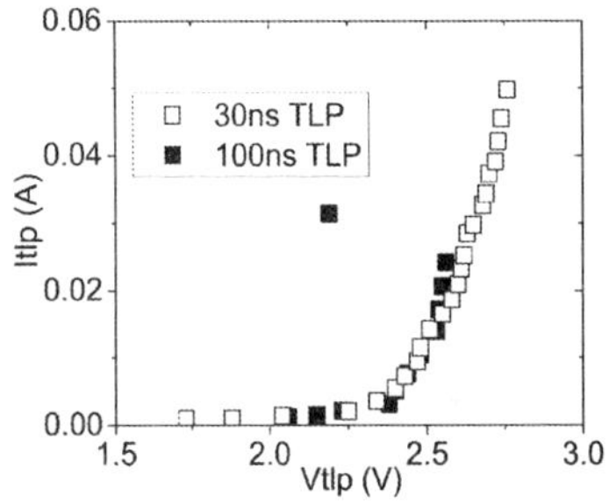

Figure 9: 100ns/30ns TLP I-V characteristics for thin-oxide silicided NFET (W=288μm) [1ns TLP I-V data is not shown to maintain clarity]

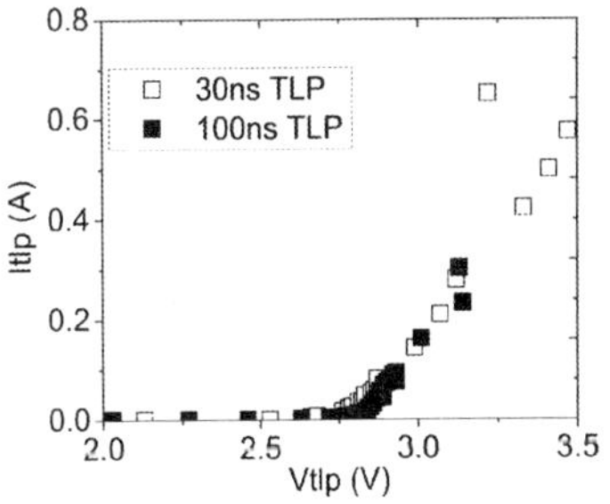

Figure 10: 100ns/30ns TLP I-V characteristics for thick-oxide silicided NFET (W=288μm) [1ns TLP I-V data is not shown to maintain clarity]

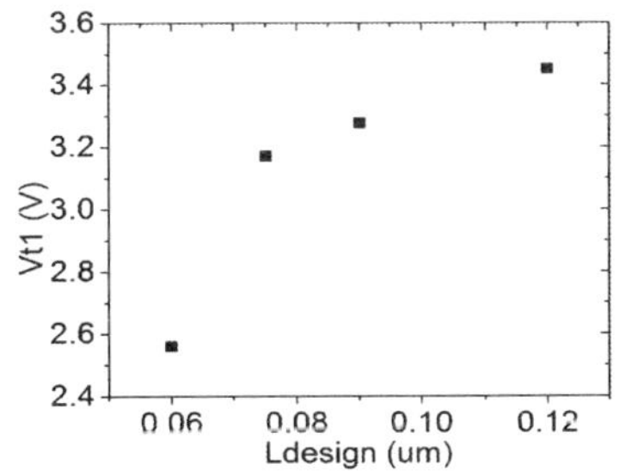

Figure 11: Variation of trigger voltage (Vt_1) with Ldesign for thin-oxide silicided NFET for 100ns TLP.

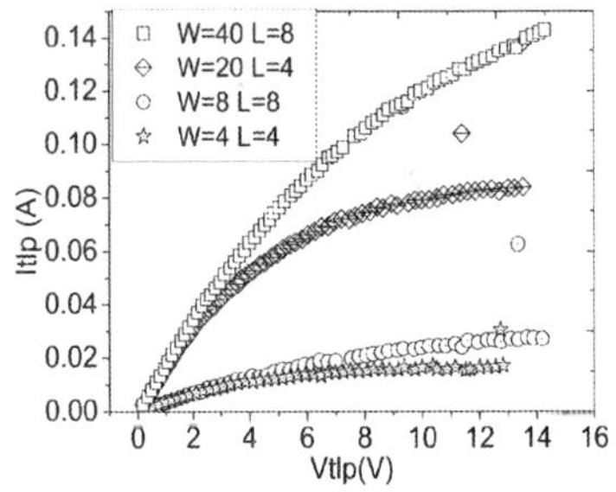

Figure 12: 100ns TLP I-V characteristics for BR (floating PC) of different width and lengths (W/L dimensions in μm)

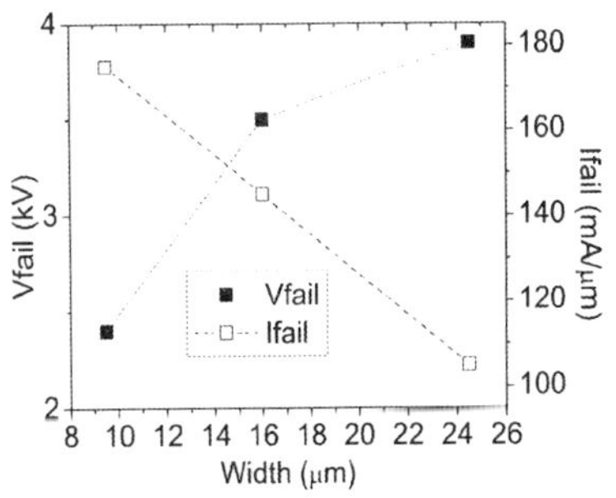

Figure 13: Variation of V_{fail} and I_{fail} with the M1 width for M1 length = 1500 μm as extracted from the HBM measurement.

Table 1: Important ESD parameters for SG and DG silicided NFETs

ESD Parameters (based on 100ns TLP)	Thin-oxide silicided NFET (Lnom=0.06μm)	Thick-oxide silicided NFET (Lnom=0.12μm)
Trigger Voltage (Vt1)	2.3V	2.7V
Failure Current (It2)	≤0.1 mA/μm	1.1mA/μm
Gate-oxide breakdown voltage	~3.75V	~6.3V

Table2: BEOL ESD data based on HBM measurement

BEOL structure	Required W/L in μm	Required Number	Vfail
M1	12.5/1500		0.24kV/μm
M2	10/1500		0.25kV/μm
M3	10/1500		0.25kV/μm
CA		100 for 1.0kV	2.24V per CA
Via		160 for 2.5kV	15.6V per Via

Transport and Leakage in Super-Critical Thickness Strained Silicon Directly on Insulator MOSFETs with Strained Si Thickness up to 135 nm

I. Åberg[1], Z. Cheng[2], T. A. Langdo[2], I. Lauer[1], A. Lochtefeld[2], D.A. Antoniadis[1], and J.L. Hoyt[1]
[1]MIT Microsystems Technology Laboratories, Cambridge, MA
[2]AmberWave Systems Corporation, Salem, NH
Email: iaberg@mit.edu, Tel: (617) 452-3609

Introduction

Strained Si directly on insulator (SSOI) has been suggested as a candidate to enhance transport in deeply scaled ultra-thin body fully-depleted silicon-on-insulator (FD-SOI) technogies [1-3]. To evaluate SSOI as a candidate for partially depleted applications, using thicker Si films (PD-SOI), issues relating to epitaxial growth and device processing of films with strained Si thickness (T_{Si}) greater than the critical thickness, must be addressed. Previous work has indicated that FD-SSOI is more thermally stable and less prone to misfit dislocation induced off-state leakage in super-critical films than corresponding strained Si/relaxed SiGe bulk technology [4,5]. In this work, we study both FD- and PD-SSOI with aggressive T_{Si} of up to 135 nm for 14% SSOI (14% Ge equivalent strain). The SSOI MOSFET device structure is shown schematically in Fig. 1. The as-grown heterostructure film thickness was up to 10X thicker than the equilibrium critical thickness [6].

SSOI and MOSFET fabrication

14% SSOI substrates were fabricated by transfer of the strained Si film to an oxidized handle wafer by bonding and removal of all SiGe layers [7]. The buried oxide thickness is 200 nm. Samples with as grown T_{Si} of 50, 100, and 150 nm were studied. After layer transfer T_{Si} was 38, 85, and 135 nm, respectively, measured by ellipsometry. The strain of the SSOI substrates was characterized by X-ray diffraction (Fig. 2). The strain after substrate fabrication was eqivalent to the strain in fully strained Si on relaxed $Si_{0.86}Ge_{0.14}$ (lattic strain ~0.58%). This is similar to the target 14.5±0.5% Ge in the as grown relaxed buffer layers, showing that the strain was maintained during substrate fabrication for all T_{Si}. N- and p-MOSFETs were fabricated on the SSOI substrates, and on bulk-Si CZ reference wafers. In addition to the T_{Si} splits, high- and low-thermal budget splits were also made. For the high thermal budget split, LOCOS was grown by wet oxidation at 1000ºC for 30 minutes independent of T_{Si} (Fig. 1(a)). In total, the wafers received ~1 hour at 1000ºC. The field region was partially recessed prior to LOCOS, for the T_{Si}=135 nm substrates. Some of the LOCOS wafers were implanted with BF_2 (1.5×10^{13} cm^{-2} at 30 keV) to create PD-SSOI N-MOSFETs. The remaining wafers had un-doped bodies on which both N- and P-MOSFETs were fabricated. The source/drain activation was at 1030ºC for 10s for all the LOCOS wafers. Si bulk wafers were included with the LOCOS SSOI process. A low thermal split was also prepared using mesa isolation instead of LOCOS (Fig. 1(b)). The source/drain activation was 900ºC for 10s for the mesa isolated wafers. For all the wafers, 5 nm-thick gate oxides were grown at 800ºC. N+ poly-Si gates were used for both N- and P-MOSFETs.

Results

Subthreshold slopes for the T_{Si}=38 nm LOCOS split were 68 and 74 mV/dec. for P- and N-MOSFETs, respectively (Fig. 3). Electron mobilities for the 14% SSOI LOCOS devices were enhanced 1.5-1.6X compared to the Si bulk control (Fig. 4) while hole mobilities were slightly degraded (Fig. 5), consistent with earlier results on ~15% biaxially strained Si [1]. Electron and hole mobility vs. T_{Si} is shown in Fig. 6 and 7 respectively. The electron mobility at both high and low inversion charge densities (N_{inv}) remains independent of T_{Si}, for both the high thermal budget LOCOS and low thermal budget mesa splits (Fig. 6). However, the mesa devices show slightly higher mobility at high N_{inv}. For holes, mobility also remains independent of T_{Si} for the mesa devices, while mobility appears to improve slightly with T_{Si} for the LOCOS wafers, but remain below the Si bulk control (Fig. 7). For the typical W and L used to extract mobility in this work (in the range of 10 to 100 µm) it was confirmed that mobility was independent of device geometry, suggesting that local stress induced by LOCOS isolation is not likely the origin of the difference observed between LOCOS and mesa isolated devices in Fig. 7. For strained Si on bulk relaxed SiGe, off-state leakage associated with misfit dislocations is a concern. However, even for T_{Si} much larger than the critical thickness, off-state leakage at high gate-to-drain bias remains independent of T_{Si} (Fig. 8, measured on L=1 µm devices), encouraging for the application of SSOI to PD-SOI technology. Floating body PD-SSOI output characteristics are shown in Fig. 9. In Fig. 10, the electron mobility of 85 nm thick 14% PD-SSOI is compared to that of undoped 14% FD-SSOI of equal thickness. The effective field was approximated from $E_{eff}=(Q_b+Q_{inv}/2)/\varepsilon_S$, with the bulk charge obtained by $Q_b=N_a*X_{dmax}$. The electron mobility enhancement of

1.5X compared to universal mobility is maintained in 85 nm thick PD-SSOI even in the presence of high channel doping. In Fig. 11, electron mobility in highly doped PD-SSOI is independent of thickness up to T_{Si}=135 nm.

Summary

We have demonstrated that mobility in 14% SSOI is independent of the strained Si thickness, even for as grown films 10X thicker than the critical thickness. Off-state current also remains independent of T_{Si}. The successful fabrication of PD-SSOI with electron mobility enhancement maintained at 1.5X, for high channel doping and strained Si thickness up to 135 nm, was also demonstrated, showing promise for thicker film PD-SOI applications.

We thank staff and colleagues at MTL/MIT and at AmberWave Systems. This work was made possible by support from the MARCO MSD Center and the SRC.

References

[1] K. Rim, et al., IEEE IEDM, p. 49, 2003.
[2] I. Åberg, et al., 2004 Symp. on VLSI Tech. p. 52.
[3] I. Aberg, et al., IEEE EDL, 26, p. 661, 2005.
[4] I. Lauer, et al., IEEE EDL, 25, p. 83, 2004.
[5] J.G. Fiorenza, et al., Semicond. Sci. Tech., 19, p. L4, 2004.
[6] D.C. Houghton, J. Appl. Phys. 70, p.2136, 1991.
[7] T.A. Langdo, et al., SOI Conf. Dig., p211, 2002.
[8] S. Takagi, et al., Trans. Elec. Dev., 41, p. 2357, 1994.
[9] H. Nayfeh, et al., IEEE EDL, 24, p. 248, 2003.

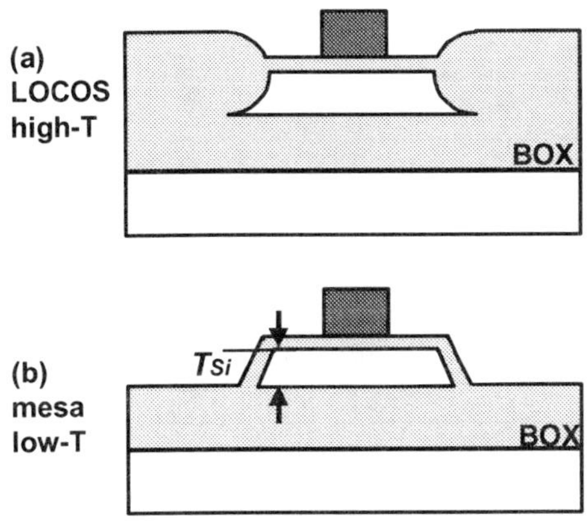

Fig. 1 14% SSOI structure and isolation/thermal splits. (a) LOCOS used 1000°C-30 min. wet oxidation, and S/D anneal was 1030°C-10 s. (b) Mesa isolation used dry etch in an SF_6 plasma. Mesa isolated wafers received S/D activation of 900°C-10s.

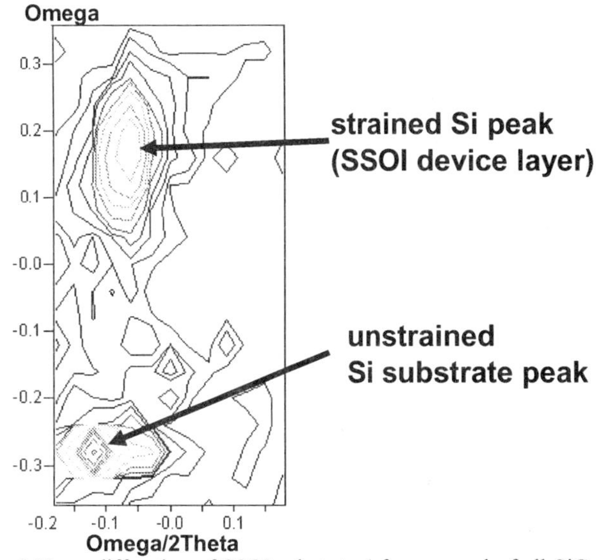

Fig. 2 X-ray diffraction of SSOI substrate (after removal of all SiGe layers, but before device fabrication), showing strain equivalent to fully strained Si on a relaxed $Si_{0.86}Ge_{0.14}$ (14% SSOI). The as grown thickness of the strained Si layer was 150 nm (135 nm after substrate fabrication). Substrates with thinner as grown thickness had the same level of strain.

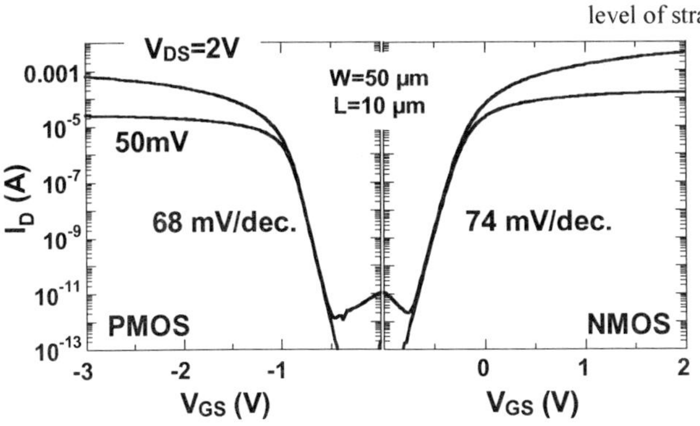

Fig. 3 I_D vs. V_{GS} for NMOS and PMOS 14% FD-SSOI (T_{Si} =38 nm). The wafers were isolated by LOCOS (Fig 1(a)). N+ poly-Si gates were used for both NMOS and PMOS.

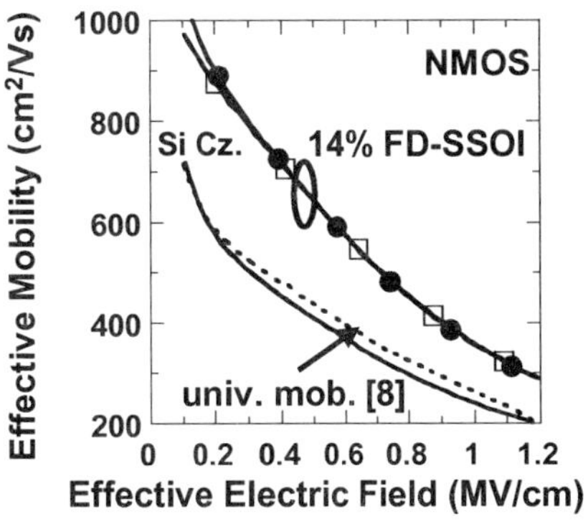

Fig. 4 NMOS effective mobility as a function of E_{eff}. μ_{eff}=L/W*I_D/($Q_{inv}V_{DS}$) was extracted by integration of C_{GC}, measuring I_D at V_{DS}=50mV. The universal mobility is indicated [8]. 14% FD-SSOI with T_{Si}=135 nm (filled symbols) has nearly identical mobility as that of T_{Si}=38 nm (open symbols).

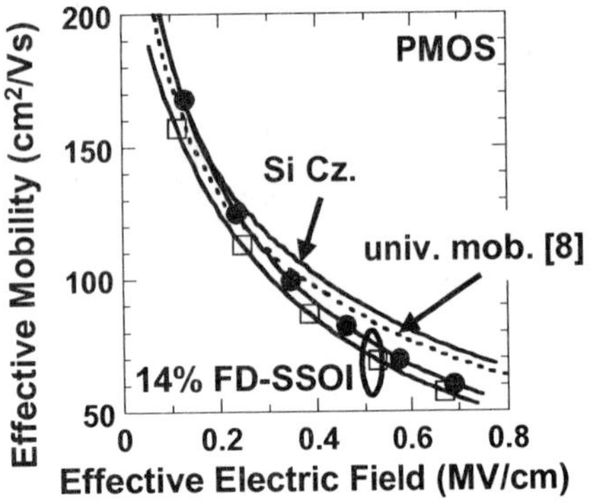

Fig. 5 PMOS effective mobility as a function of E_{eff}. 14% FD-SSOI hole mobility is slightly degraded vs. universal hole mobility [1,8].

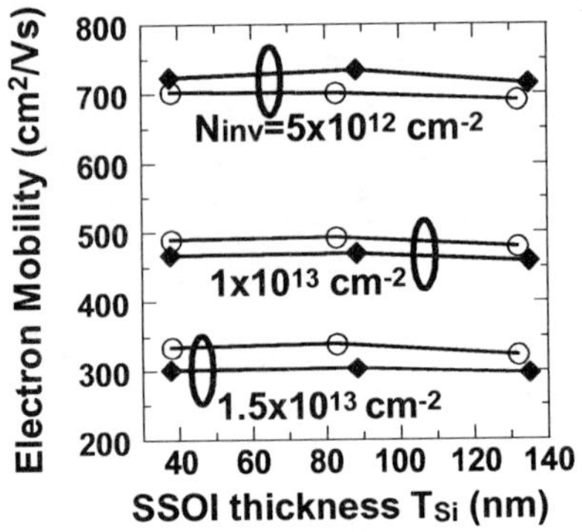

Fig. 6 Electron mobility for 14% FD-SSOI vs. T_{Si}. No thickness dependence is observed for either LOCOS (filled symbols) or mesa (open) splits.

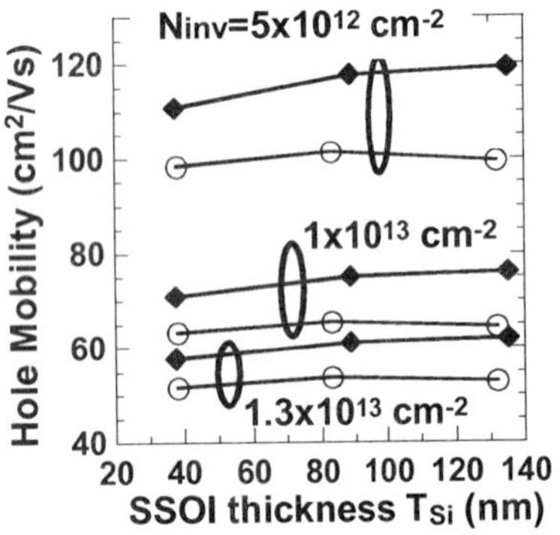

Fig. 7 Hole mobility for 14% FD-SSOI vs. T_{Si}. The LOCOS split (filled symbols) has slightly higher mobility than the mesa split (open).

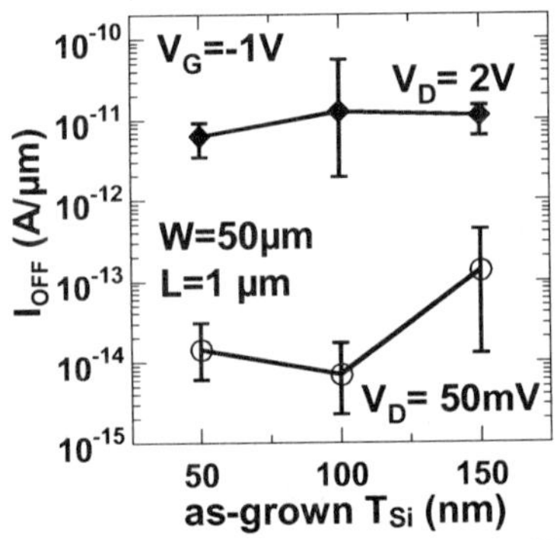

Fig. 8 I_{OFF} vs. the as grown T_{Si} for 14% FD-SSOI N-MOSFETs, measured on L=1 μm devices. Error bars indicate the range of observations. No T_{Si} dependence is observed.

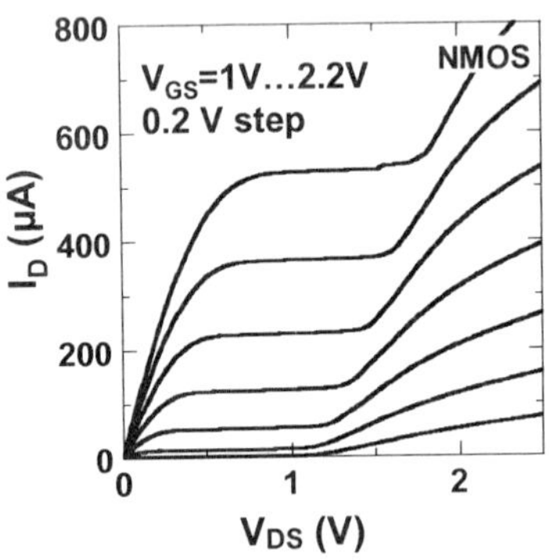

Fig. 9 Output characteristics for floating-body 14% PD-SSOI (T_{Si}=38 nm). As for unstrained PD-SOI, the kink effect is observed at high drain bias.

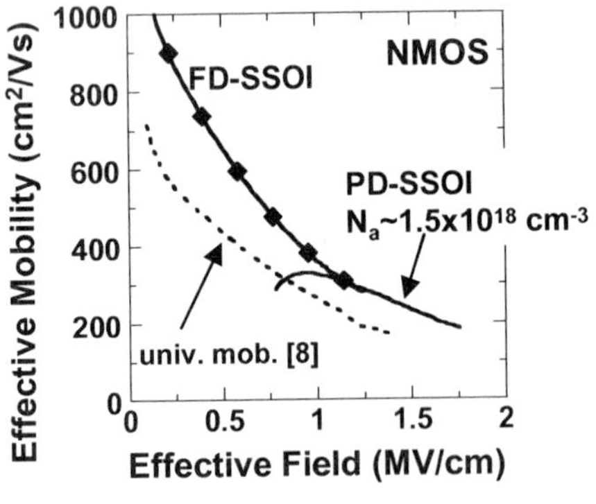

Fig. 10 Electron mobility vs. effective field. The 85 nm thick PD-SSOI mobility (line) is enhanced by 1.5X compared to universal mobility (dotted line). FD-SSOI of similar thickness (filled symbols) has equal enhancement factor, similar to findings in bulk strained Si on relaxed SiGe [9]. The doping level of the PD-SSOI was estimated by SIMS and threshold voltage measurements.

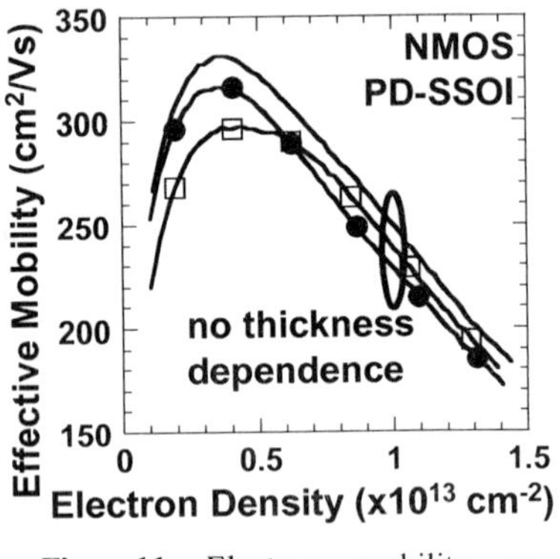

Fig. 11 Electron mobility vs. inversion electron charge density for 14% PD-SSOI. Similar mobility is observed for T_{Si}=38 nm (open symbols), 85 nm (line) and 135 nm (filled symbols).

Edge-Defined 90nm TFTs with Adjustable V_T in a 3-D Compatible Process

Jawad Nasrullah,*† James B. Burr,† G. Leonard Tyler,* Yoshio Nishi*
*Dept. of Electrical Engineering, Stanford University, Stanford, CA 94305
Telephone: (650) 725–1638, Email:jawad@ieee.org
†Transmeta Corporation, 3990 Freedom Circle, Santa Clara, CA 95054

Introduction

Threshold voltage (V_T) adjustability through back-gate bias (V_B) facilitates power management in VLSI by providing a mechanism for balancing leakage and switching energies [1]. Dynamic threshold voltage control with back-gate bias has been demonstrated in SOI [2]; it is especially important for 3-D stacking where heat builds up [3]. We implement short-channel thin-film transistors (TFTs) in solid-phase crystallized polycrystalline silicon (poly-Si) with dynamically adjustable V_T, in a 3-D compatible process. Aggressive scaling of device features is achieved via a low-temperature-compatible edge-defined patterning process [4]. Edge-defined lithography is attractive for 3-D integration, since it offers a very cost-effective solution for the multiple CD mask layers in stacked devices. This work focuses on the processing technology that achieves device V_T adjustability in second and higher layers of short-channel stacked devices. The device mobility and adjusted subthreshold behavior reported here are, however, comparable to those in other TFT work [5].

Process and Device Details

Fig.1 shows key steps of edge-defined TFT fabrication. The TFT's active area is built in a 50nm-thick deposited amorphous silicon (a-Si) channel that is later crystallized using rapid thermal annealing (RTA). An underlying-blanket, boron-doped Si film is used as a global back-gate electrode, separated from the channel by a 50-nm thick deposited back-gate-oxide film. A blanket film of 12nm deposited oxide is used for front-gate dielectric after active-area patterning. An in-situ phosphorus-doped a-Si film is deposited and patterned into 90nm-length gates, Fig.1(a). The device gates are fabricated with an edge-defined patterning process as detailed in Ref. [4]. First, a spacer is defined around a sacrificial registration block. The sacrificial block is then stripped away, leaving the outline of the registration as a spacer. The spacer width is defined by the spacer-film thickness, and controls the final gate length (L_g) by acting as the mask, Fig.1(b). A second pattern is superimposed at this stage with a patterned photoresist, enabling fabrication of edge-defined and photo-defined features at the same level. The photoresist and spacer are then used to pattern gates of the transistors, as shown in Fig.1(c). The TFT channels are undoped. Arsenic is implanted to create sources and drains, whereas the gates are deposited in-situ doped a-Si. After opening up contact holes, the devices are subjected to a 900°C spike RTA to activate dopants and to convert the channel a-Si to poly-Si form. Contact metal is a 50nm titanium film followed by a 450nm aluminum film. Lastly, the devices are annealed at 350°C in forming gas for an hour.

Results

Fig.2(a) is a micrograph of a typical device, comprised of three 90nm edge-defined front gates. These three gates are defined by two edge-registration polygons. A gate contact pad is superimposed, to make electrical connection to the gate. Fig.2(b) is a close-up of a 90nm gate stack fabricated using a similar edge-defined patterning process. Note the rounded spacer mask on top of the patterned gate, and the sidewall variations [4]. Fig.2(c) shows the device structural details.

Fig.3 shows the device performance adjustability of a typical single-gate device with the application of back-gate bias. Fig.3(a) presents the I_D–V_G family of curves for this device with values of V_B ranging from −6V to +6V. For poly-Si TFTs, the value of subthreshold slope is typically 300–400mV/decade, which is higher than the ideal value of 60mV/decade. Fig.3(b) presents the I_D–V_G family of curves for this device at V_D=1.5V. Comparison of Fig.3(a) with Fig.3(b) shows that I_{off} is 2–3 orders of magnitude higher for V_D=1.5V. The dependence of the turn-off behavior on the drain voltage results in a lower threshold voltage; hence I_D increases. Figs.3(c)–(f) show the behavior of four specific parameters vs. V_B, based on the device I_D–V_G curves for V_D=0.3 and 1.5V, described above. Fig.3(c) gives the change of V_T with V_B. Fig.3(d) shows the change in subthreshold slope, SS, with V_B. Subthreshold slope decreases with decreasing values of V_B due to better front-gate control for lower values of V_B. Fig.3(e) shows the change of the on-state current, I_{on}, with V_B for a fixed gate potential of 3.3V. On current decreases with decreasing values of V_B due to increase in V_T. For V_D=1.5V, I_{on} decreases more rapidly with reduction in V_B than is the case where V_D=0.3V. Fig.3(f) shows the off current, I_{off}, change with V_B for a fixed front-gate potential of 0V. Off current decreases due to decreasing values of V_B and increasing V_T.

Fig.4(a) shows the family of I_D–V_D curves with V_B =5V and 0V for a L_g =90nm device. These curves do not show saturation of the drain current with increasing V_D, suggesting control of the channel conductivity by the drain similar to that from the gate. We attribute this to series resistance, uncontrolled inward diffusion of dopant atoms from the source and the drain resulting in shorter electrical channel length, and the formation of unwanted resistive current paths beyond the control of the front-gate. In a quadruple-front-gate device, the I_D–V_D curves look more saturated as seen in Fig.4(b); but its turn-off behavior is similar to that for a single-front-gate transistor, Fig.4(c).

Fig.5 shows the back-gate bias behavior of device parameters for ≈20°C, 60°C, and 110°C. Fig.5(a) shows V_T change with V_B. Higher temperatures result in lower V_T. Subthreshold slope decreases with lower temperature, consistent with traditional device behavior, Fig.5(b). Fig.5(c) shows I_{on}

0-7803-9212-4/05/$20.00 ©2005 IEEE

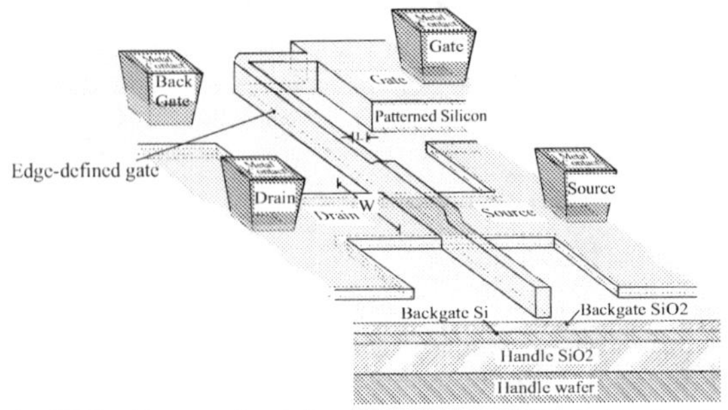

(a) TFT with one edge-defined front gate and blanket back-gate electrode. Optical lithographys is used for patterning features other than the front gate.

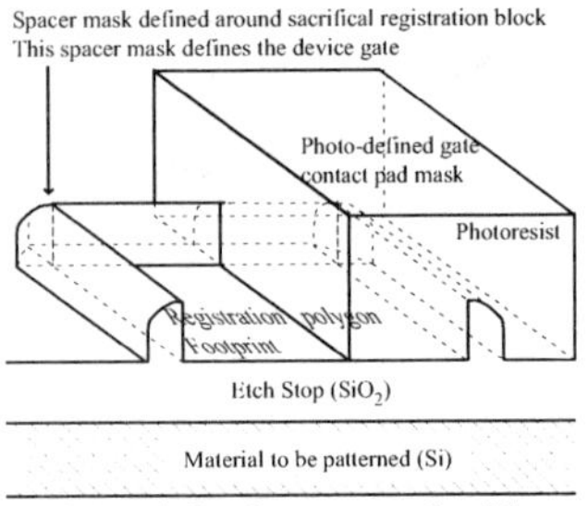

(b) Edge-defined spacer mask with superimposed photo mask.

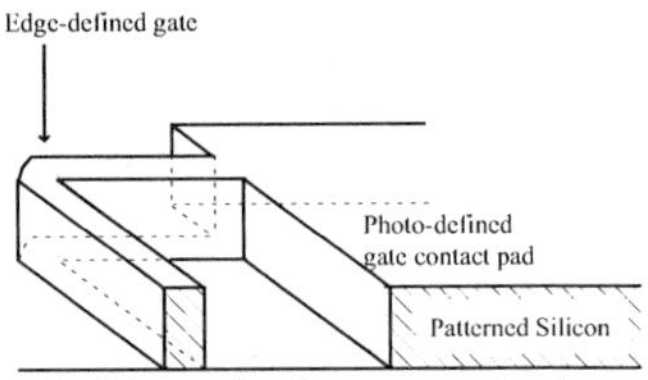

(c) Edge-defined gate and photo-defined gate contact pad.

Fig. 1. Device process summary.

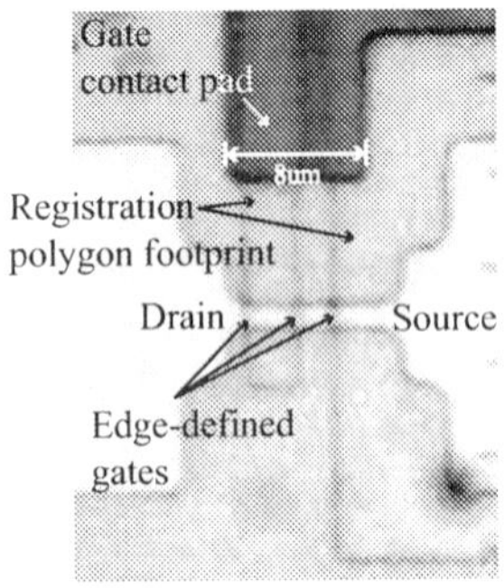

(a) Micrograph of a triple-front-gate TFT.

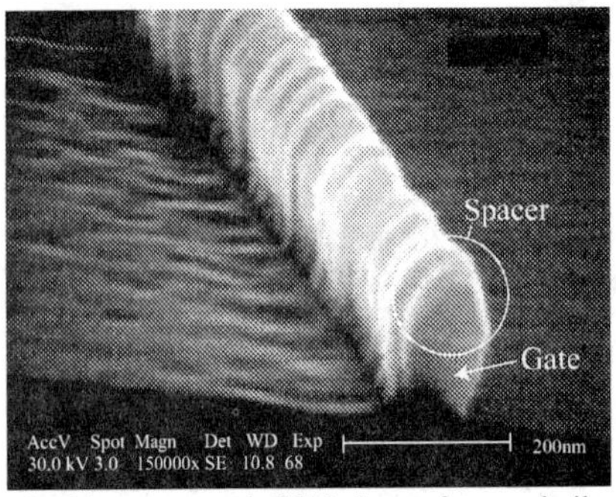

(b) 90nm gate stack SEM picture from a similar lithography experiment. scale=200nm

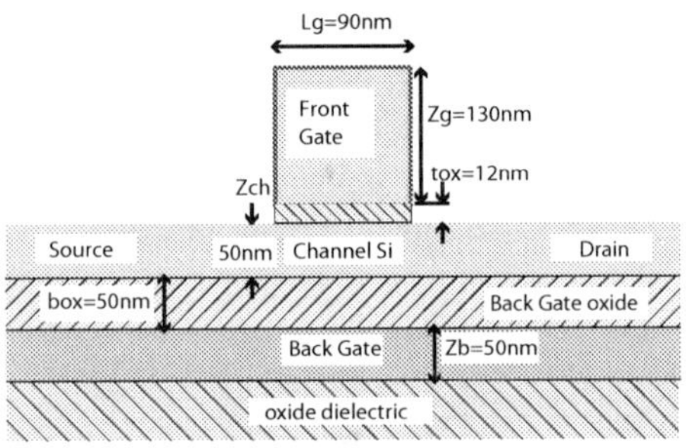

(c) TFT cross section.

Fig. 2. Device Pictures.

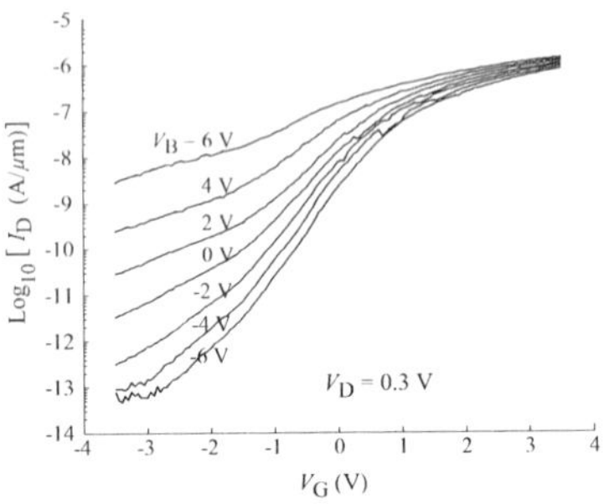

(a) I_D–V_G curves for V_D=0.3 V.

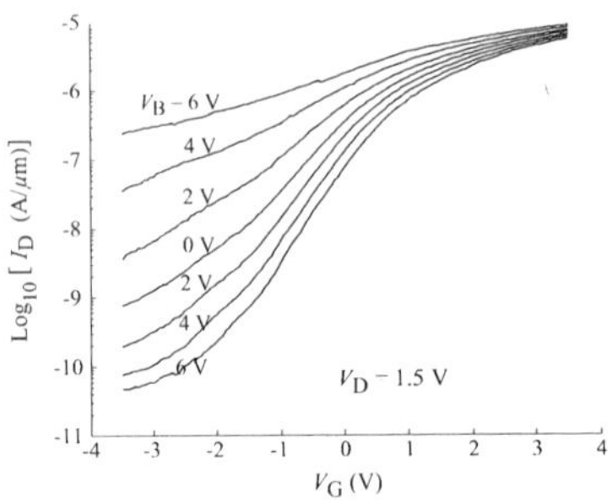

(b) I_D–V_G curves for V_D=1.5 V.

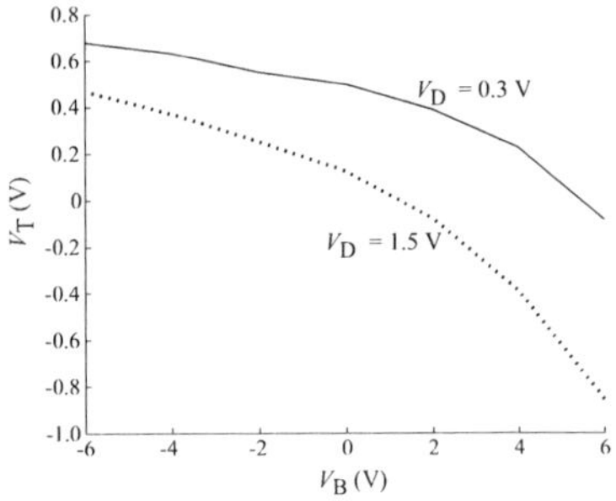

(c) Threshold Voltage vs. back-gate bias.

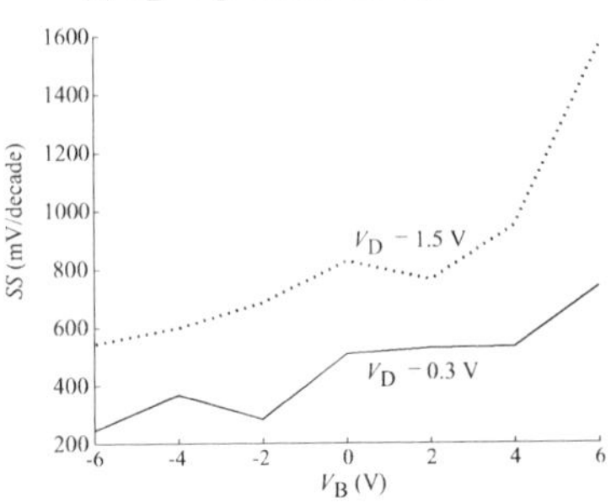

(d) Subthreshold slope vs. back-gate bias.

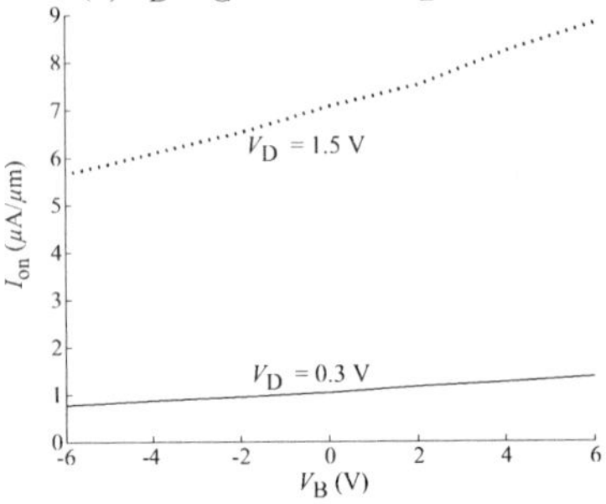

(e) On current vs. back-gate bias.

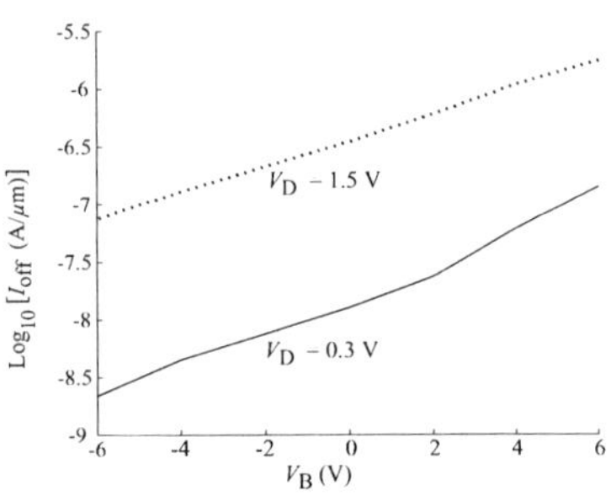

(f) Off current vs. back-gate bias.

Fig. 3. Effect of applied back-gate bias on the electrical behavior of a transistor with L_g=90nm, W_g=1500nm at ≈20°C. V_S=0.

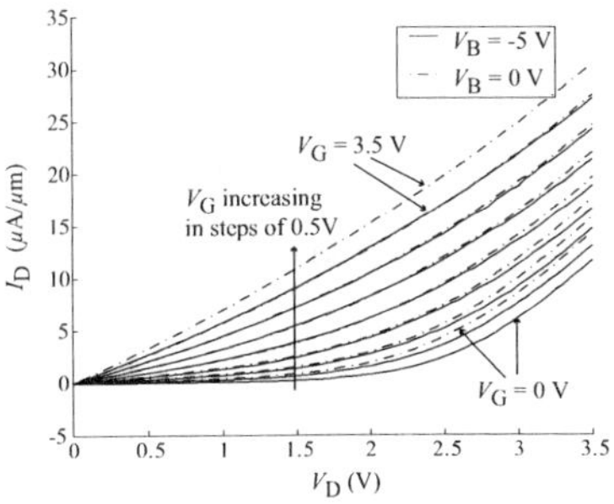

(a) I_D–V_D curves, L_g=90nm.

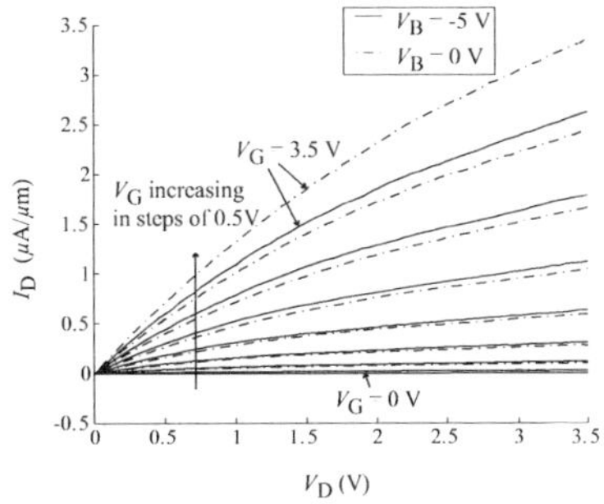

(b) I_D–V_D curves, L_g=4×90nm.

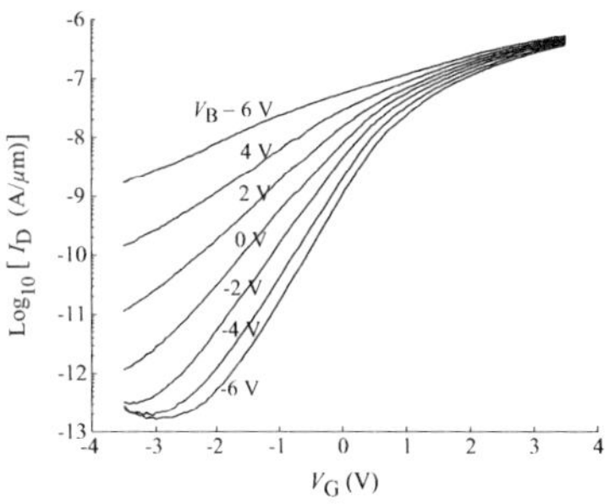

(c) I_D–V_G curves, L_g=4×90nm, V_D=0.3V.

Fig. 4. Device curves showing the effect of applied back-gate bias (V_B) at 20°C. V_S=0.

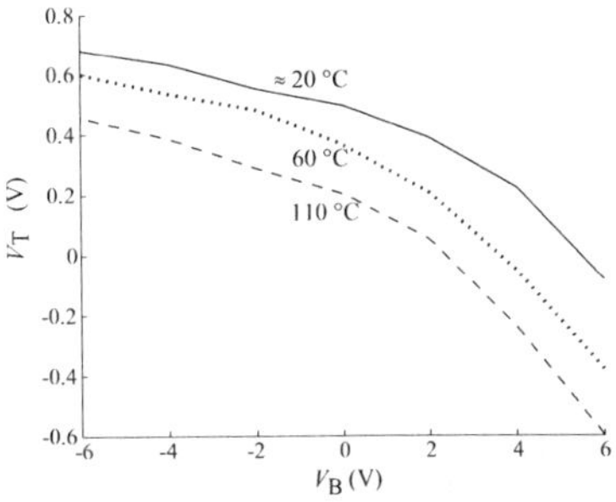

(a) Threshold voltage vs. back-gate bias.

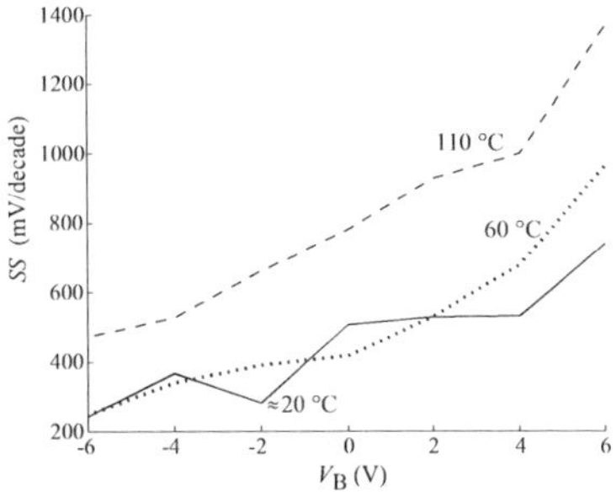

(b) Subthreshold slope vs. back-gate bias.

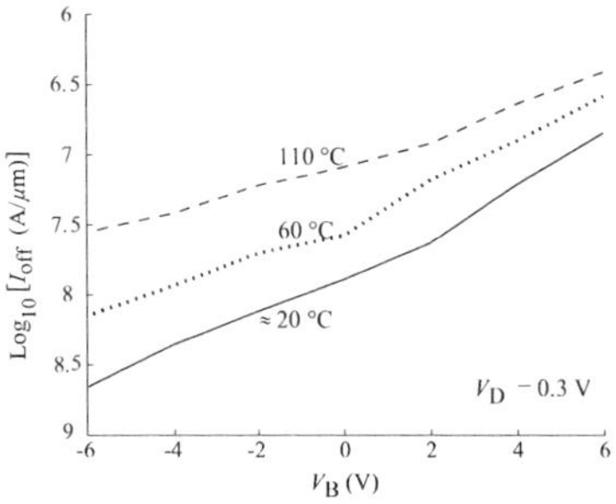

(c) On-current vs. back-gate bias.

Fig. 5. Effect of temperature on a device with L_g=90nm. V_S=0.

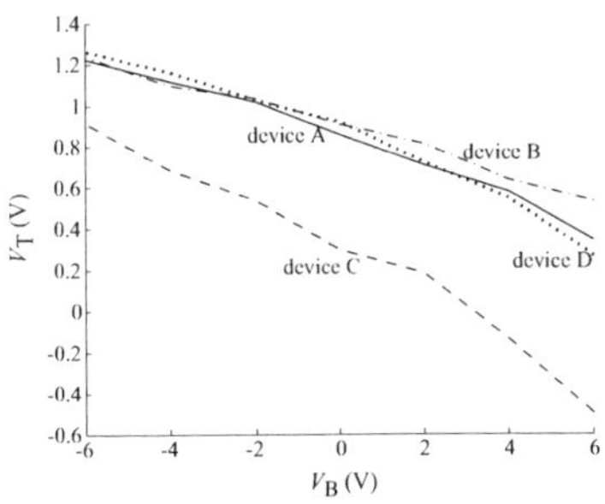

(a) Threshold voltage vs. back-gate bias.

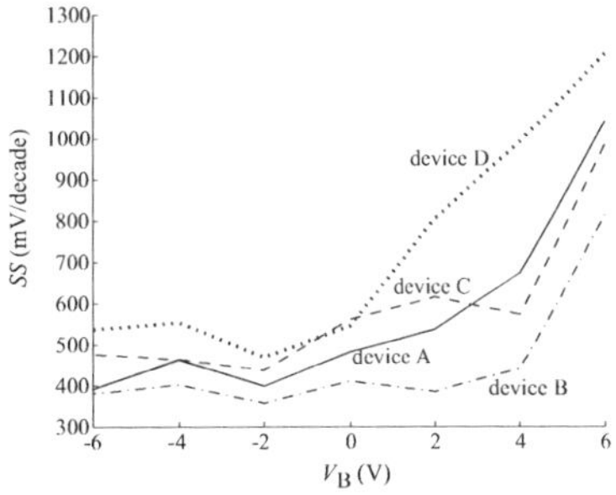

(b) Subthreshold slope vs. back-gate bias.

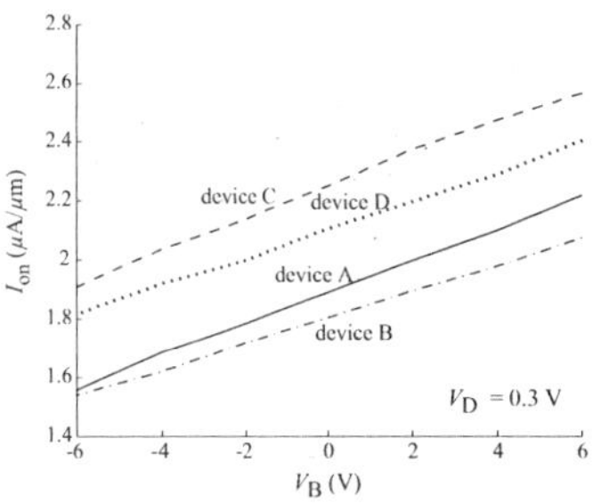

(c) On-current vs. back-gate bias.

Fig. 6. Device-performance local variations for four identical devices in close proximity with L_g=90nm, W_g=2000nm, and V_D=0.3V at 20°C. V_S=0.

change with V_B. On current decreases with lower values of V_B due to increasing V_T. However, with higher temperature I_{on} increases. Carrier mobility is already lower than *c*-Si due to poly-Si grain boundaries in the channel; therefore the degradation in mobility with temperature is not apparent in I_{on} behavior as would be the case with *c*-Si channel devices. Instead, reduction in V_T helps to increase I_{on} with high temperature.

Fig.6 shows device performance variations with V_B for four similar devices a few hundred microns apart. Fig.6(a) shows the change in V_T with V_B; note that for devices A, B, and D V_T can be shifted by approximately 70 mV/V of V_B. Device C, however, has 300mV and lower threshold voltage as compared to the other three devices. This indicates that devices A, B, and D have similar turn-on behavior, whereas device C is different. Figs.6(b)&(c) respectively show the variations in SS and I_{on} for these four devices. The device variations are expected [6], and we observed them. These device performance variations come from randomly oriented grain boundaries and enhanced dopant diffusion along grain boundaries. Since the TFT gate length is comparable to the grain size, device variations are pronounced due to less grain-effect averaging. The grain boundaries perpendicular to the channel result in increased V_T and lowered mobility, whereas the ones parallel to the channel degrade subthreshold performance. The enhanced diffusion along grain boundaries results in channel shortening, hence in increased on-current.

Conclusions

Thin-film transistors of L_g =90nm with insulated back gates, compatible with 3-D stacking, have shown V_T adjustability. Threshold voltages of these devices are adjustable by about 100mV per volt of back-gate bias. Threshold-voltage variations are observed in these devices, and are plausibly attributed to poly-Si grain-boundary related charge traps and to channel-length shortening.

Acknowledgment

The authors acknowledge helpful discussions with A. Matthews, J. McVittie, E. Perozziello, and V. Svilan.

References

[1] V. Svilan et. al., in *ISLPED*, pp. 268–272, Aug. 2000.
[2] I. Y. Yang et. al., *IEEE T. Electron Dev.*, vol. 44, pp. 822–831, May 1997.
[3] S. J. Souri et. al., in *Proc. 37th Conf. on Design Automation*, 2000.
[4] J. Nasrullah et. al., in *IEEE Conf. on Nanotech.*, pp. 502–505, Aug. 2003.
[5] R. Watts and J. Lee, *Elect. Dev. Lett.*, vol. 14, pp. 515–517, Nov. 1993.
[6] P. Walker et. al., *IEEE T. Electron Dev.*, vol. 51, pp. 212–219, Feb. 2004.

TiN metal gate thickness influence on Fully Depleted SOI MOSFETs physical and electrical properties

J. Widiez*, M. Vinet, T. Poiroux, P. Holliger, B. Previtali, P. Grosgeorges, M. Mouis** and S. Deleonibus

CEA/DRT-LETI - 17, Rue des Martyrs 38054 Grenoble Cedex 9, France, email: julie.widiez@cea.fr

*also with STMicroelectronics, 850 Rue J. Monnet 38926 Crolles, France **IMEP, ENSERG, 38016 Grenoble, France.

Abstract

In this paper, we report the influence of the TiN metal gate thickness on fully depleted (FD) silicon-on-insulator (SOI) MOS transistors with SiO_2 and HfO_2 gate dielectrics. The threshold voltage as well as the Electrical Oxide Thickness (EOT), the gate leakage current and the channel mobility are impacted. Smaller the TiN layer is, better are the transistors characteristics. This is due to a lower charge effect when the TiN thickness decreases.

Introduction

Fully Depleted (FD) Silicon-On-Insulator (SOI) MOSFETs are an attractive choice for sub-20nm gate-length scaling. As undoped channel are used to avoid dopant fluctuation effects in such ultra-thin SOI devices, metal gates become necessary to adjust the threshold voltage. In addition, the use of metal gates offers advantages, which include the suppression of polysilicon depletion effect and a good compatibility with high-κ gate dielectrics. TiN material is a very attractive candidate as metal gate because of its thermal stability, low resistivity and high CMOS-process compatibility [1]. In this paper we study the impact of the TiN metal gate thickness on threshold voltage, EOT, gate leakage current and on channel mobility.

Experiment

The devices were fabricated on SOI wafers with an initial silicon thickness of 50nm thinned down to 15nm +/- 2nm and a buried oxide thickness of 100nm. The transistor channel was left undoped with a p-type concentration ($Na=10^{15}cm^{-3}$). The various gate dielectrics are: 2nm rapid thermal oxide (RTO) or 3nm HfO_2. Thin films of HfO_2 were deposited by Atomic Layer Deposition (ALD) with a ($HfCl_4$,H_2O) chemistry. Post deposition annealing was performed at 600°C in N_2 followed by the metal gate deposition. CVD TiN layer using $TiCl_4$ and NH_3 as precursors is deposited at 680°C and encapsulated with a 50nm highly-doped polysilicon layer. Three TiN thicknesses are processed: 5nm, 10nm and 40nm either on SiO_2 and HfO_2 dielectrics. The poly-Si layer is dry etched followed by a chemical etching to eliminate the TiN layer. The process ended up following a standard SOI technology: extension implantation, spacers' formations, raised source/drain epitaxy and nickel salicidation. Poly-Si/SiO_2 transistors were fabricated as references.

Results and discussion

Devices characteristics

Fig. 1 shows I_D-V_G electrical characteristics for long channel p- and n-FDSOI devices with 3 different thicknesses of TiN gates on HfO_2. 5nm and 10nm TiN gates devices show perfect subthreshold characteristics with a subthreshold slope of 60mV/dec. On the contrary, 40nm TiN transistors have a subthreshold slope of 64mV/dec and show strong gate leakage at high V_D which severely degrades the off-state current. A positive 230mV threshold voltage shift is observed between the 5nm and the 40nm TiN metal gate transistor. This may be due to negative charges. On SiO_2 gate dielectric, the threshold voltage also decreases when the TiN thickness decreases (Table 1). A threshold voltage of 0.28V is obtained (Table 1) with the 5nm TiN metal gate transistor on HfO_2, a value suitable for Low Operating Power (LOP) nMOSFETs.

The gate capacitance curves (Fig.2) show an oxide re-growth when the TiN thickness increases. The linear dependance between the Equivalent Oxide Thickness (EOT) and the TiN thickness is the same on SiO_2 and on HfO_2 (Fig. 3): a 11Å EOT increase is observed for the 40nm TiN gate transistors. Moreover, for a superior EOT, these devices exhibit the highest gate leakage current value (Fig. 4). These degradations may be due to the chlorine (Cl) diffusion into the gate oxide (SIMS profiles presented in Fig. 5 & 6). Cl atoms diffusion in the dielectric oxide (SiO_2 or HfO_2) is all the more important as the metal gate is thick. The high electro-negativity of Cl atoms can decrease the oxide dielectric constant or free oxygen atoms can diffuse to the SiO_2/Si interface inducing gate oxide re-growth [2]. Moreover, the Cl diffusion in the dielectric can degrade its properties which can explain the higher gate leakage for thicker metal gate [3].

Long channel mobility

The electron and hole effective mobilities (μ) are extracted from the split-CV technique (Fig. 7 & 8). The same trend is observed for both carriers: the mobility is improved when the metal gate thickness is minimized. At the maximum mobility (μ_{max}), a 108% (63%) electron mobility improvement is obtained by reducing the TiN thickness from 40nm to 5nm on HfO_2 and SiO_2 respectively. Usual mobility degradation using high-κ gate dielectric is observed but its value highly depends on the initial metal gate thickness: at μ_{max}, 22% degradation is observed for the 5nm TiN thickness, 36% for the 10nm and 56% for the 40nm. To understand the mobility difference, temperature dependence was examined (Fig. 9). The low-field μ plot versus the temperature (T from 460K to 125K) evidences that Coulomb scattering is the main mechanism responsible of the mobility degradation in the 40nm TiN gate devices (Fig. 10). As the temperature decreases, μ strongly increases down to 200K for the 5nm TiN devices, showing that Phonon scattering is the main mechanism. For the 40nm TiN devices, μ rapidly saturates and decreases. From previous results, we can speculate that Coulomb scattering is associated with chlorine atoms diffusion.

Conclusion

FDSOI devices have been fabricated with various TiN metal gate thicknesses on SiO_2 and HfO_2. The study demonstrates the importance of reducing the metal gate thickness to enhance the transistor performance. EOT, gate leakage current and mobility are maximized thought a reduction of the Coulomb scattering mechanism.

0-7803-9212-4/05/$20.00 ©2005 IEEE

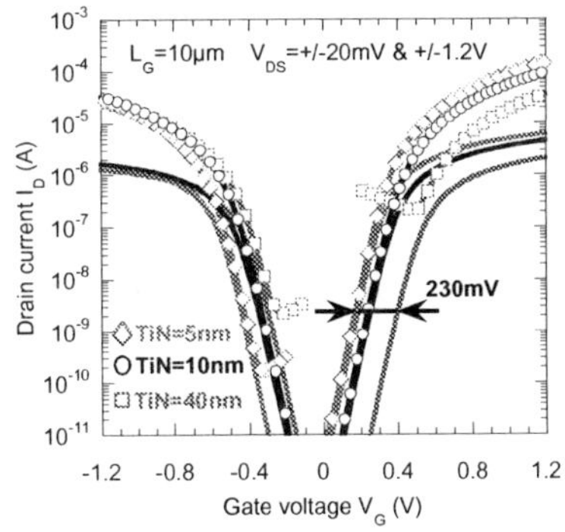

Fig. 1: I_{DS} versus V_G at V_{DS}=+/-20mV and +/-1.2V in n- and p-FDSOI MOSFETs with 3 TiN thicknesses: 5, 10 and 40nm.

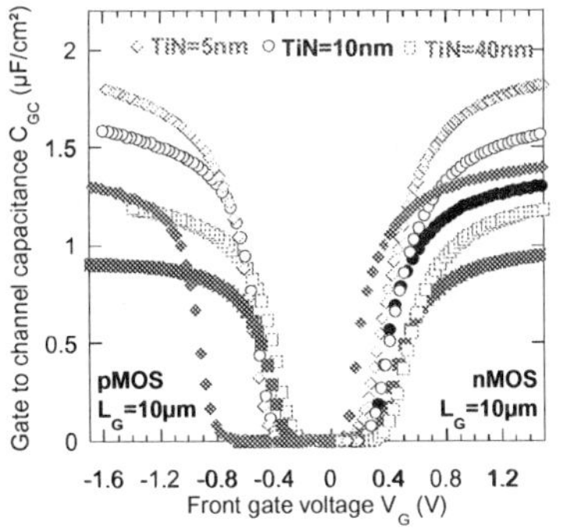

Fig. 2 : Capacitance versus front gate voltage for different TiN thicknesses on HfO_2 (empty symbols) and on SiO_2 (full symbols).

SiO_2 – t_{phys}=2nm					
gate		Poly-Si	TiN	TiN	TiN
TiN thickness		**0**	**5**	**10**	**40**
N	EOT	20	20	22	31
	V_{th}(V)	-0.2	**0.18**	0.39	0.44
P	EOT	20	22		34
	V_{th}(V)	-1	-0.78		-0.47
HfO_2 – t_{phys}=3nm					
gate			TiN	TiN	TiN
TiN thickness			**5**	**10**	**40**
N	EOT		15	17	24
	V_{th}(V)		**0.28**	0.37	0.51
P	EOT		15	17	25
	V_{th}(V)		-0.54	-0.47	-0.37

Table 1: Extracted values of the EOT (Å) and the threshold voltage (V_{th}) for the different nMOS and pMOS gate stacks on SiO_2 and HfO_2.

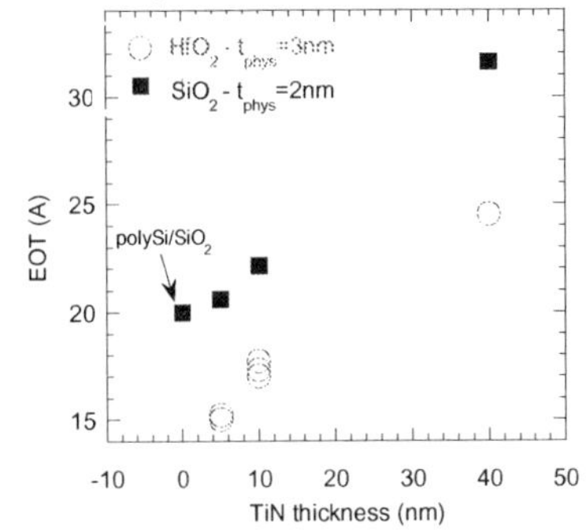

Fig. 3: EOT(Å) versus the TiN thicknesses. 0nm corresponds to the poly-Si gate.

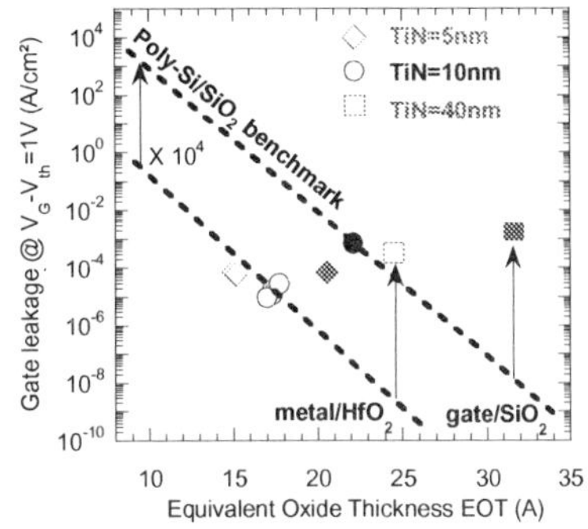

Fig. 4: Gate leakage versus EOT for poly-Si and TiN gates on HfO_2 (empty symbols) and on SiO_2 (full symbols) for nMOSFETs.

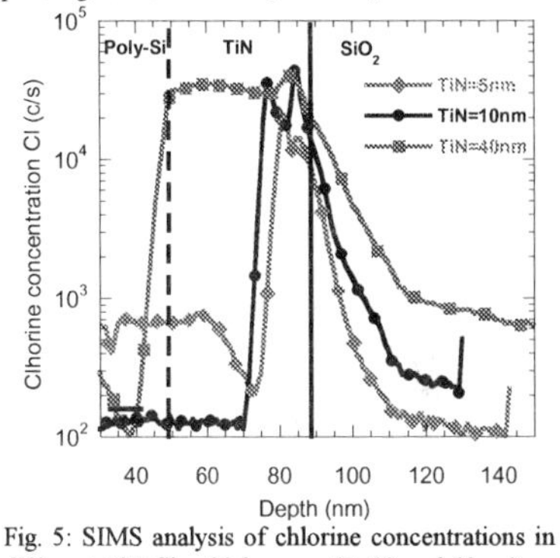

Fig. 5: SIMS analysis of chlorine concentrations in different TiN film thicknesses (5, 10 and 40nm) on SiO_2 gate dielectric.

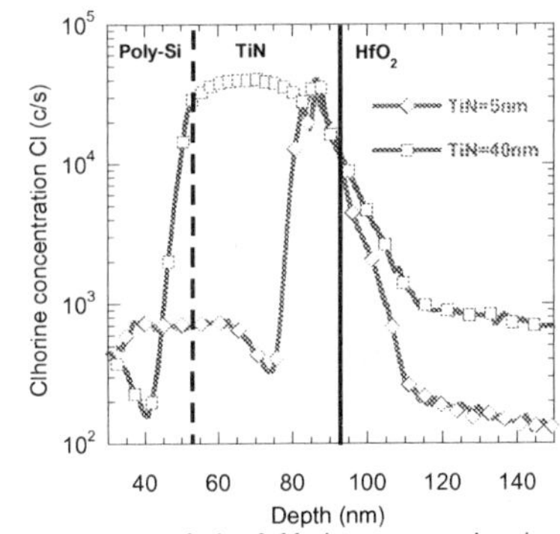

Fig. 6: SIMS analysis of chlorine concentrations in TiN film on HfO_2 dielectric in 5nm and 40nm TiN films.

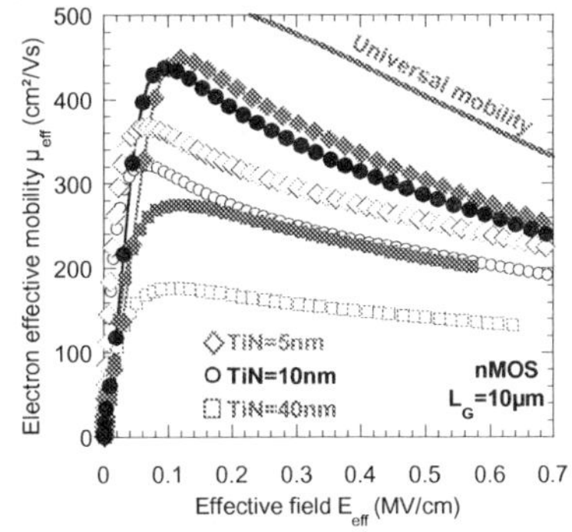

Fig. 7: Electron effective mobility versus effective field on HfO_2 (empty symbols) and on SiO_2 (full symbols) for 3 thicknesses of TiN (5, 10 and 40nm).

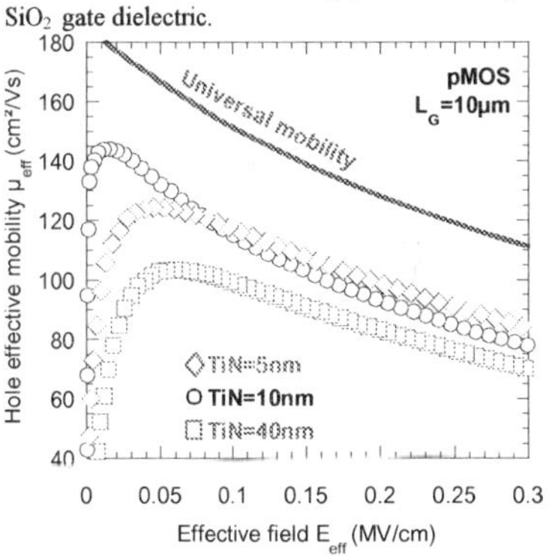

Fig. 8: Hole effective mobility versus effective field on HfO_2 for 3 thicknesses of TiN (5, 10 and 40nm).

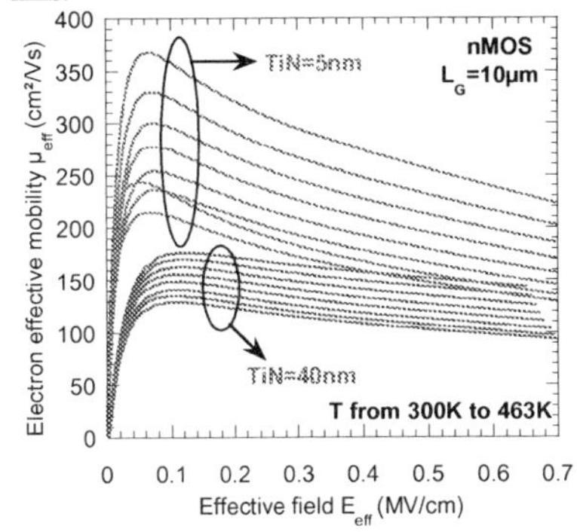

Fig. 9: Electron effective mobility versus effective field on HfO_2 for different temperatures (300K to 463K) for 2 TiN thicknesses (5 and 40nm).

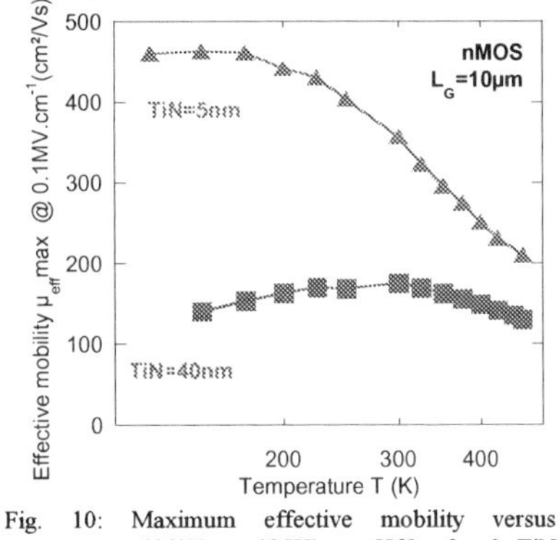

Fig. 10: Maximum effective mobility versus temperature (460K to 125K) on HfO_2 for 2 TiN thicknesses (5 and 40nm).

References

[1] H. Wakabayashi *et al.*, *IEDM Tech. Digest*, p.447, 1996.
[2] S. Maîtrejean *et al.*, *MRS Fall Meeting*, 2003.
[3] M.Moriwaki *et al.*, *JJAP*, 40, p.2679, 2001.

Acknowledgements

The authors would like to thank CEA-LETI pilot lines. This study was partly funded by the Advanced Devices Program of the Alliance (STMicroelectronics, Philips and Freescale Semiconductors).

A New Leakage Mechanism in SOI Smart Power Technologies

S. Schwantes, A. Heid, F. Dietz, M. Graf and V. Dudek
ATMEL Germany, Theresienstrasse 2, 74072 Heilbronn, Germany
Email: stefan.schwantes@hno.atmel.com, Tel: ++49 7131 67 3268

Introduction

Smart power SOI technologies are becoming popular because of better isolation. The SOI technology introduces a new electrode, the substrate electrode referred to as back gate. Several papers have reported back gate induced leakage current [1,2]. This work reports a new leakage mechanism induced by the interaction of the neighbouring device and that back gate. Moreover, possible countermeasures for a 0.8 μm 80V SOI technology are presented.

High-voltage devices

Fig. 1 depicts the stadium structure of a p-channel DMOS transistor (PDMOST) with shorted source and body. This is a common mode of operation. The PDMOST shows excessive leakage current for increasing source, body and gate voltage while the drain and the back gate remain at 0V. The neighbouring box was grounded, too. This is likely to happen, when the PDMOST is turned off to block large voltages while the adjacent device is a turned-on NMOS. The measured drain current is represented by the black squares in fig. 2. This is caused by a depletion layer at the buried oxide (BOX) and the trench interface. If this depletion layer hits the source, punch-through occurs as illustrated by the simulated depletion charge and hole current in fig. 3. The 800nm wide, polysilicon filled trench with 150nm liner oxide per side can suppress tunnelling currents, but it can not shield the electrical impact of the neighbouring device. This leakage mechanism can be suppressed by an additional 800 nm wide n+ ring separating the source from the trench's depletion zone (fig.4). The new device does not show leakage current below the operation voltage of 80V. However, 11% more area is required for a device with 20 μm channel width.

Low-voltage devices

Besides the PDMOST the low-voltage CMOS part is affected by the operation of the neighbouring device, too. Fig. 5 plots the drain leakage current of the 5V NMOS and PMOS. Both were turned off by an appropriate gate voltage and weakly biased at the drain (0.1V). Leakage current is observed if the potential difference V_{NB} between the neighbouring box and the body of the device exceeds a certain limit which shows a strong dependence on the back gate voltage. This potential difference could be caused by a turned-off PDMOST in case of the 5V NMOS and a turned-on NMOS in case of the 5V PMOS. Similar to the PDMOST, the leakage is caused by a depletion layer induced by the neighbouring box potential. The depletion layer connects the source and the drain regions and causes punch-through (Fig. 6). The depletion charge at the BOX interface induced by the back gate potential modulates the width of the depletion zone at the trench sidewall and, therefore, the threshold of the leakage current (fig.7). The threshold of this current is shifted after 1000s stress with V_{NB}=80V for the NMOS and V_{NB}=-80V for the PMOS (Fig. 8). After stress, the PMOS will cease to function whereas the leakage threshold of the NMOS has been increased. Hot-hole injection into the polysilicon of the trench for bulk technologies was reported by [3]. However, the threshold shift here can only be explained by electrons trapped into the polysilicon jeopardising the reliability of circuits. Therefore, the polysilicon filled trench was replaced by a TEOS filled trench that screens the effect of the neighbouring device and shows no deterioration after stressing (Fig. 9).

Conclusions

The proximity of high-voltage and low-voltage devices endangers the functionality of circuits with highly area optimised design rules. This can be avoided by a polysilicon free trench module without any additional area requirements.

References

[1] K.Yallup et al., "Back gate effects in thick film SOI CMOS devices", *IEEE SOI Conf.*, p. 49, 1991.

[2] Y. Hayasaki et al., "Back-Channel effect on SOI CMOS for high-voltage power ICs", *ISPSD*, p. 337, 1997.

[3] B. Elattari et al, "Impact of charging on breakdown in deep trench isolation structures", *ESSDERC*, p.513, 2003.

0-7803-9212-4/05/$20.00 ©2005 IEEE

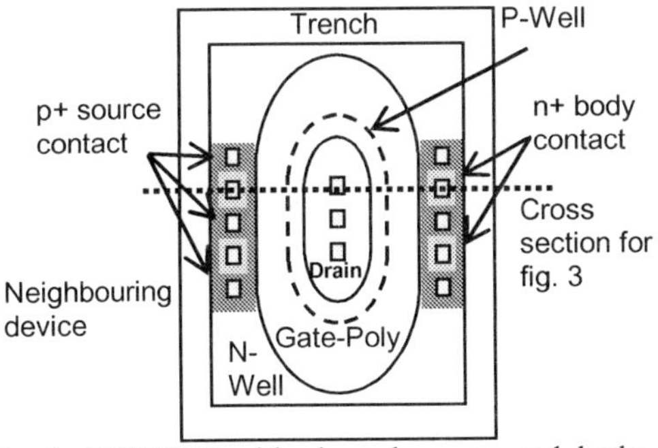

Fig. 1. PDMOST with shorted source and body electrodes isolated by a deep trench from the adjacent devices.

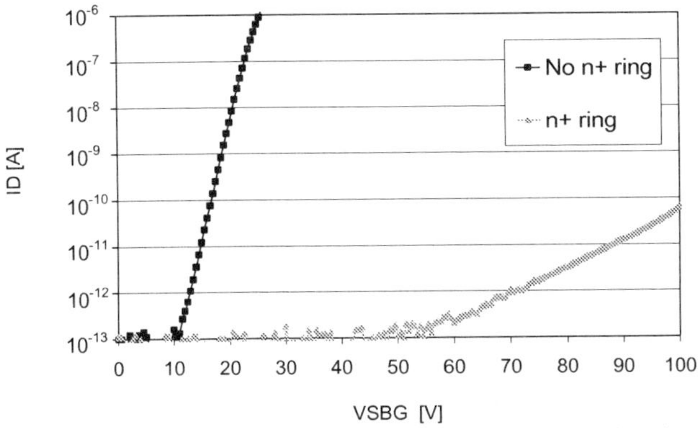

Fig. 2. Drain current of the turned-off PDMOST as a function of the source, body and gate voltage for the device shown in fig. 1 (black squares) and fig. 4 (grey triangles).

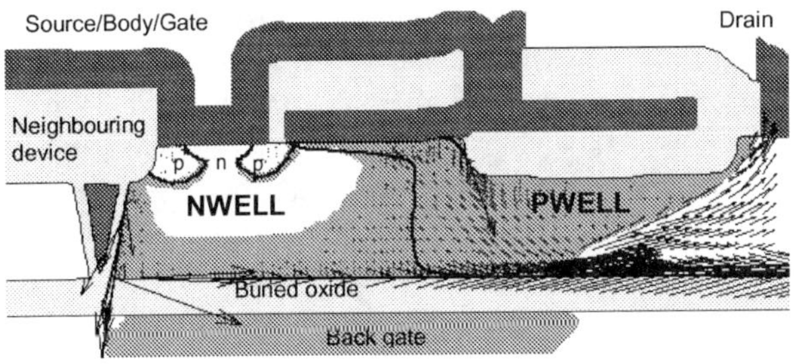

Fig. 3. Simulated depletion charge and hole current of the PDMOST. The depletion regions of the buried oxide interface and the trench merge with the p+ source contact causing a punch-through current. (V_{SBG}=20V, V_D=V_{NB}=V_{BG}=0V)

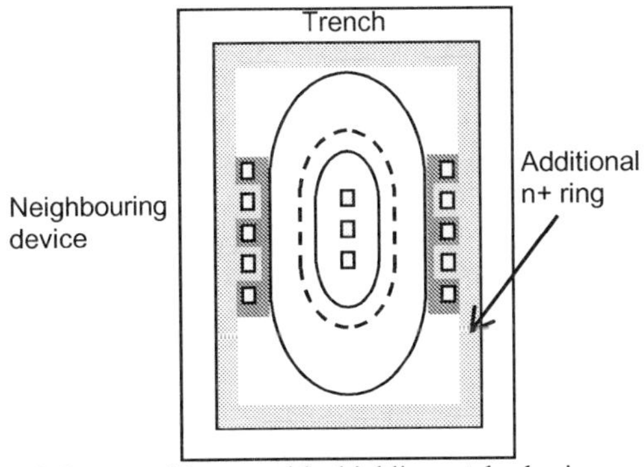

Fig. 4. Improved layout with shielding n+ body ring.

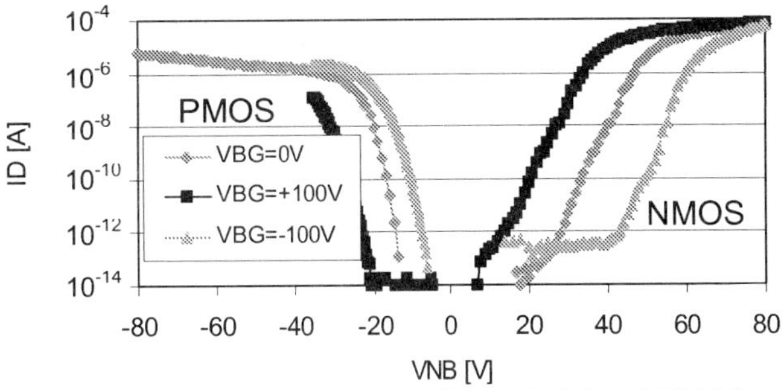

Fig. 5. Leakage of the turned-off 5V NMOS and PMOS vs. neighbouring box potential for different back gate voltages.

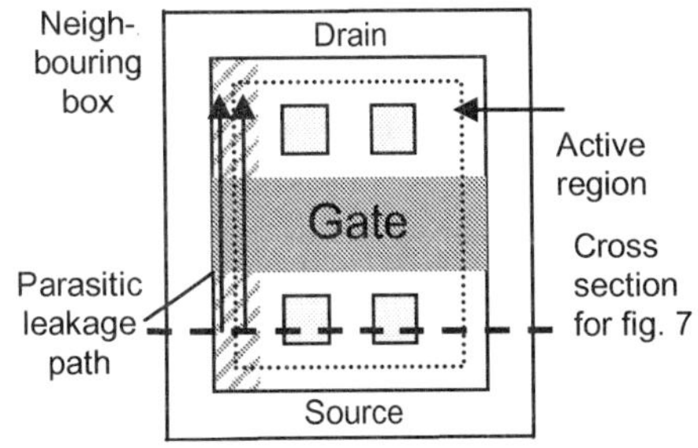

Fig. 6. Parasitic leakage current path for the 5V device.

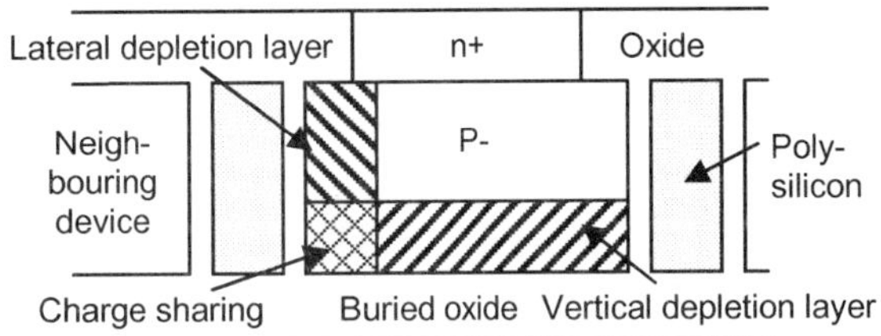

Fig. 7. Interaction of the BOX and the trench depletion layer.

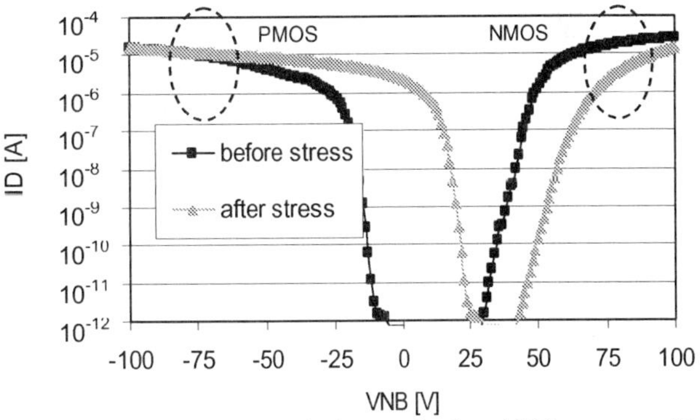

Fig. 8. Leakage current before and after 1000s stress with V_{NB}=80 V for the NMOS and V_{NB}=-80V for the PMOS.

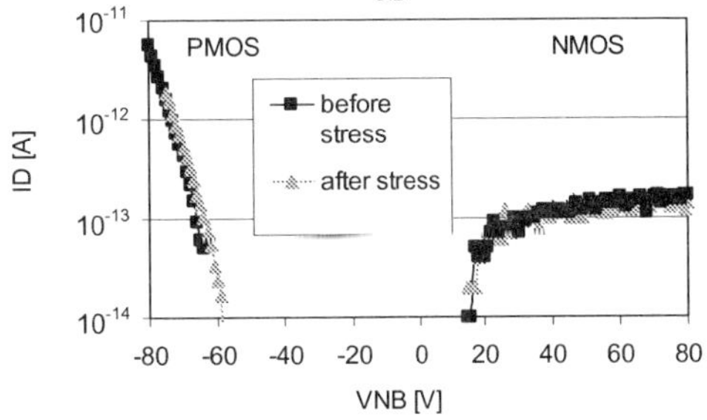

Fig. 9. Leakage current for the TEOS filled trench technology.

Experimental characterization of source-to-drain tunneling in 10nm SOI devices

J. Lolivier, X. Jehl[1], Q.Rafhay, T. Poiroux, M.Vinet, B.Previtali, M.Sanquer[1], F. Balestra[2], S. Deleonibus
CEA/DRT-LETI, 17 rue des Martyrs, 38054 Grenoble Cedex 9, France
[1] CEA/DRFMC, 17 rue des Martyrs, 38054 Grenoble Cedex 9, France,
[2] IMEP–ENSERG, 23 avenue des martyrs, BP 257, 38016 Grenoble, France

Introduction

The ITRS roadmap foresees the production of transistors with gate length (L_g) down to 10nm in 2016. The aggressive scaling of MOSFET requires new architectures such as Fully Depleted Silicon-On-Insulator (FD SOI) or even MultiGate devices in order to ensure an excellent electrostatic control. For all these architectures, at ultra-short gate length, the off-state leakage, partly determined by quantum tunnelling between source and drain, could threaten MOSFET scalability [1].
This work deals with the electrical characterization down to 4K of FD SOI MOSFET with a physical gate length down to 10nm. Temperature measurements are used to highlight source to drain tunneling: which is evidenced at room temperature for the first time. Finally resonnant tunneling effect is observed.

Device description

The device fabricated owns a 1.5nm gate oxide and a 60nm thick polysilicon. Ebeam lithography allows to design gate lengths down to 10nm. The final device is presented on a TEM cross-section (Fig.1) [2]. Channel thickness, obtained either by electrical extraction [3] or ellipsometric measurements, reaches value as low as 8nm. The effective channel length estimated with technological simulation tools (BCA, [4]) is expected to be 5nm shorter than the gate length (Fig.2) which means that for L_g=10nm the effective channel length is 5nm. This value is theoretically sufficient to observe source to drain tunnelling at room temperature [5].

Source to drain tunneling

As the tunnelling current is mainly dependent on the width and height of the potential barrier between source and drain, whereas the thermal current depends especially on barrier height and temperature, the temperature was used to discriminate between the different physical current contributions [6].
The use of an ultra-thin film and of BF_2 pocket implantation allows well controlled short channel effects down to L_G=15nm (Fig.3). For the longest devices (L_g=1μm to 45nm and T_{Si}=14nm), the substherhold Swing (SS) exhibits a perfect linear decrease with the temperature as expected (Fig.4). The same trend is also observed for L_g=20nm (L_{eff}=15nm) except below 100K where SS saturates. At low temperature, the linear trend of the SS as a function of temperature is suppressed indicating that the tunnelling current exceeds the thermal current.
For the 10nm gate length devices (L_{eff}=5nm), the off-state leakage current is dramatically increased (Fig.5). Moreover the pocket implantation and thus high coulomb scattering prevents the drain current to increase as the temperature decreases, contrary to the other gate lengths. We have reported SS as a function of temperature for three different devices taken on the same die: we do not any more observe any linear behavior for these devices, and source to drain tunnelling is thus evidenced at room temperature (Fig.6). The gate lithography fluctuations (σ=2.5nm) accounts for the SS dispersion.

Resonant tunneling effect

The same measurements were performed at low temperature with a smaller V_g step and V_d=10mV. The conductance curve exhibits high peaks at various gate voltage (gain of 2 decades for V_G=-0.208V) (Fig.7). The peaks height increases as the temperature decreases (Fig.8). This phenomenon is related to resonant tunnelling effect, via energy states linked to channel impurities [7]. These impurities can be isolated atoms introduced in the silicon during the extensions or pockets implantations. The peaks height is all the more important that the impurity atom lies in the middle of the channel. To fit the experimental curves the coupling coefficient was α_c=0.115 which is in good agreement with our short effective gate length as compared to the silicon thickness.

Conclusion

SOI devices with gate length down to 10nm were fabricated, resulting in 5nm effective gate length transistors. Source to drain tunnelling was assumed to be predominant in the off-state current even at room temperature. Low temperature measurements have also highlighted resonant tunnelling via impurities.

0-7803-9212-4/05/$20.00 ©2005 IEEE

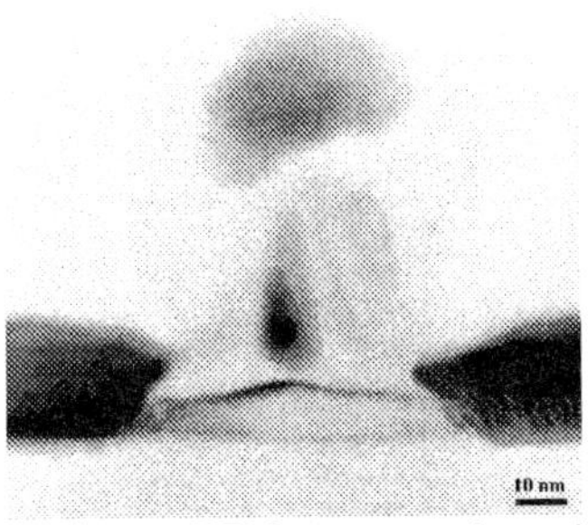

Fig1: SOI FD device TEM cross section with L_g=10nm

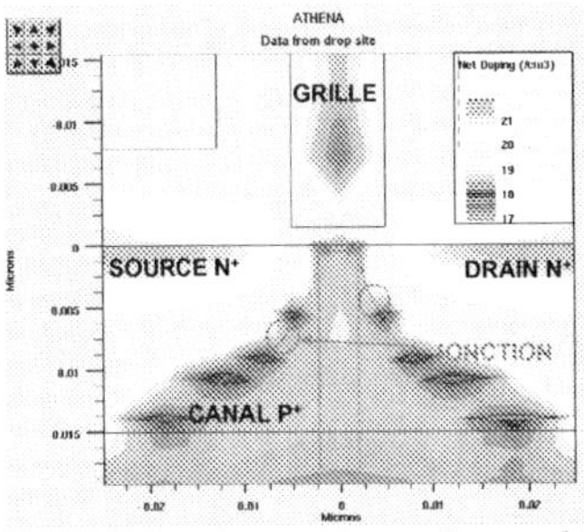

Fig2: Simulated L_{eff} for a 10nm physical L_g transistor

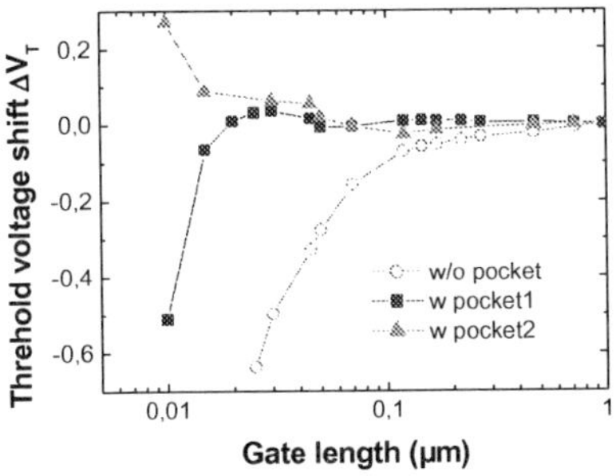

Fig3: SOI FD Threshold voltage shift versus gate length for devices w/o pockets implantation, V_d = 50mV.

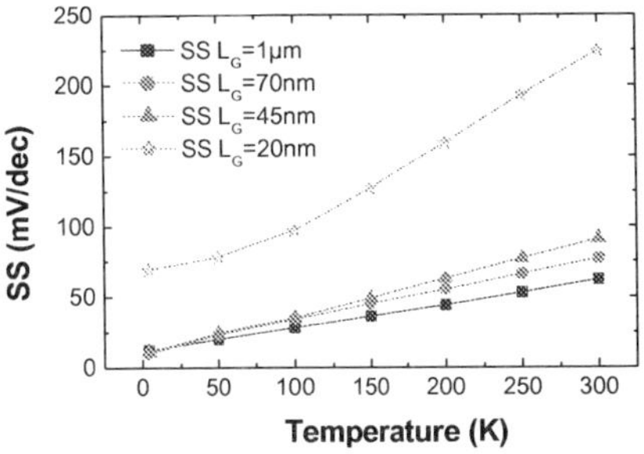

Fig4: SS as a function of temperature for various gate length (L_g =1µm, 70nm, 45nm and 20nm), V_d=50mV.

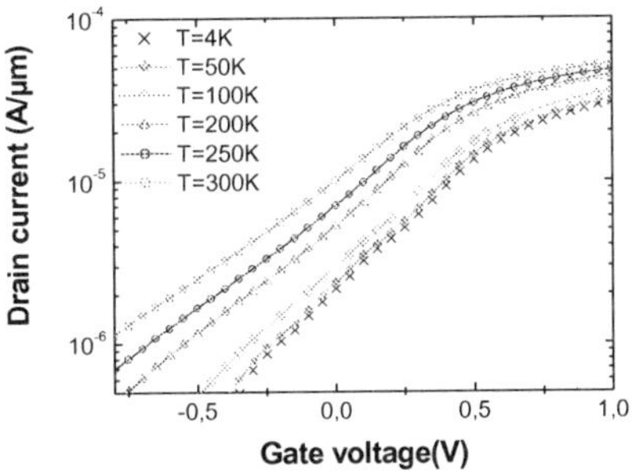

Fig5: Drain current versus gate voltage for L_g =10nm at various temperature, V_d=50mV.

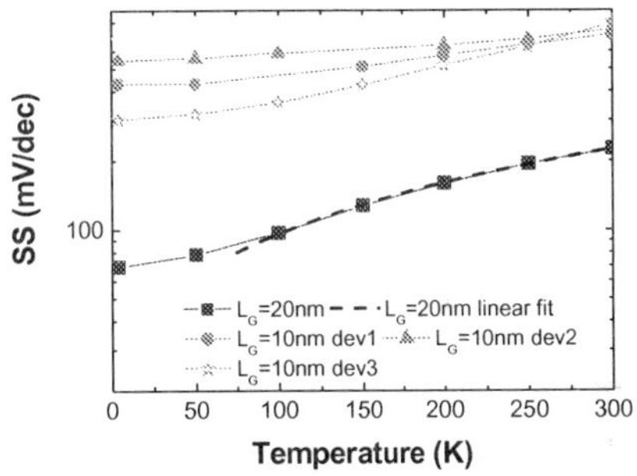

Fig6: SS as a function of temperature for L_g=20nm and 10nm (three different devices). V_d=50mV, V_g step=10mV.

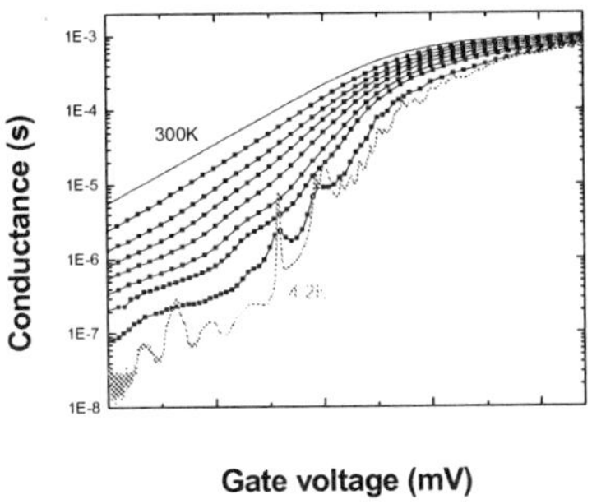

Fig7: $I_d(V_g)$ for LG=10nm and V_g step=1mV.

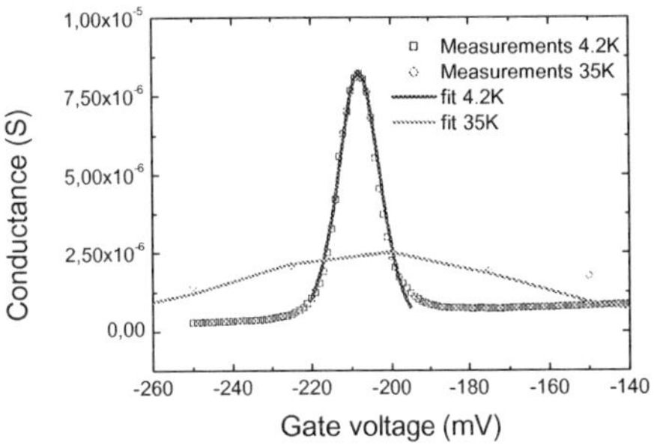

Fig8: Conductance overshoot measurements and fit.

References

[1] F. Boeuf *et al*, IEEE Trans. Nanotech. 3, n°1, p105, 2004.
[2] J. Lolivier *et al*, Int SOI conference, p17, 2004.
[3] T. Poiroux *et al*, Int SOI conference, p73, 2004.
[4] Silvaco user's manual, 2003.
[5] M. Bescond *et al*, Proc. of ESSDERC, p395, 2003.
[6] H. Kawaura *et al.*, Appl. Phys. Lett. 76, n°25, p3810, 2000.
[7] M. Sanquer *et al*, Phys.Rev.B 61, n°11, p7249, 2000.

Session #3

CIRCUIT TECHNIQUES & APPLICATIONS

CO-CHAIRS

Richard Brown

University of Utah

Ted Houston

Texas Instruments

CIRCUIT TECHNIQUES AND APPLICATIONS Session 3

Design and Evaluation of Basic Analog Circuits in an Emerging MuGFET Technology

G. Knoblinger[1], F. Kuttner[1], A. Marshall[2], C. Russ[3], P. Haibach[3], P. Patruno[4], T. Schulz[3,5], W. Xiong[2,5], M. Gostkowski[5], K. Schruefer[3], C.R. Cleavelin[2]

[1] Infineon Technologies Austria AG, Siemensstrasse 2, 9500 Villach, Austria
[2] Texas Instruments Inc., 13560 North Central Expressway, Dallas, TX, USA
[3] Infineon Technologies AG, Balanstrasse 73, 81541 Munich, Germany
[4] SOITEC, Parc Technologique des Fontaines, 38190 Bernin, France
[5] ATDF Advanced Technolgy Development Facility Inc., Austin, TX, USA
Email: Gerhard.Knoblinger@infineon.com , Tel: +43 4242 305 6534

Introduction

Multi-gate MOSFET (MuGFET) e.g. FinFET, TriGate devices (Fig. 1,2) are the most promising candidates for beyond 45nm technology CMOS nodes [1,2]. For future SoC solutions in these technologies the ability to realize also analog building blocks is of utmost importance. Up to now only a few publications are available concerning the perspective of FinFETs for analog applications [3] and no reports and measurement results can be found about the realization of analog circuits with these advanced devices.

In this work we demonstrate for the first time the design and realization of basic analog circuits (low voltage bandgap, Miller op amp and current reference) with FinFET devices, including measurement results.

Circuit design and measurement results

The available technology offers double-gate devices with a fin height of 88nm and a width of 55nm with 2nm physical gate oxide thickness and a minimum channel length of 45nm, a silicided poly resistor with 10Ω/sq. and one metal layer [4]. To improve the output conductance, the presented analog circuits are using devices with a channel length of 250nm. Fig. 3 shows typical I/V curves for this device in the linear and saturation region.

For the design of the low voltage bandgap a pn-junction is realized with a so called fin lubistor (gated diode). Fig.4 shows the layout and the I/V behavior of this structure with gate connected to the cathode. For the I/V measurements 500 fins are switched in parallel. This gives an IS=1.6e-19A for one fin. This value is also used for the diode model in the circuit simulation.

The schematic of the realized low voltage bandgap [5] is shown in Fig.5. The difference compared to the design procedure with planar devices is the width scaling in discrete steps of one fin (2*88nm), but for most analog circuits this restriction is not really a problem. Fig.6 shows the schematic and micro photograph of the Miller op amp and Fig.7 the schematic of the Vth current reference. Fig.8 demonstrates the functionality of the bandgap with a Vout vs. VDD plot. It can be seen, that the regulation loop begins to work at around VDD=1.1V, and above VDD=1.2V the output voltage is almost independent of the supply voltage. Fig.9 shows the temperature dependence of the bandgap at VDD=1.5V. In Fig.10 the output current Iout vs. Vout for the Vth current reference is shown. An output resistance of rout=5.5MΩ can be extracted. The most important analog building block is a Miller op amp and Fig.11 shows the gain vs. frequency behavior of the realized op amp, configured as a 20dB inverting amplifier. In addition the low frequency open loop gain was measured and a value of $A0 \approx 45$dB was reached. This is comparable with typical open loop gains achievable in deep sub-micron planar CMOS technologies.

Conclusion

The successful design and realization of the most important analog circuits in a MuGFET technology with no additional analog specific process steps or devices is presented for the first time. This opens the possibility for future SoC designs also on these advanced technologies.

References

[1] B. Yu et al., Proc. of the IEDM 2002, p. 251-254.
[2] S.E. Thompson et al., IEEE Trans. on Semiconductor Manufacturing, 2005, p. 26-36
[3] V. Kilchytska et al, Proc. of the ESSDERC 2004, p. 65-68
[4] W. Xiong et al., IEEE Electr. Dev. Letters, 2004, p. 541-543
[5] H. Banba et al, IEEE Journal of Solid-State Circuits., 1999, p. 670-674.

0-7803-9212-4/05/$20.00 ©2005 IEEE

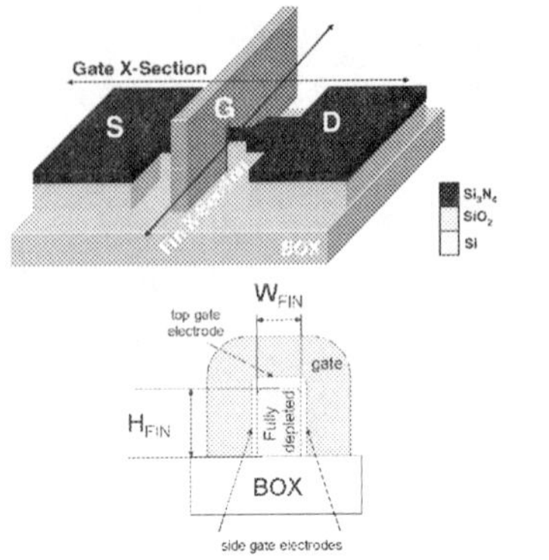

Fig.1: 3-D visualization of a multi-gate (MuGFET, FinFET) device and fin cross section.

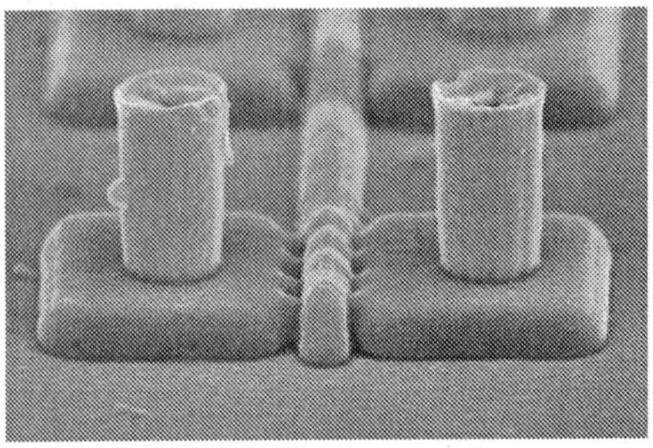

Fig.2: SEM of a 4 fin structure including S/D area and contacts.

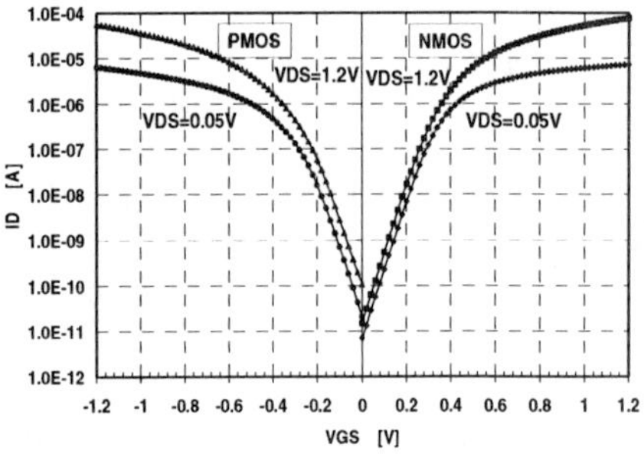

Fig.3: ID vs. VGS for the NMOS and PMOS FinFET in linear and saturation region for an analog device with L=250nm and two fins in parallel.

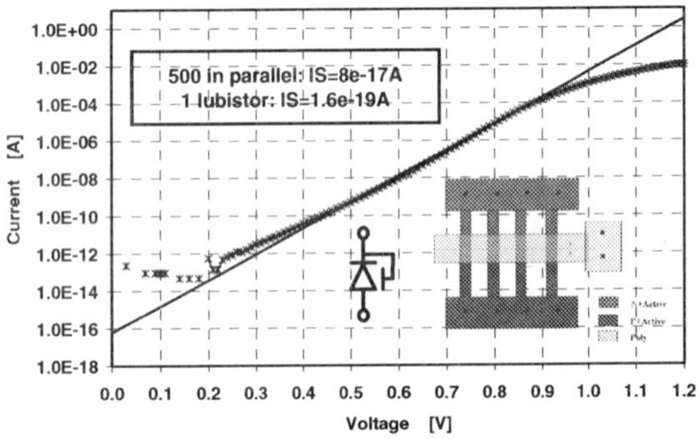

Fig.4: Layout, symbol and I/V characteristics of the fin lubistor (gated diode).

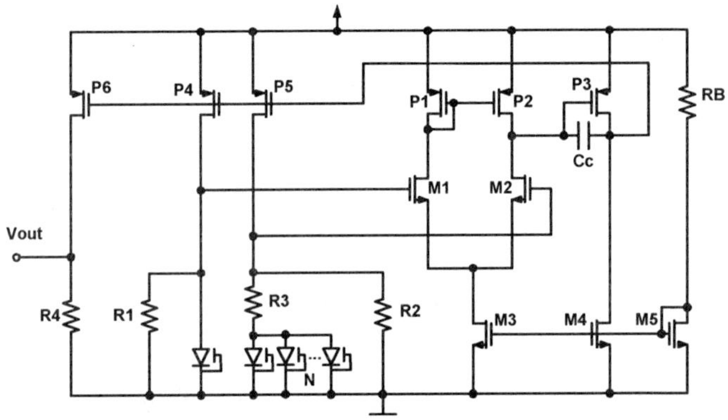

Fig.5: Schematic of the low voltage bandgap circuit.

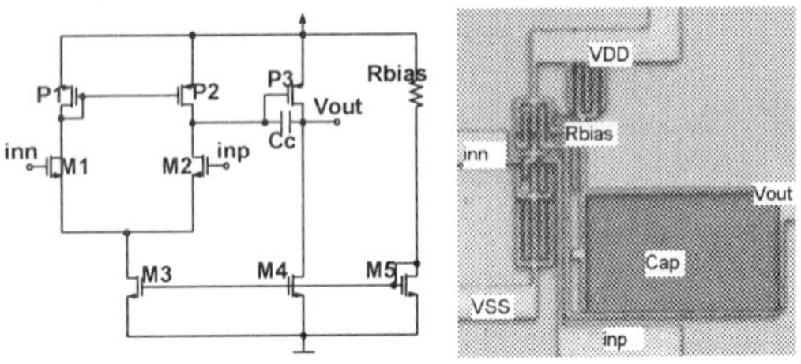

Fig.6: Schematic and micro photograph of the Miller op amp

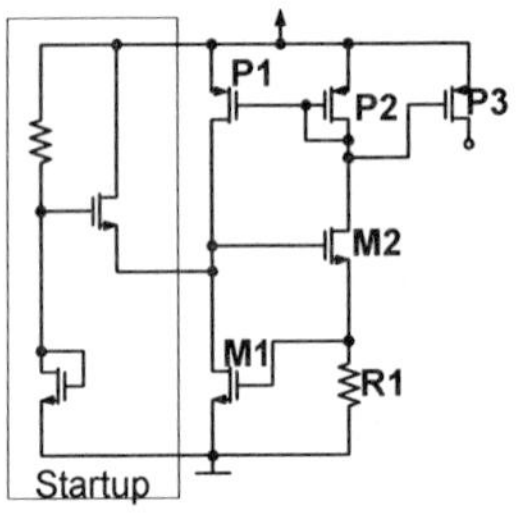

Fig.7: Schematic of the Vth current reference incl. the start up circuit

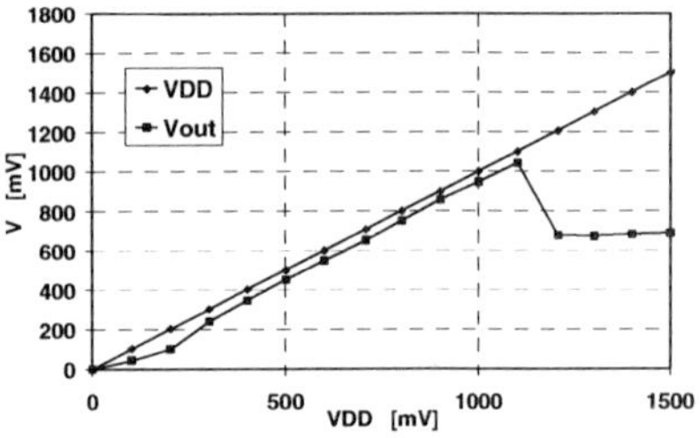

Fig.8: Vout vs. power supply voltage (VDD) of the bandgap reference at room temp.

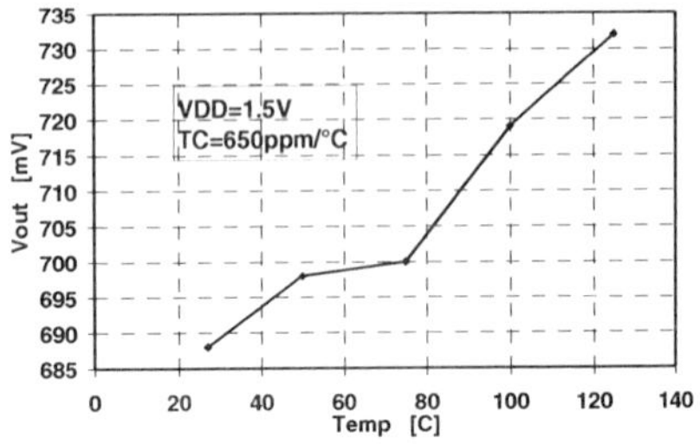

Fig.9: Temperature dependence of the bandgap reference

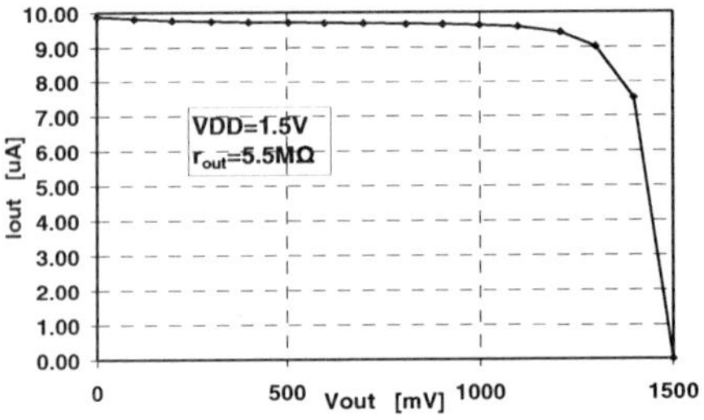

Fig.10: Iout vs. Vout of the Vth current reference

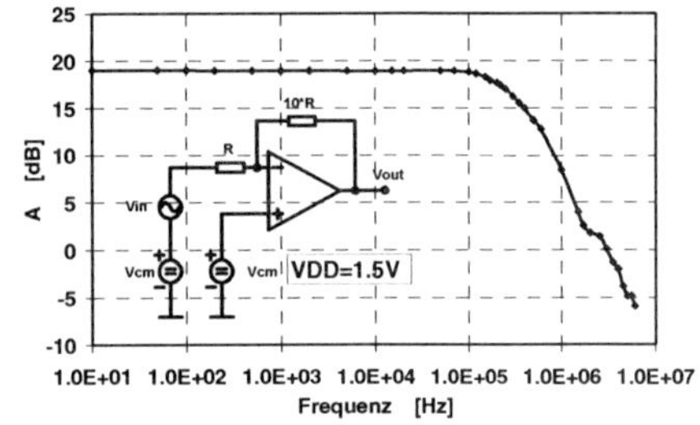

Fig.11: Gain vs. frequency of the inverting op amp circuit (target: A=20dB).

A 1V, Ka Band Prescaler with V_T Control in 90nm CMOS SOI

Razvan Ionita[1], Mihai A.T. Sanduleanu[2], Eduard Stikvoort[2] and Andrei Vladimirescu[1,3]
[1] Institut Superieur d'Electronique de Paris - ISEP, 21 rue d'Assas 75270 Paris Cedex 06, France
[2] Philips Research Eindhoven, Prof Holstlaan 4, 5656 AA, Eindhoven, The Netherlands
[3] University of California at Berkeley, BWRC, 2208 Allston Way, Berkeley, CA 94704, USA
Email: razvan.ionita@isep.fr, mihai.sanduleanu@philips.com, eduard.stikvoort@philips.com, andrei@eecs.berkeley.edu, Tel: (+33) 1.49.54.52.66

Introduction

The conventional circuit of the frequency divider is based on the Source Coupled Logic (SCL) or CML family which is the fastest logic family in CMOS. At low supply voltages, the SCL fails to work properly for reason of the stacking of transistors. The static frequency divider of Plouchart et al. [1] is implemented in 120nm CMOS SOI by using the well-known CML latch. It has a maximum operating frequency of 25GHz for a supply voltage of 1V. Some other CMOS prescaler topologies are published in literature [2,3] above 10GHz.

This paper deals with a less conventional frequency divide-by-two circuit based on a D-latch that operates at low power supply voltage. The V_T control via body-biasing technique allows enhancing the frequency performance of the prescaler and provides a tuning mechanism of the sensitivity curve for optimal operation with respect to frequency and process variations.

The circuit is realized in a partially depleted (PD) 90nm CMOS SOI process. The current consumption of the prescaler core without buffers is 60mA from a 1V supply. On-wafer measurements with differential GSSG RF probes show correct operation from 8GHz to 28GHz and a shift in the minimum sensitivity of more than 6GHz when operating the devices in reverse to forward body-bias.

Divider Topology

The frequency divide-by-two circuit is based on a cascade of two D-latches. The D-latch with V_T control is shown in Fig. 1. The core of the D-latch is the SR latch formed by transistors M1…M4. The set and reset functions are implemented with NOR transistors M5, M6, M7 and M8, M9, M10 respectively. This D-latch works with the R-NMOS type of logic and has the advantage of providing the same delay from CK to Q output and D to Q output ($t_{d\_CK \rightarrow Q} = t_{d\_D \rightarrow Q}$). The D-latch has differential D inputs and differential Q outputs with almost rail-to-rail swing.

The D-latch has a single-ended clock input but the total prescaler needs a differential input clock. Therefore, a differential input / output clock buffer is used. This leads to the circuit implementation shown in Fig. 2. The input and output buffers are SCL type. In the input buffer (see Fig. 3) shunt-peaking, stack inductors, in series with 50Ω resistors boost the input bandwidth. The common-mode biasing voltages at the input of the prescaler are derived from a resistive divider. As the output buffers are operating at half the input frequency, they are 50Ω differential buffers without inductors.

The V_T control of the NMOS transistors is realized with body-biasing. A DC signal applied at the input NB controls the body of the transistors. Applying a negative or positive DC voltage to the transistor body, results in a change of the threshold voltage and the transistor transconductance. Consequently, a proportional frequency shift of the input power sensitivity of the frequency divider is produced. The shift is to the left in the frequency domain for forward body-bias (FBB), and to the right for a reverse body-biasing (RBB) as sketched in Fig. 4.

Measured results

The circuit was implemented in a 90nm PD CMOS SOI process with seven metal layers: six metal layers and one thick on top. Body-contacted transistors are used for all SOI transistors. The layout of the D-latch is depicted in Fig. 5. Extra attention was payed for symmetry and the dimensioning of the interconnect for speed and maximum current density. The active area of the prescaler, including input and output 50Ω buffers,

0-7803-9212-4/05/$20.00 ©2005 IEEE

is $350 \times 400\mu m^2$. The chip photomicrograph is shown in Fig. 6. For the nominal condition ($V_{SB} = 0$), the prescaler has an operation range of 9 to 27GHz with the highest sensitivity at 23GHz. At 1MHz offset from the carrier, the phase noise at the output of the divider is –127dBc/Hz (see Fig. 7).

Fig. 8 presents the output spectrum of the the prescaler when a 24GHz clock signal is applied at the input. This frequency is close to the maximum sensitivity input frequency for nominal biasing with $V_{SB} = 0$. The spectrum of the 12GHz output signal is seen together with the input clock, after combining the two signals at the spectrum analyzer input. The power consumption of the divider is 60mW @ 1V supply voltage. The minimum input power sensitivity is improved by applying a positive voltage to the body of NMOS transitors (FBB) as shown in Fig. 9. The maximum operating frequency of the prescaler is increased up to 28GHz for negative body voltages (RBB).

Conclusion

This paper presents a static frequency divider in a 90nm PD CMOS SOI process. The divider uses a novel D-latch topology and has an operation range of 8 to 28GHz with maximum sensitivity tuning of ±3GHz around 22GHz. The D-latches were implemented with NMOS transistors in R-NMOS logic. A new method is proposed for tuning the sensitivity curve of the prescaler by controlling the threshold voltage of the transistors. The V_T spread due to process variations is compensated too. The V_T control shows an improvement of the prescaler sensitivity with forward body-biasing voltages and an increase of the frequency range with reverse body-biasing voltages. At maximum operating frequency, the power consumption of the divider is 60mW (1V supply voltage) and the active area, including buffers, is $350 \times 400\mu m^2$.

References:

[1] J.-O. Plouchart et al., RFIC Symposium, pp. 329-332, IEEE, 2003.
[2] H.D. Wohlmuth et al, RFIC Symposium, pp. 231-334, IEEE, 2002.
[3] H.D. Wohlmuth et al, Proc. of ESSCIRC, pp. 823-826, IEEE, 2003.

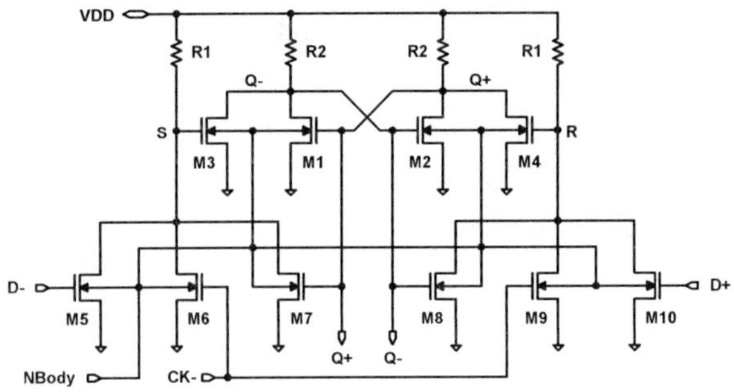

Fig. 1. D-latch with V_T control.

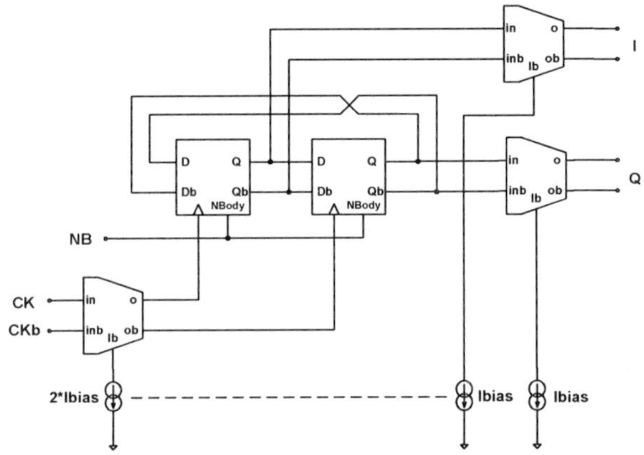

Fig. 2. Static prescaler with body voltage control.

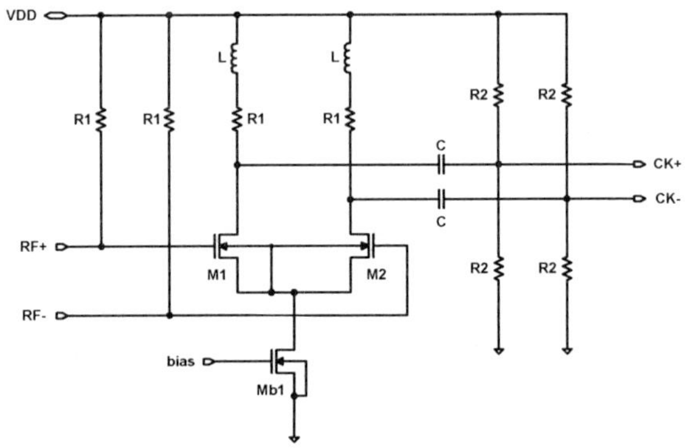

Fig. 3: Input clock buffer with shunt peaking.

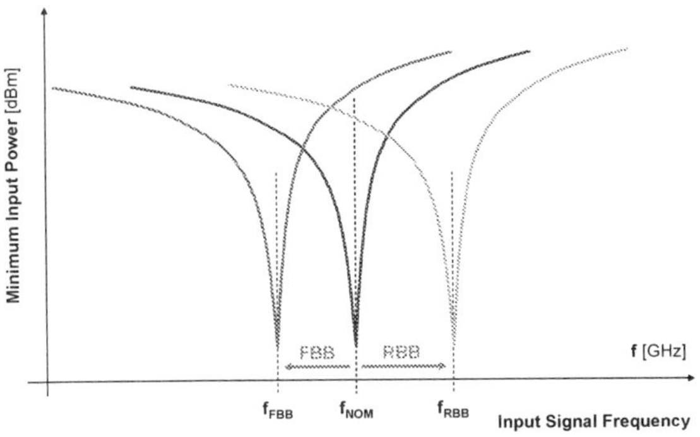

Fig. 4: FBB and RBB effects in the input power sensitivity of the prescaler with body control.

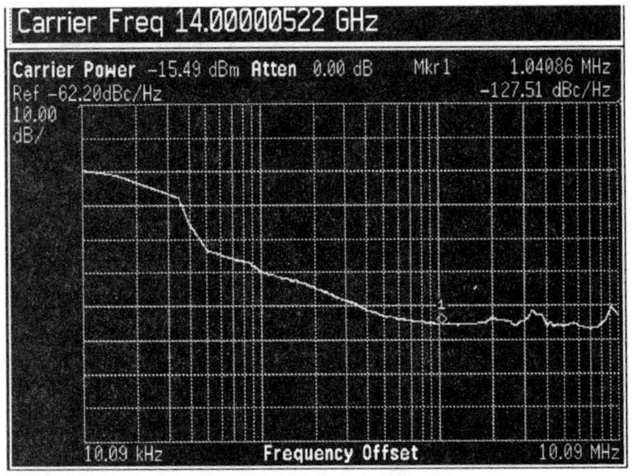

Fig. 7. Phase noise at the divider output.

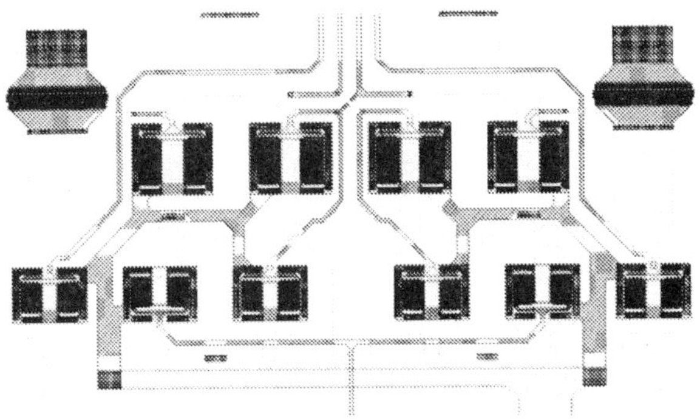

Fig. 5. Layout of the D-latch.

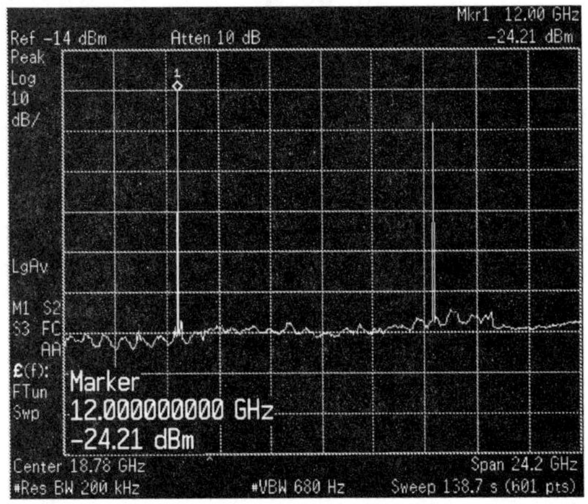

Fig. 8. Output spectrum of the prescaler for 24GHz clock input signal.

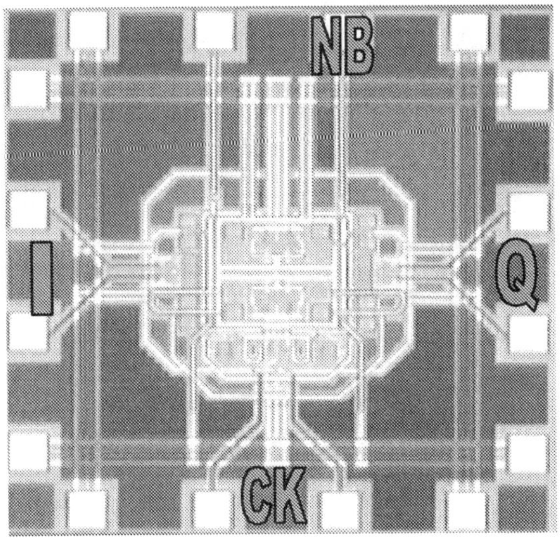

Fig. 6. Chip photo.

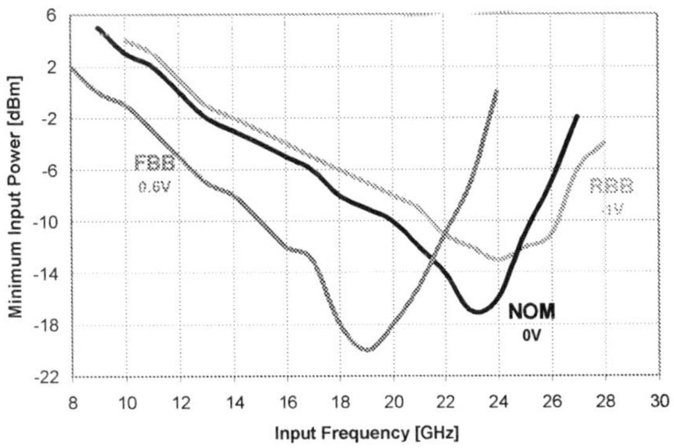

Fig. 9. Measured input power sensitivity curves.

2005 IEEE International SOI Conference

Floating Body Effects on the RF Performance of FDSOI RF Amplifiers

C.L. Chen, R.T. Chang,* P.W. Wyatt, C.K. Chen, D-R Yost, J.M. Knecht, and C.L. Keast
Lincoln Laboratory, Massachusetts Institute of Technology, Lexington, MA
*Currently at Atheros Communications, Sunnyvale, CA
Email: clchen@ll.mit.edu, Tel: (781) 981-4507

Abstract

The impact of body contacts on the floating body effect of fully depleted (FD) SOI at high frequencies is studied. It is found that the floating body effect is negligibly small for FDSOI FETs and the body contact (BC) increased parasitic capacitance and degraded the performance. We show that the linearity of an X-band amplifier, fabricated with a 180-nm FDSOI technology, is unchanged by the BC in continuous wave and pulse mode operations.

Introduction

It has been shown that the floating body can significantly change the device parameters of a partially depleted (PD) SOI FET at DC and high frequencies [1]. As a result, the linearity, which becomes increasingly important in communication and radar systems, of analog or RF amplifiers, can degrade without a body contact (BC). However, in addition to increased device size and layout complexity, the BC can also introduce undesirable parasitics that degrade the RF performance. Although a thinner silicon layer in fully depleted (FD) SOI offers some mitigation, it is not clear if BC is still required to eliminate the floating body effect.

In this work, we evaluate the effect of the body contact on RF characteristics of FDSOI FETs and compare the linearity of actual X-band amplifiers operated in the continuous wave (CW) and pulse mode.

Device and Amplifier Fabrication

The process is a 180-nm fully depleted SOI technology with metal T-gate structures and high-resistivity substrates [2]. The thickness of the SOI is 45 nm. Maximum f_T and f_{max} for the n-FET are in the 70 GHz range and measured minimum noise figure was 0.5 dB at 10 GHz. The top Al metal thickness is 2 μm.

Experiment Results

The BC design of a two 10-μm-wide gate-finger nFET is shown in Fig. 1. The p^+ implanted contacts were made in the middle and both ends of the poly-Si gate on the source side. The total gate width was increased to 11.5 μm per finger to maintain the same 10-μm effective source width.

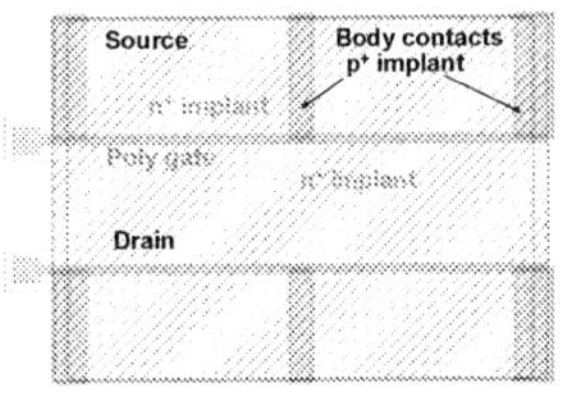

Fig. 1 nMOSFET with source side body contact.

The I_{ds}-V_{ds} plots for nFETs with gate length, L_g, of 180 and 500 nm are shown in Fig. 2. The kinks caused by the floating body effect are clearly shown in the 500-nm FET without BC. Although smaller for the 180-nm FETs, the kinks are still present without BC and this is confirmed by measured output conductance. The I_{ds} is slightly higher with BC, probably because the effective drain width is larger than 10-μm source width.

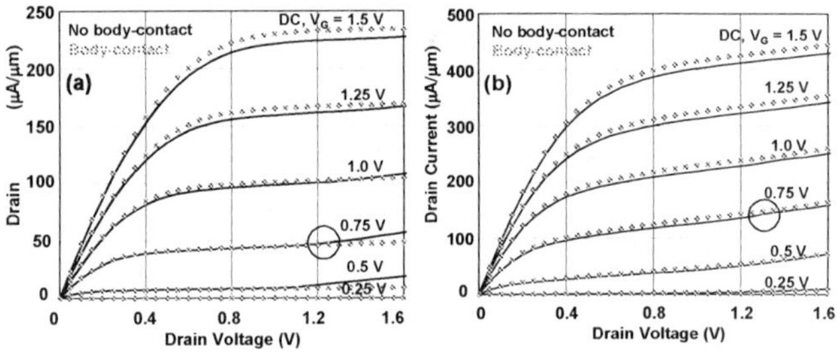

Fig. 2 DC characteristics of nMOSFETs. (a) L_g = 500 nm. (b) L_g = 180 nm.

The scattering (S) parameters of the FETs were measured for RF characterization. The kinks in the I_{ds}-V_{ds} plots suggest that the output conductance is most likely to be affected by the floating body effect. Figure 3 shows the output S parameter, S_{22}, of nFETs without BC biased at drain voltages above and below the kinks, exhibiting no difference in RF output resistance despite the large difference in DC output resistance.

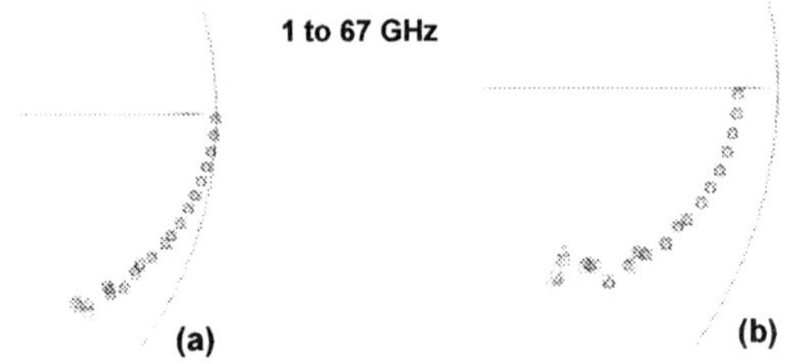

Fig. 3 Measured S_{22} of no-BC nMOSFETs with gate length of (a) 500 nm and (b) 180 nm. The gate bias is 0.5 V and each plot includes 5 sets of data with V_{ds} from 0.9 to 1.3 V in 0.1-V steps.

The kink observed in the IV plots is caused by change in the body potential, and that potential generally cannot follow the high-frequency signal [3]. The lack of output conductance variation with V_{ds}, which is observed in PDSOI [1], suggests that the capacitive body coupling is also negligible with a thin fully depleted body even without the BC.

The input S_{11}, plotted in Fig. 4 for 180-nm nFETs, shows higher input capacitance of the device with body contacts. The input capacitances, estimated from S_{11}, are

This work was supported by the Defense Advanced Research Projects Agency under Air Force Contract FA8721-05-C-0002. Opinions, interpretations, conclusions, and recommendations are those of the authors, and are not necessarily endorsed by the United States Government.

0-7803-9212-4/05/$20.00 ©2005 IEEE

34.9 and 37.6 fF for FETs without and with body contacts, respectively. Calculated DC gate capacitance is 30.3 fF for the same gate area without body contacts. One probable explanation is that the actual gate area of FET with body contact is larger by keeping the same effective conducting channel width. The S_{22} data suggests that FETs with BC also have slightly higher output capacitance. The large gate area may also explain why S_{12}, a measure of feedback from the drain to the gate, is always higher for FETs with BC. The difference is approximately 1.2 dB over the entire frequency range.

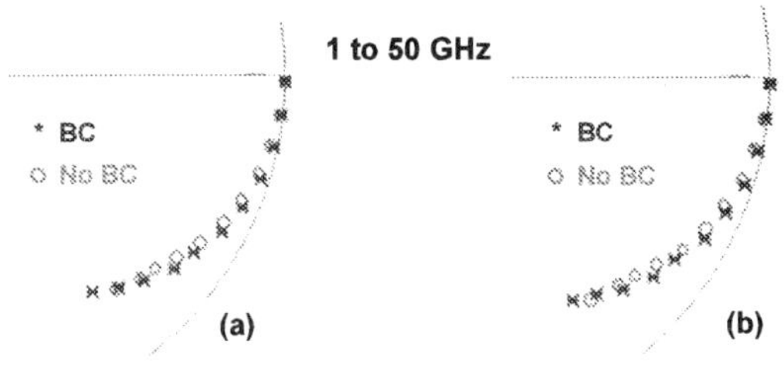

Fig. 4 Measured S_{11} of an nMOSFET with 180-nm gate length at V_{gs} = 0.75 V. (a) V_{ds} = 1.5 V. (b) V_{ds} = 0.9 V.

The plots of h_{21} and the maximum gain of adjacent FETs with and without BC are shown in Fig. 5. Because of the higher capacitance, the gain of devices with BC is always slightly lower than that without it. Similar characteristics were observed for pMOSFETs and measured f_T and f_{max} are listed in Table I. Note that the biases chosen here are not for the highest speed.

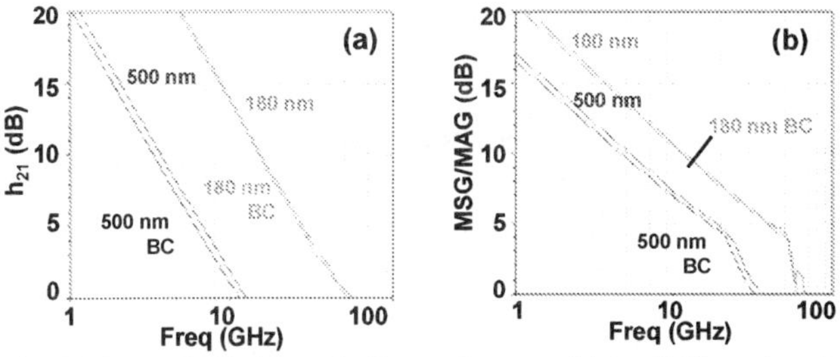

Fig. 5 Measured (a) h_{21} and (b) maximum stable/available gain of nMOSFETs at V_{ds} = 1.5 V and V_{gs} = 0.75 V.

Table I. Measured results at V_{ds}=1.5 V, V_{gs}=0.75 V

	nFET f_T	nFET f_{max}	pFET f_T	pFET f_{max}
180 nm	51	66	28	56
180 nm BC	50	60	26	52
500 nm	12	34	2.7	9.0
500 nm BC	11	31	2.1	7.0

Equivalent circuit models are used to better understand the effect of body contact. Figure 6 shows the DC output resistance, R_{out}, of 500-nm 2 × 10 n-MOSFETs with and without BC. The sharp drop of R_{out} indicates the onset of the kink. The equivalent circuits were derived for V_{ds} = 0.9 V and 1.3 V while fixing V_{gs} = 0.75 V to capture the impact of the kink.

A simple equivalent circuit shown in Fig. 7 is used. The C_{ds} and R_{ds}, representing the drain to source coupling via drain/source junctions, have no effects in DC measurement. In the p^+-implanted body-contact region, there is only one pn junction from the drain to the channel and this coupling is expected to be stronger.

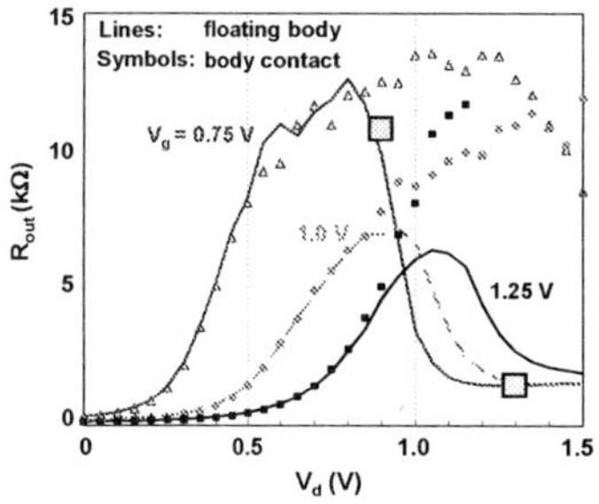

Fig. 6 DC output resistance of 500-nm 2 × 10 n-MOSFETs with and without body contacts. The two shaded squares show the DC R_{out} at the biases used for the equivalent circuit modeling.

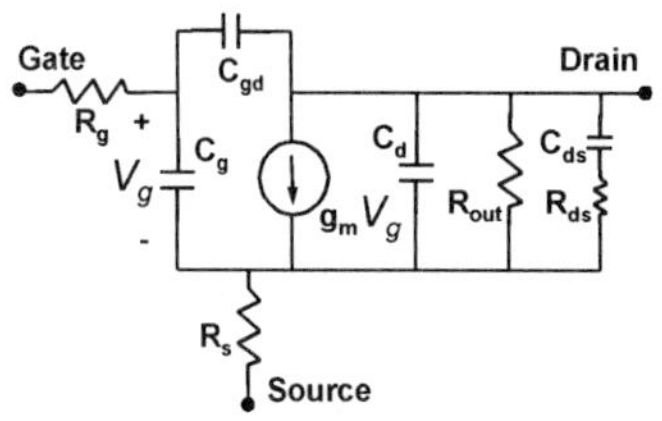

Fig. 7 Equivalent circuit for the FET.

The values of the components in the equivalent circuit were optimized to best match the measured S parameters from 1 to 67 GHz. Measured and calculated from equivalent circuit results for the 500-nm FET without BC are shown in Fig. 8.

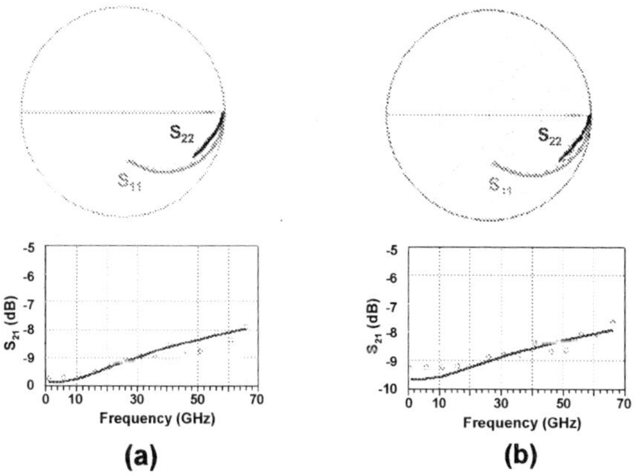

Fig. 8 Measured (circles) and calculated (lines) results of 500-nm nFET without BC. (a) V_{ds} = 0.9 V. (b) V_{ds} = 1.3 V.

Table II Component values of the equivalent circuit

	V_{ds} = 0.9 V 500-nm	V_{ds} = 0.9 V 500-nm BC	V_{ds} = 1.3 V 500-nm	V_{ds} = 1.3 V 500-nm BC
C_g (fF)	47.0	49.0	47.0	49.0
g_m (mS)	3.8	3.9	3.9	4.0
R_{out} (kΩ)	6.0	8.0	6.0	8.0
C_d (fF)	0.8	1.0	0.8	1.0
C_{ds} (fF)	7.0	8.0	7.0	8.0
R_{ds} (kΩ)	1.8	1.2	1.8	1.2
C_{gd} (fF)	14.0	15.5	14.0	15.5

Good match is achieved for this FET and the one with BC as well. The values of selected components are listed in Table II. Further increasing the R_{out} values in the equivalent circuit does not affect simulated S parameters and therefore, R_{out} values could be higher and are difficult to be accurately determined. For the FET without BC, the same R_{out} value in the equivalent circuit for V_{ds} below and above the kink is in contrast to the sharp drop of the DC values, marked by the two shaded boxes in Fig. 6. The body contact raises both g_m and gate capacitance values because of the wider effective gate width. However, the amount of capacitance increase is larger than that of g_m. The body contact also adds additional coupling from drain to source through the p^+-implanted region. As a result, both f_T and f_{max} are lower for the FET with BC.

Because of the increase of DC output conductance at higher V_{ds}, one possible impact of the floating body effect would be on the linearity of RF amplifiers. Although the equivalent circuit suggests that the kink is negligible at RF, the small-signal S-parameter measurement does not provide information on linearity. To evaluate the effects of the floating body on RF circuits, an X-band amplifier was designed and fabricated for this study.

As shown in the circuit diagram in Fig. 9, the amplifier uses a cascode configuration with twelve 180-nm 5-μm-wide gate fingers. Inductors and capacitors are used for impedance matching and all the DC bias circuits are also integrated. Except for the source-side body-contacts (SBC) to the FET, two otherwise identical amplifiers were fabricated side by side for evaluation. Because of the narrow gate fingers, there are only two body contacts for each finger and the actual gate width was increased to maintain the 5 μm source width.

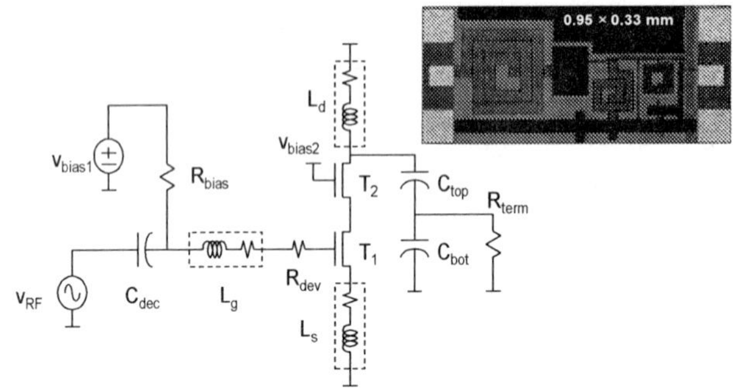

Fig. 9 Schematic diagram and photo of the X-band amplifier.

The amplifier without body contacts (NBC) has 8.2 dB of gain peaked at 9.3 GHz with 1.3 GHz of bandwidth. The two-tone continuous wave test was used to evaluate the linearity. The two signals were centered at 9.35 GHz and separated by 10 KHz. Measured fundamental and third-order intermodulation product (IP3) are plotted in Fig. 10. The output power at the interception point, OIP3, is 1.2 and 0.8 dBm for NBC and SBC amplifiers, respectively. The 0.4 dBm difference is within the range of typical process variations and confirms our expectation that the there is no difference in linearity when the two amplifiers are operated in the CW mode. The gain in the linear region of the NBC amplifier is approximately 0.9 dB higher than the SBC amplifier. Because floating body effects are more likely at higher drain voltage, a bias voltage as high as 5 V, which is much higher than designed 1.5 V, was applied and there is still no difference in OIP3 between the two amplifiers.

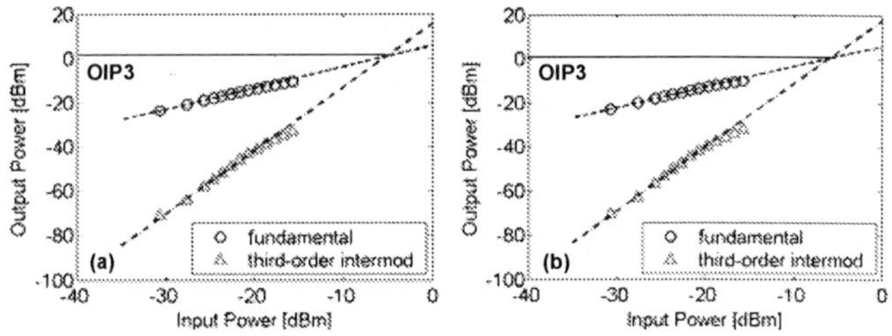

Fig. 10 Measured fundamental and third-order intermodulation product power of the X-band amplifier. (a) With body contact. (b) Without body contact.

The next set of measurement used input waveforms with modulated pulses. The large signal change during the on and off cycles may create a signal-dependent gain, resulting from the dynamic bias by the charge present in the floating body. To characterize any such effect, signals with pulse widths from 50 to 500 ns and repetition rates from 12.5 to 1000 kHz were applied. The repetition rates are low enough to have AC kink effects [3]. Across all these conditions, the 1-dB power compression of the two amplifiers track to within 0.7 dBm, showing no apparent floating body effect.

Conclusion

The small-signal S-parameters and the equivalent circuits suggest that the floating-body effects, observed in DC I-V measurement for FETs without body contacts, are negligibly small at RF. This is consistent with the fact that there is no apparent difference in linearity for X-band amplifiers with and without body contacts. Although the extent may vary with different body contact designs, our results showed that the added parasitics of the body contact can degrade the RF performance. Except maybe for low-frequency noise considerations, this work suggests that body contacts are not necessary for FDSOI FETs for RF circuit applications.

References

1. O. Rozeau, et al., "Impact of floating body and BS-tied architectures on SOI MOSFET's radio-frequencies performances," Dig. 2000 SOI Conf., pp. 124-125.
2. C-L. Chen, et al., "High-performance fully depleted SOI RF CMOS," IEEE Elect. Dev Letts., vol. 24, no. 4, 2003, pp. 251-253.
3. Y-C. Tseng, et al., "AC floating body effects and the resultant analog circuits issues in submicron floating body and body-grounded SOI MOSFETs," IEEE Trans. Electron Devices, vol. 46, no. 8, 1999, pp. 1685-1692.

A COMPARISON OF 10 GHZ FREQUENCY DIVIDERS IN BULK AND SOI 0.13 μm CMOS TECHNOLOGIES

Alexandre Engelstein, Jean-Michel Fournier*, Vincent Knopik**, Christine Raynaud****

*Institut de Microélectronique, Electromagnétisme et Photonique (IMEP),
23, rue des Martyrs 38016 GRENOBLE
**STMicroelectronics 850, rue Jean Monnet - BP 16 F-38926 CROLLES
*** LETI-CEA, 17, rue des Martyrs 38054 GRENOBLE
E-mail: alexandre.engelstein@insa-lyon.fr

ABSTRACT

The goal of this paper is to focus on the advantages of SOI technologies for high speed digital circuits for RF application, through the study of the consumption of 10 GHz frequency dividers. Dynamic and static structures are implemented in Bulk and SOI CMOS 0.13 μm technologies, and the measured consumptions are compared. As the capacitance of the drain-source diffusions and interconnections are reduced in SOI because of the BOX, dynamic consumption reductions of 20 % and 25 % are measured between Bulk and SOI technologies, respectively for the dynamic and static structure.

1. INTRODUCTION

During the last 10 years, the SOI technologies have proved to be a solution for high-speed digital applications (PowerPC 750* [1], AMD Athlon 64 [2]). The dynamical threshold voltage reduction due to the floating body increases the switching speed of digital circuits [3,4], and the buried oxide layer (BOX) reduces parasitic capacitances allowing energy saving at a fixed frequency. Some digital circuits in RF systems, such as RF frequency dividers, can take advantage of these properties to increase their performances. In order to quantify this technological improvement, a dynamic and a static frequency dividers 2:1 are designed in both Bulk and SOI 0.13 μm CMOS technologies and their consumption are compared at 10 GHz. The same design architecture and transistors dimensions have been used for bulk and SOI; only well contacts have been suppressed in SOI case. In the section 2, we present the structure of the two dividers compatible with low supply voltage (1.2-V). We present in section 3 the estimated capacitance reduction between the two technologies, which fits the measurements presented in section 4.

2. FREQUENCY DIVIDER STRUCTURES

2.1 Dynamic divider

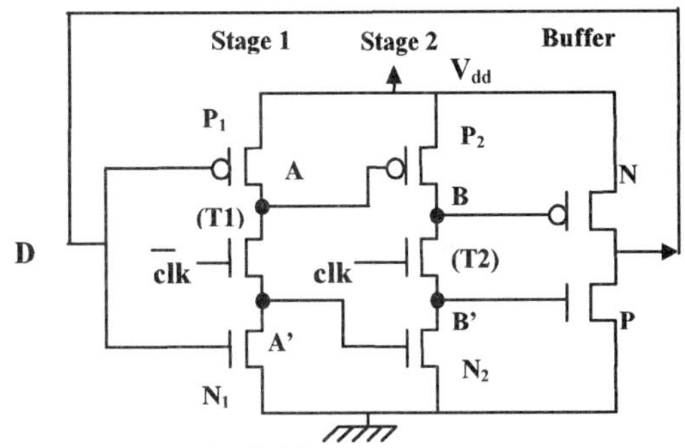

Figure 1. Dynamic divider.

A classical structure is used for the dynamic divider (Fig. 1). The 2 inverter stages (P_1-N_1) and P_2-N_2) are switched from memory to transparent state by switches T1 and T2. The memory function is done by the capacitances at the A, A', B, B' nodes. As there is no current generator neither loads, this circuit works rail-to-rail and is well-suited for low voltage applications. The NMOS and PMOS sizes are respectively 4/0.13μm and 8/0.13 μm.

2.2 Static divider

The structure comes from [5]. It is designed with two latches master/slave presented Fig.2.

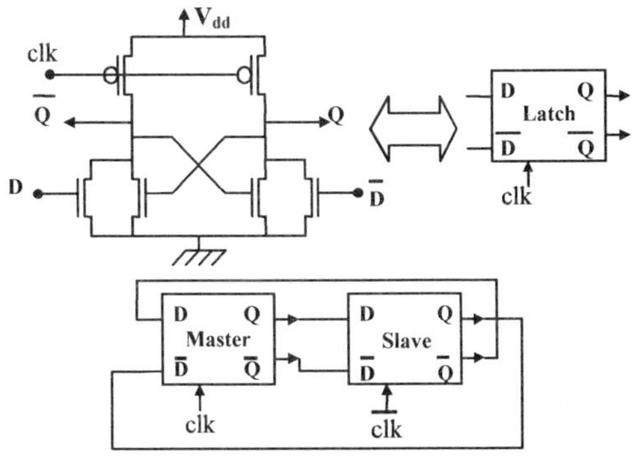

Figure 2. Static divider

0-7803-9212-4/05/$20.00 ©2005 IEEE

The signal is rail-to-rail and the structure of the latch stacks only 2 transistors. This divider is then well-suited for low voltage applications. When the clock changes from V_{dd} to 0, the latch is in a transparency mode during the switching time, and then stays in memory mode during half a clock period. In this memory mode, the nodes Q and $\overline{Q}$ become insensitive to D and $\overline{D}$ voltage variation. When the clk = Vdd, the latch is in inactive mode, and $Q = \overline{Q} = 0$. Figure 3 shows the evolution of the node voltage of a latch.

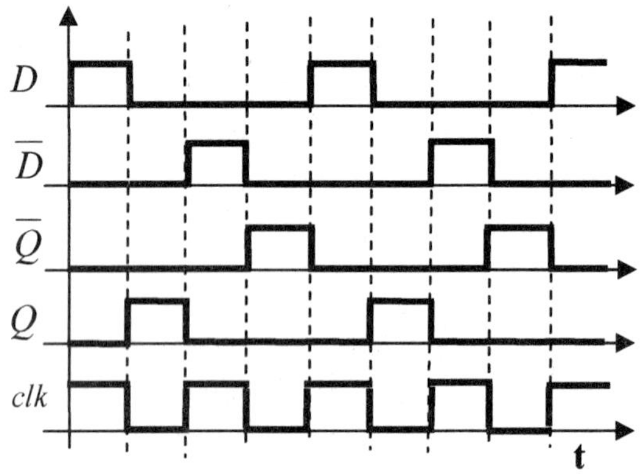

Figure 3 : evolution of a latch node voltage

This structure is called static because of the static consumption when the latch is in memory mode. In inactive mode however, there is no static consumption, the current sources pMOS being switched off. The simulations of the bulk circuits show a static consumption participation of 25 % of the global consumption at 10 GHz. This static consumption is identical in SOI technology because of the close V_{th} in the two technologies, so the capacitance reduction in SOI only impacts on 75 % of the global consumption reduction. The NMOS and PMOS sizes are respectively 2/0.13 and 6/0.13.

3. LOAD CAPACITANCE ESTIMATION

The load capacitances at the critical nodes (A, A', B, B' and D for the dynamic divider, the drain nodes for the static divider) are divided in 3 types: the gate, the diffusion and the interconnect capacitances.

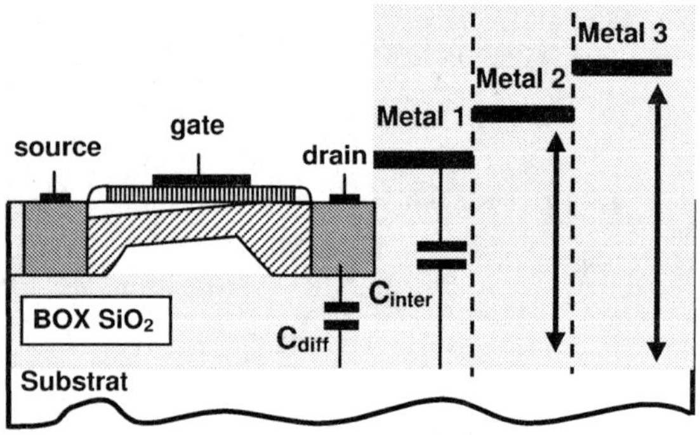

Figure 4 : the tree types of load capacitance

The gate capacitances are determined by the gate oxide thickness and are the same for both technologies. With the BOX thickness of 400 nm, the diffusion capacitances C_{diff} are reduced by a factor of 4-5 (figure 4). The BOX also reduces the interconnect capacitance (C_{inter}) by increasing the thickness of the SiO_2 dielectric between the metallization layer and the substrate. The reduction depends of the level of metallization layer. From technological data, we have obtained reduction factor of 1.6, 1.3 and 1.15 respectively for the first, second and third layer of metallization. Beyond this level, the reduction factor is negligible. As we have mainly used low level interconnect, we consider an average of 1.5 reduction factor for the interconnect capacitances.

	Gate capacitance	Diffusion capacitance	Interconnect capacitance
Dynamic divider (Bulk)	55%	10%	35%
Static divider (Bulk)	23%	12%	65%
SOI reduction factor	1	4-5	~1.5
Dynamic divider load reduction = Dynamic divider consumption reduction			**20 %**
Static divider load reduction			**33 %**
Static divider consumption reduction at 10 GHz			**33 % * 75 % = 25 %**

Table 1 Contribution of bulk load capacitance and estimation of load capacitance reduction in SOI

In table 1 we take into account the factor 75 % to estimate the global consumption reduction from the total load capacitance of the static divider. It's important to notice that the threshold voltages V_t's have almost the same values between the two technologies. Then, the direct short-circuit currents throughout N and P devices during transition which are depending on V_{th}'s have roughly the same values. So the current reduction is representative of capacitances reduction between Bulk and SOI processes.

4. MEASUREMENTS

The dividers are packaged with a 10 GHz VCO providing the differential clock and with a buffer for the 50 Ohm measurements. The buffer capacitance is taken into account for the capacitance estimation in the last section. In order to take into account only the dividers consumption, the dividers have separated supply voltages from VCO and buffer. Fig 5 and 6 show the comparison of Bulk and SOI circuits' current consumption. The frequency shift in the SOI version is caused by the reduction of parasitic capacitance in the VCO, especially by removing the pattern ground shield (PGS) of the inductance, which increases its oscillation frequency.

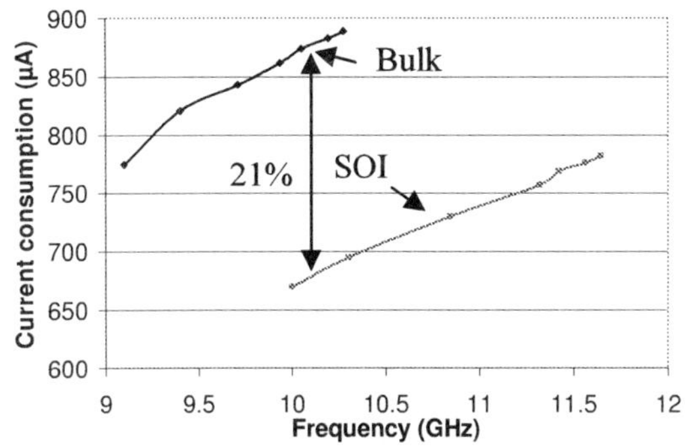

Figure 5. Comparison of Bulk and SOI dynamic divider current consumption

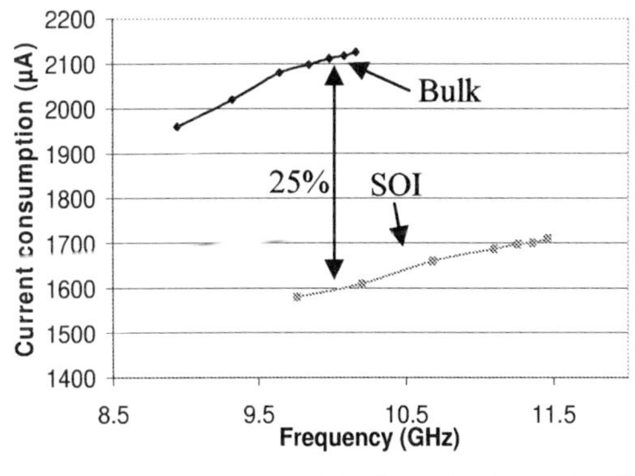

Figure 6. Comparison of Bulk and SOI static divider current consumption

The measurements show a current reduction of 25% and 21% respectively for the static and dynamic divider at 10 GHz. These results are quite consistent with the estimations given in table 1.

Fig 7 presents the simulated results of the input sensitivity for the two SOI frequency dividers. They both show self oscillating frequencies around 8 GHz and a minimum input signal from the VCO of -5 dBV is needed between 9.5 and 11.5 GHz. With the high input load of the frequency dividers, the VCO output signal is -4 dBV, which is sufficient for the system to work between 9.5-11.5 GHz.

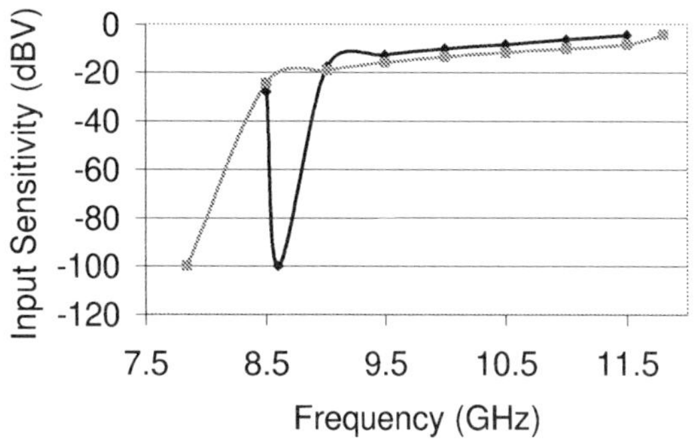

Figure 7. input sensitivity for dynamic and static dividers

5. CONCLUSION

The advantage of the SOI on Bulk technology for high speed digital circuits used in RF blocks is shown in this paper. The comparison of current consumption between 10 GHz frequency dividers designed in both technologies has been presented. The diffusion capacitances are reduced by factors of 4-5, and considering the back-end dimensions, the interconnect capacitances are reduced by an estimated average factor of 1.5. The current consumption measurements confirm this estimation, and show a consumption reduction of 25 % and 21 % respectively for both static and dynamic dividers.

6. REFERENCES

[1] http://www.research.ibm.com/journal/rd/462/shahidi.html

[2] http://www.amd.com/us-en/Processors/ProductInformation/0,,30_118_9486_9490,00.html

[3] Jacques Gautier, Keith A. Jenkins, Jack Y.-C. Sun, "Transient effect in floating body SOI NMOSFETs", *IEEE SOI conference 1995*

[4] Daniel L. Stasiak, F. Mounes-Toussi, Salvatore N. Storino, "A 440-ps 64-bit Adder in 1.5-V/0.18 µm Partially Depleted SOI technology", *IEEE journal of solid state circuits, vol.36, no.10, October 2001*

[5] Razavi B., Lee K.F., Yan R.H., "Design of high-speed, low power frequency dividers and phase locked loops in deep submicron CMOS", *IEEE journal of solid state circuits, vol.30, no.2, February 1995*

Active Body-Biasing Control Technique for Bootstrap Pass-Transistor Logic on PD-SOI at 0.5V-VDD

M. Iijima, M. Kitamura, K. Hamada, K. Fukuoka, M. Numa, A. Tada†, and S. Maegawa†
Kobe University, 1-1 Rokko-dai, Nada, Kobe 657-8501, Japan
†Renesas Technology, 4-1 Mizuhara, Itami, Hyogo 664-0005, Japan
numa@kobe-u.ac.jp, +81(78) 803-6089
†maegawa.shigeto@renesas.com, +81(72) 784-7324

Abstract

In this paper, we propose an Active Body-biasing Controlled (ABC)-Bootstrap PTL (Pass-Transistor Logic) on PD-SOI at 0.5 V-V_{DD} for ultra low power design. Applying active body-biasing to Bootstrap PTL is the key for higher performance without output voltage loss by boosting gate voltage with coupling capacitance between source and body. Lowering V_{th} by body biasing also contributes for high speed operation. Experimental results have shown improvement in both delay time and power consumption.

Introduction

Aggressively scaled-down CMOS technology faces the speed-improvement degradation due to increase in power consumption. Lowering threshold voltage causes the drastic increase in leakage power. The low power design style is moving from coarse-grained approaches to fine-grained ones for ultra low power operation [1].

SOI MOSFET has been attracting attention as a new device technology for improvement in both speed and power consumption in recent years [2]. In addition, PD-SOI is able to control the threshold voltage (V_{th}) by changing body voltage. Although VTCMOS (Variable Threshold CMOS) [3], well known as one of dynamic V_{th} control techniques, controls V_{th} at block level, PD-SOI is suitable for the control of V_{th} at transistor level because the body region of transistor is separated each other [4]. DTMOS (Dynamic Threshold MOS) [5], also known as ABC (Active Body-biasing Control) technique, enables transistors to turn on quickly without increase in leakage power by lowering V_{th} only if needed. However, DTMOS cannot be used with supply voltage (V_{DD}) over 0.6 V due to PN current from source / drain to substrate.

In order to achieve ultra low power operation, lowering supply voltage has been one of the most effective approaches since dynamic power consumption decreases in proportion to the square of V_{DD}. Lowering V_{DD}, however, causes degradation in performance seriously. Thus we employ the dynamic body control technique at 0.5 V-V_{DD}. Scaling V_{DD} down to 0.5 V enables the application of ABC-technique in SOI circuits without any additional circuits for voltage level conversion.

ABC-Bootstrap PTL

The Pass-Transistor Logic (PTL), such as LEAP (LEAn integration with Pass-transistors) [6], is suitable for low power design because PTL circuit is synthesized with fewer number of transistors than CMOS logic. In addition, the low parasitic junction capacitance with SOI devices between silicon substrate and source / drain is effective to speed up PTL circuits composed of plural nMOSFET's connected in series. The single-rail type PTL, however, has the disadvantage of voltage loss at the high-level output. Even though the output voltage V_o is expected to be $V_o = V_{DD}$, the output voltage is actually able to rise only up to $V_o = (V_{DD} - V_{th})$, which causes increase in short circuit current and degradation of driving power.

The use of Bootstrap PTL [7] has been discussed to overcome the issue of output voltage loss and degradation of speed. The key point with Bootstrap PTL is that the drain voltage of nMOS pass-transistor is able to rise up to V_{DD} without using pull-up pMOS by boosting the gate voltage higher than V_{DD}. A Bootstrap PTL configuration consists of two nMOS's T1 and T2 as shown in Fig. 1 (a). nMOS T1 is a pass-transistor for

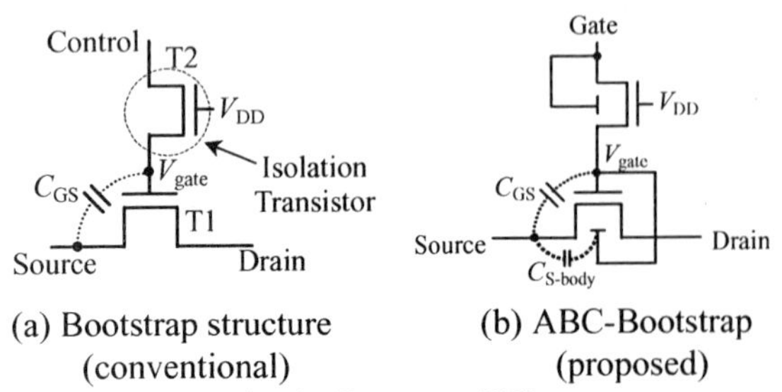

Fig. 1: Bootstrap PTL.

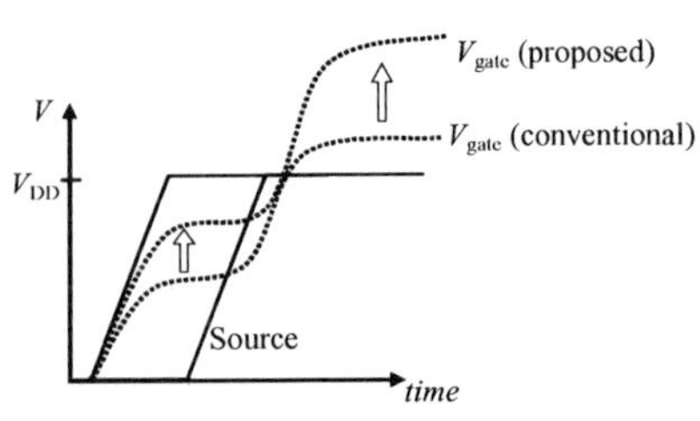

Fig. 2: Boosting effect of Gate voltage.

data propagation and T2, called "isolation transistor," ensures capacitive coupling in T1 between source and gate, so that the gate voltage of T1 rises up higher than V_{DD}. The capacitance C_{GS} between source and gate helps to boost gate voltage when the source signal begins to rise as shown in Fig. 2.

We propose ABC-Bootstrap PTL shown in Fig. 1 (b) to improve speed and driving power with conventional Bootstrap PTL by the following two points:

i) Speed-up by forward body-biasing higher than V_{DD}.
ii) Make use of capacitance $C_{S\text{-}body}$ between source and body to boost gate voltage more.

The ABC-Bootstrap PTL strongly contributes to speed-up during pull-up transition because the gate voltage raised up higher than V_{DD}. This is effective to lower V_{th} than conventional Bootstrap PTL. In contrast to the conventional approaches using multiple-supply voltages for forward body-biasing, single-V_{DD} is enogh to do so by ABC-Bootstrap PTL. Moreover, ABC-Bootstrap PTL offers improvement in boosting effect of gate voltage as shown in Fig. 2. Only the coupling capacitance C_{GS} between source and gate is used to boost the gate voltage in conventional Bootstrap PTL. On the other hand, the coupling capacitance $C_{S\text{-}body}$ between source and body is also used to boost the gate voltage in ABC-Bootstrap PTL. This is the key idea with our technique to improve both speed and driving power.

Simulation Results

We have performed SPICE simulation with BSIM3-based SOI transistor model. The process rule is 0.18 μm PD-SOI and the supply voltage is 0.5 V. The threshold voltages with nMOS and pMOS are set to $V_{th\text{-}n} = 0.24$ V, $V_{th\text{-}p} = -0.34$ V, respectively. ABC-Bootstrap PTL has been applied to XOR gate [7] and to 4-bit MCC Adder [8].

In Fig. 3 shows the trade-offs between delay and standby leakage power when V_{th} changes from 240 mV to 310 mV at 10 mV step. In terms of delay time, ABC-Bootstrap PTL shows the better result with speed degradation caused by higher V_{th}. In case of conventional Bootstrap PTL, delay time increases rapidly for higher V_{th}. The delay time with conventional Bootstrap PTL is 3.3 times as long as that of nMOS-only PTL when V_{th} is 310 mV. Therefore, ABC-Bootstrap PTL shortens delay time effectively and avoids the speed degradation caused by high V_{th}. Each circuit dissipated almost the same standby leakage power as long as V_{th} is kept unchanged. ABC-SOI contributes to speed up without increasing standby leakage power because ABC-SOI operates like body-tied SOI at standby mode.

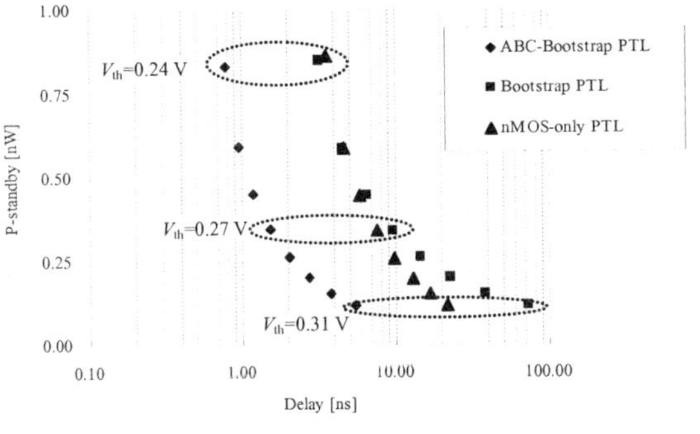

Fig. 3: Variation of delay and standby power for several V_{th}.

Table 1: Power saving effect by ABC-Bootstrap PTL

circuit		V_{DD}	Delay [ns]	Active Power [nW]	Standby Power [nW]
XOR	conv.	0.56 V	1.54	28.4	1.01
	proposed	0.50 V		22.5 (0.79)	0.82 (0.81)
4bit-MCC Adder	conv.	0.63 V	9.02	447	12.8
	proposed	0.50 V		266 (0.60)	8.14 (0.63)

( ): proposed / conv.

We have compared the power consumption of XOR gate, 4-bit MCC Adder after adjusting proper supply voltages in order to achieve same operation speed. As shown in Table 1, both active power and standby leakage power have been decreased effectively. ABC-Bootstrap PTL saves active power by 21 %, 40 % for XOR gate, 4-bit MCC Adder, respectively. Moreover, the standby leakage power is also reduced by 19 %, 37 %, respectively.

Conclusion

We have proposed ABC (Active Body-biasing Controlled) -Bootstrap PTL with PD-SOI for high-speed operation even at 0.5 V-V_{DD}. One of the advanteges is speed-up by forward body-biasing higher than V_{DD}, and the other is making use of the coupling capacitance $C_{S\text{-}body}$ between source and body to boost gate voltage higher than conventional Bootstrap PTL. As a result, our technique has succeeded in the operation of no less than 4 pass-transistors connected in series at 0.5V-V_{DD}.

References

[1] T. Sakurai, European Solid-State Circuits Conference, pp. 11-18, 2004.
[2] G. Shahidi, et al., Proc. IEEE Int'l SOI Conf., pp. 1-4, 1999.
[3] T. Kuroda et al., IEEE J. Solid State Circuits, vol. 31, no. 11, pp. 1770-1779, 1996.
[4] Y. Hirano et al., IEDM Tech. Dig., pp. 467, 2000.
[5] F. Assaderaghi et al., IEDM Tech. Dig., pp. 809-812, 1994.
[6] K. Yano et al., IEEE J. Solid State Circuits, vol. 31, no.6, pp. 792-803, 1996.
[7] K. Fujii et al., IEICE Trans. Electron., vol. E86-C, no. 4, pp. 604-611, 2003.
[8] N. H. E. Weste, K. Eshraghian, Principles of CMOS VLSI Design, Addison - Wesley Publishing Company, 1985.

State of the art integrated millimeter wave passive components and circuits in advanced thin SOI CMOS technology on High Resistivity substrate

F. Gianesello[1], D. Gloria[1], C. Raynaud[1], S. Montusclat[1], S. Boret[1], C. Clément[1], C. Tinella[1], Ph. Benech[2], J.M. Fournier[2] and G. Dambrine[3]
[1]STMicroelectronics, FTM, 850 avenue Jean Monnet, 38926 Crolles
[2]IMEP UMR 5130, 23 rue des martyrs- BP. 257-38016 Grenoble
[3]IEMN,cité scientifique, avenue Poincarré, BP69, 59662 Villeneuve d'Ascq
Email: frederic.gianesello@st.com

Abstract

In this paper, a comparison between transmission line (TL) integrated in High Resistivity (HR) Silicon On Insulator technology (SOI), standard CMOS and InP technologies is made. State of the art performances are reported on HR SOI with loss propagation of about 0.4 dB/mm@40 GHz and < 1 dB/mm@100 GHz. These results demonstrate that using HR SOI wafer suppressed substrate losses (like for III-V technology). In addtion, model have been developed for the described TL. To illustrate these results two microwave passive circuits, a 80 GHz coupler and a 80-100 GHz band-pass filter with both 2.5dB insertion losses (for the HR SOI version) have been realized in standard bulk CMOS and HR SOI technologies for comparison and modelization purpose.

Introduction

Today, advanced 65 nm CMOS node offer Ft higher than 150 GHz [1] allowing new millimeter wave applications on silicon. To address such applications, silicon technology would have to offer high quality integrated passive components. Unfortunately, this goal would be hard to reach using standard bulk technologies because of their associated lossy substrate HR. On the contrary, SOI technology is fully compatible with HR ($\rho > 1$ KΩcm). Previous works have demonstrated potential gain if using HR SOI concerning isolation [2] and improvement of integrated inductor performances [3]. However, few works have been reported at millimeter frequencies. In this study, integrated coplanar transmission line (CPW) and circuits integrated on standard STMicroelectronics SOI are described and compared to similar CMOS and also InP technologies.

CPW lines description and performances

At microwave frequencies TL is a key component and has to be characterized and modeled cautiously. CPW is prefered at millimeter range since it has better quality factor than microstrip and it is less sensitive to process variation. Moreover, it will make no sense to use microstrip on HR since the substrate is shielded. CPW is widely used and flexible (Fig.1). To compare performances, three key parameters are extracted thanks to relation given in [4]: Characteristic impedance, Attenuation constant and Propagation constant. Thus, 50 Ohms characteristic impedance lines have been realized on bulk and HR SOI by using top level metallization and exploring different factors shape. The best results obtained on standard bulk process and on HR SOI are compared on Fig.2. We decided to stack all the metallization on HR SOI in order to reduce metallic losses and to evaluate substrate losses. The dimensions of the line are the same as those commonly used for W-band design on InP [5].The performances obtained for this stacked line could be compared to state of the art structure in GaAs or InP technologies (Fig.3). This result demonstrates the insulating behaviour of the HR Unibond substrate used in STMicroelectronics SOI technology, even if the resistivity is only 1 KΩcm.

Millimeter wave passive circuits results: SOI HR vs bulk and InP

To illustrate the performances of the TL described above, two passive circuits have been studied. First one is a stub based band-pass 80-100 GHz filter (Fig.4) using CPW TL realized in top level metallization. Second one is a 94 GHz InP based hybrid coupler which has been blind ported (Fig.5) on HR SOI by using the stacked CPW TL since the stacked TL does not work on bulk because of substrate losses. Measurements have been performed on wafer up to 110 GHz using an Agilent HP8510XF vector network analyzer. For the filter, the SOI version has 2.5dB insertion losses in band whereas the bulk version has 5dB (Fig.6). Moreover, simulations achieved using the developed TL model accord well with these results. For the coupler, on SOI version a shift has been observed in frequency and resonates at 80 GHz (Fig.7) due to an increase of capacitance at the underpath level (we can not use air bridge), but the losses are similar to InP around 2.5dB.

Conclusion

For the first time, a commercially available standard 130 nm SOI CMOS technology on 200 mm HR Unibond substrate shows passive components performances at millimeter frequencies equivalent to III-V technologies. Moreover, HR SOI technology offers advantages in term of speed and consumption for the digital side. As a consequence, HR SOI seems to be a good candidate to address incoming wireless technology (WiMax, 60 GHz WLAN, ...) in order to offer low cost, low power SOCs required by mass market applications.

0-7803-9212-4/05/$20.00 ©2005 IEEE

References

[1] L.F. Tiemeijer et Al., IEDM, 2005 (to be published)
[2] J.P. Raskin, Electron Devices, pp. 2252,dec. 1997
[3] Jonghae Kim et Al., IEEE MTT-S, vol. 1, 2003
[4] S. Montusclat et Al., IEEE-JINA, pp. 120, 2004.
[5] S. Boret, PhD Thesis, IEMN, 1999.

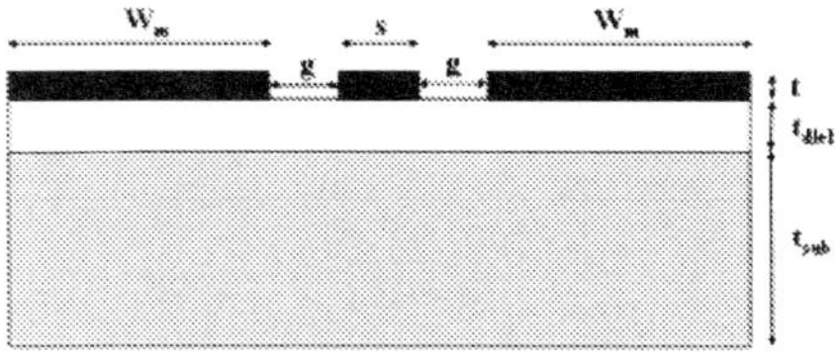

Figure 1: CPW structures cross section

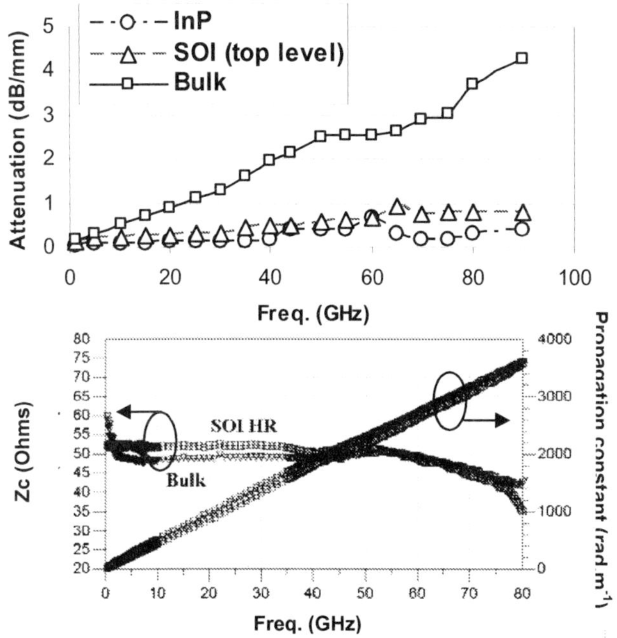

Figure 2: Comparison between measured 50 Ohms CPW TL performances, realized by using top level metallizations, in bulk, HR SOI and a reference on InP.

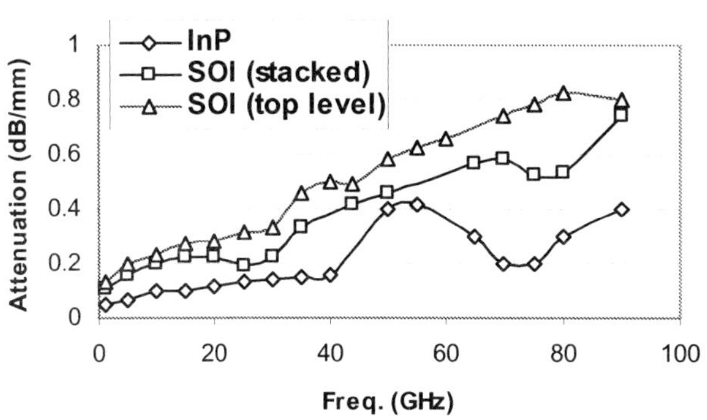

Figure 3: Comparison between measured 50 Ohms CPW TL performances, realized by stacking all the metallizations in HR SOI compared to the same structure realized on InP and reference CPW TL using top level metallization on HR SOI.

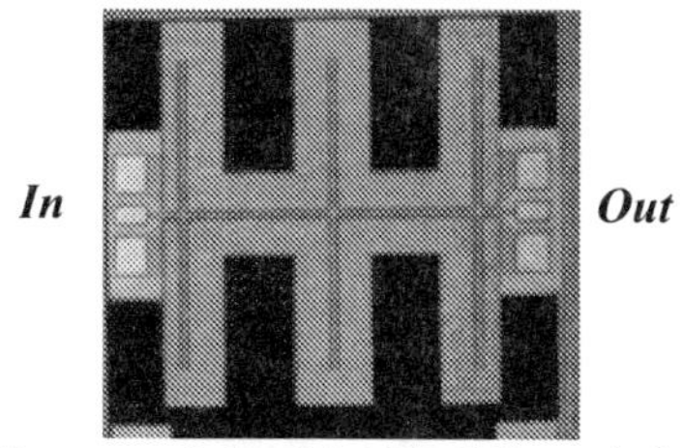

Figure 4: Layout of the 80-100 GHz stub based band-pass filter

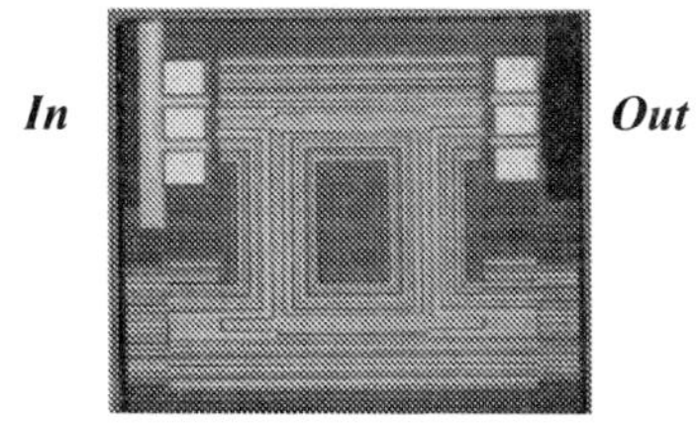

Figure 5: Layout of the 80 GHz hybrid coupler

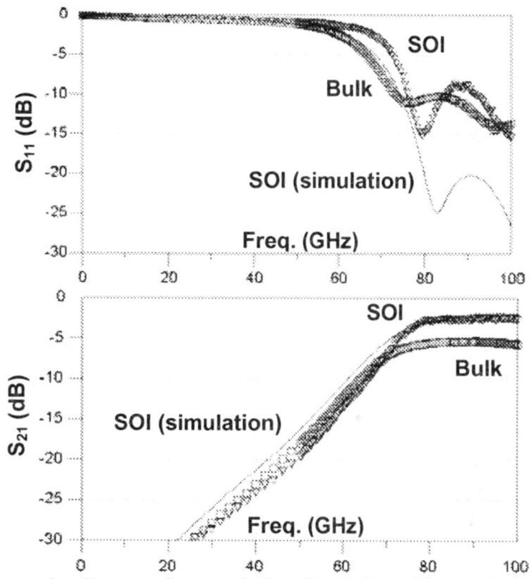

Figure 6: Comparison of the 80-100 GHz stub based band-pass filter performances: bulk vs. HR SOI

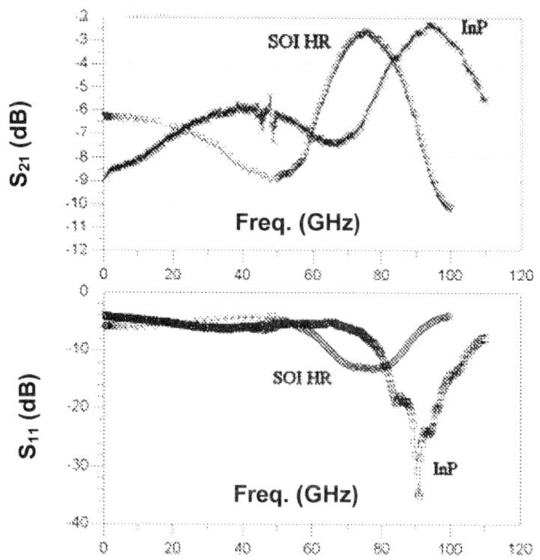

Figure7: Comparison of the hybrid coupler measured performances, for the same layout: HR SOI vs. InP

Session #4

POSTER SESSION

CHAIR

Hector Sanchez

Freescale Semiconductor

POSTERS — Session 4

NOTE: Poster numbers were chosen to associate the posters by technical content with the oral sessions. Poster, P3.1, means the poster, P, is the first poster similar in technical content to the papers presented in session 3; P3.2 is the second poster with similar content to session 3.

Electrical Characterization of Ultra-Thin SOI Films : Comparison of the Pseudo-MOSFET and Hg-FET Techniques

F. Allibert,[1] N. Bresson,[1,2] K. Bellatreche,[1] C. Maunand-Tussot,[1] and S. Cristoloveanu[2]
[1] Soitec S.A., Chemin de Franques, Bernin, 38926 Crolles, France
[2] IMEP, ENSERG, BP 257, 38016 Grenoble Cedex 1, France.
Email: frederic.allibert@soitec.fr, Tel: (+33) 438-92-1746

Abstract As the MOSFETs dimensions are scaled down, following the ITRS roadmap, the need for SOI wafers with ultra-thin Si films (UTF) becomes acute. Characterization of these wafers with simple, process-independent, and fast turnaround methods is very important. In this paper, we present for the first time 3 key aspects: (i) properties of UTF down to 10 nm thickness, (ii) comparison of thinning techniques (sacrificial oxidation vs. SC1), and (iii) comparison of Pseudo-MOSFET and Hg-FET methods for UTF.

Measurement The Pseudo-MOSFET and Hg-FET techniques have been thoroughly described for application on relatively thick SOI wafers [1,2]. These two methods enable the characterization of the BOX, film/BOX interface, and electron/hole channels. The parameter extraction is done as in a MOSFET. Typical $I_D(V_G)$ and $g_m(V_G)$ curves are shown on Figures 1 and 2. The main difference between the two techniques resides in the source/drain contacts. For Hg-FET the geometrical factor ($f_g \equiv W/L$ in regular FETs) is well defined (concentric circular contacts) [3] but a surface treatment of the wafer is necessary prior to measurement, resulting in a time dependency of the extracted parameters. The measurement takes several hours. On the other hand, for the point contacts of the Pseudo-MOSFET, f_g must be defined by simulation or calibration [4], but no surface treatment is required and the measurement is much faster.

Questions have been raised concerning the application of these techniques to ultra-thin SOI layers [5,6]. Based on systematic measurements, we discuss the impact of the film thickness (T_{Si}) and thinning methods on extracted values of carrier mobility (μ), threshold (V_T) and flat-band (V_{FB}) voltages.

Sample preparation To insure a differentiation of the measurement effects from the possible impact of thinning on the material quality, the samples were prepared using two different methods.

Method 1: 8 wafers were thinned by thermal sacrificial oxidation down to 73, 50, 37, 36, 24, 21 and 20 nm.

Method 2: Ten wafers were thermally thinned down to 82 nm, then further thinned to 66, 43, 32, 21, 11 and 10 nm by wet etching in a SC1 solution.

Results and discussion Figure 3 presents the variation of the flat-band and threshold voltage with T_{Si} for chemically thinned wafers as measured with both techniques. Figure 4 complements the results by showing that similar data are obtained when comparing wafers thinned chemically and thermally. The increase in V_T and V_{FB} for UTF is usually associated with an increase in D_{it} and Q_{BOX} respectively, due to oxidation induced defects. This explanation does not stand here since SC1 thinning is not expected to change D_{it} and Q_{BOX}. We explain this $V_{T,FB}$ increase by the proximity of the top film surface, that contains numerous dangling bonds acting as traps, to the conduction layer [6]. A coupling model has been developed.

The impact of T_{Si} on mobility is shown on Figures 5 (electrons) and 6 (holes). In both cases, a drop is observed in the mobility values extracted with the Pseudo-MOSFET while those extracted with the Hg-FET remain constant down to 11 nm. Two phenomena can be invoked. First, the electric field profile changes as T_{Si} decreases, thus leading to an increasing influence of the second order mobility reduction factor (θ_2) related to interface scattering. The effect of θ_2 is more easily corrected by the Hg-FET technique (constant and higher mobility) than by the Pseudo-MOSFET (degraded mobility). Second, the geometrical factor is well defined for the Hg-FET and does not change much with T_{Si}. In Pseudo-MOSFETs, f_g, defined by the distribution of the current lines, can vary with T_{Si} [7], leading to an apparent mobility degradation. The mobility change in Hg-FET with extremely thin film (10 nm) suggests that other effects, such as quantization, may come into play.

Conclusion Though the extraction of the electrical parameters becomes harder for UTF SOI, we have demonstrated the possibility to characterize *in situ* SOI wafers down to 10 nm. There is no impact of film thickness on carrier mobility and oxide charge but extraction model adaptations become necessary.

Acknowledgements The authors thank N. Baumel and F. Poncet from Soitec for their help.

Reference

[1] S Cristoloveanu et al, IEEE T.E.D. 47, p.1018, 2000.
[2] H.J. Hovel, Sol. Stat. Elect. 47 (2003) 1311–1333.

[3] D. Munteanu et al, El. & Sol. St. Lett. 2 1999 p.242.
[4] S. Williams et al, Microelec. Eng. 15 (1991) p.195.
[5] F. Allibert et al, IEEE Int. SOI Conf., 2002.
[6] N. Bresson et al, Proc. ECS spring met. 2005.
[7] K. Komiya et al, IEEE T.E.D. 52, N°3, 2005, p.406.

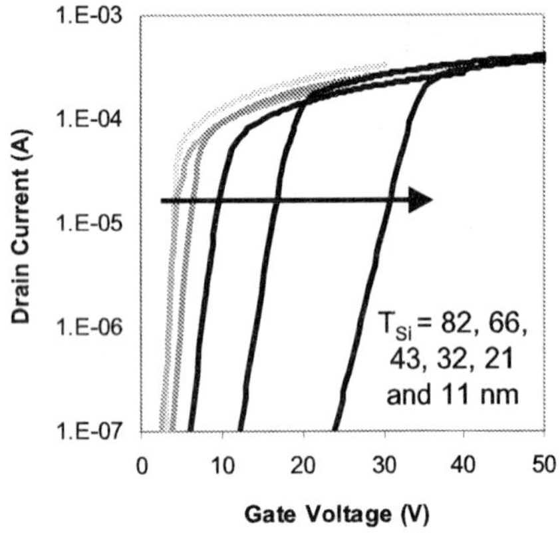

Figure 1: $I_D(V_G)$ characteristics obtained with the Hg-FET technique for various film thickness T_{Si}.

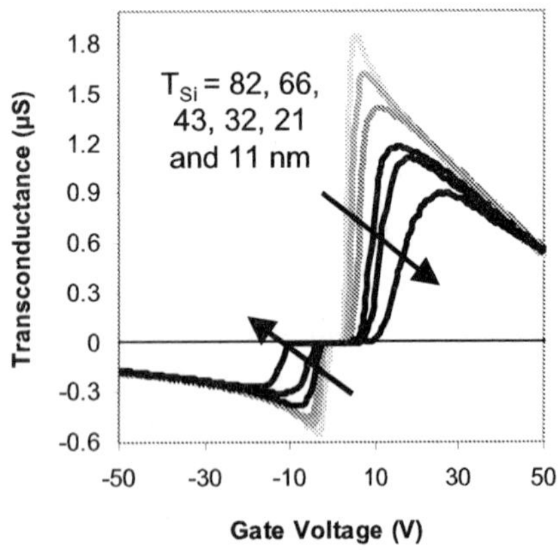

Figure 2: $g_m(V_G)$ characteristics obtained with the Pseudo-MOSFET technique for various T_{Si}.

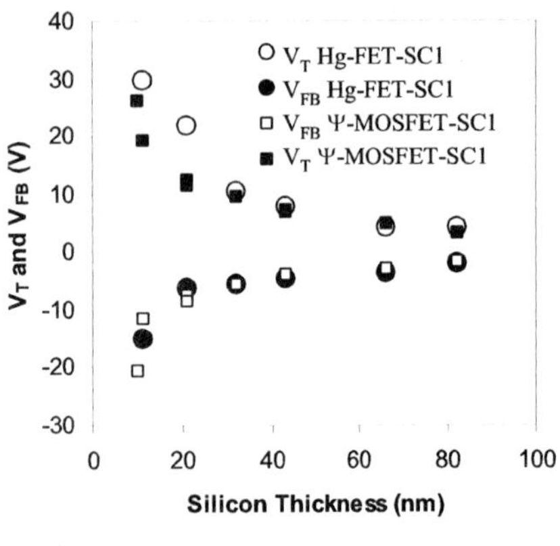

Figure 3: Variation with T_{Si} of V_T and V_{FB} values extracted by Hg-FET and Pseudo-MOSFET.

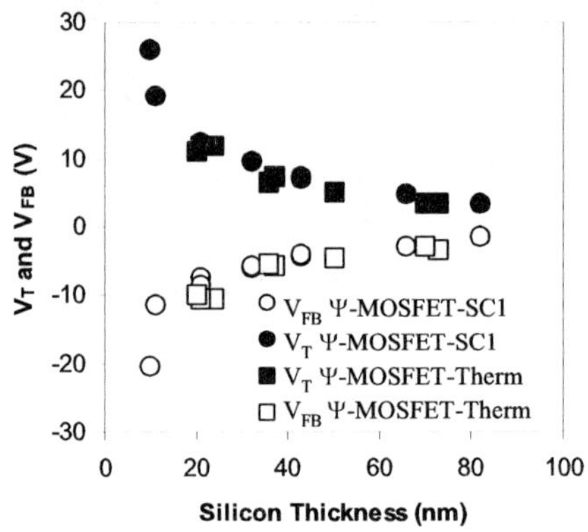

Figure 4: Comparison of V_T and V_{FB} values extracted by Pseudo-MOSFET for various T_{Si}.

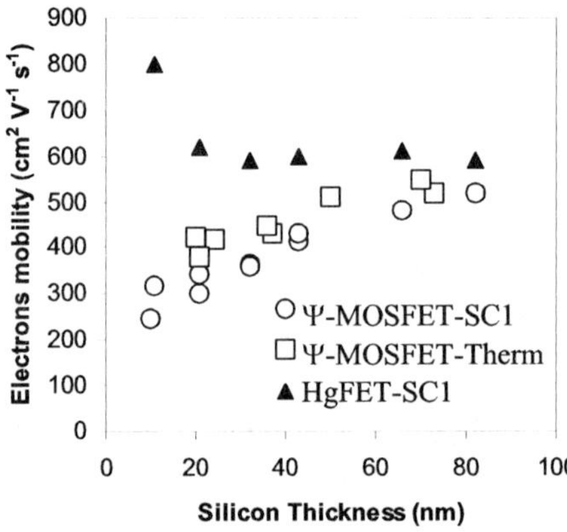

Figure 5: Comparison of electron mobility values measured by Hg-FET and Pseudo-MOSFET for various T_{Si} and thinning methods.

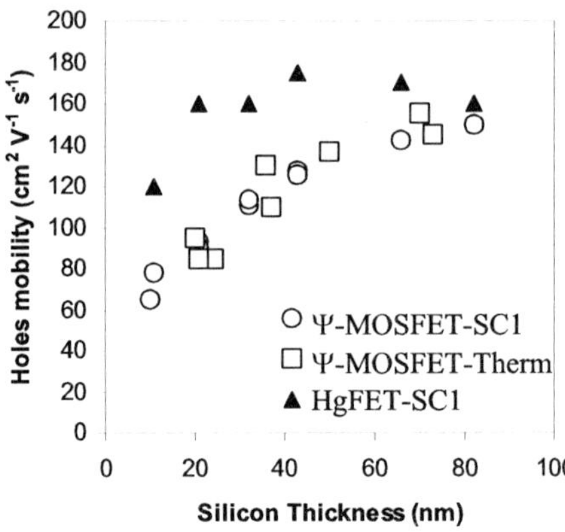

Figure 6: Comparison of hole mobility values measured by Hg-FET and Pseudo-MOSFET for various T_{Si} and thinning methods.

Impact of Random Dopant Induced Fluctuations on Sub-15nm UTB SOI 6T SRAM Cells

K. Samsudin*, B. Cheng, A. R. Brown, S. Roy and A. Asenov
Device Modelling Group, Dept. of Electronics and Electrical Engineering
University of Glasgow, Glasgow G12 8LT, Scotland, UK.
*Email: K.Samsudin@elec.gla.ac.uk, Tel: +44 141 330 2964

Abstract: The CMOS scaling increases the impact of intrinsic parameter fluctuation on the yield and functionality of SRAM. A statistical circuit simulation framework which can fully capture intrinsic parameter fluctuation information into the compact model has been developed. The impact of discrete random dopants in the source and drain regions on 6T SRAM cells has been investigated for well scaled Ultra Thin Body (UTB) SOI MOSFETs with physical channel length in the range of 10nm to 5nm.

Introduction: Around the 90nm technology node, intrinsic parameter fluctuations will drasticaly reduce the noise margin and overall speed in SRAM based on conventional MOSFETs [1]. SRAMs based on UTB SOI MOSFETs will outperform the conventional MOSFETs one due to superior electrostatic integrity, significant reduction of collective junction capacitance on the bitline and improved intrinsic parameter fluctuations. Although UTB device can operate without dopant within the channel region there are necessarily discrete random dopants in the source/drain regions that will introduce fluctuations in the effective channel length and access resistance at nanoscale dimensions.

Simulation Methodology: The physical parameters of carefully scaled UTB SOI MOSFETs are summarised in Table 1. In order to capture statistical variation associated with the discrete random dopants in the source and drain regions as illustrated in Fig. 1, an ensemble of 200 macroscopically identical, but microscopically different devices is created and simulated using the Glasgow 3D 'atomistic' drift diffusion simulator [2].

An improved two-stage statistical parameter extraction methodology [3] is employed to capture all device fluctuation information obtained from the 3D 'atomistic' simulator into a representative set of BSIMSOI-FD compact models library. Fig. 2(a) illustrates the quality of BSIMSOI extraction from 3 extreme cases of the physically simulated I_D-V_G characteristics. The static transfer characteristics of the 6T SRAM cells is depicted in Fig. 2(b). Transistors in the SRAM circuit are randomly selected from the compact models library during SPICE simulation of 1, 2 and 3 cell ratio.

Results and Discussions: Figure 3 illustrates the distribution of the static noise margin (SNM) due to discrete random doping for SRAM cells with different cell ratios, corresponding to each channel length being investigated here. It clearly follows the expected trend that increasing cell ratio improves cell stability although it delivers less improvement of the SNM with smaller channel length. The SRAMs based on 5nm transistors could not achieve the required yield even at cell ratio 3, while 10nm and 7.5nm requires at least a cell ratio of 2 and 3 respectively. Typical write operation stability considering quasi-static operation is shown in Table 3. The relative read operation speed of the different cells is obtained by measuring the discharge time for the bitline as shown in Fig. 4 and Table 4. As a measure of the write times, the worst-case flip time of the bitlines was recorded as shown in Table 5. Roughly the discrete random dopants will cause 40% performance difference for read operation and less then 5% difference for write operation.

Conclusions: UTB SOI MOSFETs could extend the benefits of SRAM scaling beyond the 65nm technology node [5]. However, novel device architectures such as Double Gate MOSFETs are required after the 10nm channel length mark. It is important that each potential sources of intrinsic parameter fluctuation is carefully studied to fully map their impact on the corresponding SRAM cell generations.

References

[1] B. Cheng, S. Roy, and A. Asenov, "The impact of random doping effects on CMOS SRAM cell," in *ESSCIRC*, 2004.

[2] A. R. Brown, F. Adamu-Lema, and A. Asenov, "Intrinsic parameter fluctuations in nanometre scale thin-body SOI devices introduced by interface roughness," *Superlattices and Microstructures*, vol. 34, pp. 283–291, 2003.

[3] K. Samsudin, B. Cheng, *et al.*, "Impact of body thickness fluctuation in nanometre scale UTB SOI MOSFETs on SRAM cell functionality," in *6th European Conference on Ultimate Integration of Silicon*, (Bologna, Italy), pp. 45–48, April 2005.

[4] P. Stolk, H. Tuinhout, *et al.*, "CMOS device optimization for mixed-signal technologies," *Tech. Digest IEDM*, pp. 215–218, 2001.

[5] "International Technology Roadmap for Semiconductors," 2003.

0-7803-9212-4/05/$20.00 ©2005 IEEE

Channel Length (nm)	10	7.5	5
Gate Oxide Thickness, T_{ox} (nm)	0.67	0.5	0.33
Body Thickness, T_{si}(nm)	2.5	2.25	2.0
Buried Oxide Thickness, T_{box}(nm)	50		
Channel Doping, N_a (cm^{-3})	1e14		
source-drain Doping, $N_{s/d}$ (cm^{-3})	2e20		

Table 1: Generic device parameters considered.

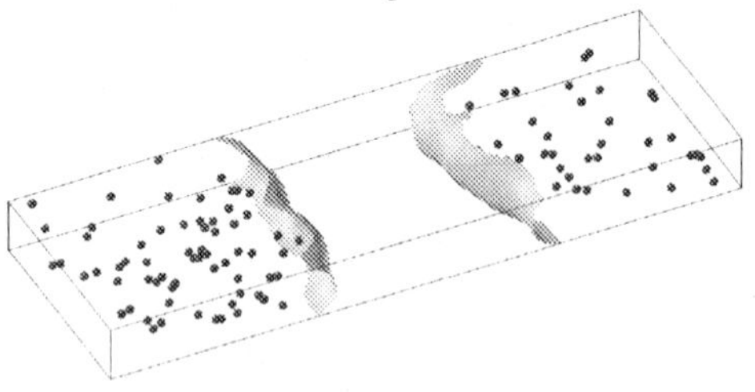

Fig. 1: Location of discrete random dopants in the source/drain regions and corresponding variation of effective channel length in the silicon body of a 10nm UTB SOI MOSFET.

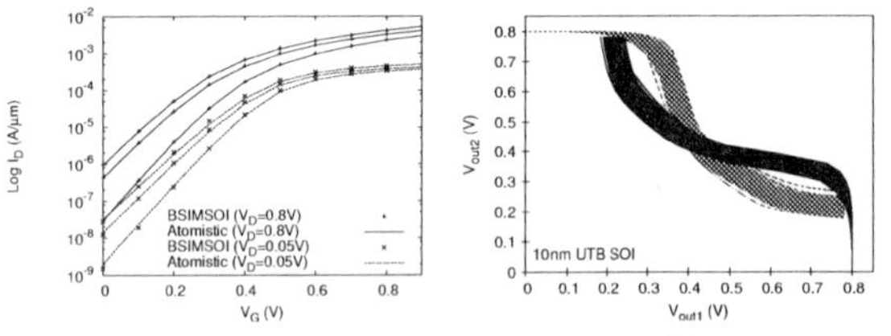

(a) (b)

Fig. 2: (a) Compact model calibration of gate characteristics for 3 extreme fluctuation cases of 10nm channel length. All characteristics are from nMOSFET UTB SOI. (b) Static transfer characteristics of 200 statistical 6T SRAM cells utilising 10nm UTB MOSFET with a cell ratio of 1.

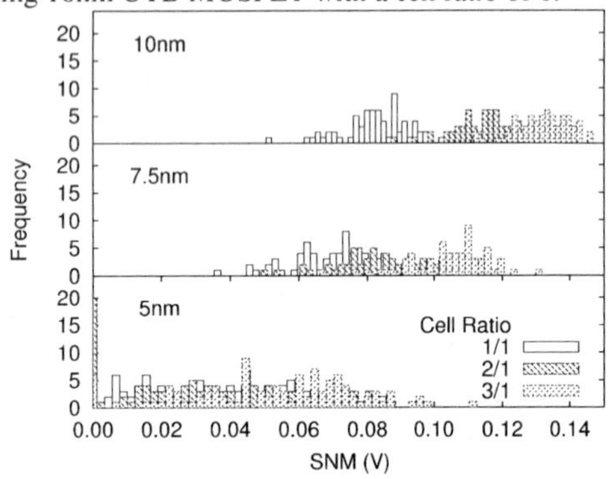

Fig. 3: Static Noise Margin Distributions.

	Cell Ratio	Mean, μ(mV)	SD, σ(mV)	σ/μ(%)
10nm	1	81.9	10.1	12.3
	2	115.4	7.8	6.7
	3	132.9	7.1	5.3
7.5nm	1	70.2	11.5	16.5
	2	89.8	12.3	13.7
	3	106.2	10.7	10.1
5nm	1	34.0	22.1	65.0
	2	37.5	23.9	63.7
	3	53.8	23.5	43.6

Table 2: Static Noise Margin distributions during read operation. The Static Noise Margin (SNM) is the minimum DC noise voltage needed to flip the cell state. As a guideline, μ-6σ is required to exceed 4% of the supply voltage to achieve 90% yield on 1Mbit SRAM [4].

	Cell Ratio	Mean, μ(mV)	SD, σ(mV)	σ/μ(%)
10nm	1	542	17	3.2
	2	586	18	3.1
	3	612	19	3.1
7.5nm	1	473	27	5.7
	2	512	26	5.1
	3	534	27	5.0
5nm	1	451	57	12.6
	2	485	58	12.0
	3	505	56	0.11

Table 3: Switch Point Voltage distributions . The write operation stability could be measured by switch point voltage which is defined as the bit line voltage that will cause data to begin to change under a write operation.

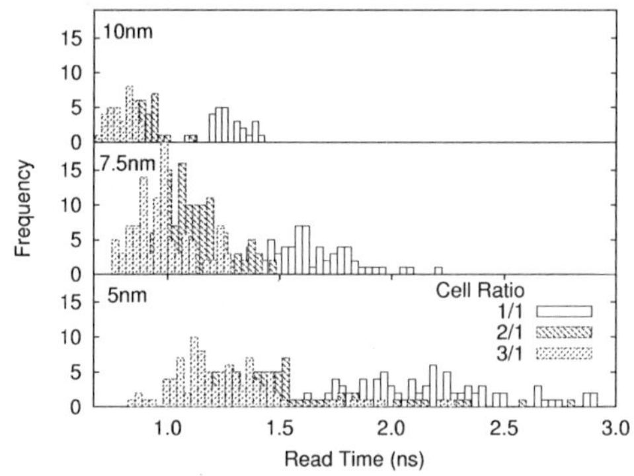

Fig. 4: SRAM read discharge time distribution.

	Cell Ratio	Mean, μ(ns)	SD, σ(ns)	σ/μ(%)
10nm	1	1.28	0.073	5.7
	2	0.91	0.053	5.8
	3	0.79	0.048	6.2
7.5nm	1	1.65	0.196	11.8
	2	1.14	0.129	11.2
	3	0.98	0.114	11.6
5nm	1	2.09	0.366	17.5
	2	1.49	0.299	20.0
	3	1.28	0.257	20.1

Table 4: SRAM read discharge time distribution. The read discharge time is measured when one of the bitlines is reduced to half of the supply voltage after the wordline is raised.

	Cell Ratio	Mean, μ(ps)	SD, σ(ps)	σ/μ(%)
10nm	1	231.4	3.91	1.69
	2	234.0	4.29	1.83
	3	236.1	4.66	1.97
7.5nm	1	241.8	7.77	3.21
	2	244.8	8.46	3.46
	3	247.3	9.30	3.76
5nm	1	255.8	21.82	8.53
	2	260.3	25.30	9.72
	3	265.0	29.91	11.29

Table 5: SRAM Write time distribution. The flip time was measured from the time the wordline was raised after placing an opposite value on the bitlines to when the latter of the two storage nodes reached 90% of its final value.

Characterization of Ultra-Thin SOI MOSFETs by Coupling Effect between Front and Back Interfaces

A. Ohata[1], S. Cristoloveanu[1], M. Cassé[2], A. Vandooren[3], and F. Daugé[1]
[1] IMEP ENSERG, B.P. 257, 38016 Grenoble, France, [2]CEA-LETI, 17 rue des Martyrs, 38054 Grenoble, France
[3]Freescale semiconductor, 850 rue J. Monnet, 38926 Crolles, France
Tel: +33-4-76-85-6039, email: ohata@enserg.fr

Abstract By investigating the coupling effect between the front and back channels of ultra-thin SOI-MOSFETs, the mobility enhancement by volume inversion, the channel separation between the front and back channels, and the impact of the SOI thickness are discussed. As a result, it is shown that a careful check is necessary for confirming which channel (front or back) is inverted in a short device because the channel separation is difficult in transconductance curves. Furthermore, it is shown that the SOI thickness depends on the device size and location, leading to a complex behavior.

Introduction: The electrical characteristics of thin-SOI MOSFETs have attracted attention because the SOI thickness is approaching the transition region from the classical regime to the quantum regime. In the quantum regime, a thin-SOI film becomes a quantum well, with quasi-flat potential and full volume inversion. Investigating the coupling effect between the front and back interfaces would provide useful information, allowing us to distinguish whether or not the device operation is in quantum mode. In this paper, we discuss the coupling characteristics as well as new issues caused by the ultra-thin SOI film.

Volume Inversion[1]**:** The devices used in this study are SOI n-MOSFETs with a silicon film thickness of 15 nm (except in the raised source and drain regions), a 2 nm thick gate oxide and a poly-Si gate. The doping concentration in the SOI body is approximately 10^{17} cm^{-3}. Figures 1 show the transconductance (G_m) of the two channels (front and back) measured for various substrate and gate biases. The lateral shift of these curves, that is the coupling effect, occurs due to the decrease in the threshold voltage V_{gth} and V_{subth} with increasing V_{sub} and V_g, respectively. The plateau is due to the inversion layer at the opposite interface. The mobility at G_m peak increases with V_{sub} in Fig.1(a) (V_g in Fig.1(b)) because the electric field decreases. However, once the opposite channel is inverted by applying a large positive V_{sub} in Fig.1(a) (with V_g in Fig.1 (b)), the main peak becomes constant. This implies that the mobility does not increase due to the inversion layer at the opposite interface. It can be concluded that these devices do not benefit from mobility enhancement by volume inversion. This fact is consistent with the result obtained by evaluating the thickness dependence of the mobility[2]: the SOI film is not thin enough to make the wave function spread into the entire body, from one interface to the opposite interface.

Channel Separation: We have already reported that G_m curve in a short channel has a single peak[3] as shown in Fig.2, although a plateau appears in long channels (Figs.1). By simulation, we revealed that the series resistance can mask the formation of the two separate channels. We now provide more direct evidence. Figure 3 shows capacitance versus V_g curves for various V_{sub}. Under a large V_{sub}, a plateau appears due to the back channel creation and the capacitance value at the plateau is determined by $(1/C_{si}+1/C_{ox})^{-1}$[4]. Thus, altough it is impossible to observe the front and back channel creation separately in G_m curves, it becomes evident that the back and front channels form independently.

Figure 4 shows G_m curves featuring a single peak, for various substrate biases in a short, undoped SOI MOSFET with 9-nm thickness. Figure 5 shows $V_{gth}(V_{sub})$ and $V_g(V_{subth})$. An important feature is that the back channel is inverted even at V_{sub} = 0 V due to the sweeping of V_g and thereby the peak of G_m at V_{sub} = 0 V in Fig.4 is due to the *back channel*. The front channel peak is strongly reduced by the series resistance and totally masked. Thus, in order to discuss the electrical characteristics in short-channel, ultra-thin-SOI, it is necessary to investigate the coupling curve that confirms which channel is prevailing.

Thickness Variation: Figure 6 shows $V_{gth}(V_{sub})$ and $V_g(V_{subth})$ with various channel lengths. Although the slopes of $V_{gth}(V_{sub})$ are equal, the slopes of $V_g(V_{subth})$, which are governed by the film thickness[5], clearly depend on the channel length. Figure 7(a) and (b) show V_{gth} and V_{subth} vs. SOI thickness determined by the slope of $V_g(V_{subth})$. The SOI thickness dependence of V_{gth}, for the same channel length, follows the theory[5]. We show that the channel length dependence of the slope is actually due to the SOI thickness variation. Devices made by a different process, such as 15 nm SOI thick MOSFETs with a high-k gate insulator, exhibit the same features. We show that thickness variations are induced by the process at the wafer level and during device-island patterning.

Conclusion: By investigating the coupling effect between the front and back channels of ultra-thin SOI-MOSFETs, we discussed (i) the possibility of the mobility enhancement by volume inversion, (ii) the channel separation, and (iii) the overall impact of the SOI thickness. We found it necessary to check which channel (front or back) is inverted in a short device because the channel separation is difficult in G_m curves. Furthermore, the SOI thickness depends on the device size and location, leading to a complex behavior.

References

[1] F. Balestra *et al.*, *IEEE Electron Device Lett.*, 8, p410, 1987.
[2] A. Ohata *et al.*, *Proc. ESSDERC*, p.109, 2004.
[3] A. Ohata *et al.*, *INFOS 2005*, in press.
[4] J. Chen, *et al.*, *IEEE Trans.* ED-39, p2346, 1992.
[5] Lim and Fossum, *IEEE Trans.* ED-30, p.1244, 1983.

0-7803-9212-4/05/$20.00 ©2005 IEEE

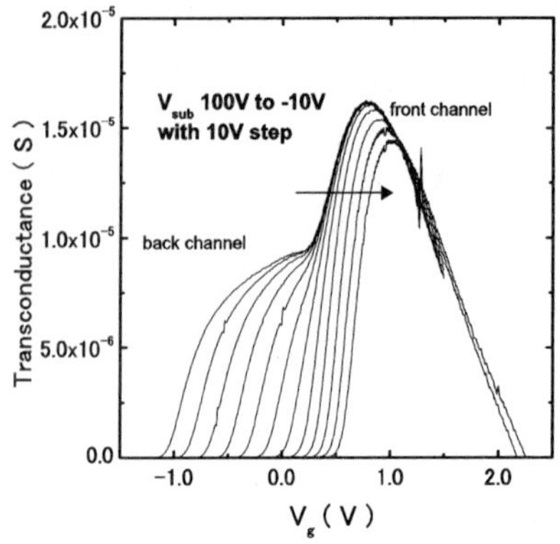

Fig.1(a): Front channel transconductance vs. gate voltage with the substrate bias as a parameter in a long, 15 nm thick SOI MOSFET (V_d = 50 mV, W = 100 μm, L = 100 μm).

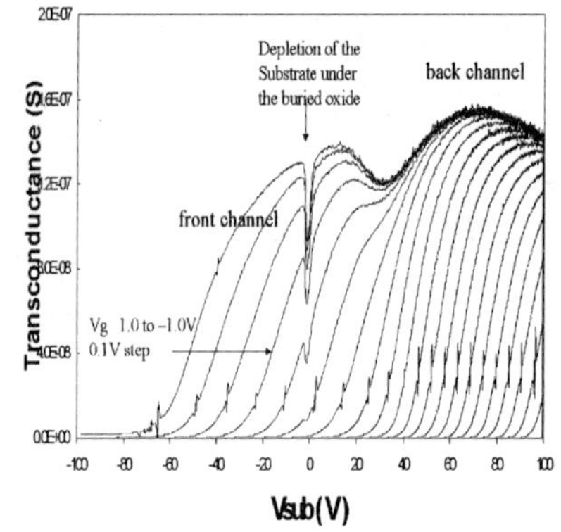

Fig.1(b): Back channel transconductance vs. substrate voltage with the gate bias as a parameter (V_d = 50 mV, W = 100 μm, L = 100 μm).

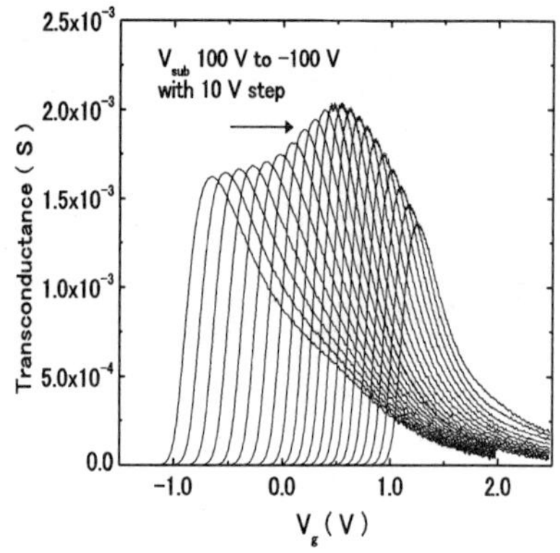

Fig.2: Front channel transconductance vs. gate voltage with the substrate bias as a parameter in a short, 15 nm thick MOSFET (V_d = 50 mV, W = 100 μm, L = 0.5 μm).

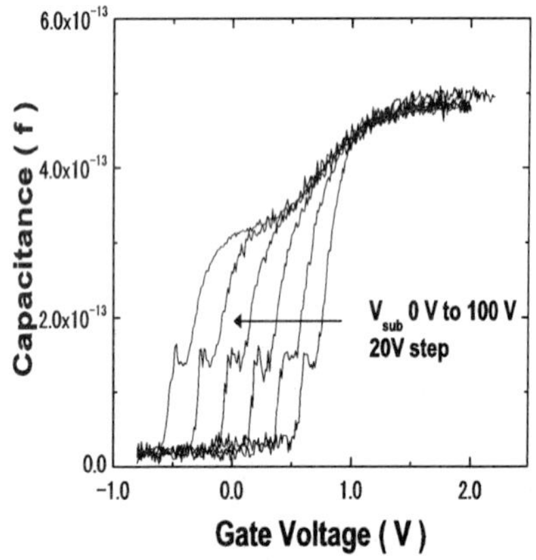

Fig.3: Capacitance versus gate voltage with the substrate voltage as a parameter (W = 100 μm, L = 0.5 μm).

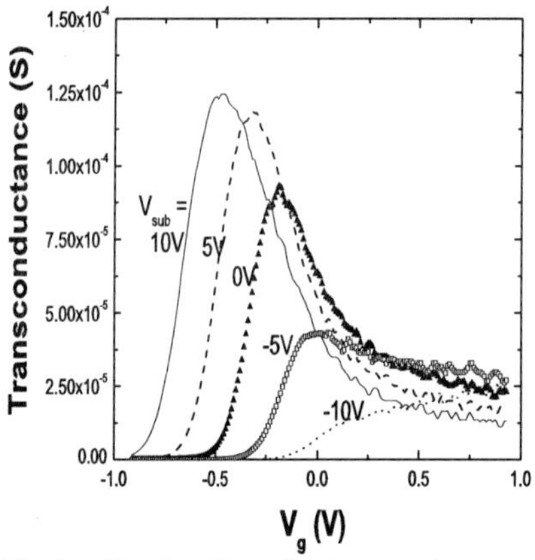

Fig.4: Front channel transconductance versus gate voltage with the substrate bias as a parameter in a 9 nm thick SOI MOSFET (V_d = 50mV, W = 10 μm, L = 0.075 μm). The gate oxide thickness is 4.5 nm.

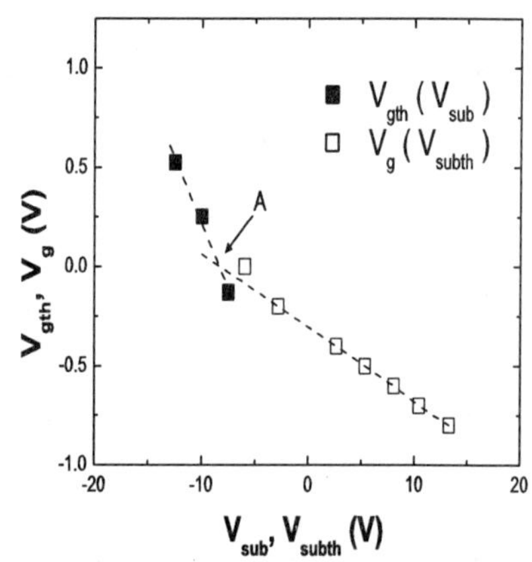

Fig.5: Front channel threshold voltage V_{gth} vs. substrate voltage V_{sub} and back channel threshold voltage V_{subth} vs. gate voltage V_g. At point A, both the front and back channels are simultaneously inverted.

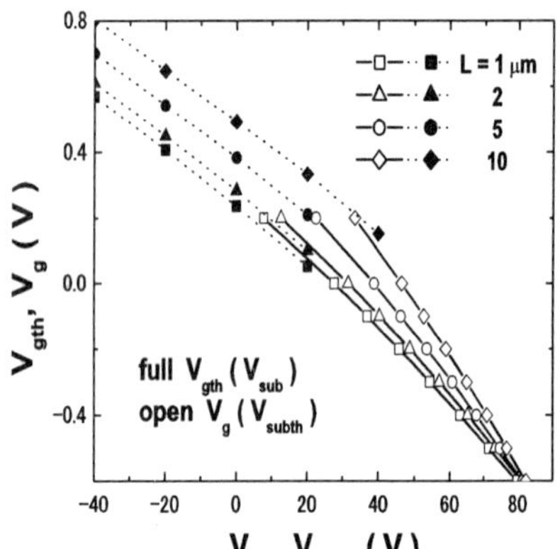

Fig. 6: Front channel threshold voltage V_{gth} vs. substrate voltage V_{sub} and back channel threshold voltage V_{subth} vs. gate voltage V_g. W = 25 μm, L= 1, 2, 5, 10 μm. The slopes of V_g (V_{subth}) increase with the channel length whereas the slopes of V_{gth} (V_{sub}) are constant.

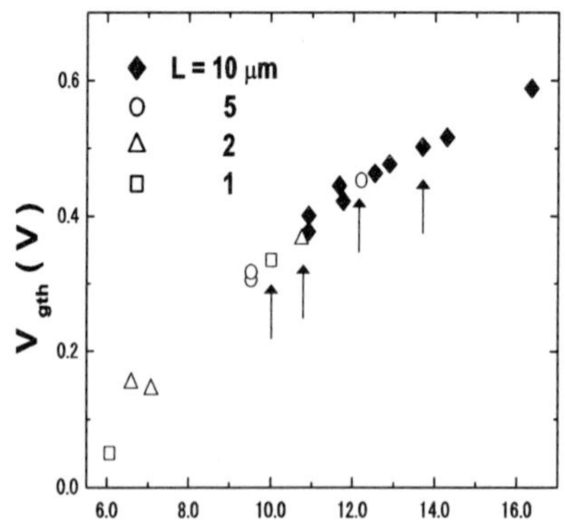

Fig. 7(a): Front channel threshold voltage V_{gth} versus SOI thickness. The data pointed by arrows are obtained on devices located in the same area, which have a common drain.

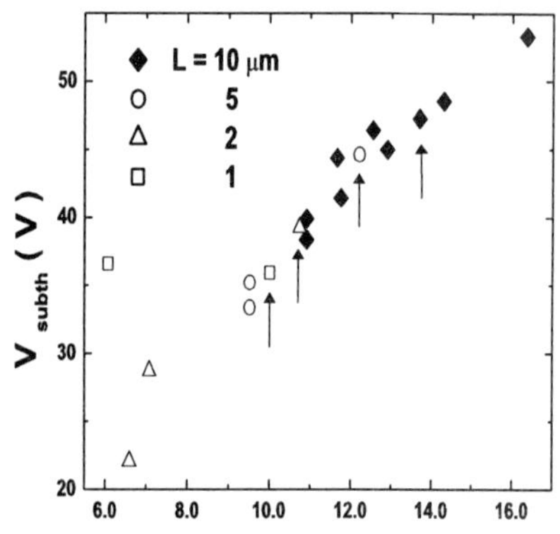

Fig. 7(b): Back channel threshold voltage V_{subth} versus SOI thickness. The data pointed by arrows are obtained on devices located in the same area, which have a common drain.

Wideband characterization of body-accessed PD SOI MOSFETs with multiport measurements

D. Lederer[1], O. Rozeau[2] and J.-P. Raskin[1],
[1]Microwave Laboratory of the Université Catholique de Louvain,
Place du Levant 3, 1348 Louvain-la-Neuve, BELGIUM
Email: lederer@emic.ucl.ac.be, Tel: +32 (0)10 47 81 05
[2]LETI-CEA, Rue des Martyrs 17, 38054 Grenoble CEDEX 9, France

Introduction: Body-accessed partially depleted (PD) SOI devices are known to provide high DC performance due to the suppression of floating body effects [1]. However, due to the large value of body resistance (R_{be}) this higher performance is lost at radio frequencies (RF) [2], [3]. Indeed, at RF voltage drops across R_{be} prevent the body from being a cold node, thereby inducing transconductance and output conductance degradations in Dynamic Threshold (DT) MOSFETs [3]. As lower R_{be} values reject these effects to higher frequencies, it is important to reduce R_{be}. Accurate characterization of R_{be} is therefore crucial to assess the efficiency of the body contact in a given technology. It can also be used to validate compact models. However in previous works only rough estimations of R_{be} could be given due to a lack of external body connection [2], [3]. In this work we present an accurate method for completely characterizing the body terminal of PD SOI devices (including R_{be}), based on three-port RF measurements. The third port is connected to the body of the devices.

Devices and measurement setup: The devices originate from a 0.13 µm SOI technology from ST Microelectronics, Grenoble. Different geometries (varying number of fingers (N_f) or finger width (W_f)) and body connections (either to the gate (DTMOSFET) or to a third RF access) were considered. In all cases the body was accessed at both sides of the gate in order to reduce R_{be}. The S-parameters were measured with a multiport Rhode&Schwartz VNA up to 8 GHz and were de-embedded with a 3-port open subtraction method.

Results and discussion: The modeling of the RF PD SOI devices was based on the small signal equivalent circuit presented in Figure 1. It can be seen that both intrinsic (subscript *i*) and extrinsic (subscript *e*) elements were considered between each pair of electrodes, including the body (**B**). By extrinsic, the authors mean all parasitics that could not be removed during the de-embedding step. The modeling of the body node was therefore achieved by considering all extrinsisic capacitances (C_{bge}, C_{bse} and C_{bde}), its access resistance (R_{be}), its intrinsic body-gate (C_{bgi}) as well as body-source (C_{bsi}) and body-drain (C_{bdi}) junction capacitances, and its intrinsic body-source junction resistance (R_{jbsi}). The body-drain junction resistance was neglected due to reverse biasing of this junction when devices are biased in saturation. The extraction of R_{be} was performed by analyzing the output admittance seen from the body node terminal (Y_{33}). Indeed, as shown in Figure 2 the value of $Im(Y_{33})/(2\pi f)$ clearly exhibits a pole-zero pair dependence, which is typical of the simple *R-C* network included in Figure 2. In this circuit, the high frequency (C_{HF}) value of $Im(Y_{33})/(2*\pi f)$ is simply the sum of all extrinsic capacitances seen from the body terminal (i.e., $C_{HF} = C_{be} = C_{bse} + C_{bde} + C_{bge}$) while its low frequency value (C_{LF}) is the sum of all intrinsic and extrinsic capacitances seen from the body terminal (ie, $C_{LF} = C_{bsi} + C_{bdi} + C_{bgi} + C_{be}$). In these conditions, it is immediate to see that the pole expression is given by

$$f_p = \frac{1}{R_{be}(C_{LF} - C_{HF})} = \frac{1}{R_{be}(C_{bsi} + C_{bgi} + C_{bdi})} \quad (1)$$

and is therefore strongly dependent on the value of R_{be}. By fitting the *R-C* network on the measurements this value could then be extracted. Figure 3 shows the value of R_{be} obtained for devices with varying N_f (with constant $N_f.W_f$ product) and W_f (with constant N_f) values. It is seen that (1) R_{be} closely decreases as $(1/N_f)^2$ (similarly to R_{ge} [4]) while (2) a linear dependence with W_f is observed. These two trends agree with predictions made by the scalable BSIMSOI model [5], further supporting the validity of the extraction method. The extrapolated value of R_{be} obtained for $W_f = 0$ (~150 Ω) therefore provides a good estimation of the parasitic resistance associated with the body interconnects (R_{bout}) outside the active region. The figure shows that its contribution is not negligible with regards to the overall value of R_{be}. Indeed normalized values of R_{be} and R_{bout} were found to be respectively close to 21 kΩ /µm/finger and 2.25 kΩ/finger, assuming only one contact per finger. The body node characterization was further achieved by extracting (C_{bge}, C_{bgi}) and (C_{bde}, C_{bdi}) from the high frequency and low frequency parts of $Im(Y_{31})$ and $Im(Y_{32})$, respectively. The body-source capacitances were then obtained from C_{LF} and C_{HF} and the back gate transconductance (g_{mbi}) was given by $Re(Y_{23})$. The value of R_{jbsi} could theoretically be extracted from $Re(Y_{33})$ =

0-7803-9212-4/05/$20.00 ©2005 IEEE

$(R_{be} + R_{jbsi})^{-1}$ at low frequencies but was too high (>100 kΩ) to be accurately measured. It was therefore assumed to be infinitely high in the model. The rest of the device parameters were obtained with the cold-FET method [4], while R_{se}, R_{de} and R_{ge} were obtained with a method depicted in [6]. Figures 4a and b show that an accurate modeling of both real and imaginary parts of the Y parameters is obtained up to ~ 4 GHz for a 4 μm-wide device with N_f= 15, V_g = 0.6 V, V_d = 1.2 V, V_b = 0.6 V. The model was then further tested by connecting the body to the gate terminal, forming a two-port network and compared with measurement results obtained on a DTMOSFET with the same geometry. It is seen in Figures 4a and b that a good agreement is obtained between measured and simulated data, despite a small output conductance difference observed for the DTMOSFET, which could be due to a subtle bias shift.

Conclusion: We have presented in this work an original method based on 3-port RF measurements to accurately extract the body resistance (R_{be}) in body-accessed PD SOI MOSFETs. This method can be used to assess the validity of compact models such as BSIMSOI. The RF body access also enabled a precise characterization of intrinsic and extrinsic body capacitances. A complete 3-port model was then derived and further validated on DT MOSFET measurements by connecting the body to the gate terminal.

Acknowledgement: The authors thank N. Fel from CEA-DAM, Paris, for designing the devices and C. Raynaud from ST Microlectronics, Crolles, for supplying the devices. This work has been funded by the European project MEDEA +T206.

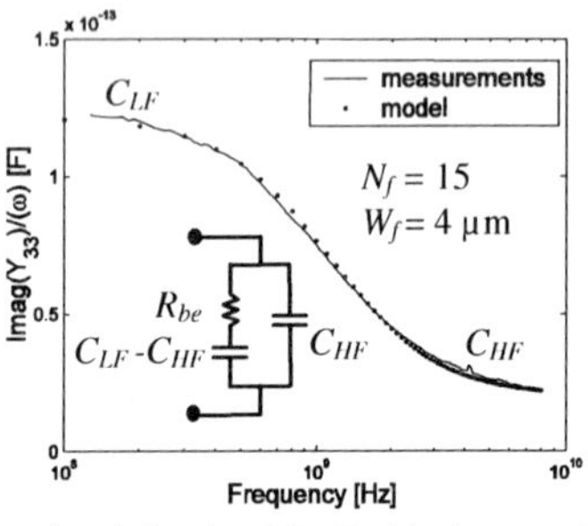

Figure2: Measured and simulated Im(Y_{33})/ω data as a function of f for V_g = V_b = 0.6 V and V_d = 1.2 V, inset: R-C model used.

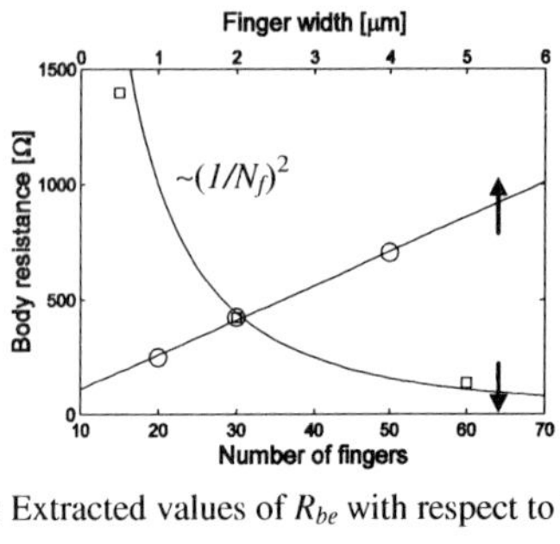

Figure 3: Extracted values of R_{be} with respect to W_f and N_f.

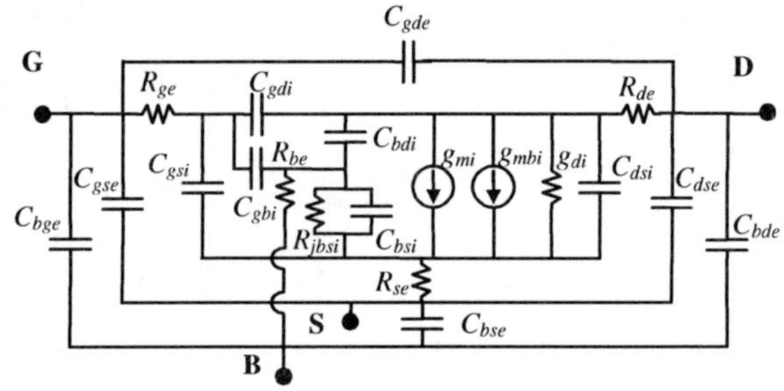

Figure 1: Complete small signal equivalent circuit of the 3-port devices, including the external body node.

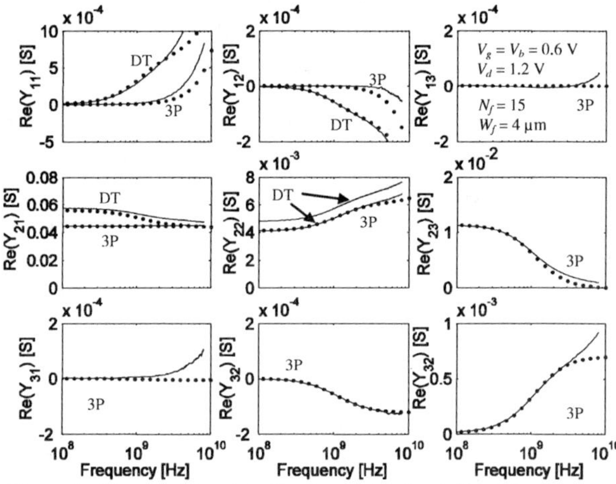

Figure 4a: Measured (straight lines) and simulated (dots) Re(Y_{ij}) for the 3-port device (3P) as well as for the DT MOSFET (DT) with body connected to gate.

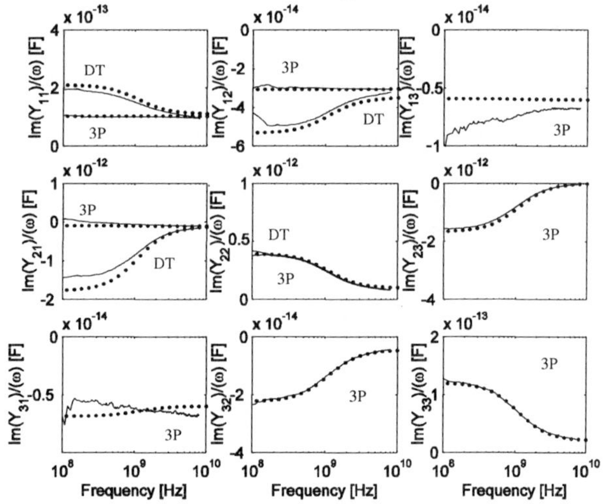

Figure 4b: Measured (straight lines) and simulated (dots) Im(Y_{ij})/ω for the 3-port device (3P) as well as for the DT MOSFET (DT) with body connected to gate.

References

[1] J.-P. Colinge, Kluwer Academic Publishers, 1997.
[2] D. Lederer *et al.*, *2003 IEEE Int. SOI. Conf.*, 76-77.
[3] M.Dehan *et al.*, *SSE* 49(2005), 67-72.
[4] J.-P. Raskin *et al*, *IEEE TED* 45(1998), 1017-1025.
[5] www-device.eecs.berkeley.edu/~bsimsoi.
[6] A. Bracale *et al.*, Analog Integrated Circuits and Signal Processing, 25(2000), 159-167.

Unusual Gate Current Transient Behavior in SOI MOSFETs

M. Bawedin[1], S. Cristoloveanu[2] and D. Flandre[1]
[1]Microelectronics Lab.(DICE), UCL, Louvain-La-Neuve, Belgium
[2]IMEP, ENSERG, BP 257, 38016 Grenoble, France

Abstract. We report unusual gate current variations with gate bias and time in accumulation regime. Measurements and simulations show the importance of the time-dependent floating body potential in both partially and fully depleted SOI MOSFETs. These variations during front-gate voltage scan and the resulting transient gate current are discussed as a combination of several mechanisms such as capacitive coupling, SRH generation, band-to-band tunneling and impact ionization. The gate current behavior is important for EEPROMs and depends on the device geometry and operating conditions.

Introduction

The transient variation of the drain current when the gate bias is switched between two logic levels is well documented [1-3]. Less attention has been paid to the transient gate current variation, although this is a critical issue for memory writing and erasing. In this paper, we discuss non-conventional gate current characteristics that are induced by the transient variation of the body potential.

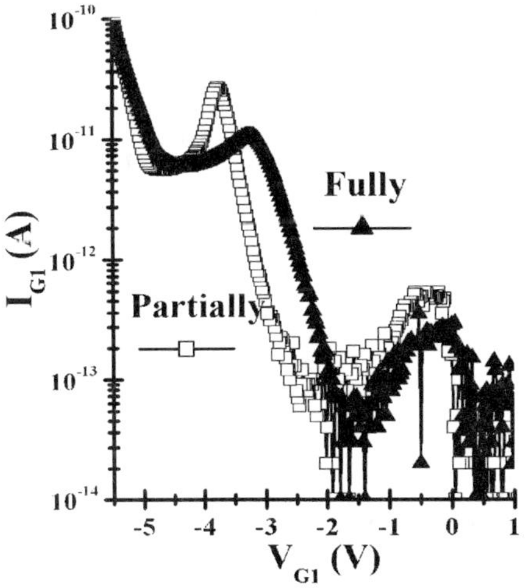

Figure 1: Measured front gate current I_{G1} versus voltage V_{G1} for FD and PD SOI nMOSFETs. Scan speed is 1V/s from positive to negative V_{G1} values. Gate width and length are 100μm. Drain, source and back gate are grounded, the body floating.

Detailed measurements were performed on partially depleted (PD) devices, in which the body contact availability allows probing the body potential, but similar behavior was observed on fully depleted (FD) devices (Fig.1). Test devices were from two different SOI technologies: PD SOI n-MOSFETs (from ST-Microelectronics) with 4.5 nm thick gate oxide and 100 nm film, and FD SOI nMOSFETs (from LETI) with 4.5 nm and 40nm thick gate oxide and film respectively.

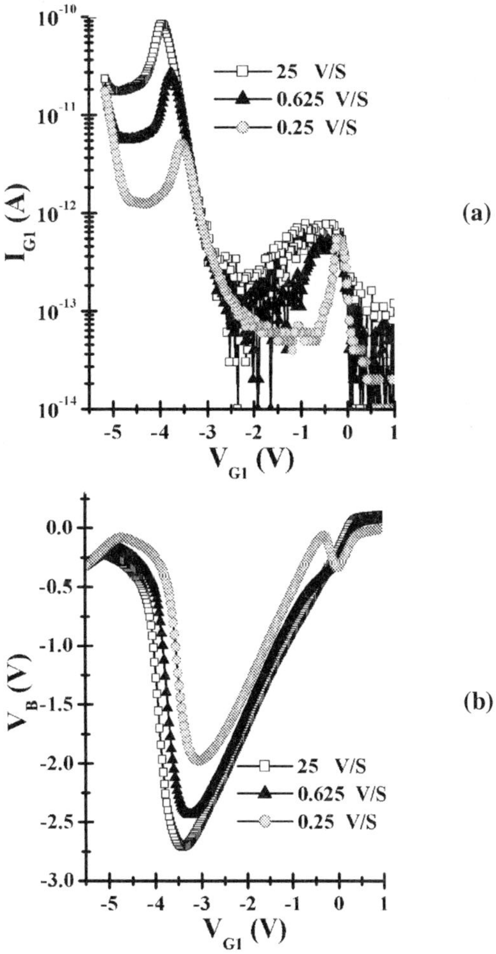

Figure 2: Measured (a) I_{G1} and (b) related body potential V_B variations for various V_{G1} scan speeds (PD nMOSFET of Fig. 1).

0-7803-9212-4/05/$20.00 ©2005 IEEE

Experimental results

The transient gate current I_{G1} can show unexpected, non-monotonic variations as the front interface moves from inversion (or depletion) into accumulation (i.e. positive-to- negative front gate voltage scan for nMOSFETs, Fig.1). A first current peak is observed close to the expected flat-band voltage (V_{G1}= -0.5V, Fig.1). A second, larger peak appears for more negative bias ($V_{G1} \approx$ -3 V). The I_{G1} peaks are followed by significant drops although $|V_{G1}|$ keeps increasing. The position and magnitude of I_{G1} peaks and valleys depend on several technological parameters such as oxide and silicon film thickness, gate dimensions and doping.

The V_{G1} scan speed significantly affects the position and level of the current peaks (Fig.2a). We relate the I_{G1} signature to the transient variations of the body potential V_B. Figure 2b shows that the second peak of I_{G1} corresponds to a remarkable drop in V_B. Faster scans lead to more accentuated features in I_{G1} (peaks) and V_B (valleys). Nevertheless, for this device, the I_{G1} peak values are pinned when the scan speed exceeds 25V/s.

Measurements with grounded body contact show the difference between the `transient' and `pure tunneling' components of I_{G1} (Fig.3). The current peaks are totally suppressed with grounded body contact in Fig.3, revealing the actual Fowler-Nordheim current [4] for V_{G1}<-4.5V and no significant direct tunneling for less negative V_{G1}.

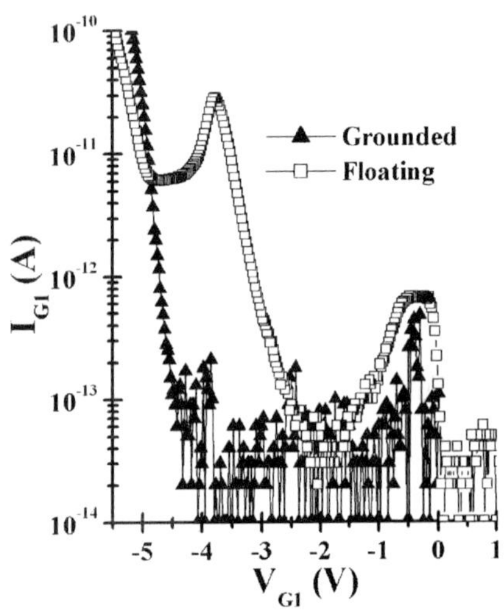

Figure 3: Measured I_{G1} with body contact grounded and floating versus V_{G1} (1V/s scan speed, PD nMOSFET of Fig. 1).

Model and Discussion

The gate current peaks are directly associated to the transient variations of V_B and V_{G1} following :

$$\frac{I_{G1\_C}}{C_1} = \frac{\Delta(V_{G1}(t) - V_B(t))}{\Delta t} \quad (1),$$

where C_1 is an empirical coupling coefficient which may differ from the oxide capacitance.

Measurements (Fig.4.a) and 2D numerical simulations [5] (Fig.4.b) of I_{G1} and I_{G1_C}/C_1 demonstrate that Eq.(1) accurately reproduces the transient gate current behavior.

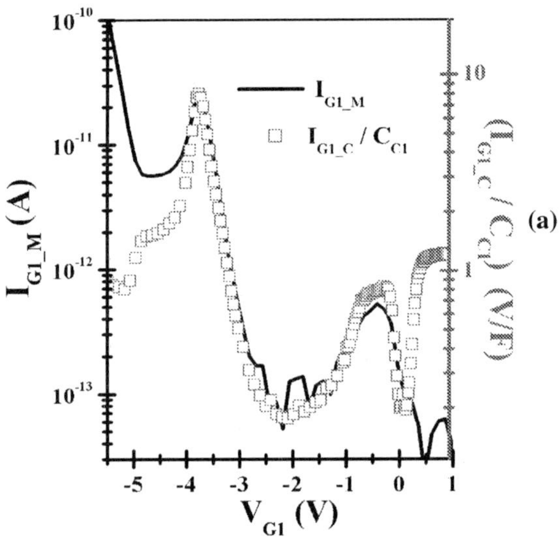

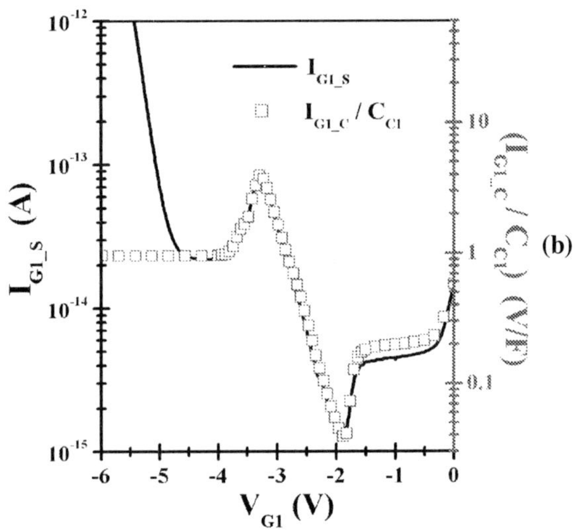

Figure 4. Calculated $\Delta(V_{G1}-V_B)/\Delta t$ compared to the corresponding (a) measured I_{G1_M} and (b) simulated I_{G1_S} front gate currents. 1V/s scan speed, experimental PD nMOSFET of Fig. 1, simulation with ISE tools for a nMOSFET with T_{OX1}=6nm, T_{SI}= 80nm, W x L= 1 x 4μm^2, N_C=5x10^{15}cm^{-3}.

In -4V<V_{G1}<0V range, I_{G1} is dominated by the $V_B(V_{G1})$ variations following Eq.(1). Note that the measured V_B is an average value over the silicon film and can differ from the interface value used to calculate Eq.(1) from simulation. For very negative V_{G1} (< -4.5V, Fig.4), the large increases of I_{G1} denotes the onset of Fowler-Nordheim tunneling, which rules out equation 1.

The detailed discussion of the operating modes (inversion, depletion, or accumulation) and dominant mechanisms is based on the simulated carriers concentrations and body potential at front interface. Several regions can be identified:

1) **-2V <V_{G1}< 0V:** the I_{G1_C}/C_1 ratio is small and decreases with V_{G1} (Fig.4.b). Indeed, the front accumulation layer can not be completed due to the lack of hole supply (Fig.5) and V_B drops by capacitive coupling through the depletion region. But negative V_B values gradually reinforce the carrier generation rate via SRH process.

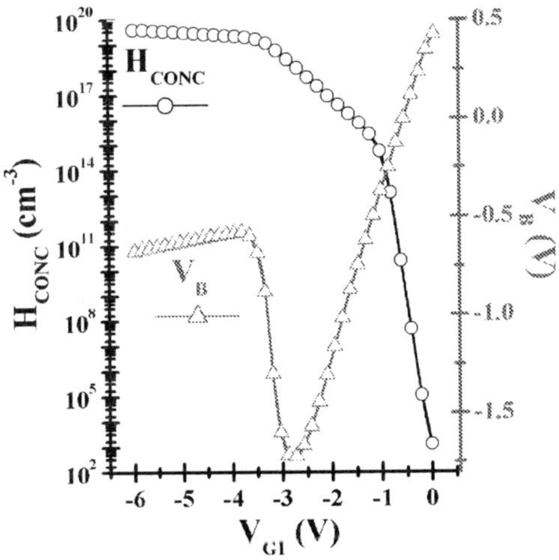

Figure 5. Simulated majority carrier concentration H_{CONC} and body potential V_B next to the interface versus V_{G1}. Same case as fig. 4b.

2) **-3.2V <V_{G1}< -2V:** I_{G1} increases and reaches a maximum value (V_{G1}=-3.2V, Fig.4.b). The hole concentration is indeed considerably enhanced leading to accumulation at the front interface which lessens the V_B variation with V_{G1}, increasing the I_{G1_C}/C_1. Moreover, when the electric field at the junctions becomes strong enough to enable band-to-band tunneling to prevail over SRH generation, the V_B evolution reverses (V_{G1} = -2.8V, Fig.5) yielding the peak of Eq.(1) value.

3) **-4V <V_{G1}< -3.2V:** the I_{G1_C}/C_1 ratio decreases again as V_B steadily increases towards the equilibrium value (Fig.5). The junction electric field at the front interface is large and turns on the impact ionization which dominates the band-to-band tunneling. As the accumulation layer is completed, V_B decreases again with V_{G1} by coupling through the oxide capacitance, yielding a I_{G1_C}/C_1 plateau value.

4) **V_{G1} < -4.5V:** Finally, the onset of Fowler-Nordheim tunneling increases the gate current while the I_{G1}/C_1 ratio keeps stable.

The details of these mechanisms will be discussed as a function of the device dimensions and doping.

Conclusion

Interesting features of the gate current in time domain are revealed and explained for the first time. This complex behavior, due to floating body effects even in FD MOSFETs, has escaped attention mainly because it occurs in accumulation and does not affect the operation of digital SOI circuits. However, it can modify the charging and discharging capability of SOI EEPROMs [6, 7] and presumably get accentuated by gate oxide scaling.

References

[1] D.E. Ioannou et al, IEEE Electron Device Lett, vol. 11(9), 409-411 (1990).

[2] D. Munteanu, D.A. Weiser et al, IEEE Trans. Electron Devices, vol. 45(8), 1678-1683 (1998).

[3] S. Cristoloveanu, S.S. Li, Kluwer Academic Publishers, Boston (1995).

[4] R. Laffont et al, ESSDERC '03. 16-18 Sept. 71-74 (2003).

[5] Dessis, Release 8.0, ISE.

[6] A. Chun-Keung Chang et al, IEEE Electron Device Lett, vol. 24(2), 75-77 (2003).

[7] Min-Hwa et al, IEEE Int. SOI Conf., 3-5 Oct., 129-130, 1995.

An investigation of Temperature Effects on CPW and MSL on SOI Substrate for RF Applications

M. Si Moussa[1], C. Pavageau[2,3], D. Lederer[1], L. Picheta[2], F. Danneville[2], J. Russat[3], N. Fel[3], J.-P. Raskin[1] and D. Vanhoenacker-Janvier[1]

[1] EMIC, Microwave Laboratory– Université catholique de Louvain –Belgium.
E-mail: simoussa@emic.ucl.ac.be, Fax: +32 10 47 87 05
[2] IEMN – UMR CNRS 8520, Villeneuve d'Ascq, France.
[3]CEA, Bruyères-Le-Châtel, France.

Introduction

One of the main markets for Silicon-on-Silicium (SOI) devices is the high-temperature applications. In the last decade, the technology advances to deep submicron to improve device performances in term of cut-off frequency. Practical application of integrated circuits requires the consideration of a wide temperature range, and transmission lines are widely used in MMIC's as interconnects and matching networks. Therefore, there is a need to investigate the performances of the transmission line structures on SOI substrate through temperature as a function of frequency.

In the next paragraphs, the behaviour of 50 Ω microstrip (MSL) and coplanar waveguide (CPW) transmission line topologies versus temperature is presented.

Experiment

One of the main challenges in silicon process is not anymore high performance transistors but ultra low loss transmission lines for achieving higher frequency and gain performances [1]. Transmission lines can be implemented using MS or CPW lines. Fig. 1 shows their structure. For the MSL, metal1 and 2 were used for the ground plane and metal6 for the signal line. The CPW is implemented on metal6 layer. Both configurations on 130 nm SOI process were modelled through full-wave EM simulations with HFSS Ansoft software environment. Correctness of line description in HFSS was ascertained by comparison with S-parameters measurements of MS lines with an excellent agreement between simulation and measurements [2].

Results and Discussion

Fig. 2 shows the attenuation coefficient of CPW lines made on standard (20 Ω.cm) and high resistivity (5000 Ω.cm) SOI wafers, as well as SOS substrate. The CPWs were designed to obtain a characteristic impedance close to 50 Ω.

By comparing, the attenuation coefficient (α) of the two lines made on SOI, the importance of the substrate losses is highlighted. The substrate losses dominate in the case of standard resistivity substrate. As we can see in Fig. 2, the losses at room temperature are strongly lowered when a HR substrate is used (α around 0.3 dB/mm over 0 – 30 GHz).

We note also that CPWs made on HR SOI or SOS exhibits better performances than MSL for impedances close to 50 Ω. For the HR SOI and SOS, the difference is small and is mainly function of the conductor resistance.

MSL do not see the substrate, but suffer from higher losses than CPW lines on HR substrates due to small conductor width. Indeed, the properties of MS lines are independent on the substrate resistivity used, because the ground plane avoids coupling between the signal and the substrate. But due to the dielectric layer (Fig. 1) between the signal conductor and the ground plane (~ 3 μm) a MSL of 50 Ω is characterized by a small W (~7 μm). Thus, the conductor losses are dominant.

On the other hand, MSL even without Alucap layer is still better than CPW on standard resistivity. Thanks to Alucap which allows a reduction in losses of 0.35dB/mm at 20 GHz is obtained.

In order to study the behaviour versus temperature, the measurements were performed over the temperature range from 25 to 250°C. Using a classical TRL de-embedding method, the lineic losses were extracted as shown in Fig. 3, the measurements done on the MSL show an increase of 80% of the lineic losses with respect to temperature which is due to the positive temperature coefficient of the metallic strip. The temperature dependence of resistivity is often represented by the empirical relationship: $\rho(T) = \rho_0 + kT$

where ρ_0 is the resistivity at a reference temperature, usually room temperature, and k is the temperature coefficient.

The same measurements setup was also applied to the CPW on standard and high resistivity (STD and HR) SOI substrates.

For CPW, the increase of the losses is more important on HR compared to STD, but its performance is still interesting at 250°C. Compared to a 50 Ω MSL, the CPW_HR shows 50% less loss, because CPW structures enable using wider lines to achieve the same

0-7803-9212-4/05/$20.00 ©2005 IEEE

characteristic impedance which allows reducing the metallic losses.

Lineic losses of CPW_HR, CPW_STD and MSL were plotted in Fig. 5 with respect to temperature.

We note a decrease of the losses for the CPW_STD, till 200°C due to the mobility reduction of free carriers in the substrate at higher temperature [3]. Above 200°C, the losses start to increase for both CPW_STD and CPW_HR. This is due to the increase of the metallic losses on the one hand, and the degradation of the substrate resistivity on the other hand for the HR substrate. For the MSL, there is a linear increase of α as expected, because the conductor losses are dominant, as shown earlier. Fig. 5 shows also that at room temperature, CPW_HR is the most interesting structure. But at 250°C, the losses are quite close for any of the 3 studied transmission lines.

Conclusion

Losses of MSL and CPW made on HR and STD SOI wafers were analyzed with respect to temperature.

MSL allows the use of STD substrate because the back ground plane shields the Si substrate. MSL can then be an interesting topology if the losses can be lowered to the same level than CPW made on HR SOI or SOS.

For the CPW, the losses are of two kinds: conductor losses, due to the metal resistivity, and substrate losses, due to the coupling between the line and the substrate.

These results demonstrate the feasibility and practical applicability of different passive structures in circuits design for both room and high temperature applications.

Due to the rapid increase in the number of metallic interconnects, the top level metals will be situated further away from the SOI substrate as the technology scales, thus reducing the substrate losses for CPW and widen the metallic strip for MSL . In a near future, 12 metal levels will be available and will enable using 5 times wider strips for MSL and thus reduce the losses by a factor of 3 making MSL as a very promising structure for RF design for the next technological node.

References

[1] M. Si Moussa et al. IEEE RFIC. June 2005. (Accepted)
[2] C. Pavageau et al. IEEE MTT-S, June 2005. (Accepted)
[3] D. Lederer et al. NATO Science Series Elsevier, 2005, pp. 192-196.

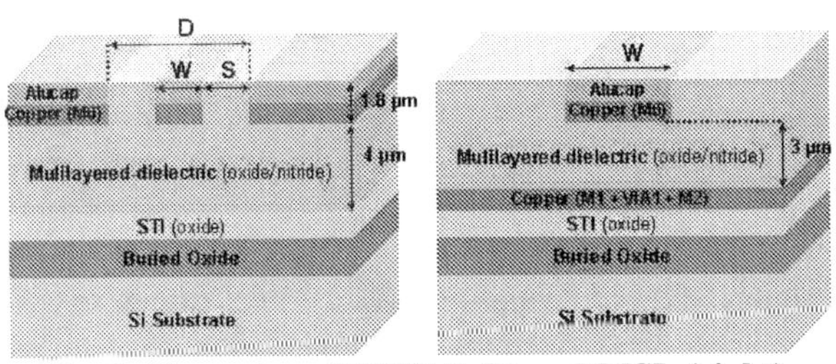

Fig. 1: Geometry of CPW (left) and MSL (right).

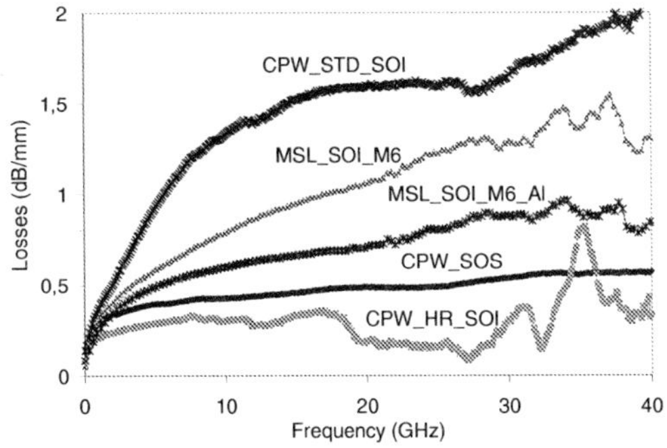

Fig. 2: Measured lineic losses of CPW on STD, HR and SOS @ room temperature.

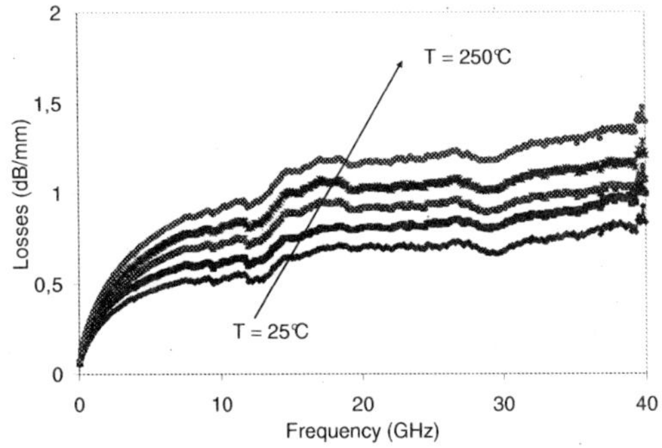

Fig.3: Measured lineic losses vs. frequency and temperature (W=7 µm).

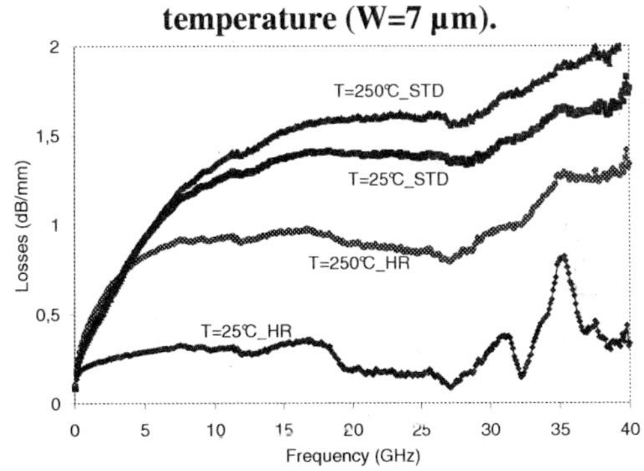

Fig. 4: Measured lineic losses vs. frequency of CPW on STD and HR (W=40 µm, S=24 µm).

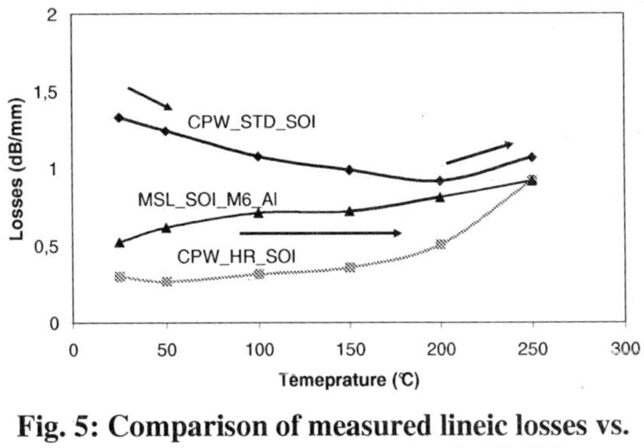

Fig. 5: Comparison of measured lineic losses vs. temperature of CPW & MSL @ 10 GHz.

Saturation Threshold Voltage Degradation in Deep-Submicrometer Fully Depleted SOI nMOSFETs Operating in Cryogenic Environments

M. A. Pavanello[1,2], J. A. Martino[1,2], E. Simoen[3] and C. Claeys[3,4]

[1] Centro Universitário da FEI, Av. Humberto de A. C. Branco, 3972, 09850-901 – São Bernardo do Campo – Brazil
e-mail: pavanello@fei.edu.br

[2] LSI, Universidade de Sao Paulo, Av. Prof. Luciano Gualberto, trav.3 n. 158, 05508-900 – Sao Paulo – Brazil

[3] IMEC, Kapeldreef 75, B 3001 Leuven, Belgium

[4] E.E. Dept., KU Leuven, Kasteelpark Arenberg 10, B-3001 Leuven, Belgium

1. Introduction

One of the most popular ways to verify the occurrence of short-channel effects is to monitor the threshold voltage (V_T) variation with drain bias (V_{DS}) rise. The presence of floating-body and the increment of lateral electric field when shrinking the dimensions hardly affects the V_T reduction in fully depleted (FD) SOI transistors, in contrast to bulk transistors where the V_T reduction with channel length (L) occurs only due to the junction barrier lowering at the source[1]. In bulk MOSFETs the temperature reduction contributes to improve the V_T degradation with V_{DS}.

In this work, additionally to the contribution of applied drain voltage on V_T reduction we demonstrate that halo implanted deep-submicrometer fully depleted (FD) SOI nMOSFETs suffer another degrading mechanism when operating in cryogenic environments.

2. Measurements

The experiments were performed in FD SOI nMOSFETs from a 0.1μm technology with a gate nitrided oxide 2.5 nm thick (t_{oxf}) and 30 nm thick silicon film (t_{Si}) made in standard ELTRAN wafer with a buried oxide thickness of 100 nm and optimized for room temperature operation. Devices presenting channel length (L) of 0.13 μm and channel width (W) of 10 μm. Halo implantation has been performed to control short-channel effects. The halo region length is 65 nm. The power supply bias voltage is 1.5 V.

The temperature has been decreased from 300 K down to 95 K and the V_T has been extracted using the double derivative technique.

Figure 1 presents the V_T difference for two drain bias, i.e. $\Delta V_T=V_T(V_{DS}=0.1V)-V_T(V_{DS}=1.5\ V)$, measured in the whole studied temperature range. Decreasing the temperature there is a practically linear increase on ΔV_T variation from about 170 mV at 300 K up to 240 mV at 95 K. This increase on V_T difference does not occur in bulk MOSFETs with similar processing characteristics and dimensions wheres the ΔV_T decreases from 52 mV at 300 K down to 49 mV at 95 K. Increasing L to 0.8μm the ΔV_T reduces to only 13 mV.

The dependence of ΔV_T variation with V_{GB} is presented in figure 2. Reducing the applied V_{GB} increases the ΔV_T from 70 mV up to 100 mV.

3. Analysis and Discussion

In order to get a physical insight on the measured results two-dimensional numerical simulations[2] were performed in 0.13 μm long transistors. Instead of the nitrided gate oxide we used 3 nm silicon dioxide without the tunneling model. A uniform boron doping level (N_{AF}) of 10^{18} cm^{-3} has been used as for this channel length the halo regions overlap. The interface trap densities at the oxide interfaces were neglected in order to dissociate the ΔV_T induced by V_{DS} to that given by trap charging at lower T.

According to ref. [1], the ΔV_T with V_{DS} can be expressed as:

$$\Delta V_T \approx -\left(\frac{S}{2.3}\right)(\beta+1)\left(1+\gamma\frac{n}{m}\frac{I_0}{I_{J0}}\right)(M-1) \quad (1)$$

where S is the subthreshold slope, β is the gain of the parasitic bipolar transistor, n is the ideality factor of the body-to-source junction, m is a constant related to the subthreshold slope (m=1+γ), I_0 is the channel current at $V_{GF}=V_T$, I_{J0} is the diode reverse saturation current and M is the multiplication factor.

Looking to the experimental results the contribution of the term I_0/I_{J0} with T lowering is negligible as when increasing L, and consequently reducing the contribution of β and (M-1) on eqn. (1), the ΔV_T vs.T can be neglected.

The reason for the enhanced ΔV_T is that both (M-1) and β increase at lower T (Table 1), simultaneously increasing the amount of generated holes and the contributed bipolar current. As the impact ionization rate increases at lower T [3], the holes generated accumulate near the source-junction modifying the threshold voltage in saturation. In addition, the impact ionization carriers bias the base of parasitic bipolar structure, also contributing to increase the ΔV_T at low T. The V_{GB} reduction increases the source-junction barrier for the generated holes which causes the larger ΔV_T observed in the experiments.

In order to explore the ΔV_T evolution with T figure 3 studies the influence of N_{AF} corresponding to three scenarios: the halo implanted as measured, the no halo implanted ($N_{AF}=5.5 \times 10^{17}$ cm^{-3}) and the low doped one ($N_{AF}=1 \times 10^{17}$ cm^{-3}).

First, the slope of the ΔV_T x T curve for halo implanted is similar to that obtained from figure 1, validating the simulation analysis. Removing the halo implantation makes the ΔV_T x T behavior nearly temperature independent and a hard reduction on N_{AF} changes the slope on this curve, similarly to observed for bulk transistors. In the latter case the ΔV_T is larger at room temperature due to the source barrier lowering. These results indicate that careful halo design has to be performed in FD SOI to overcome this ΔV_T x T degradation.

Figure 4 presents the impact of scaling t_{oxf} (filled symbols) and t_{Si} (opened symbol) on ΔV_T *vs.* T behavior.

The reduction on t_{oxf} improves the gate control and the occurrence of SCE, and the ΔV_T x T degradation. Similar results were obtained turning on tunneling models. Scaling t_{Si} also contributes not only by the better ΔV_T but also improves the ΔV_T x T degradation. A thicker t_{Si} increases β, worsening the ΔV_T at lower T.

4. Conclusion

This work demonstrated that the saturation threshold voltage of short-channel SOI nMOSFETs degrades as the temperature is reduced. The V_T degradation at low temperatures is caused by a combination between the floating-body and impact ionization effects, increasing the gain of the parasitic bipolar structure. The halo or pocket implantation contributes for the enhanced V_T degradation at lower temperatures. The absence of halo efficiently reduces both the V_T variation and the V_T degradation in cryogenic operation. A negative biasing of the back gate tends to enhance the V_T degradation with temperature.

References

[1] A. O. Adan *et al.*, *IEEE T-.ED*, v. 46, pp. 729, 1999.

[2] Atlas Program, Silvaco International, 2002.

[3] F. Balestra and G. Ghibaudo, *Device and Circuit Cryogenic Operation for Low Temperature Electronics*, Kluwer Academic Publishers, Boston, 2001.

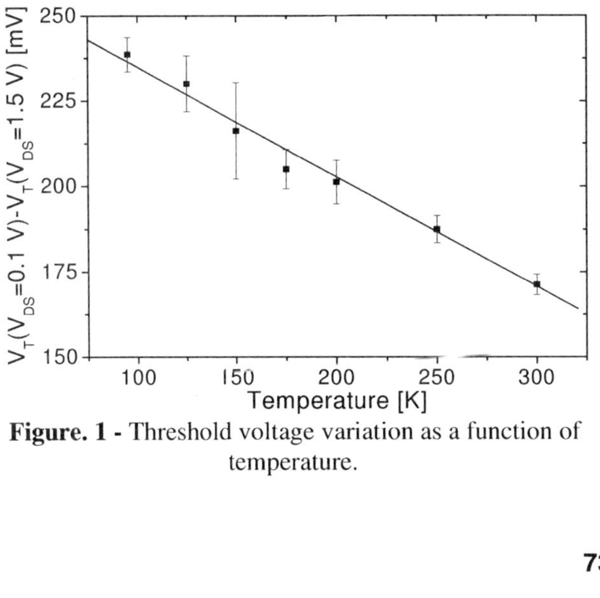

Figure. 1 - Threshold voltage variation as a function of temperature.

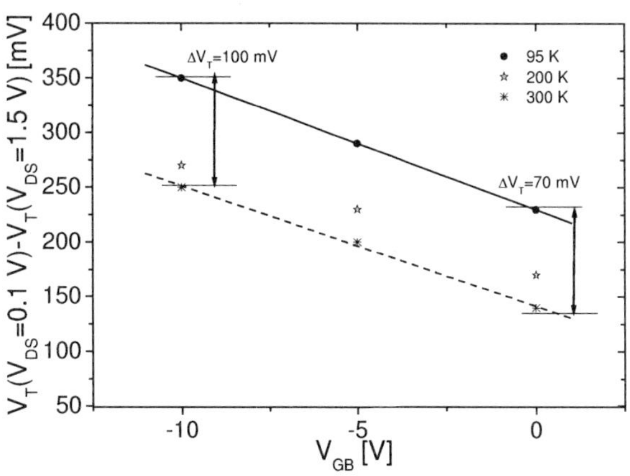

Figure 2 - Threshold voltage variation as a function of applied V_{GB} in three different temperatures.

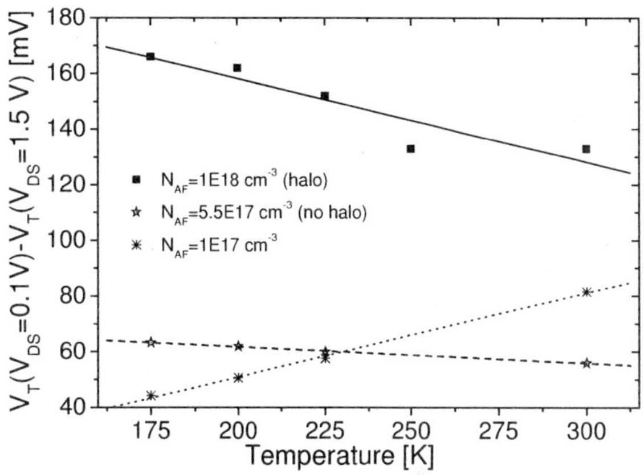

Figure 3 - Simulated V_T variation with T as a function of N_{AF}.

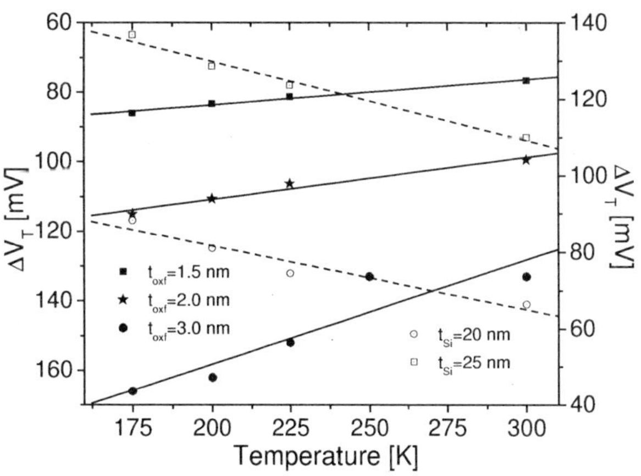

Figure 4 - Simulated V_T variation with T as a function of t_{oxf} and t_{Si}.

Table 1: Simulated (M-1) and β factors as a function of T, extracted at $V_{GF}=V_T$ and $V_{DS}=1.5$ V.

TEMPERATURE [K]	(M-1) [x10^-5]	β [x10^3]
300	3.2	29.1
200	3.4	29.4
175	3.7	29.5

2005 IEEE International SOI Conference

Dynamic Circuit Techniques Using Independently Controlled Double-Gate Devices

[1]J.B. Kuang, [2]K. Kim, [2]C.T. Chuang, [1]H.C. Ngo, and [1]K.J. Nowka
IBM Research Division
[1]Austin Research Laboratory, TX 78758 [2]T.J. Watson Research Center, Yorktown Heights, NY 10598

Introduction

Increased leakage in scaled technologies limits the robustness of dynamic circuits, especially the wide OR-style dynamic gates commonly used in high-speed designs. Strong keepers are needed for the precharge state or after the completion of evaluation to compensate for leakage and hold the correct state of the dynamic node. Strong keepers, however, cause aggravated contention and speed degradation during evaluation. Previously, a conditional keeper technique was proposed where only a fraction of the keeper strength is turned on at the onset of evaluation phase while the full strength is enabled after a delay time [1]. Alternatively, a programmable keeper technique where the effective keeper width tracked the on-die leakage was proposed to compensate for die-to-die variation [2]. Nevertheless, employment of extra devices increases the logic gate area as well as the capacitance at the dynamic node. Charge sharing is another concern which causes voltage droop in the dynamic node and degrades the noise margin. A common method to prevent the charge sharing effect consists of charging the intermediate node in a stacked configuration to full rail before evaluation. While the technique is effective, the intermediate node precharge device adds capacitance to that node and increases the circuit area. To address these issues with continued device scaling requires diligent technology circuit co-design efforts.

Circuit Design with Double-Gate Devices

Independent biasing of the front and back gate in double-gate (DG) technologies can be exploited to boost the performance and reduce the number of transistors in implementing logic functions [3-4]. In this paper, keeper schemes utilizing independent front and back gate control in DG devices are presented to achieve improved speed, noise margin, and reliability in dynamic circuits. These schemes provide the conditional keeper function with fewer devices, effectively reducing area and capacitance. Also introduced is a method that utilizes the back-gate device of a logic transistor as the precharge device for the intermediate stacked node, to prevent charge sharing, thereby reducing the capacitance and area of the intermediate node precharge device. Finally, a static keeper and footer method using the asymmetrical DG devices is described taking advantage of the unique front and back channel current characteristics.

Conditional Keepers

Fig. 1 shows the schematic diagram of a conditional keeper using independently biased DG PFET. The "slow mode" pin can be either a test mode signal that preserves the dynamic node state during low-frequency debug, or an at-speed delayed clock that turns on the front gate keeper after a successful evaluation. Figs 2 (a) and (b) depict the device structures for asymmetrical and symmetrical DG devices, respectively.

For symmetric DG devices, the strengths of the front and back gate are equal when only one gate is turned on. When both channels are on (DG mode), the total current in the front and back channel increases to more than two times higher, compared with the one-gate-on case due to the ideal subthreshold slope in the DG mode of operation. At lower V_{DD}'s, the DG mode current improvement becomes larger as the effect of gate-to-gate coupling becomes more significant [5]. Thus, this dynamic circuit technique is viable for voltage scaling, considering the timing, current drive, and device strength requirements. Two alternative circuit configurations are shown in Figs 3 (a) and (b), both of which delay the turn-on of the front gate keeper device. These schemes utilize a single DG PFET to perform the functions of both weak keeper and strong delayed-on keeper, thus reducing the capacitance and area associated with multiple keeper devices to reduce contention, improve speed, area, noise immunity, and circuit robustness.

The unique features (Fig. 4) of asymmetrical DG device structure can be preferentially utilized in circuit design. When only the back gate is biased and the front gate is grounded, the back channel current is more than one-order-of-magnitude lower than the predominant front-channel current due to the ~1V higher threshold voltage for p+ gate [5]. When both the front and back channel are turned on, the front-channel current is enhanced by approximately 2 times at V_{DD}=1V, due to gate-to-gate coupling, compared with the case when only the front gate is on. In the circuit of Fig. 1 for the asymmetrical DG devices, the front channel with the off back gate serves as a weak keeper. The back gate is turned on to increase the front channel current only when a strong keeper is desired. Thus, the circuit provides the keeper function with its strength conditionally modulated by the back gate. Figs 5 (a) and (b) show the MEDICI [6] simulation results for the fast and slow mode, with heavy and light output load, respectively. A significant slow-down of the evaluate edge in the slow mode operation is observed, thereby demonstrating the performance advantage of the proposed method. The difference between the slow and fast mode operation is larger in the case of light output load because of the more pronounced difference in effective drive current and hence the transitional slews.

Charge Sharing Mitigation

Fig. 6 depicts a scheme, using symmetric DG devices, where the back gate (BG) of transistor A1 is used as the precharge device for the intermediate stacked node "int" to prevent charge sharing. The scheme reduces the capacitance and area of the intermediate node pre-charge device, thus effectively circumvents charge sharing and conserves area with the intervention of signal "clock_b". The insert of Fig. 6 compares the dynamic node "dyn" waveforms, under a severe charge sharing condition during evaluation, with and without the anti-charge sharing back gate device. The use of anti-charge sharing back gate device prevents the collapse of dynamic node voltage and catastrophic logic fault.

Static Keepers and Footers

If a dynamic stage uses a static keeper instead of the feedback half latch, the leakage current through the always-on keeper would be unacceptably high for conventional bulk silicon, PD/SOI, or symmetrical DG technologies. In contrast, the asymmetrical DG PFET incorporates the function of two PFETs: the front gate with a strong front channel current can be connected to the clock to perform the reset function while the weak back gate, at a 1/10th - 1/20th drive strength, can be used as a static keeper without excessive leakage current. Fig. 7 shows a domino stage using asymmetrical DG devices where p1 replaces both the precharge and keeper device. The front gate precharge device of p1 is connected to the clock with only half the gate loading and no performance penalty. The back gate keeper device of p1 is tied to constant (hard wired for all modes, Fig. 7) or conditional (only in active mode to further reduce leakage, Fig. 8) ground. When the front gate is off during evaluation, the back gate becomes a keeper. MEDICI simulation results showing well behaved state transitioning are presented in Fig. 9. It is observed that $I_{on(n1)}/I_{on\_strong(p1)}=3.85$; $I_{on\_strong(p1)}/I_{on\_weak(p1)}=14.50$; $I_{on\_weak(p1)}/I_{off(n1)}=1.84\times10^4$. The circuit functions with a wide noise margin as shown in two examples of noise event simulations (Fig. 10) for both precharge and evaluate. The static keeper can maintain or restore the desired V_{DD} level. During burn-in and debug mode, there is an additional option of keeping clock high. Thus, the circuit functions as a pseudo-NMOS gate at low frequencies.

Similarly, when designing a footer device, we can exploit the asymmetrical DG device feature: the front gate with strong coupling to the front channel determines the on/off state; and the back gate with weak coupling to the channel controls the strength of the front channel. When the footer front and back gate are clock tied (Fig. 7), faster reset and lower leakage are achieved for the on and off state, respectively. Alternatively, the footer back gate can be tied to constant or conditional V_{DD} for clock load reduction (Fig. 8).

Conclusion

In summary, conditional keeper, charge sharing prevention, and clock load reduction techniques for symmetrical and asymmetrical DG devices have been presented. Performance benefit, noise immunity, area and power efficiency can be achieved when technology features are judiciously utilized in the design of dynamic circuits.

This work is supported in part by the DARPA contract NBCH30390004.

[1] A. Alvandpour, et al., "A Conditional Keeper Technique for Sub-0.13μm Wide Dynamic Gates", Dig. Tech. Papers, Symp. VLSI Circuits, 2001, pp. 29-30.

[2] C. H. Kim, et al., "A Process Variation Compensating Technique for Sub-90 nm Dynamic Circuits," Dig. Tech. Papers, Symp. VLSI Circuits, 2003, pp. 205-206.

[3] M. H. Chiang, et al., "Novel High-Density Low-Power High-Performance Double-Gate Logic Techniques," IEEE Intl SOI Conf., 2004, pp. 122-123.

[4] H. Mahmoodi, et al., "High performance and low power domino logic using independent gate control in double-gate SOI MOSFETs," IEEE Intl SOI Conf., 2004, pp. 67-68.

[5] K. Kim and J. G. Fossum, "Double-gate CMOS: symmetrical versus asymmetrical gate devices," IEEE Trans. Electron Devices, vol. 48, 2001, pp. 294-299.

[6] Taurus-MEDICI, Industry-standard device simulation tool, Mountain View, CA, Synopsis, Inc., 2003.

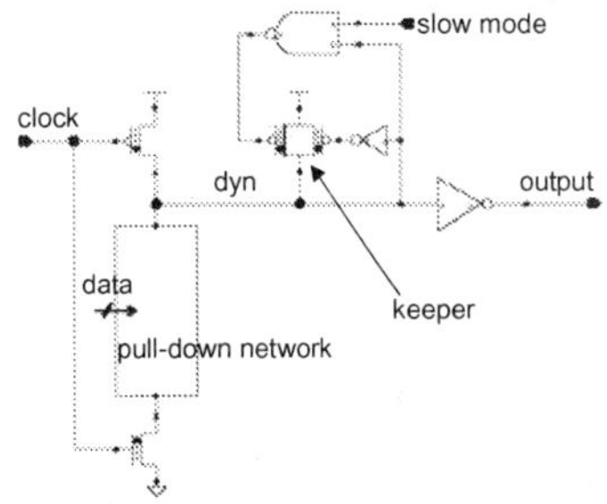

Fig. 1 Conditional keeper using independently controlled front and back gate double-gate PFET. The "slow mode" pin can be an at-speed clock or a low-frequency control signal.

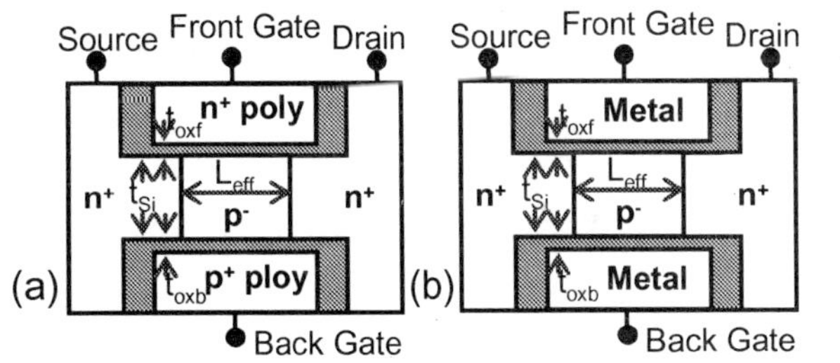

Fig. 2 Double-gate device cross sections (L_{eff} = 25 nm, t_{oxf} = toxb = 1 nm, t_{Si} = 10 nm): (a) asymmetrical double-gate NFET. For PFET, the front gate is p+ and the back gate is n+. (b) symmetrical double-gate NFET.

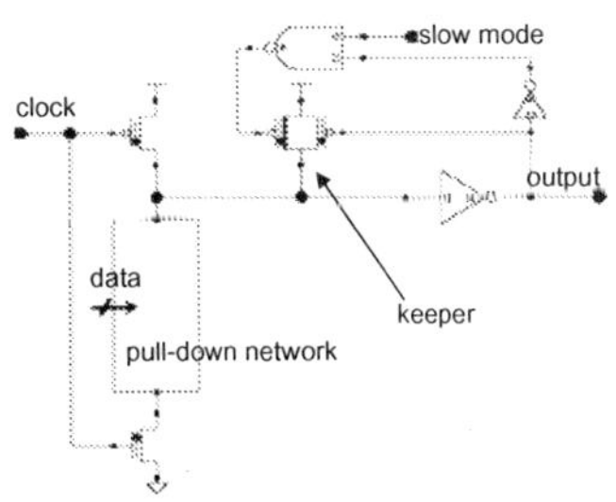

Fig. 3 (a) Alternative conditional keeper topology using a DG PFET

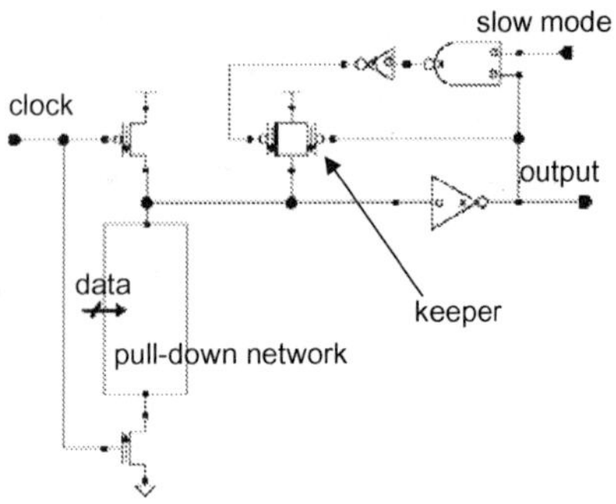

Fig. 3 (b) Alternative conditional keeper topology using a DG PFET

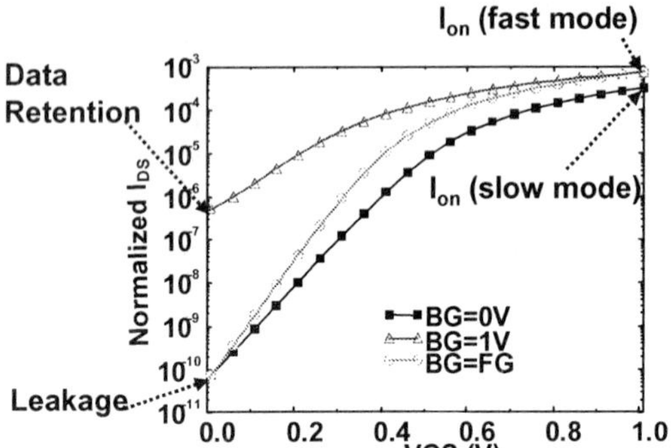

Fig. 4 MEDICI-predicted drain current versus front gate voltage for asymmetrical DG NFET for three different bias conditions. The gate terminal can be connected flexibly in design to meet the drive strength, leakage reduction, or data retention requirements.

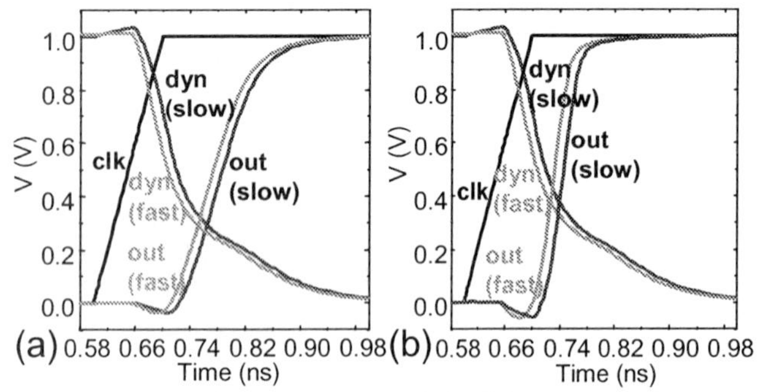

Fig. 5 MEDICI simulation for (a) heavily loaded output: clock-up to dyn-down = 55 ps (fast mode), 70 ps (slow mode), and clock-up to output-up = 123 ps (fast mode), 130 ps (slow mode), and (b) lightly loaded output: clock-up to output-up = 83 ps (fast mode), 100 ps (slow mode).

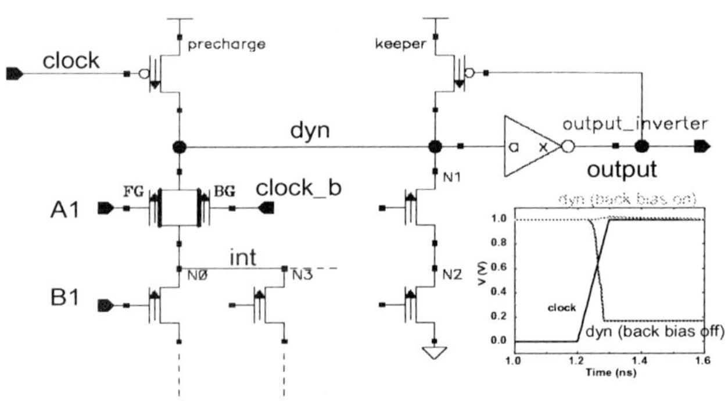

Fig. 6 Schematic diagram showing the precharge of internal stacked node "int" using back gate of the logic transistor A1. The insert shows the well behaved and collapsed dynamic node waveforms with and without the intervention of the back gate anti-charge sharing device, respectively.

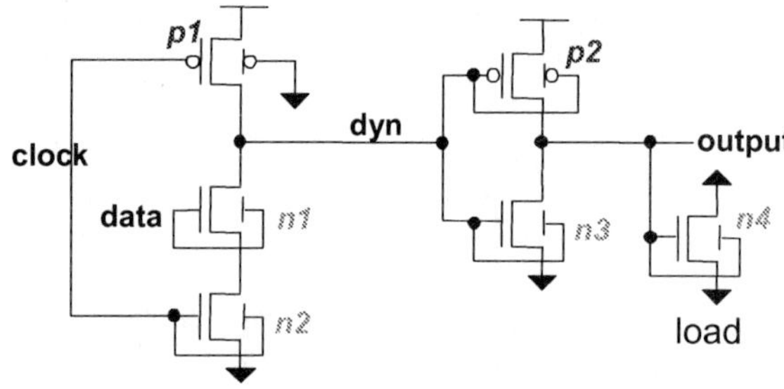

Fig. 7 Schematic diagram for a shared precharge/keeper implementation using asymmetric DG devices, where the front gate of p1 is the precharge device and back gate of p1 is the weak always-on static keeper

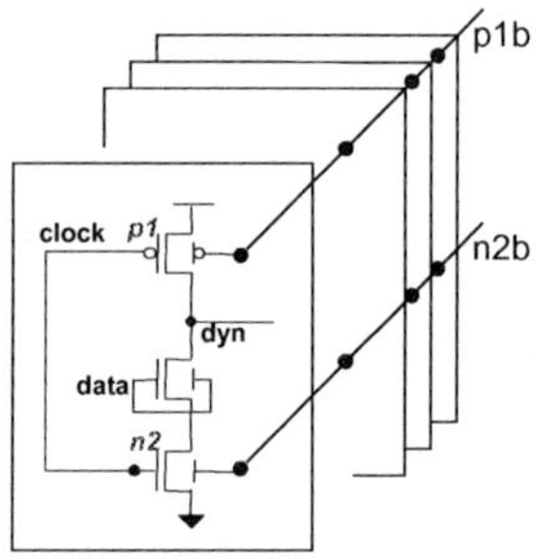

Fig. 8 Precharge/keeper device back gate control for leakage power and/or clock load reduction. The common node p1b is at ground and V_{DD} for active and standby mode, respectively. The common node n2b is at V_{DD} and ground for the active and standby mode, respectively.

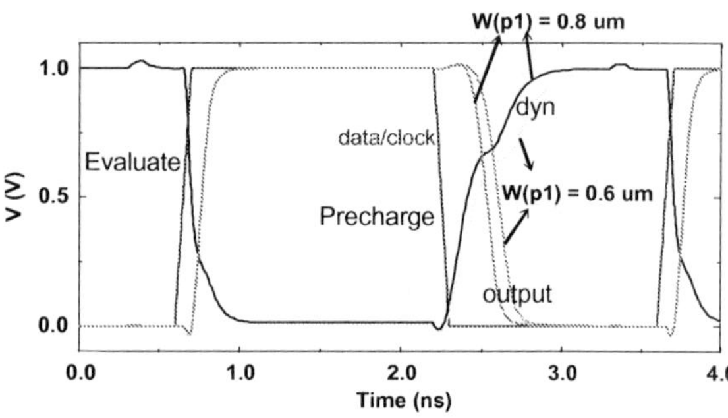

Fig. 9 MEDICI results showing the waveforms for data, clock, dyn, and output nodes for two precharge/keeper device (p1) widths of 0.6 and 0.8μm. Data and clock arrival times are assumed to be perfectly aligned in simulation.

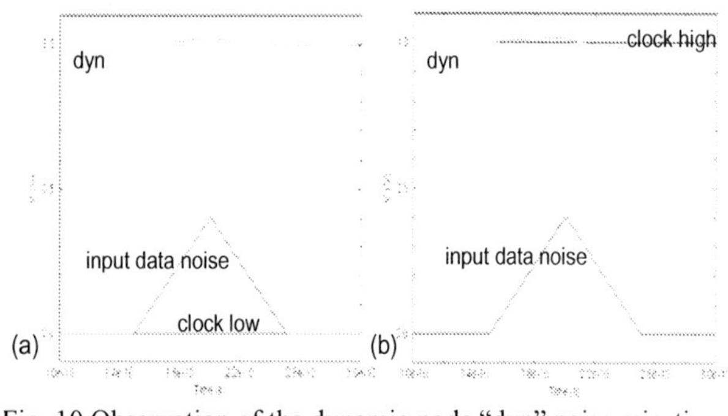

Fig. 10 Observation of the dynamic node "dyn" noise rejection behavior for the circuit of Fig. 7 during the (a) precharge and (b) evaluate interval, respectively. Node "dyn" recovers to V_{DD} in both cases in the presence of the precharge/static keeper device.

Circuit to Measure High Speed Pulse I-V Characteristics with Only DC I/O's

Mark B. Ketchen, Manjul Bhushan* and Keith A. Jenkins
IBM Thomas J. Watson Research Center and *IBM STG Division
1101 Kitchawan Road, Yorktown Heights, NY 10598

Abstract

We have implemented a new approach for measuring the pulse I-V characteristics of a set of nominally identical MOSFETs. In this scheme, all high speed signals are contained entirely inside the test structure itself. Only DC inputs and outputs are required for direct measurement of both DC and AC I-V characteristics at frequencies up to the GHz range

Introduction

PDSOI technology presents unique characterization challenges as a result of features such as switching history dependent circuit delays and frequency dependent MOSFET I-V characteristics. DC I-V characteristics are intrinsically different from those existing under high speed switching conditions due to self-heating and floating body effects. While accurate high speed characteristics can be obtained with standard pulse I-V techniques[1], this approach is not suitable for in-line testing to provide rapid feedback for device development and model-to-hardware correlation on an ongoing basis. Recently, an in-line monitor for switching history measurements using only DC I/O's was successfully introduced [2]. In a complementary fashion, this paper describes a scheme for in-line high speed I-V characterization that requires only existing in-line parametric testers with DC I/O's.

Description of In-line Pulse IV Circuit

The basic principle of this new measurement technique is illustrated in Figure 1 for the case of NFETs being the devices under test. A continuous sequence of non-overlapping pulses is generated sequentially from a ring oscillator and applied to the gates of m NFETs under test as indicated in Figure 1. The NFETs are connected in parallel and the measured current is averaged over all NFETs, each with a 100/m % duty cycle. The ring oscillator consists of nm nominally identical stages and is divided up into m equal segments, each n stages in length, where n is an even number. The output from the XNOR across each segment is a train of pulses of width T/m and period T, where T=mnt and t is the delay per stage of the ring. The ring oscillator and XNORs share a common power supply (VDD) and a common ground. The pulse trains from all m segments are non-overlapping, of magnitude VDD, and configured such that at any given time one and only one pulse is present. The output from each XNOR drives an inverter

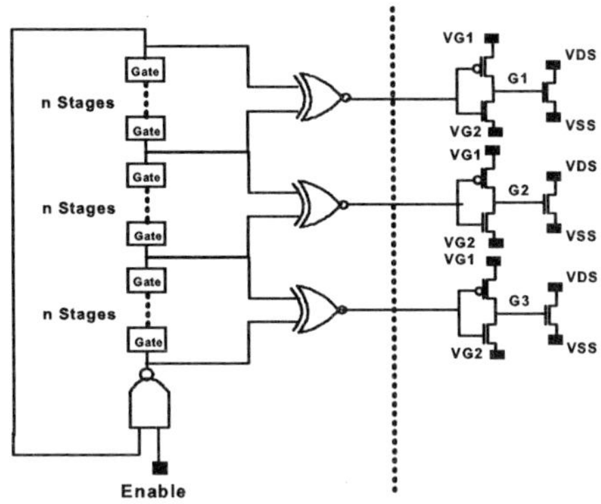

Figure 1: Circuit scheme for applying non-overlapping pulses to the gates of NFETs connected in parallel.

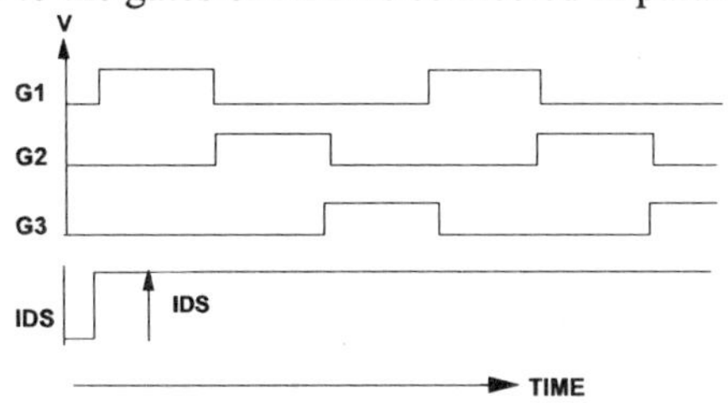

Figure 2: Pulse inputs to the gates of the three NFETs in Fig. 1 and measured IDS as a function of time.

powered by an independent power supply of amplitude VGS (= VG2-VG1). The output of each such inverter in turn drives an NFET under test with its drain held at VDS and its source at VSS. The m NFETs under test share common VDS and VSS supplies.

Consider the case where VG1=VDD, VDS=VDD, and VSS, VG2=0(ground) with m = 10. Each NFET is on and delivering IDS 10% of the time and is off the remaining 90% of the time. The current drawn from the VDS supply is constant and equal to that of a single NFET under test. To the external DC current meter, this arrangement provides a constant current which can be measured accurately as indicated in Figure 2. Since each NFET is only on 10% of the time the self-heating, with a time constant of 100's of ns [3], is 1/10 that of a similar device under conventional DC test conditions, and rendered insignificant. Furthermore the floating body potential of the FETs under test will be very nearly that of similar devices in a high duty factor steady state operation in a microprocessor, such as a clock buffer. If the ring oscillator VDD is set to 0, then the output of the XNORs will all be 0, and the outputs of all the intermediate inverters applied to the gates of

0-7803-9212-4/05/$20.00 ©2005 IEEE

all the FETs under test will be VDD. The IDS then measured as the DC current from the VDS supply is the standard IDS for the 10 FETs in parallel as is conventionally measured in a DC in-line test. This circuit thus allows one to directly measure and compare for the same devices the standard DC IDS (with self-heating and DC body effect) with the IDS under representative use conditions (minimal self-heating and steady state pulse body effect). This approach has the additional advantage of providing data averaged over m nominally identical devices.

Macro Implementation and Results

The initial implementation of this circuit is a 2500 μm x 200 μm macro designed for 90 nm PD SOI technology (Figure 3). It has two ring oscillators, one servicing two groups of m=10 NFETs and the other, two groups of m=10 PFETs. For this design n=100 and t =10 ps. VG2 is tied to the ring oscillator ground so that VG1=VGS. The design required careful attention to detail to ensure pulse integrity and precise temporal alignment. Low Vt FETs were incorporated in the inverters preceding the FETs under test to extend the operating range to of order 0.5 < VGS/VDD < 1.3 with an overall error due to misalignment and skew of less than 1% in the true IDS(AC) compared with that measured. The ring oscillator signal is fed to a frequency divider so that stage switching delay can be correlated with FET characteristics.

Figure 4 presents representative data from an experimental 90 nm technology showing IDS versus VGS in arbitrary units for VDS=1.0V. As anticipated the difference increases with increasing VGS. Figure 5 replots this data as the fractional difference between IDS(AC) and IDS(DC) over the range of 0.7 < VGS/VDD < 1.0 and also shows the PFET response over the same range. For the PFET there is a crossover at about 0.8V indicating that the body effect is dominating and has a sign opposite that of the thermal effect at low VGS. Conventional pulsed I-V data from an experimental wafer is shown in Figure 6. The NFETs show about 10% increase in IDS(AC) from its DC value at VGS=1.2V in both cases. The PFET again exhibits a region with IDS(AC) >IDS(DC).

Figure 3: Physical layout of the macro for measuring DC and AC characteristics.

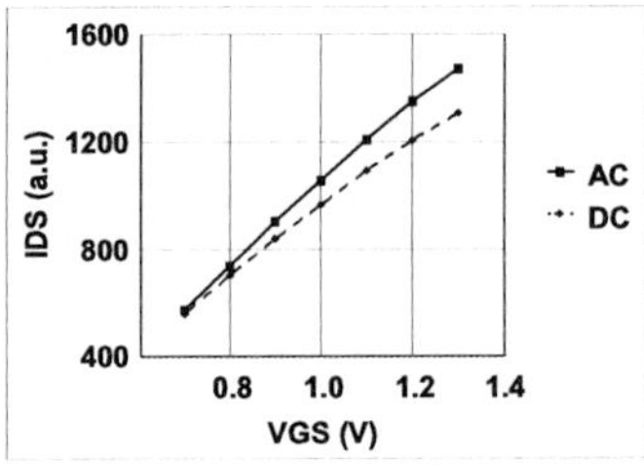

Figure 4: Comparison of DC and AC IDS for an NFET at VDS=1.0V.

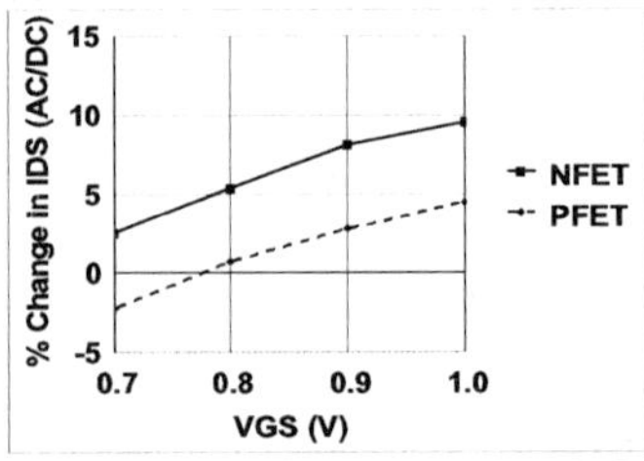

Figure 5: % change in delay (AC/DC) for NFET and PFET.

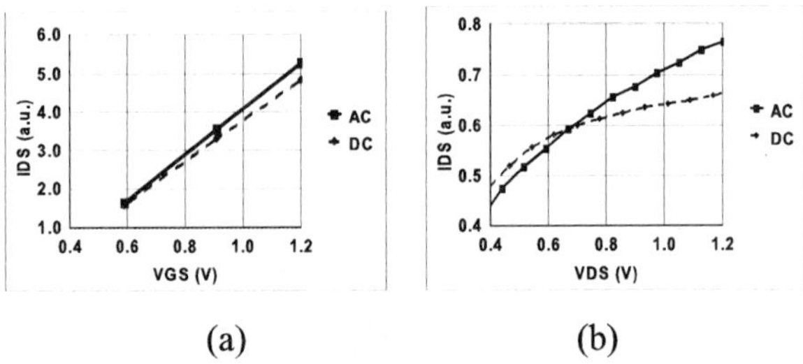

Figure 6: Conventional pulse I-V data: (a) IDS vs. VGS characteristics of a NFET at VDS=1.0V and (b) IDS vs.VDS characteristics of a PFET at VGS=1.2V and showing a region IDS(DC)>IDS(AC).

Conclusion

We have developed and implemented a circuit scheme for enabling rapid, in-line measurement of high speed pulse as well as DC I-V characteristics of a set of nominally identical MOSFETs.

Acknowledgments

We gratefully acknowledge the support of many individuals in IBM Research and STG Division for their assistance.

References

[1] Keith A. Jenkins, J. Y.-C. Sun, and J. Gautier, "Characteristics of SOI FET's Under Pulsed Conditions", IEEE Trans. on El. Dev., **44**, p. 1923 1997.

[2] D. J. Pearson, M. B. Ketchen and M. Bhushan, "Technique for Rapid, In-Line Characterization of Switching History in Partially Depleted SOI Technologies", Proc. of 2004 SOI Conference, pp. 148-150, 2004.

[3] S. Polonsky and K.A. Jenkins, "Time-Resolved Measurements of Self-Heating in SOI and Strained Silicon MOSFETs Using Photon Emission Microscopy," IEEE Electron Device Letters, **25**, pp. 208-210 (2004).

An Integrated Balun for Dual-Band LC tank VCO in 130nm CMOS/SOI

L. Geynet, E. de Foucauld, D. Cartalade and G. Jacquemod.
CEA/Leti, 17 rue des Martyrs 38054 Grenoble Cedex 9, FRANCE
Email: lionel.geynet@cea.fr, Tel: (33) 438.782.479

Introduction

In radio-frequency integrated circuits, tranformers have been an essential component for many years. They are currently used for single to differential conversion particularly for LNA and PA circuits [1]. For a LNA, the noise figure is directly related to the quality factor of the balun. The higher Q is, the lower noise figure would be. In this paper, we present a symmetric parallel plate coupled line balun for dual-band LC tank VCO. It has been fabricated in a 130nm CMOS/SOI technology from STMicroelectronics. It has been realised with six metal layers on a SOI high resistivity substrate. The metal windings of the primary and secondary are in parallel to form coupled lines. Both the primary and secondary are three turns inductors, composed of three metal layers. The primary is formed by the three upper metal layers so as to decrease resistivity and therefore increase Q, whereas secondary uses three downer metal layers. Due to high resistivity substrate SOI technology, we are able to make a high-Q balun (S21>-2dB), this can improve the feasibility of new architectures such as switched LC tanks. This study presents the advantages of such SOI balun for wireless applications, particularly for dual-band VCO using switched LC tank. In a first part, we present the design, the measurement results and a model of this transformer, in a second one, dual-band VCO applications and comparison with others switched resonator structures are proposed.

Design, Results and Models

As shown on the photograph (Fig.1), both inductances are hexagonal and multi-turns. All metal layers have the same resistivity (60mΩ/sq) except the top layer (22mΩ/sq) and the bottom layer (75mΩ/sq). The primary quality factor is therefore better than secondary. All inductors metal layers are shunted together so as to reduce serial resistances. Space between inductor wires has been widened to minimize capacitive coupling.

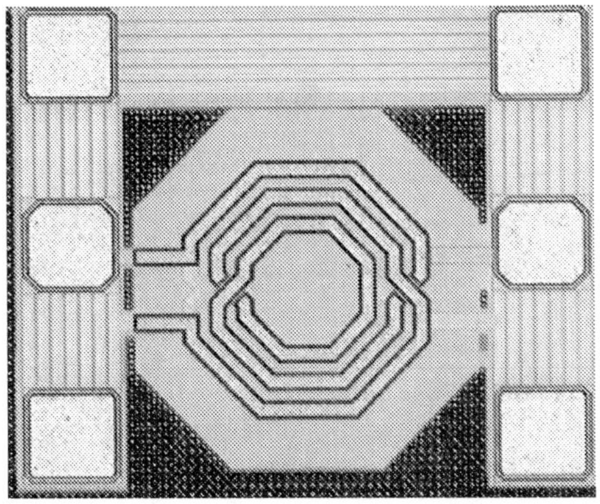

Figure 1: Photograph of the balun

This circuit has been measured using a four-ports network analyser where only two ports were used. For swiched LC tank applications we only need to study the primary characteristics. The secondary is shorted or opened so as to modify the inductance value of the primary as proved in Equation (1).

$$\begin{cases} Vp = j.Lp.\omega.Ip + j.M.\omega.Is \\ Vs = j.Ls.\omega.Is + j.M.\omega.Ip \end{cases} \quad (1)$$

If open circuit at secondary:

$$Lprimary = \frac{\mathrm{Im}(z11)}{\omega} = Lp \quad (2)$$

If short circuit at secondary:

$$Lprimary = \frac{\mathrm{Im}(z11)}{\omega} = Lp - \frac{M^2}{Ls} \quad (3)$$

Simulations have been realised using Momentum electromagnetic simulator. Results are presented in figures 2 and 3. Measurement results are very close to simulation, our substrate modeling is therefore validated.

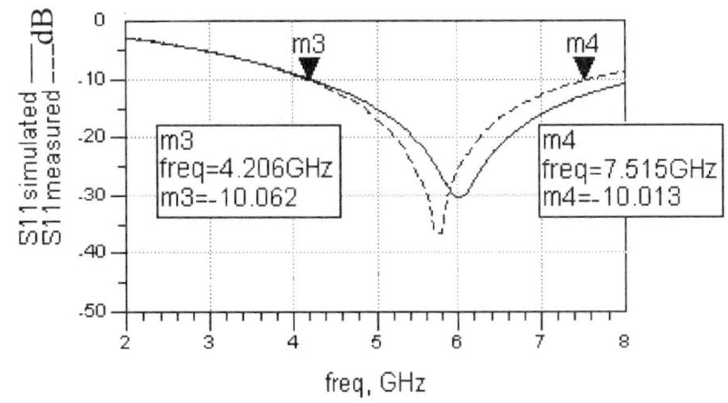

Figure 2: Measured and simulated return loss

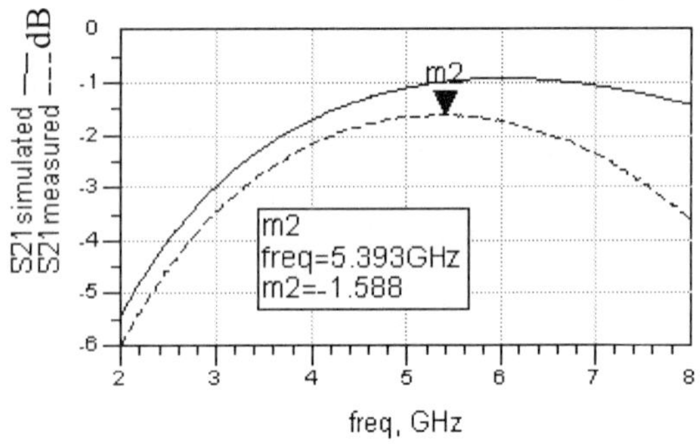

Figure 3: Measured and simulated insertion loss

Differences at high frequencies are explained by the simulator limitation about metal thickness, therefore line-to-line capacitive coupling. S parameters shows that S11 is lower than -10dB on a wide frequency band from 4 GHz to 7.7 GHz. The insertion loss is about -1.5dB at 5 GHz. The proposed balun has a strong coupling coefficient (85%) that induces a strong mutual inductance which produces an important variation of the primary inductance value. By shifting one inductor but staying symmetric, we are able to change the coupling coefficient and the primary inductance value (Fig.4.).

Switched LC tanks Applications

From these results, we have extracted a RLC simulation model of the balun to study phase noise performance in a dual-band VCO. This VCO has been realised and measurement results are expected for end of august. A Dual-Band VCO needs to commute between two oscillation frequencies without changing phase noise performance and increasing current consumption too much. By switching with a MOSFET at secondary we are able to change the primary inductance value, therefore modifying VCO oscillation frequency. Two classical switched LC tanks (Fig.5.) have been studied to compare their performance [2], one in bulk technology and

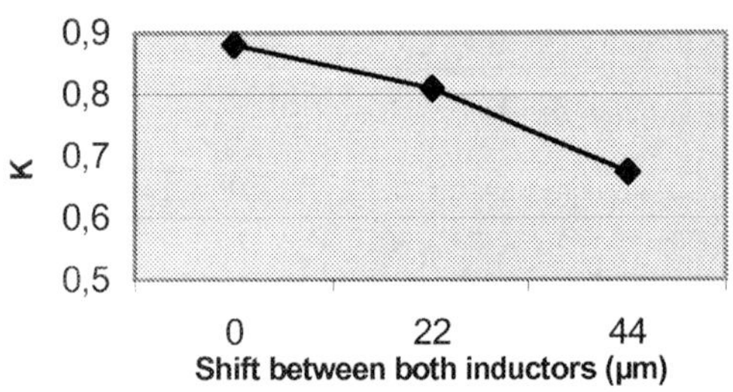

Figure 4: Coupling variation vs. shift

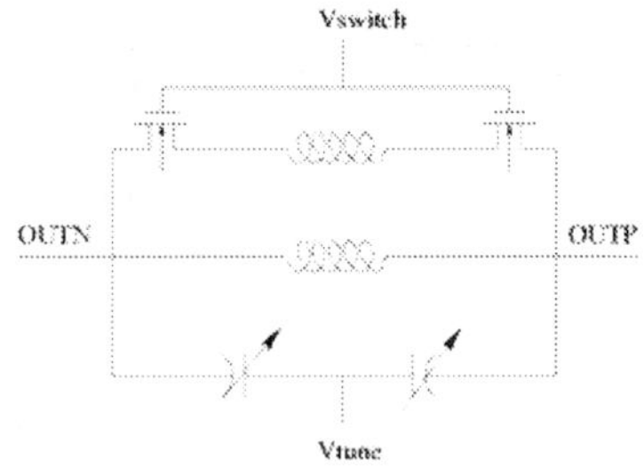

Figure 5: Classical switched tank

LC tank Structures		Switch Off	Switch On
Quality Factor	Simulation Bulk Classical Tank	6	3.7
	Measures SOI Classical Tank	9.47	6.25
	Our Structure with Balun SOI	5.7	4

Table 1: Quality Factor comparison

other one in high resistivity SOI technology. Best phase noise performance of a VCO are ensured by good quality factor and little MOSFET flicker noise. By switching on the MOSFETs, noise from channel is added and tank serial resistance is increased. Quality factor [3] and therefore phase noise are degraded (Table 1). Moreover, current consumption has to be increased in the ratio of 1.5 to ensure oscillation.

Conclusion

The feasibility of a switched LC tank for multi-standards applications has been proved. With our balun structure, the MOSFET (and its noise) is isolated at secondary, therefore Q-factor is less decreased and flicker noise doesn't contribute to phase noise. This new structure has lower phase noise than classical bulk tank particularly when switched on. Moreover this new switched tank occupies only one inductor area (290µm*190µm).

References

[1] H.Y. D. Yang, L. Zhang, and J. A. Castaneda, Design and Analysis of a Multi-Layer Transformer Balun for Silicon RF Integrated Circuits, IEEE RFIC Symposium, 2002.

[2] Seong-Mo Yim and Kenneth K. O, Demonstration of a Switched Resonator Concept in a Dual-Band Monolithic CMOS LC-Tuned VCO, IEEE Custom Integrated Circuits Conference, 2001.

[3] Kenneth 0, Estimation Methods for Quality Factors of Inductors Fabricated in Silicon Integrated Circuit Process Technologies, IEEE Journal of Solid-State Circuits, Vol. 33, NO. 8, August 1998.

Ultra-Low-Power, High-Performance, Dynamic-Threshold Digital Circuits in the FlexFET Independently-Double-Gated SOI CMOS Technology

D. Hackler*, K. DeGregorio*, S. Parke*+
*American Semiconductor, Inc., Boise, ID
+Boise State University, Boise, ID

Introduction

The ITRS roadmap projects that double-gated transistors will be needed in the future, in order to scale CMOS to the 45nm node. However, flexible, dynamic threshold control is possible now with existing independently-double-gated (IDG) CMOS technologies, and is highly desirable for ultra-low-power (ULP) SoC designs at the 180nm, 130nm, and 90nm nodes [1]. By varying the bottom gate voltage of the FlexFET IDG MOSFET from -0.5V to +0.5V, standby power can be dynamically changed over **ten orders** of magnitude, while the transistor/circuit performance can be changed by 70%. Minimally sized transistors may be used to achieve ULP in standby, while dynamic Vt adjustment is used to achieve high-performance when active. This paper demonstrates IDG FlexFET CMOS in static CMOS ring oscillators, while the advantages of applying IDG-CMOS to dynamic domino CMOS logic circuits have recently been shown as well [1]. IDG-CMOS has also recently been applied to several exciting new analog/MS/RF circuit applications, such as a single transistor mixer [5]

Device Description

FlexFET is a new SOI IDG-CMOS technology [2-4] that features a gate trench etched through thick SD regions into which an implanted JFET bottom gate and mid-gap TiN MOS topgate are self-aligned, as shown in Fig. 1. The independent top and bottom gates are contacted at opposite sides of the channel by a damascene tungsten local interconnect (LI) that is embedded in the isolation region between devices. This results in compact, planar connections to all four transistor terminals. Individual transistors can be connected as single gate (SG), double gate (DG), or (IDG) as required. Fig. 2 shows an SEM topview of an IDG inverter ring oscillator (taken after LI mask). All NMOS bottomgates are connected to NBG and all PMOS bottomgates are connected to PBG. Fig. 4 shows the schematics of the three inverter types investigated. The SG and IDG types can both operate to VDD=2.5V, while the DG type can only operate to VDD=1.0V. Fig. 3 shows that a measured bottomgate threshold control of >0.85V/V was achieved. This high degree of control means that only a small bottomgate voltage range of -0.5V to +0.5V is required to achieve the desired ultra-low-power and high performance targets. No additional area was required to make the bottomgate contacts.

Process Description

The FlexFET process uses a relatively thick 200nm SIMOX SOI layer into which SD's are implanted. Gate trenches are then etched through the SD's. The channel thickness is set by the energy of the bottomgate implant into the bottom of the gate trench, not by the SOI thickness. A raised SD structure is achieved without any epi growth thermal budget. FlexFET is a "gate-last" process, allowing high-K or ferroelectric gate dielectrics and metal topgates to be deposited and planarized into the gate trench, thus avoiding any damaging plasma gate etches or subsequent high-temperature processing. Ultrashallow SDE's are formed by disposable doped glass sidewall spacers. PECVD PSG & BSG conformal spacers with 5% dopant concentration were created on the gate trench sidewalls, used to drive in the SDE's, and then removed and replaced by nitride spacers. These nitride spacers permit 180nm channels to be created with relaxed 350nm lithography. A 4.5nm oxynitride and a 30nm TiN layer form the topgate stack. Only 8 masks were used through M1. The recessed channel & bottomgate structure maintains an ultra-thin 20nm channel while allowing 200nm thick SD's for lower SD series resistance. These unique device structural features make the FlexFET transistor highly scalable.

Results & Discussion

Fig. 5 shows the IDG-CMOS propagation delay as a function of both NBG and PBG. The slowest condition (44ps) is for both bottomgates "off" at -0.5V, while the fastest condition (26ps) is for both bottomgates "on" at +0.5V. The table in Fig. 6 shows that the threshold voltage can be varied by 0.85V, resulting in the standby current reduced to 10^{-16}A/um at -0.5V, while the drive current is increased to 850uA/um at +0.5V. Essentially all static & dynamic digital circuits, as well as analog/MS/RF circuits can benefit from IDG operation.

References

[1] H. Mahmoodi, et. al., 2004 IEEE SOI Conf., pp. 67-68, Oct. 2004.
[2] S. Parke, et. al., 2004 IEEE SOI Conf., pp. 104-105, Oct. 2004.
[3] S. Parke, et. al., 2005 IEEE WMED, pp. 35-37, April 2005.
[4] S. Parke, et. al., 2005 IEEE Aerospace Conf., 7.0902, Mar. 2005.
[5] L. Mathew, et. al., 2004 IEEE SOI Conf., pp. 187-189, Oct. 2004.

0-7803-9212-4/05/$20.00 ©2005 IEEE

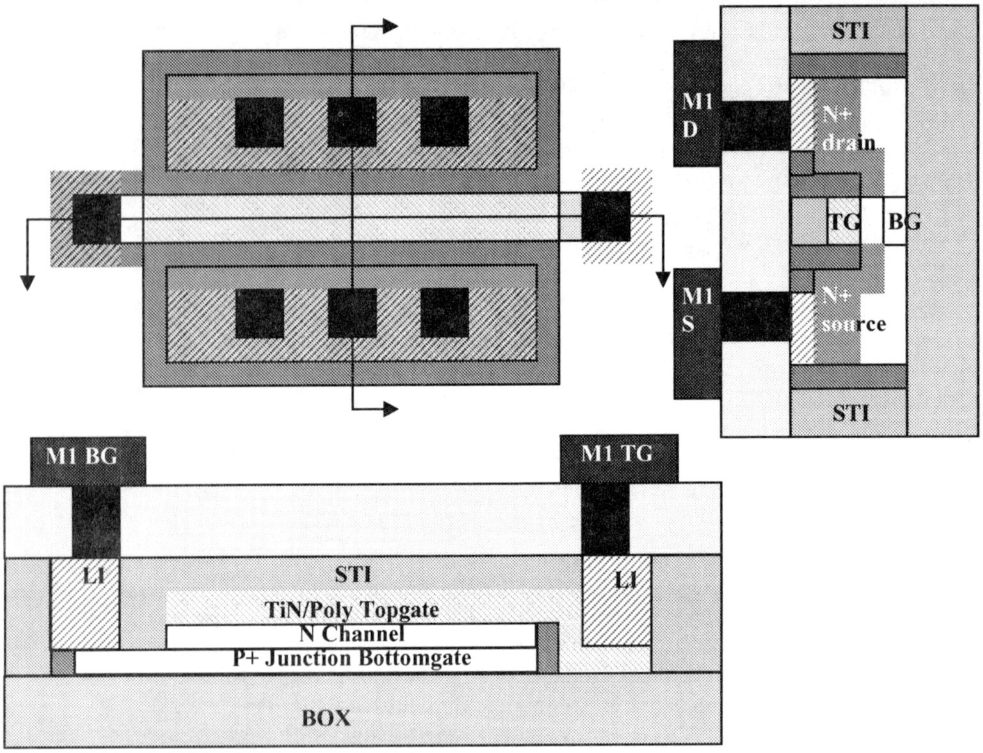

Fig. 1 FlexFET NMOS Top View and X & Y Section Drawings

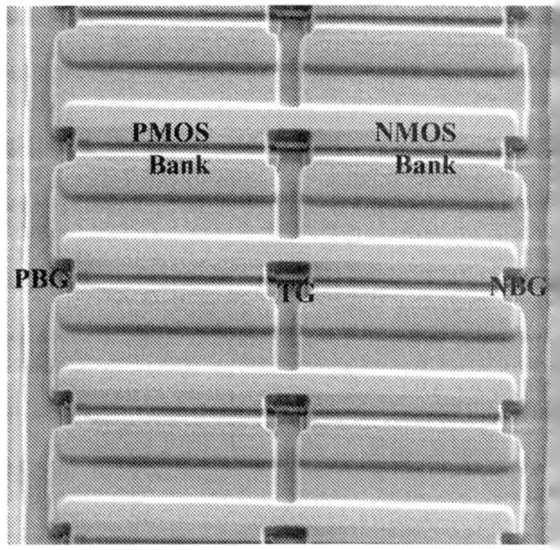

Fig. 2 FlexFET IDG-CMOS Inverter Oscillator SEM

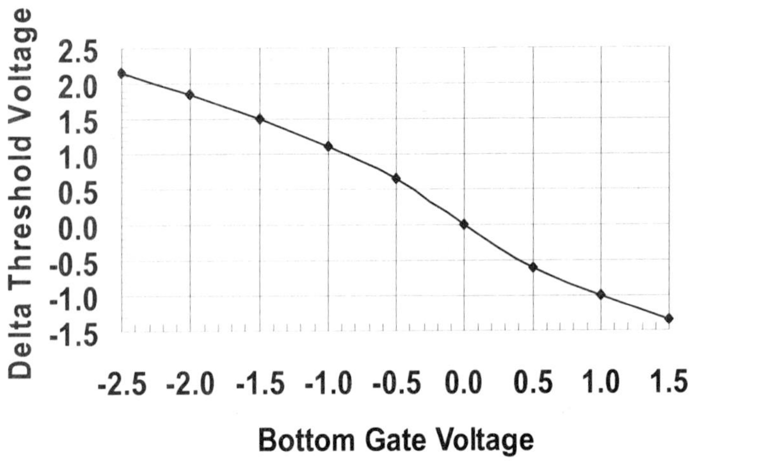

Fig. 3 Measured NMOS Vt shift vs. BG Voltage, showing >0.85V/V

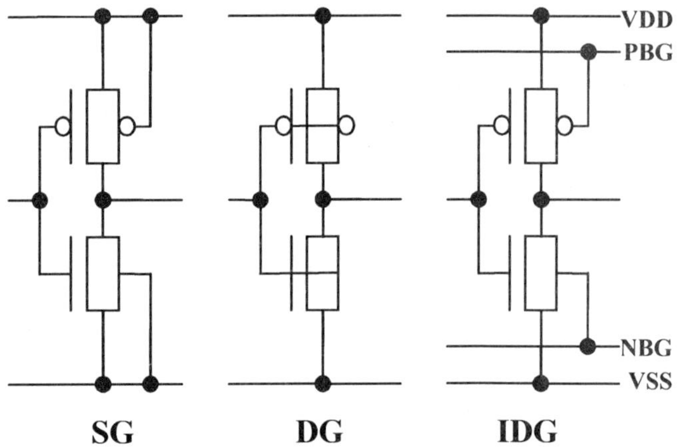

Fig. 4 SG, DG, IDG Ring Oscillator Schematics

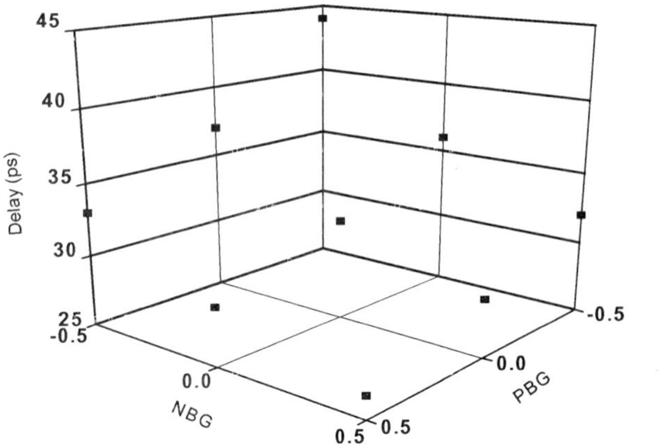

Fig. 5 IDG-CMOS Propagation Delay vs. BG Voltages

		Slow ULP	Nominal	Fast HP
NBG & PBG	V	-0.5	0	+0.5
VT	V	0.85	0.425	0
Ioff	uA/um	1.0E-10	1.0E-05	1
Pstandby	uW/um	2.5E-10	2.5E-05	2.5
Ion (NMOS)	uA/um	520	700	850
Pactive	uW/um	191	263	323
Prop Delay	psec	44	32	26
PDP	fJ/um	8.4	8.4	8.4

Fig. 6 180nm FlexFET IDG-CMOS Performance Table

2005 IEEE International SOI Conference

Independent Gate Skewed Logic in Double-Gate SOI Technology

Tamer Cakici, Hamid Mahmoodi, Saibal Mukhopadhyay and Kaushik Roy
School of Electrical and Computer Engineering, Purdue University, West Lafayette, IN, USA

Introduction:

Due to excellent control of short channel effects, Double-Gate SOI (*DGSOI*) devices have emerged as the device of choice for circuit design in sub-50nm regimes [1]. Reduced short channel effects in *DGSOI* alleviate the issue of leakage power. However, the overall power including the dynamic power still imposes a great challenge in future high performance designs. Therefore, there is a need for innovative low power techniques in *DGSOI* technologies. Double-gate SOI technologies with Independent front and back Gate control (*IGSOI*) have been developed recently [2,3]. We exploit *IGSOI* in noncritical paths to reduce the strength of the non-critical transistors below the minimum size. Such a mode of operation reduces capacitive loading on critical transistors, allowing them to be smaller in size for any target delay. In this paper, we propose an independent gate skewed logic style which achieves high performance and low power by exploiting the dynamic V_t and lower capacitive loading properties of *IGSOI* devices.

Device Characteristics:

In this work, we have designed symmetric double-gate SOI devices for 50nm and 35nm gate lengths (Fig. 1). The devices use metal gates with workfunctions that are within the range of reported manufacturable workfunctions [4]. The device and circuit simulations are done using TAURUS-DS [5]. In thin SOI devices, the quantization of electron energy in the body has a strong impact on V_t and gate capacitance, and hence, we have considered this quantum effect in our simulations.

Independent Gate Skewing:

In a precharge-evaluate logic, relative strengths of pull-up and pull-down transistors are modified in favor of a particular transition direction to achieve better delay. This modification, known as "skewing", increases I_{ON} of critical transistors and changes trip points in favor of the critical transition. Hence, skewing reduces the critical delay. We propose a new way of skewing through use of *IGSOI* (IG skewing in Fig. 2). In Fig. 2, significant skewing is achieved by connecting the back gate of the *PMOS* transistor to V_{DD}. Further skewing can be achieved by upsizing critical transistors. Skewing changes the trip point of a gate in favor of the critical transition.

As observed from Fig. 2, *IG* skewing changes the trip point in favor of the preferred transition (evaluation transition, in this case high-to-low at the output). For any given *NMOS* size, the *IG* skewed inverter has a smaller trip point. In other words, in the *IG* skewing case, a target trip point (skew) can be achieved with a smaller *NMOS* transistor. Hence, compared to *DG* skewing, *IG* skewing results in less switching capacitance, and therefore, less switching energy for the inverter at any target delay (as shown in Fig. 3(b)). The issue, however, is that as the evaluation delay reduces by skewing (Fig. 3(c)), the delay of the other transition direction (precharge delay) increases (Fig. 3(d)). Depending on the application, the precharge delay may not be acceptable beyond some skew, limiting delay reduction on the critical path. In the *IG* skewing, however, skewing can be dynamically changed for any preferred transition (dynamic *IG* skewing in Fig. 3(a)). For the inverter example in Fig. 3, in dynamic *IG* skewed gate, the back gate of the *PMOS* is connected to the clock signal (*CLK*) instead of V_{DD}. *CLK* is set to high and low, if the gate is in evaluation and precharge phase, respectively. Dynamic *IG* skewing significantly reduces the precharge delay (Fig. 3(d)). By dynamic *IG* skewing, the precharge delay is reduced to the precharge delay of the *DG* skewed inverter at any *NMOS* size.

This work is supported in part by Semiconductor Research Corp. (#1078.001), Gigascale System Research Center (MARCO), Intel and IBM Corp.

In the *IG* skewed circuit, the skewing is achieved by control of the strength (effective width and V_t) of the noncritical transistor through back gate bias. The strength of a transistor can be represented by its I_{ON}. The ratio of I_{ON} of double-gate drive (I_{ONDG}) to that of single-gate drive (I_{ONSG}) for a double gate transistor is greater than 2 since $|V_{tSG}|>|V_{tDG}|$, as shown in Fig. 4. This ratio increases by device scaling due to the increase in V_{dd}-$|V_t|$ ratio of *DG* to *SG* as the device is scaled down. Hence, the effectiveness of the proposed *IG* skewing technique improves with technology scaling. An increase in the I_{ON} ratio provides an opportunity to downsize the critical transistor in *IG* skewed circuits. Smaller sized critical transistors result in reduced power dissipation.

Independent Gate Skewed Logic Style:

Based on the proposed *IG* skewing, a new skewed *CMOS* logic style is presented to improve performance and reduce power dissipation. We have used a chain of four inverters to demonstrate the proposed skewed logic style as shown in Fig. 5. As the gates are more skewed, the evaluation delay reduces; however, the precharge delay increases. The precharge delay should be as small as the evaluation delay in order to maximize the performance. This can be achieved by selectively inserting *PMOS* precharge transistors on logic stages along with *NMOS* footers to avoid short circuit current (Fig. 5). As the evaluation delay decreases by skewing, the precharge delay has to catch up, which in turn requires upsizing the precharge transistors and/or inserting precharge transistors on more number of stages. To minimize power dissipation for any target delay, the number, location, and size of precharge transistors should be optimally chosen. The overall delay can be reduced by more skewing (less evaluation delay) and insertion of larger and/or more precharge transistors (less precharge delay). This results in an increase in the overall capacitance of the circuit and hence, larger energy dissipation.

In the *IG* skewed logic (Fig. 5(b)), back gates of noncritical transistors are connected to clock signals (*CLK* or *CKB*) to dynamically change the skew of the gates for the preferred transition in both evaluation and precharge phases. In the evaluation phase when *CLK* is high, back gates of *PMOS* and *NMOS* noncritical transistors are connected to V_{DD} and *GND*, respectively, providing significant skewing. Moreover, in the precharge phase when *CLK* is low, the back gates of these transistors are connected to *GND* and V_{DD}, respectively, resulting in faster precharging. Faster precharging helps in reducing the size of the inserted precharge transistors. The *IG* skewed logic achieves the same skewing as *DG* skewed logic with smaller sizes (less gate capacitance) for critical transistors, and therefore less power dissipation for a given target delay (Fig. 6). In the 50nm node, the *IG* skewed circuit shows savings of 22% in total energy per switching. Moreover, the delay is reduced by 8.5%. Since the *DG* to *SG* I_{ON} ratio increases with technology scaling (Fig. 4), the energy savings of this logic style also increases with scaling (Fig. 6). In the 35nm node, the proposed technique shows a reduction of 25% in total energy per switching and 9.5% in delay.

Conclusion:

Independent gate control of double gate SOI devices [2,3] can be effectively exploited to improve performance and reduce power in sub-50nm circuits. In this paper, we have proposed a skewed logic style using independent gate operation of double gate SOI devices.

References:

[1] *L. Chang et al.*, Proc. of IEEE, vol. 91, pp. 1860–1873, Nov. 2003
[2] *D. Fried et al.*, IEEE EDL, vol. 24, pp. 592-594, Sep. 2003
[3] *L. Mathew, et al.*, IEEE SOI Conf., pp. 187-189, Oct. 2004
[4] *C. Cabral et al.*, Symp. on VLSI Tech., pp.184-185, 2004
[5] *Taurus-DS*, V2004.09, Synopsys Inc.

0-7803-9212-4/05/$20.00 ©2005 IEEE

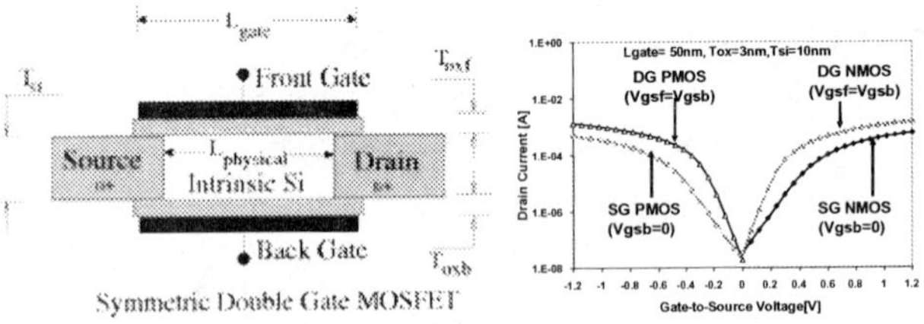

NMOS			PMOS		
Tech. Node	50nm	35nm	Tech. Node	50nm	35nm
L_{gate} (nm)	50	35	L_{gate} (nm)	50	35
$L_{physical}$ (nm)	40	28	$L_{physical}$(nm)	40	28
$T_{oxf}=T_{oxb}$ (nm)	3	2	$T_{oxf}=T_{oxb}$(nm)	3	2
T_{si} (nm)	10	7	T_{si}(nm)	10	7
Vdd (V)	1.2	1.1	Vdd (V)	1.2	1.1
Workfunc. (eV)	4.55	4.51	Workfunc. (eV)	4.89	4.93
DG Vt (V)	0.28	0.3	DG Vt (V)	0.32	0.34
SG Vt (V)	0.37	0.39	SG Vt (V)	0.41	0.43
I_{ON} (μA/μm)	1470	1850	I_{ON} (μA/μm)	1290	1630
I_{OFF} (nA/μm)	31.9	30.1	I_{OFF} (nA/μm)	17.9	17.5
Sub. Swing (mV/decade)	70	70	Sub. Swing (mV/decade)	74	72
DIBL (mV/V)	55	53	DIBL (mV/V)	57	56

Fig. 1: Double Gate SOI devices and their characteristics

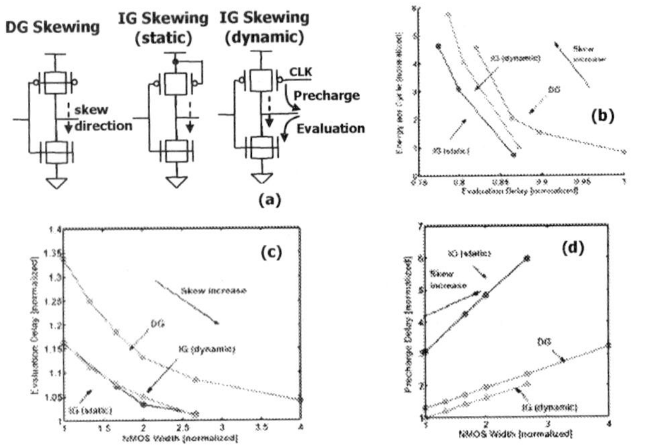

Fig. 3: Energy and delay advantage of Independent Gate (IG) skewing compared to Double Gate (DG) skewing

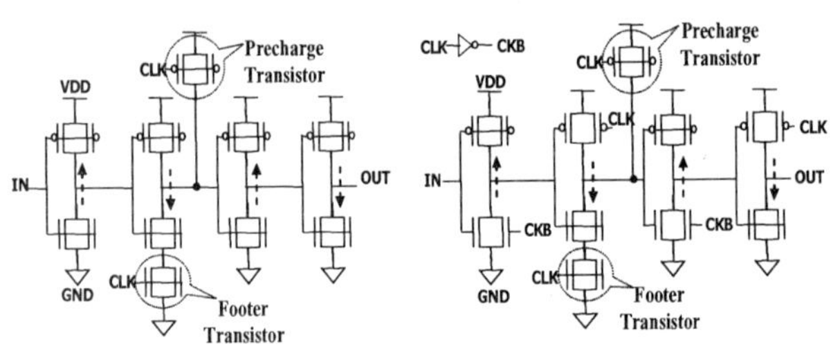

Fig. 5: Schematics of (a) DG and (b) proposed IG skewed logic styles

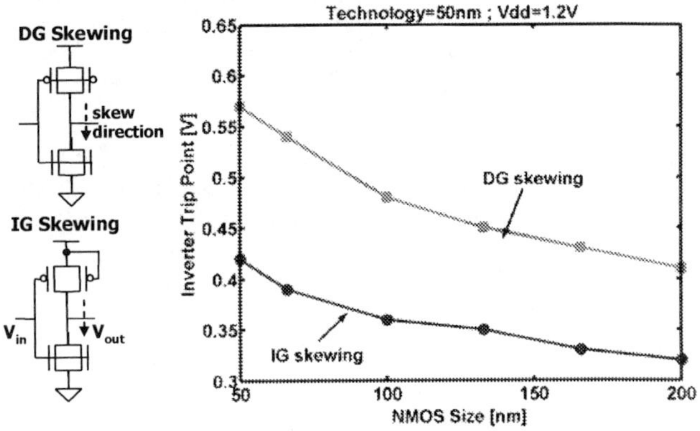

Fig. 2: Trip point of an inverter with Double Gate (DG) skewing and proposed Independent Gate (IG) skewing (PMOS width is kept minimum)

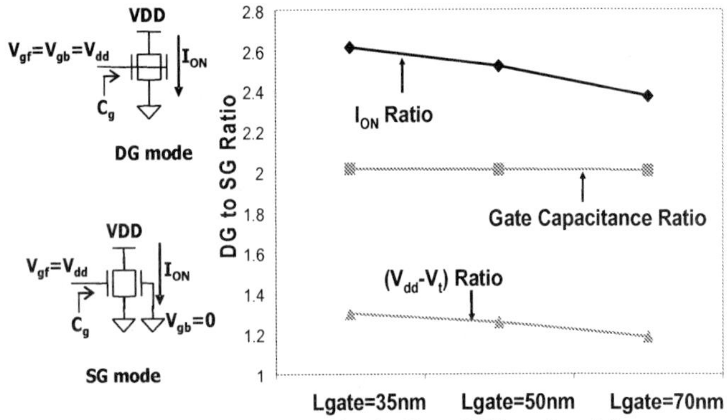

Fig. 4: Double Gate (DG) to Single Gate (SG) ratio of I_{ON}

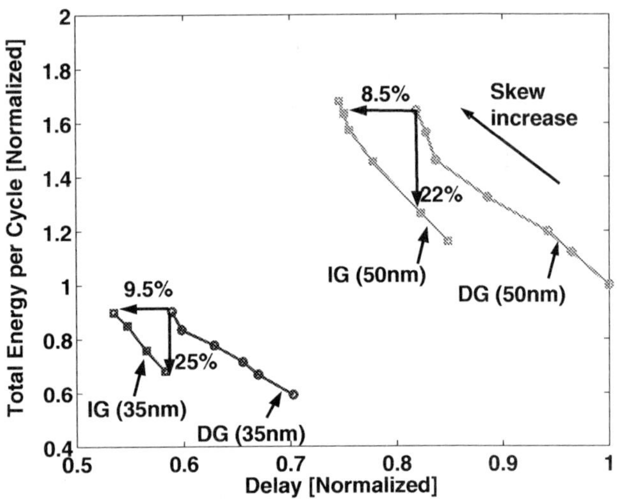

Fig. 6: Comparisons of DG and IG skewed logic styles (clock energy is included in total energy per cycle)

High-Density Logic Techniques with Reduced-Stack Double-Gate MOSFETs

Meng-Hsueh Chiang, Keunwoo Kim*, Ching-Te Chuang*, and Christophe Tretz**

National Ilan University, I-Lan 260, Taiwan, ROC

*IBM T. J. Watson Research Center, Yorktown Heights, NY 10598, USA

**IBM Engineering & Technology Services, San Jose Design Center, CA 95193, USA

Email: mhchiang@niu.edu.tw, Tel: +886-3-935-7400 ext. 653

Introduction

Because of the better short-channel effects (SCEs), double-gate (DG) technology can overcome the scaling limit of single-gated bulk-Si or SOI devices to extend CMOS scaling [1]. DG devices offer additional design flexibility where the two gates can be tied together and/or independently biased [2]. The separate biasing in DG device has recently received lots of interest as it is technologically feasible [2, 3] and advantageous for many applications [4-7]. We propose novel DG logic circuit schemes which significantly reduce the chip area and improve performance by halving the number of transistors in the stacked devices as well as parallel devices for implementing logic functions. The performance, power, and design trade-offs for logic gates are analyzed via mixed-mode two-dimensional numerical simulations [8].

Reduced Stacking in High-V_T DG Device

Our previous work [7] reduces the number of "parallel" transistors by utilizing the front- and back-channel as independently biased gates. The proposed scheme in this paper reduces the number of stacked transistors, as shown in Fig. 1 for a 2-way NAND. The symmetrical DG nFET in this study has 25 nm effective channel length, lightly-doped ($N_{body} = 10^{15}$ cm^{-3}), thin silicon film (t_{Si} = 9 nm), scaled oxide thickness ($t_{oxf} = t_{oxb}$ = 1 nm), and near-mid-gap gate work function ($\Phi_{GfS} = \Phi_{GbS}$ = 4.85 eV). Fig. 2 shows the cross section of the DG device. Fig. 3 shows MEDICI-predicted I_{DS}-V_{GS} characteristics of a high-V_T metal-gate nFET. In single-gate (SG) mode, i. e., when one gate is ground and the other gate of the nFET is biased at 0.6 V, its current is still only 4×10^{-7} A/μm, and the device is still "Off" for logic by any practical definition. When both gates are turned on (DG mode), the V_T decreases and the subthreshold slope improves significantly due to the extended gate-to-gate coupling [9], and the drive current, $I_{on(DG)}$ (I_{on} in DG mode) is almost three-order-of-magnitude higher than $I_{on(SG)}$ (I_{on} in SG mode). Thus, when both gates are turned on, the device conducts a large current and is "On" for logic. This indicates that a single high-V_T DG device could offer the functionality for two stacked devices.

Notice that in a regular-V_T symmetrical DG device, the V_T difference between DG and SG modes are relatively small. This is because once the gate voltage ramped above V_T, the channel is "On", and the inversion layer pins the surface potential and shields the gate-to-gate coupling [9]. Thus, the current in DG mode is only more than twice higher than that of SG mode, as shown for a regular V_T pFET in Fig. 4. Note that the DG pFET and the aforementioned DG nFET have the same near-mid-gap gate, implying technology simplicity. However, if we directly set the targeted I_{off} of the same 4×10^{-7} A/μm for nFET in nominal DG operation without using the V_T modulation technique via gate-to-gate coupling, it will result in imbalanced V_T's, thereby necessitating two different mid-gap gates for nFET and pFET, respectively.

For 2-way NAND (Fig. 1(c)), the two stacked nFETs in the conventional scheme can be replaced by a single high-V_T DG nFET with independently biased front/back gates. The regular-V_T DG pFET works as two parallel pFETs whereas the high-V_T nFET works as two stacked nFETs. When only one input is high, the nFET is still "Off" with I_{off} of 4×10^{-7} A/μm. When both inputs are high, the nFET works in DG-mode with large drive current and is "On". Since the nFET only switches when both inputs are high, the performance of NAND benefits from the large drive current in DG mode. 2-way NOR has the complementary structure and V_T's. The stacked pFET would be implemented using a single high-V_T double-gate pFET, and the parallel nFETs would be implemented using a single regular-V_T nFET. The proposed scheme can be implemented in any symmetrical double-gate technologies with two independently controlled gates and properly chosen multiple V_T's.

MEDICI Simulation Results and Discussion

Fig. 5 shows MEDICI-predicted waveforms for the proposed 2-way NAND with 2 transistors (Fig. 1(c)). The circuit correctly performs the NAND function. Fig. 6 shows MEDICI-predicted average delay versus V_{DD} of the proposed 2-way NAND (Fig. 1(c)) loaded with another NAND, as compared with the conventional scheme. Both the proposed and conventional schemes use the same near-mid-gap gate for nFET and pFET, as the gate work function for the conventional scheme is adjusted to achieve balanced V_T's. The proposed scheme offers 5% faster performance at V_{DD} = 0.6 V with over 2x reduction in area/capacitance due to halved transistor count and less isolation/wire. This lower capacitance speeds up the circuit. Fig. 7 shows MEDICI-predicted average delay versus V_{DD} of the proposed (unloaded) 4-way NAND, as compared with the conventional counterpart. The proposed scheme offers 19% faster performance at V_{DD} = 0.6 V. The improvement is very significant for the unloaded case due to much lower intrinsic resistance and capacitance of the proposed scheme. As the stacked heights/transistors or inputs increase, the performance advantage of the proposed scheme becomes larger due to the reduction of "ON" series resistance, as shown in Fig. 8. Due to the reduced stacking, even 6-way NAND (NAND6) could achieve acceptable performance and reliable functionality. Schematics of 6-way logic are described in Fig. 9.

Conclusions

We have presented a high-density DG logic circuit technique exploiting the unique V_T modulation effect through the extended gate-to-gate coupling in high-V_T symmetrical DG devices. The scheme reduces the number of stacked transistors (hence area/capacitance and standby/dynamic power), and improves performance. The performance improvement and power reduction are evaluated/validated using mixed-mode two-dimensional numerical simulations.

0-7803-9212-4/05/$20.00 ©2005 IEEE

Acknowledgements: M.-H. Chiang was partially supported by the National Science Council of Taiwan, ROC under contract NSC93-2215-E-197-002, and K. Kim and C. T. Chuang were partially supported by the DAPAR contract NBCH30390004 for this work.

References: [1] E. J. Nowak et al., IEEE Cir. Dev. Mag., p. 20, Jan/Feb 2004. [2] L. Mathew et al., IEEE SOI Conf. p. 187, 2004. [3] Y. X. Liu et al., IEDM, p. 986, 2003. [4] S. Mitra et al., IEEE SOI Conf., p. 177, 2003. [5] H. Mahmoodi et al., IEEE SOI Conf., p. 67, 2004. [6] A. Kumar et al., Proc. IEEE SOI Conf., p. 119, 2004. [7] M.-H. Chiang et al., Proc. IEEE SOI Conf., p. 122, 2004. [8] Taurus-MEDICI, Synopsis, Inc., 2003. [9] H.-K. Lim and J. G. Fossum, IEEE T-ED, vol. 30, pp. 1244, 1983.

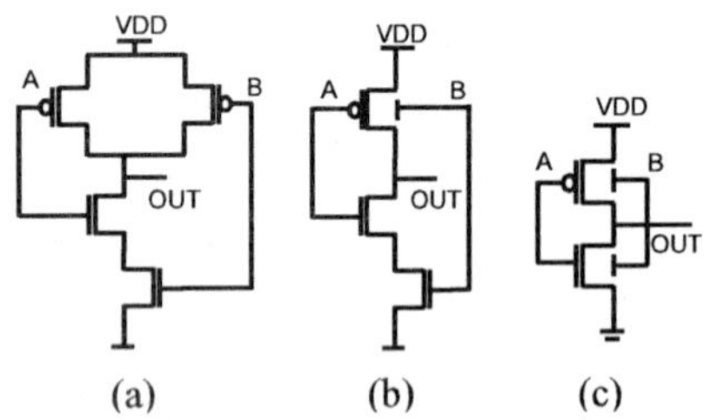

Fig. 1: Schematics of 2-way NAND for the (a) conventional, (b) halved "parallel" devices [7], and (c) proposed scheme. In the proposed scheme (c), the single nFET has high-V_t to function as "stacked" NAND logic, while pFET has regular-V_t to function as two parallel devices.

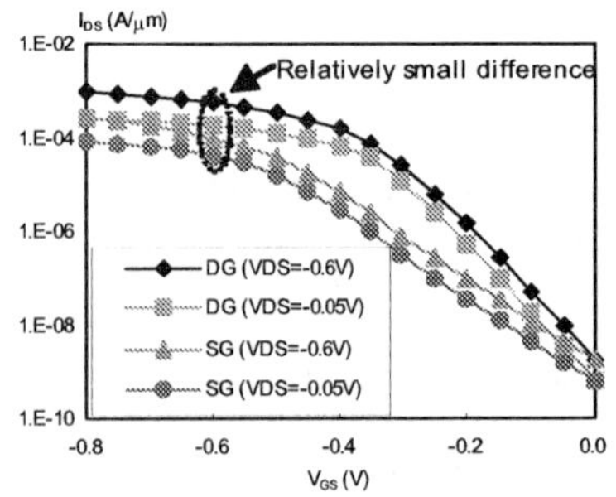

Fig 4: MEDICI-predicted I_{DS}-V_{GS} (at V_{DS} = -0.05 V and -0.6 V) of regular-V_T metal-gate pFET with lightly-doped body in Singe-Gate (SG) mode and Double-Gate (DG) mode of operation.

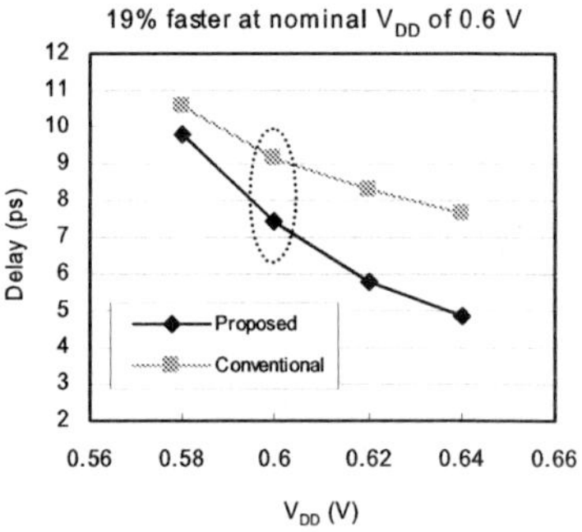

Fig. 7: MEDICI-predicted delay versus V_{DD} for the proposed unloaded 4-way NAND, compared with conventional scheme.

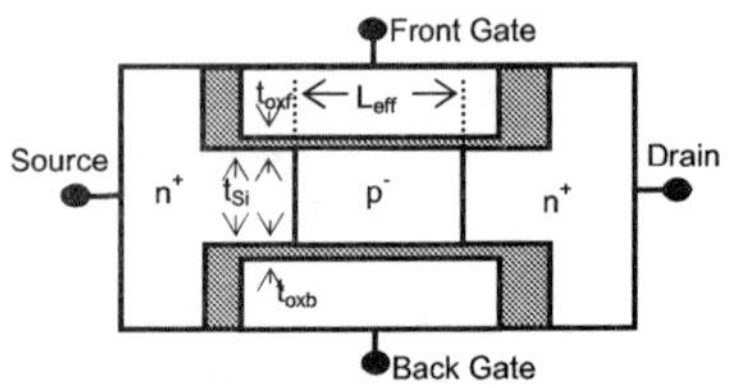

Fig. 2: Simulated symmetrical DG nFET: L_{eff} = 25 nm, $N_{body} = 10^{15}$ cm^{-3}, t_{Si} = 9 nm, $t_{oxf} = t_{oxb}$ = 1 nm), and $\Phi_{Gfs} = \Phi_{Gbs}$ = 4.85 eV.

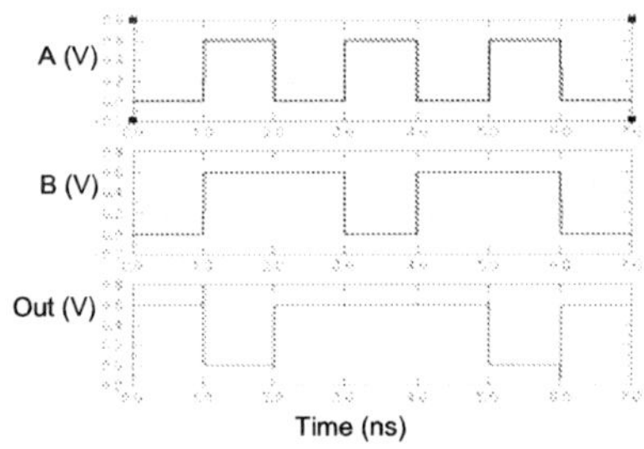

Fig. 5: MEDICI-simulated input and output waveforms for the proposed 2-way NAND scheme in Fig. 1 (c) (V_{DD} = 0.6 V).

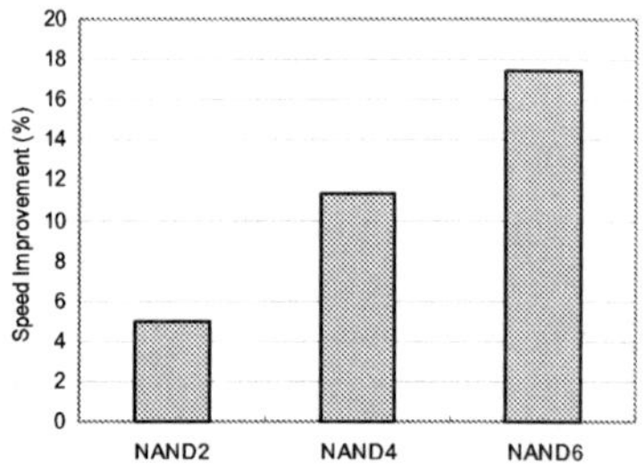

Fig. 8: Speed improvement (%) of the proposed scheme (loaded case) over conventional scheme for NAND2, NAND4, and NAND6 (V_{DD} = 0.6 V).

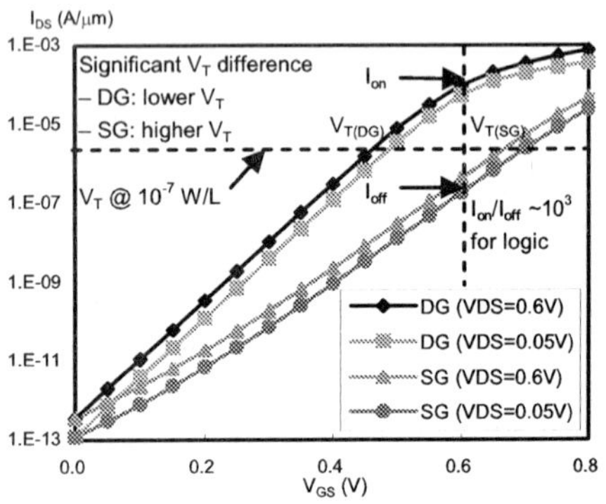

Fig. 3: MEDICI-predicted I_{DS}-V_{GS} characteristics (at V_{DS} = 0.05 V and 0.6 V) of high-V_T metal-gate nFET with lightly-doped body in Singe-Gate (SG) mode and Double-Gate (DG) mode of operations.

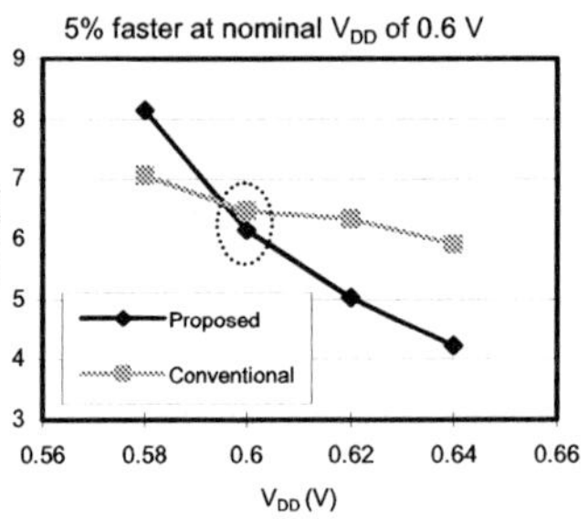

Fig. 6: MEDICI-predicted delay versus V_{DD} for the proposed (Fig. 1(c)) 2-way NAND loaded with another NAND, compared with conventional scheme (Fig. 1(a)).

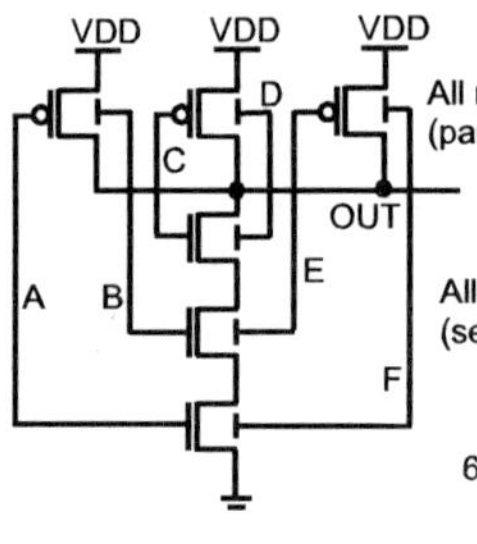

Fig. 9: Schematics of (proposed) 6-way NAND. 6-way NOR can be also implemented with 3 high V_T (series) pFETs and 3 regular V_T (parallel) nFETs.

A *Back-to-Face* Silicon Layer Stacking for Three-Dimensional Integration

C. S. Tan, K. N. Chen, A. Fan, and R. Reif
Microsystems Technology Laboratories, Massachusetts Institute of Technology
60 Vassar Street, 39-625, Cambridge, MA02139
Tel: (617) 253-0717; E-mail: tancs@mtl.mit.edu

Introduction

Three-dimensional (3-D) integration, in the form of a vertical stack of several interconnected device layers, has many performance, integration and cost advantages [1]. There are research efforts on a few possible technology options to fabricate a 3-D structure, either using recrystallization [2], [3] or wafer bonding [4]-[6]. While recrystallization finds niche applications in memory devices, thermal budget imposed by the first layer and imperfect crystallization of the second layer make this technique less attractive. Another approach to stack thin silicon device layers that is under intense research is the wafer bonding and etch-back approach. Wafer bonding can be achieved using a conductive [4] or a dielectric [5], [6] bonding medium. Silicon device layers can be stacked either in a *back-to-face* or *face-to-face* fashion. In *face-to-face* stacking, donor wafer is bonded to a substrate wafer in a face-to-face manner and etched back to the required thickness. In *back-to-face* stacking, however, the donor wafer is bonded to a temporary handle wafer for mechanical support and etched back to the required thickness. Stacking is completed by bonding the thinned silicon device layer (that is attached to the handle wafer) to a substrate wafer in a back-to-face manner and followed by handle wafer release. In this letter, we describe a *back-to-face* silicon layer stacking method using a combination of low temperature oxide wafer bonding (as a temporary bond between the donor wafer and the handle wafer) and Cu wafer bonding (as a permanent bond between the thinned silicon layer and the substrate wafer).

Wafer Preparation

The experiment was performed on blanket silicon wafers. All wafers used in this experiment were n-type 150 mm Si-(100) wafers. Two sets of wafers were prepared: One group was the handle wafers and the other was the SOI dummy wafers (as donor and substrate wafers). The handle wafers were covered with 5000 Å of thermal oxide for protection against chemical attack during donor wafer etch-back. Wet silicon etchants such as tetraethylammonium hydroxide (TMAH) is known to have excellent selectivity towards thermal oxide [7]. SOI dummy structures were prepared by growing 5000Å of thermal oxide as buried oxide (BOX) on silicon wafers followed by the deposition of 4000 Å of undoped poly-silicon at 620 °C. Since the donor and substrate wafers already have devices and interconnects fabricated on them in actual process, all subsequent process temperatures were limited to 400 °C and below. The donor wafers were coated with 1 μm of tetraethyl-orthosilicate (TEOS) source plasma-enhanced chemical-vapor-deposition (PECVD) oxide at 350 °C. PE-TEOS oxide is an attractive choice for interlayer dielectrics (ILD) and passivation layer because of its high deposition rate and low process temperature.

Low Temperature Oxide Wafer Bonding

To begin the silicon layer transfer and stacking process, the donor wafers were bonded to the handle wafers for mechanical support and ease of wafer handling. While epoxy [8] or adhesive material [9] can be used for this type of bonding, we have avoided process complexity as a result of the use of new materials and resorted to direct oxide to oxide wafer bonding which is more CMOS friendly. While the protective thermal oxide on the handle wafers possesses suitable properties for wafer bonding, PE-TEOS oxide on the SOI dummy wafers prepared at low temperature requires additional surface preparations before wafer bonding can be initiated [10], [11]. PE-TEOS oxide on SOI dummy wafers were densified in atmospheric N_2 ambient at 350 °C for 5 hours to drive away all trapped gas molecules. This step allowed degassing from the porous SiO_2, which was detrimental to the bonding, to take place prior to bonding. Surface smoothness is a very critical factor that determines successful wafer bonding. It is well known that PE-TEOS oxide has very rough surface. To reduce the surface roughness, they were chemical-mechanically polished (CMP) for 3 min after densification. The surface roughness improved from 8.53 nm to 0.48 nm as previously reported [11]. All handle wafers and donor wafers were activated by oxygen plasma followed by a 10 min piranha (H_2O_2:H_2SO_4 = 1:3) clean. This cleaning was necessary to terminate the wafer surfaces with hydroxyl (OH) groups to initiate hydrogen bonds during wafer bonding. Activated handle wafers were aligned to donor wafers in an EV Group aligner and the pairs were bonded as schematically shown in Fig. 1(a). Bonding was performed at room temperature under a contact force of 1 kN for 2 min. After bonding, the bonded wafer pairs were annealed at 300 °C in atmospheric N_2 ambient for 3 hours to enhance the bonding strength. The bonding

0-7803-9212-4/05/$20.00 ©2005 IEEE

strength is estimated to be about 2300 mJ/m^2 using the Maszara's crack opening method [12].

Donor Wafer Thinning

The next step in the flow was donor wafer thin-back as described in Fig. 1(b). The substrate of the bonded donor wafer can be thinned back using a combination of mechanical grinding and TMAH strip. Since mechanical grinding has no selectivity between silicon and oxide, the grind back was stopped at about 75 μm away from the BOX layer. The remaining 75 μm thick of silicon layer can be stripped easily using 12.5 *wt* % TMAH at 85 °C for about 100 min. Since TMAH has excellent selectivity towards oxide, the etching stops on the BOX layer. Therefore SOI wafers are attractive choice in this type of silicon layer stacking as the BOX layer serves as excellent etch stop layer.

Permanent Cu-to-Cu Wafer Bonding

The thinned silicon layer on the handle wafer was then bonded to a substrate wafer using Cu as the bonding medium as shown in Fig. 1(c). 50 nm of Ta and 300 nm of Cu were deposited on the thinned layer and the substrate wafer in an e-beam system. Ta acted as diffusion barrier. Thermo-compression bonding was done at 400 °C under a contact force of 4 kN for 1 hr in vacuum. The bonded pair was then annealed in atmospheric N_2 ambient at 400 °C for 1 hr to promote Cu inter-diffusion and grain growth for higher bonding strength. Comprehensive studies on Cu wafer bonding can be found in [13] and references therein.

Handle Wafer Release

Fig. 1(d) shows the final step in the flow, *i.e.*, the handle wafer release. This can be done in the following two ways:

1) Mechanical Grinding and TMAH Strip: The most direct way to remove the handle wafer is by mechanical grinding and TMAH strip. The handle wafer was ground to about 75 μm from the bonding interface. The remaining 75 μm of silicon layer was stripped in 12.5 *wt* % TMAH solution that stopped on the protective oxide layer. Fig. 3 is a cross-sectional SEM image showing all final layers on the stack. The close-up views of the bonding interfaces are also shown. This SEM image confirms the layer stacking described above. The resulting wiggling interface between the bottom poly-Si and the Ta layer was due to the surface roughness of the unpolished poly-Si layer. No interfacial void is observed at the bonding interface between the PE-TEOS and thermal oxides. Substantial inter-diffusion and grain growth in the Cu can be observed in the close-up SEM image at the bottom right. Note that the original Cu bonding interface has disappeared and a homogeneous of bonded Cu layer is obtained.

2) Hydrogen Induced Wafer Splitting: To minimize process damages to the Cu bonding interface as a result of mechanical grinding, hydrogen induced wafer splitting can be used to release the handle wafer. H_2^+ ions at a dose of 5 x 10^{16} cm^{-2} were implanted at an energy of 150 keV into the handle wafer prior to wafer bonding. Upon heating at the appropriate temperature, lateral micro-cracks induced by hydrogen will release the handle wafer at the location around the peak of hydrogen implant profile. Fig. 3 is the SIMS profile of the implanted hydrogen with a peak about 300 nm into the silicon handle wafer from the oxide-silicon interface. Since this process is temperature sensitive, all heat treatments prior to layer splitting were kept at 300 °C or below. Therefore, Cu wafer bonding was done at 300 °C for 1 hr to prevent premature wafer de-bonding. Hydrogen induced wafer splitting was achieved with an anneal at 400 °C for 1 hr. This step also enhanced the Cu layer bonding strength. Fig. 4 is SEM image of the stack after the handle wafer separation. We have used substrate wafer with 1 μm of PE-TEOS oxide on top. The remaining 300 nm of Si left over by handle wafer on the top of the stack can be stripped using a short TMAH dip. The close-up view shows a homogeneous Cu layer with grains that extend from one end to the other end suggesting that significant grain growth has taken place even though the thermal budget was reduced to provide a temperature window for Cu wafer bonding and hydrogen induced layer splitting.

Conclusion

We have successfully demonstrated a back-to-face ultra-thin silicon layer stacking based on low temperature wafer bonding and etch-back. This type of silicon layer stacking can be expanded to wafers with device and interconnect layers to fabricate three-dimensional integrated circuits (3-D ICs). Electrical connection between layers can be achieved by interlayer vertical via formed by bonded Cu layers.

[1] K. Banerjee, *et al.*, Proc. IEEE, 89, pp. 602-633, 2001.

[2] V. Subramaniam, *et al.*, Electron Device Lett., 20, pp. 341-343, 1999.

[3] V. W. C. Chan, *et al.*, IEEE Trans. Electron Devices, pp. 1394-1399, 2001.

[4] A. Fan, *et al.*, Electrochem. Soild-State Lett., 2(10), pp. 534-536, 1999.

[5] J.-Q. Lu, *et al.*, Proc. IEEE Int. Interconnect Tech. Conf., pp. 219-221, 2001.

[6] K. Warner, *et al.*, Proc. IEEE SOI Conf., pp. 123-124, 2002.

[7] P. H. Chen, *et al.*, Sens. Actuators A, 93, pp. 132-137, 2001.

[8] S. van der Groen, *et al.*, Int. Conf. Solid-State Sensors and Actuators, pp. 629-632, 1997.

[9] Y. Hayashi, *et al.*, Symp. VLSI Tech. Dig., pp. 95-96, 1990.

[10] C. S. Tan, *et al.*, Appl. Phys. Lett., vol 82, no 16, pp. 2649-2651, 2003.

[11] C. S. Tan, *et al.*, Electrochem. Solid State Lett., 8(1), pp. G1-G4, 2005.

[12] W. P. Maszara, *et al.*, J. Appl. Phys., vol. 64, no. 10, pp. 4943-4950, 1988.

[13] K. N. Chen, *et al.*, Electrochem. Solid State Lett., 7(1), pp. G14-G16, 2004.

* *This work was supported in part by the Defense Advanced Research Projects Agency (DARPA) on Three Dimensional Integration (3DI) which is funded at the Massachusetts Institute of Technology, through a subcontract from IBM under contract number N66001-04-C-8032. C. S. Tan thanks Applied Materials for the award of a Graduate Fellowship.*

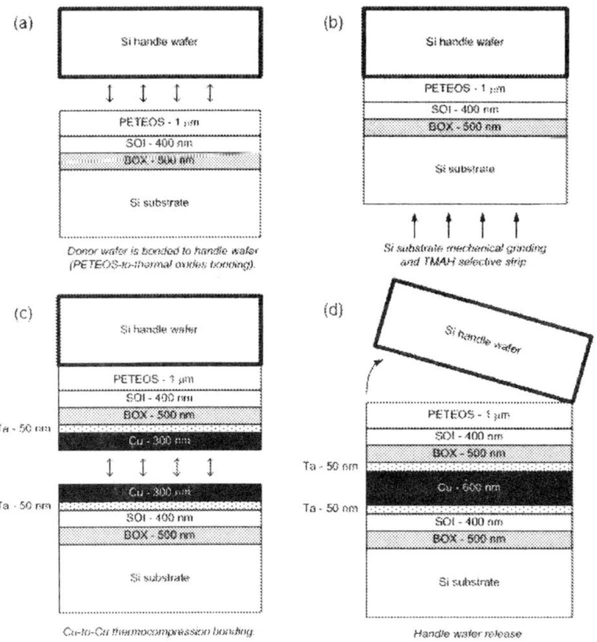

Fig. 1. Schematics of *back-to-face* silicon layer transfer and stacking for three-dimensional integration.

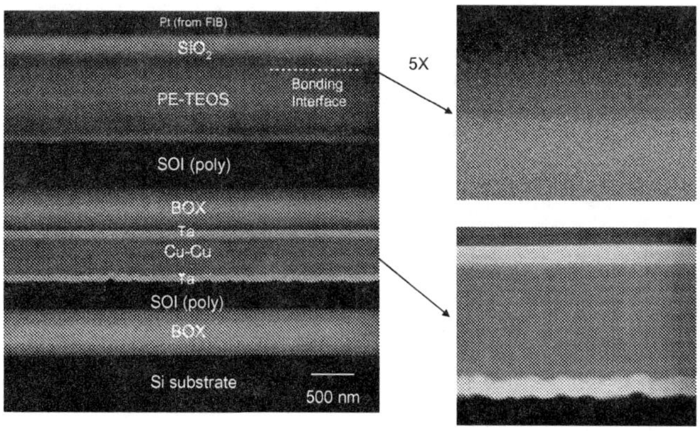

Fig. 2. SEM shows *back-to-face* silicon layers stack obtained after handle wafer grinding and TMAH strip.

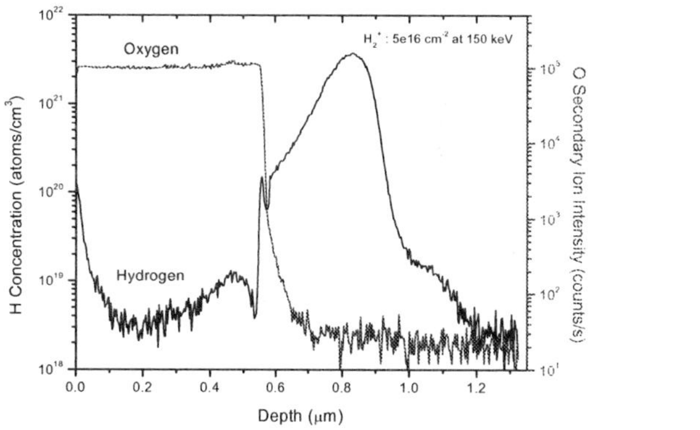

Fig. 3. SIMS profile of implanted hydrogen into the handle wafer. The hydrogen peak situated about 300 nm from the oxide-silicon interface into the silicon handle wafer.

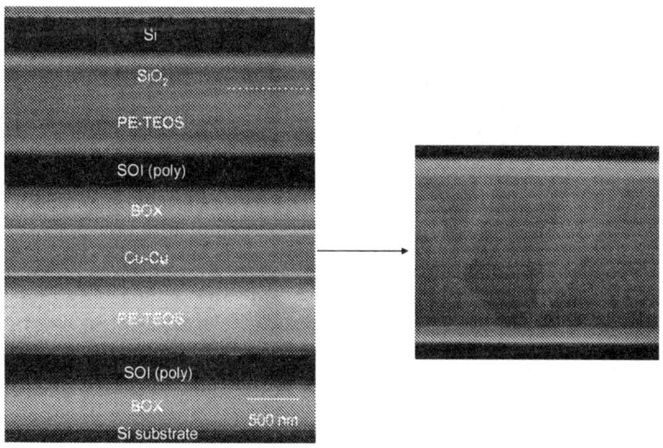

Fig. 4. SEM shows the silicon layers stack obtained after handle wafer release. Handle wafer released at the peak of the implanted hydrogen profile.

Relaxation of Strained-SOI substrates by RTA process

Kosuke Yamasaki, Daisuke Kosemura, Satoshi Tanaka, and Atsushi Ogura
School of Science and Engineering, Meiji University
1-1-1 Higashimita, Tama-ku, Kawasaki, Kanagawa 214-8571, Japan
Email: a_ogura@isc.meiji.ac.jp, Tel/Fax: +81 (44) 934-7324

Ichiro Chiba and Ryosuke Shimidzu
PHOTON Design Corporation.

Three different type strained-Si substrates, EPI, SGOI and SSOI, were evaluated by newly developed high-resolution UV-Raman spectroscopy. The structure relaxation occurred by RTA at 1050°C. The Raman peak shifted toward larger wavenumber after RTA for SSOI, indicating strain relaxation. The peak shifted toward lower wavenumber in the case of EPI due to the Ge diffusion rather than strain relaxation. The SGOI showed complex characteristics of SSOI and EPI.

Introduction

Recently, there has been considerable interest in using strained-Si film for a material in the channel area of MOSFET. Because tensile strain raises carrier mobility both for electrons and holes, therefore improves device performance. Various structures have been proposed to obtain high-quality strained-Si films. They are often combined with SOI technologies, such as SGOI (SiGe on insulator) and SSOI (strained-Si on insulator). The most important characteristic for the substrates should be the strain and its tolerance for device fabrication process. In the state-of–the-art LSI fabrication, RTA (Rapid Thermal Annealing) for impurity activation in source and drain region is the most severe process in the view point of thermal budget. Therefore, we evaluate the strain relaxation due to the RTA process.

Experiment

Structures of evaluated substrate were illustrated in Fig.1. Three types of strained-Si substrate, EPI, SGOI and SSOI, were evaluated in the experiment. The strained-Si thickness for EPI substrate was 17.5 nm with 20% Ge concentration in underlying relaxed SiGe layer. The SGOI substrate had approximately 17, 70, and 200 nm for strained-Si, SiGe, and BOX film thicknesses, respectively. Ge concentration was 20%. The SSOI had a structure with 17 nm strained-Si film directly on the 200 nm BOX layers.

Samples were annealed at 1050℃ for 1, 3, and 5 minutes using a conventional RTA apparatus. Strains in the sample before and after RTA were evaluated by using UV-Raman spectroscopy.

UV-Raman System

Figure 2 schematically illustrates a UV-Raman spectroscopy/microscopy system used in this study. We have already reported the usefulness of UV-Raman for strained-Si evaluation [1,2]. The penetration depth of UV laser is appropriate for evaluation of thin strained-Si film which is typically less than 20 nm. Moreover, the resonant effect using 364 nm excitation laser is very attractive, because strong Raman signal can be obtained without sample damage. We also improved the precision of measurement by *in-situ* calibration of wavenumber using Rayleigh scattering as a standard. It may also be possible to calibrate the optics using Hg lamp. However, slight difference of light pass and angle between calibration lamp and Raman signal may cause some degradation in repeatability. Rayleigh scattering can be a precise marker. However, it has been difficult to obtain both Reyleigh scattering and Raman signal simultaneously with keeping a high wavenumber resolution due to the limitation of the number of pixels in CCD detector. We therefore, separate the concave mirror (dual focusing mirror) in the spectrometer into two parts, one is for Reyleigh scattering and the other is for Raman signal, and successfully obtained both signal with the same CCD detector simultaneously (Fig. 3).

Results and Discussion

Figure 4 shows Raman spectra obtained from SSOI and EPI before and after RTA. Slight peak shift toward larger wavenumber was observed in SSOI after RTA, implying strain relaxation. However, peak shift toward lower wavenumber clearly appeared in EPI. This can be explained by the Ge diffusion occurring rather than strain relaxation. In both cases, peak shifts were larger

0-7803-9212-4/05/$20.00 ©2005 IEEE

than the strain variation observed in the same sample (approximately 0.2-0.3 cm^{-1}).

Peak shifts and peak width (FWHM) changes after RTA at 1050 °C for 5 minutes were summarized for all strained-Si substrates in Table 1. The FWHM of EPI increased significantly. This might reflect degradation of crystal quality. On the contrary, the FWHM of SSOI decreased. Therefore we could conclude the crystal quality was improved after RTA. We assume the strain caused by crystal imperfection was relaxed while intentionally induced strain kept unchanged. The characteristics of the peak shift and FWHM change for SGOI was complicated. This might be some kind of complex what happened on SSOI and EPI.

References

[1] S. Nakashima *et al.*, Appl. Phys. Lett., **84**, 2533 (2004).

[2] A. Ogura *et al.,* Extended Abstract of the 4th International Symposium on Advanced Science and Technology of Silicon Materials.

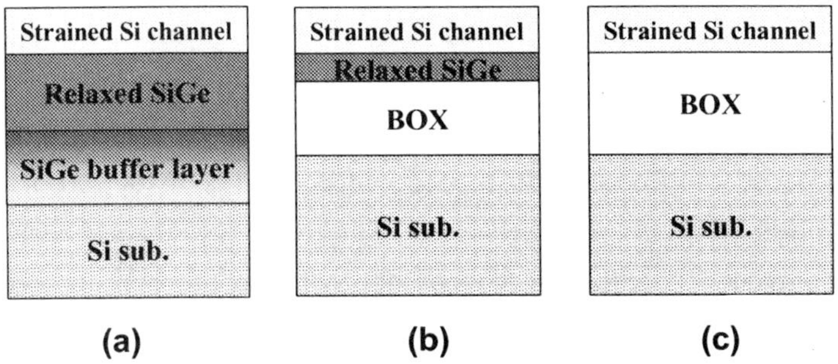

Fig. 1 Schematic structures (a) EPI, (b) SGOI and (c)SSOI substrates.

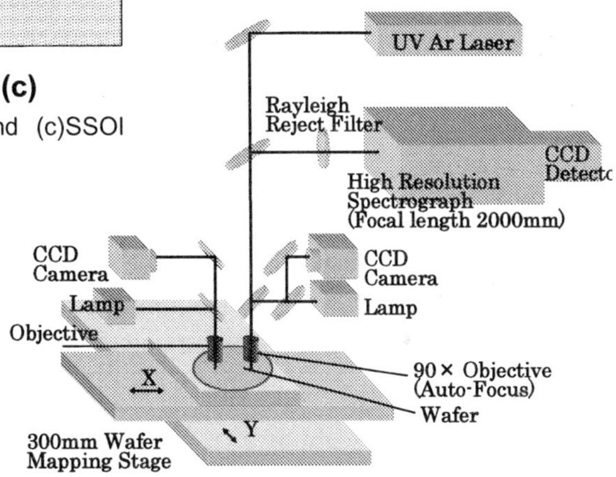

Fig.2 Schematic iluustration of UV-Raman system.

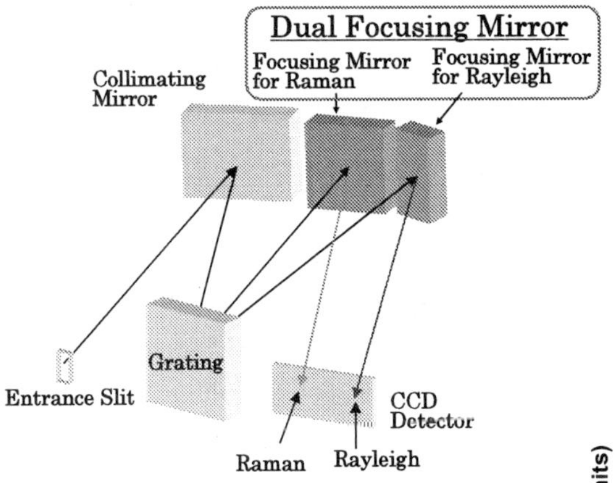

Fig. 3 Novel monocrometer using dual focusing mirror for both Raman and Rayleigh siganl.

	Peak shift (cm^{-1})	FWHM (cm^{-1})
EPI	-0.82	0.77
SGOI	-0.08	0.36
SSOI	0.39	-0.33

Tbale 1 Peak shifts and FWHM chages aftter RTA at 1050 °C for 5 minutes.

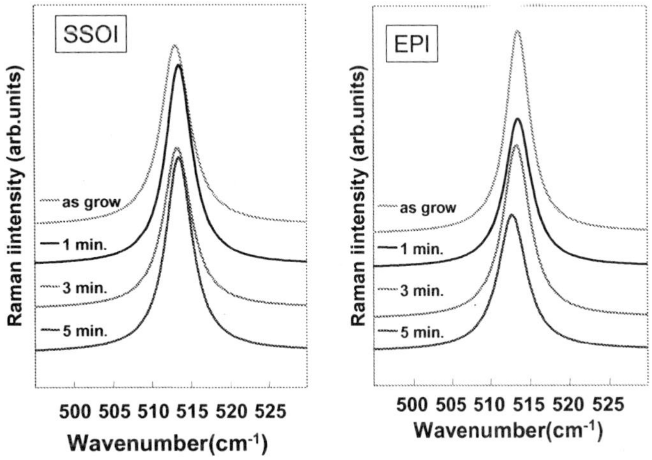

Fig. 4 Raman spectra form SSOI and EPI before and after RTA.

The Formation of Sub-micro Partial SOI Materials by SIMOX Technology

Jiayin Sun, Jing Chen, Meng Chen, Xi Wang
Shanghai Institute of Microsystem and Information Technology, Chinese Academy of Sciences, Shanghai 200050, China

The partial (or patterned) SOI technology [1, 2] is proposed one of the methods to overcome the self-heating problem in power IC and others. The conventional SOI structure suffers from self-heating effect for the very low thermal conduction of the buried oxide (BOX) layer. The patterned BOX layer can decrease the self-heating effect while maintaining good insulation between the low power circuit and the high power cell. With the development of SOI power ICs [3], the partial SOI wafers in sub-micrometer scale are expected for the further applications. In this paper, we fabricate the partial SOI materials with the sub-micro scale patterned BOX layer by mask SIMOX technology.

The starting wafers of our experiment are 100mm p-type (100) Czochralski silicon wafers. The masking patterns were fabricated on the wafers by e-beam lithography after grown of 500 nm thermal masking oxide layer. Oxygen ions ($^{16}O^{+}$) were implanted into the masked wafers with a dose of $1.8\times10^{17}cm^{2}$ by an ULVAC IM-200 ion implanter at 45keV. The as-implanted wafers were annealed at 1300°C for 5h in a flowing Ar ambient with 3% oxygen.

In this paper, we try to fabricate sub-micro windows in BOX layer. The pattern thermal oxide layer we used for mask implantation is shown in Fig.1. The typical widths of thermal oxide strips are 150nm (Fig.1 a) and ~75nm (b), and the space of each strips is 1μm. The O^{+} ions are implanted through the patterned masking oxide. The XTEM images of the SOI structures are shown in Fig.2. The widths of BOX windows are 236nm (left) and 150 nm (right). It is very interesting that the windows are about 75 nm wider than the corresponding masking strips.

In our opinions, the width of BOX layer can be influenced by two factors at least. On one hand, the implanted O^{+} ions can destroy the mask. The mask thermal oxide strips become narrower after implantation, which gets the windows narrower than the strips above. On the other hand, the O^{+} ions will be condensed into a mass during the formation of SOI structure at 1300°C. The O^{+} ions on the edge can congregate to the center of the patterned BOX layer, which gets the windows wider. It is the same for the first factor in any width of mask oxide, but it is not for the second. The ions congregation during annealing will be unobvious if the BOX layer is narrow, for the interplay of O^{+} ions on both edges.

We design the following experiment to prove our theory. The masking oxide grooves are fabricated on the bulk Si wafers (Fig.3). The typical width of grooves are 390nm (Fig.1 a), 190nm (b) and 120nm (c). The partial SOI structures after annealing are shown in Fig4 and 5, corresponding to Fig.3. The width of BOX layers are 315nm, 138nm, and 120nm, respectively. We obtain the 315nm and 138nm width BOX layer (Fig.4) while the mask grooves is 390nm and 190nm width. The BOX layers beneath are 75nm and 52nm narrower than grooves. The congregation of O^{+} ions is very obvious in this case. But we obtain the 120nm width BOX layer (Fig.5) using 120nm width mask groove. According to the theory above, it is for the broaden effect is not very obvious, and just can be counteracted by the fist factor.

So, in the fabrication of BOX windows, the space of each strips are 1μm, so the BOX layer is 1μm width, which is wide enough for the ions congregations. After annealing, the BOX layer is narrower, and then we always obtain the windows with 75nm wider.

The sub-micro partial SOI materials have been fabricated by mask SIMOX technology. We have obtained the 150 nm width window in BOX layer and 120nm width BOX layer, successfully. The congregation of O^+ ions and implantation destroy must be considered for its influence to the scale of patterns. 120nm is almost the balance scale for the two factors. If the BOX layer is wide enough, the scale of BOX patterns will be changed with 75nm, normally.

1. F. Udrea, Electronics and Communication Engineering Journal, Feb. 2000, p27-40.
2. Udrea, F.; IEEE EL, 1997, V.33 No.1035, p907-909.
3. J. Lou, ICSE2002 Proc. 2002, Penang, Malaysia, p314-318.

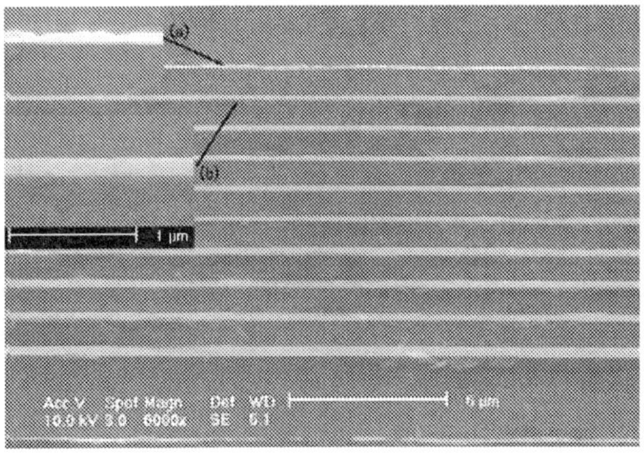

Fig.1 The SEM image of masking oxide strips

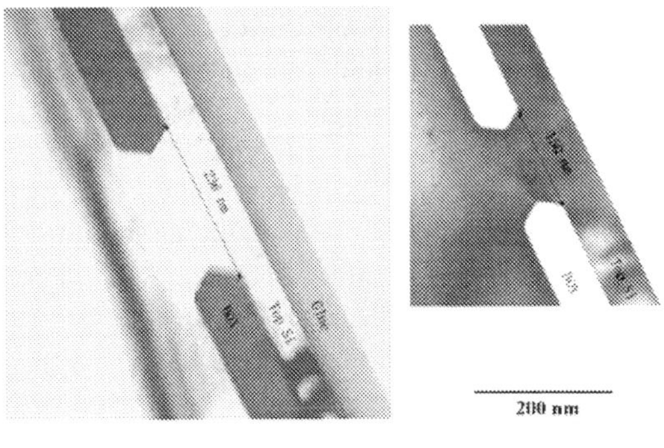

Fig.2 The XTEM image of sub-micro partial SOI structures formed by Fig.1

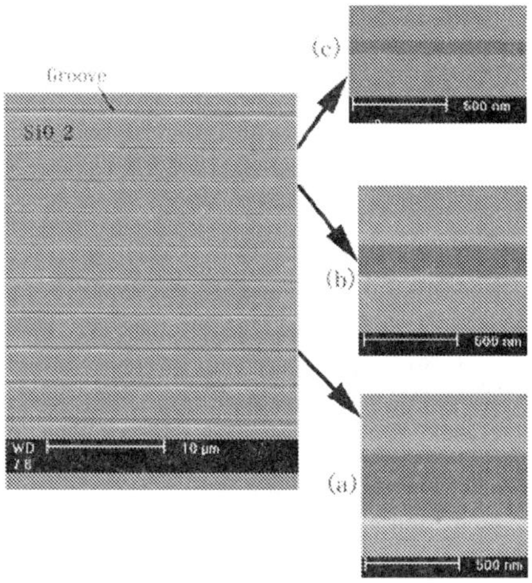

Fig.3 The SEM image of masking oxide grooves

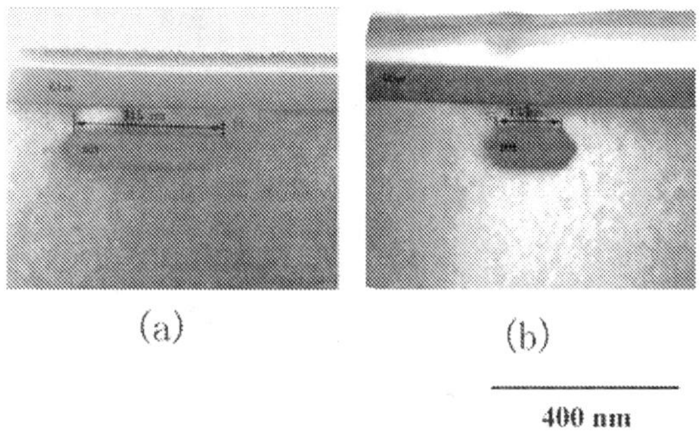

Fig4 The XTEM image of partial SOI structures formed by Fig.3 (a) and (b).

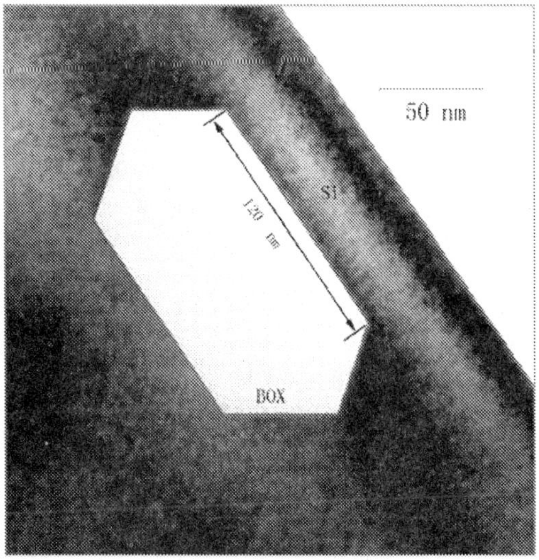

Fig5 The XTEM image of sub-micro pattial SOI structures formed by Fig.3 (c).

Lateral and Vertical Coupling Effects in MIGFETs

S. Eminente [12] Kyoung-Il Na [13] S. Cristoloveanu [1] L. Mathew[4] A. Vandooren[4]
(1) IMEP, 23 rue des Martyrs, B.P. 257, 38016 Grenoble Cedex, France, *Email: eminente@enserg.fr*
(2) ARCES Center, Bologna Italy (3) Kyungpook National University, South Corea
(4) Freescale Semiconductor Inc., Austin, USA and Crolles, France.

Abstract

We have performed measurements on double-gate FinFETs with independent gate contacts in order to study the coupling effects. In these devices, it is possible to separate the coupling between the two lateral channels and the coupling between one lateral channel and the back gate. The lateral coupling effect between the two gates is strong and can be used to tune the threshold voltage of the device whereas the vertical coupling is weaker and depends on the MIGFET size.

Introduction

The MOS scaling trend is pushing the device dimensions down into the 10 nm range. In order to overcome the scaling problems (high doping levels, low mobility, high leakage currents) various device structures have been proposed: the Double-Gate SOI FinFET is the most interesting one because of its excellent characteristics [1]. In Double-Gate devices the two gate contacts are tied together but recently a modification has been promoted: the Multiple Independent Gate FET (MIGFET) [2], is a FinFET where the two lateral gates are independent and can be controlled separately.

The shift of the threshold voltage with the substrate bias is called coupling effect. In a MIGFET we can distinguish the lateral coupling between the two lateral gates and the vertical coupling between the lateral gates and the back gate. The purpose of this work is to investigate these various types of coupling.

Device Chracteristics

The fabrication process of MIGFET devices has been presented in [2]. SOI silicon wafers were patterned into fins of about 100 nm height with two independent gate regions on the two sides of the fin: the separation between these regions is achieved through a top spacer. The gate stack is composed of a 2.4 nm oxide layer and polysilicon gate contacts. We have considered MIGFETs with fin widths of 30 and 50 nm and gate length of 80 nm. All devices are n-channel with undoped body. The BOX thickness is about 100 nm.

Lateral and Vertical Coupling

We consider the lateral coupling between the two main gates $G1$ and $G2$ with grounded back gate. Fig.1 shows the I_{DS} vs. V_{G1} curves measured for $V_{DS} = 0.1$ V, $T_{si} = 30$ nm with the opposite gate voltage V_{G2} varying from -0.8 to 0.5 V. Because of the capacitive coupling, the threshold voltage decreases while the opposite gate is moving from accumulation to inversion and the I_{DS} , G_m vs. V_{G1} curves shift to left. Figure 2 shows the threshold voltages extracted from Fig. 1 and from a MIGFET with $T_{si} = 50$ nm. For the narrower fins the V_{th} vs. V_{G2} curve is quasi-linear, and the slope is -0.35. The Lim and Fossum 1D model [3] predicts a lower slope, about -0.2. This difference confirms that, in FinFETs, 2D and even 3D model are needed [4][5]. In particular, for narrow fins, the fringing fields between the lateral gates through the BOX cannot be neglected. The lateral coupling should decrease in wider fins and indeed the slope of the curve is slightly reduced to about -0.29 for $T_{si} = 50$ nm. We now investigate the effect of the back-gate bias V_{back} in MIGFETs. Figure 3 shows the simple case of a DG MIGFETs ($V_{G1} = V_{G2}$) where the fringing fields minimize the V_{back} impact. The lateral shift and the transconductance peak degradation are smaller as compared to Fig.1. Figure 4 illustrates a 2D coupling effect: the overall variation of the threshold $V_{th}(V_{back})$ is weak for $T_{si} = 30$ nm. The slope does not depend on the opposite gate bias. The coupling is comparable to the DG operation but not totally ignorable. Instead if the fin is wider, the back-gate has a larger influence on V_{th}: when $V_{back} = 10$ V the back channel is inverted and V_{th} drops rapidly; when $V_{back} < -5$ V the potential in the bottom part of the fin is controlled only by back channel accumulation. Figure 5 and 6 show how the coupling between the lateral channels is modified by V_{back}. When $V_{back} < 10$ V the threshold voltage can be tuned but the slope is hardly altered. For $V_{back} = 10$ V the curve becomes steeper as soon as the back-channel inversion is induced by the lateral gates. Note that, when three inversion channels coexist, the extracted V_{th} is underestimated.

Conclusions

We have evaluated the various 2D coupling effects in FinFETs featuring independent gate contacts. A lateral coupling exists and permits to tune the threshold voltage for optimum device operation. Further tuning can also be provided by the back gate except in ultra narrow fins. The back-gate effect is important for optimizing the fin geometry in two respects: performance enhancement and sensitivity to radiation induced by BOX charges.

References

[1] G.K.Celler, S.Cristoloveanu *Journal of Applied Physics*, vol 93 n. 9, 2003.
[2] L. Mathew et al. , "CMOS Vertical Multiple Independent Gate Field Effect Transistor (MIGFET)", *2004 IEEE International SOI Conference Proceedings*
[3] H-K Lim, J.G. Fossum; *IEEE Trans. Electron Devices*, vol. 30 p. 1244, 1983.
[4] J. Frei et al. ; *IEEE Electron Device Letters*, vol. 25 p. 813-815, 2004.
[5] J. P. Colinge et al.; *IEEE Electron Device Letters*, vol. 24 p. 515-517, 2003.

0-7803-9212-4/05/$20.00 ©2005 IEEE

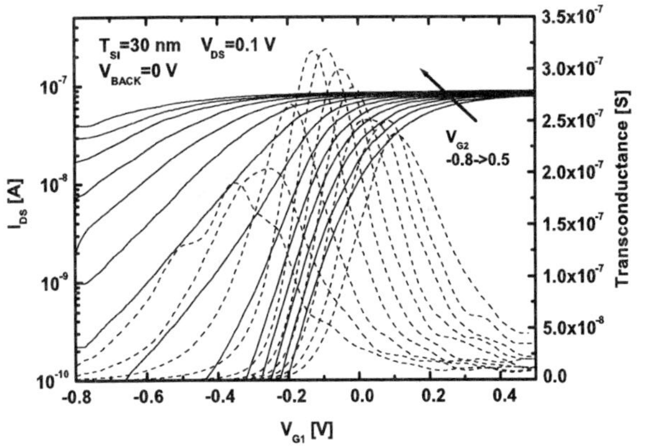

Figure 1: Drain current and transconductance versus gate1 voltage for different gate 2 bias, $T_{si} = 30$ nm.

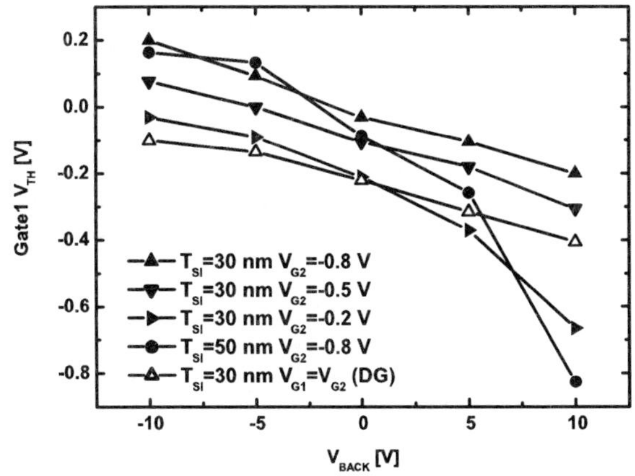

Figure 4: Threshold voltage versus back-gate bias for devices with two T_{si} and under different measurements conditions. For the narrower fin the slope does not depend on the opposite gate bias unless it is biased near inversion.

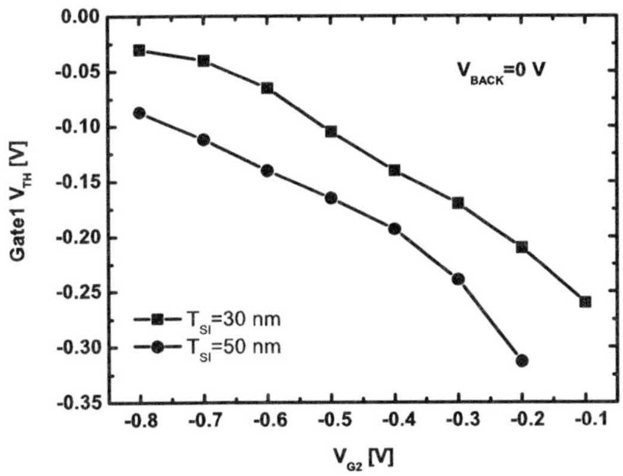

Figure 2: V_{th}, for different fin thicknesses, versus voltage at the opposite gate. V_{th} has been extracted as the V_{G1} where d^2I_{DS}/dV_{G1}^2 is maximum.

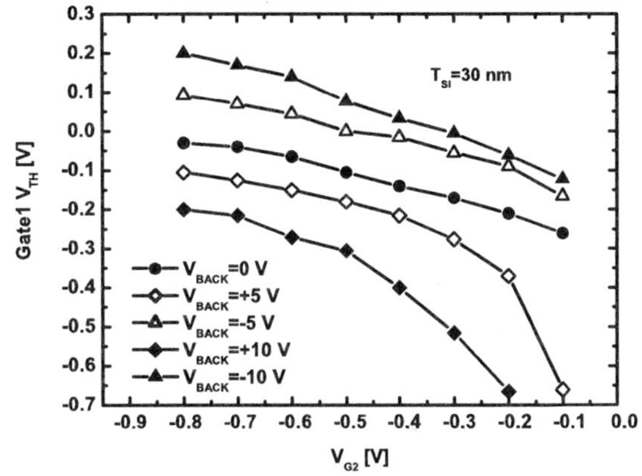

Figure 5: Threshold voltage of gate1 versus V_{G2} with different back-gate bias, $T_{si} = 30$ nm. When the back-channel is open, the extraction of the V_{th} is more difficult and underestimated.

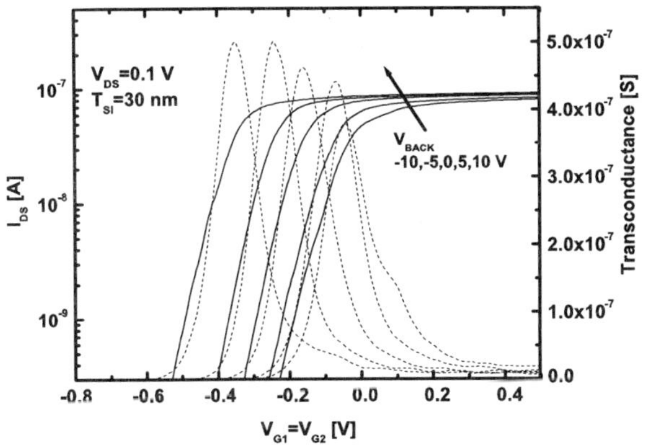

Figure 3: Drain current and transconductance in double-gate mode for different back-gate bias, $T_{si} = 30$ nm.

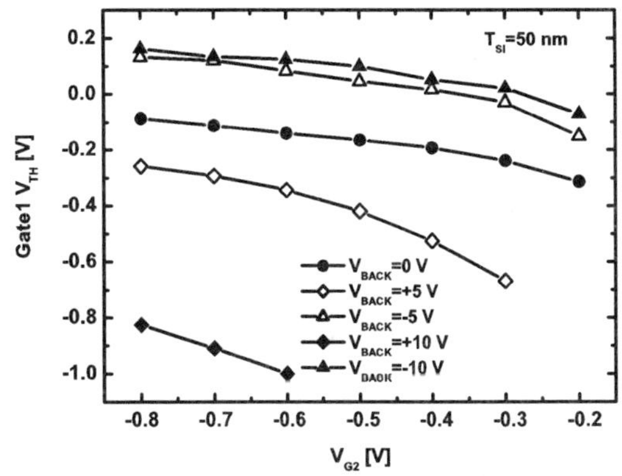

Figure 6: Threshold voltage of gate1 versus V_{G2} with different back-gate bias, $T_{si} = 50$ nm.

Device Design Considerations for Nanoscale Double and Triple Gate FinFETs

Abhinav Kranti and **G. Alastair Armstrong**

Northern Ireland Semiconductor Research Center (NISRC), School of Electrical and Electronic Engineering, Queen's University Belfast, Ashby Building, Stranmillis Road, Belfast BT9 5AH, Northern Ireland, UK.
E-mail: a.kranti@ee.qub.ac.uk, Phone: +44-2890974596, Fax: +44-2890667023

Introduction

Over the past few years, FinFETs have emerged as favorite candidates for device scaling at the end of ITRS roadmap [1-5]. Technologically FinFETs can be fabricated as Triple gate (TG) (Fig. 1(a)) or as Double gate (DG) devices. In DG FETs, the top gate is electrically isolated from the body thus resulting in the effective gate width of $2H_{fin}$ instead of $2H_{fin}+T_{fin}$ for TG FinFETs where H_{fin} and T_{fin} are the height and thickness of the fin. In the present work, based on 3D simulations, we analyze in detail the impact of (i) lateral source/drain (s/d) doping gradient, (ii) spacer width, (iii) gate workfunction (φ_m) and (iv) aspect ratio ($AR=H_{fin}/T_{fin}$) on the performance of TG and DG FinFETs and present device design guidelines for 65 nm node for HP, LOP and LSTP logic technologies. Based on the above considerations, we show that DG FinFETs show greater flexibility in selecting device and technological parameters as compared to TG FinFETs. The design guidelines proposed for 65 nm node are shown to be valid for 45 nm node for TG and DG FinFETs and a design window is presented to achieve ITRS targets for the three logic technology requirements with triple and double gate MOSFETs.

Results and Discussion

TG and DG devices have been simulated using 3D simulator, ATLAS [6]. DG FinFETs were simulated by using a thicker top gate oxide of 50 nm instead of a much thinner side gate oxide of 1.3 nm, 1.9 nm and 2.3 nm for HP, LOP and LSTP logic technologies respectively. The doping (N_a) of the p-type SOI layer was 10^{15} cm^{-3}. The spacer width (s) was varied from (0.1) L to (1.0) L, L being the gate length and s/d doping profile defined by its gradient (d) at the gate edges, was varied from 0 to 9 nm/decade, as shown in Fig. 1(b). Equivalent oxide thickness (EOT) in inversion accounting for the gate depletion, finite inversion layer capacitance and physical oxide thickness has been used as per ITRS guidelines. The simulations have been performed with drift diffusion (DD) models using CVT mobility model with default parameters. The use of energy balance (EB) model, which results in over estimation of current [7] in nanoscale regime would not influence the overall conclusions of the work.

Fig. 2-4 show the I_{on}-I_{off} graphs for HP, LOP and LSTP technologies respectively, along with the ITRS targets (shown by rectangle) for 65 nm node. At a constant s, the value at highest I_{on} denotes d=9 nm/dec while the lowest I_{on} ($I_{ds}@V_{gs}=V_{ds}$) represents d=0 nm/dec. Larger values of d along with smaller s values result in a higher I_{off} ($I_{ds}@V_{gs}=0,V_{ds}$) due to shorter effective channel length. Higher d at the gate edge not only reduces the effective channel length in weak inversion region but also draws more carriers into the spacer region near the gate edges, thus increasing I_{on} [8]. It is important to note that I_{on}-I_{off} curves sharply bend upwards at lower s values for both TG and DG FinFETs thus indicating that I_{on} does not increase much, even though I_{off} increases exponentially. Large d at lower s values cause a significant amount of dopant spill into the channel and not only results in severe SCEs but also limit I_{on} due to the parasitic series resistance. Therefore those possible combinations of s, d, φ_m and AR that satisfy the ITRS targets but also result in limiting I_{on} should be clearly avoided for optimal design of TG and DG FinFETs. Although DG devices result in higher I_{on} (mA/mm), they also achieve a higher I_{off} (nA/μm) as compared to TG devices. However, I_{off} values for DG FinFETs are still lower than the ITRS target of I_{off} for all the three logic technologies. It is more appropriate to change spacer values in an optimal design as the minimum value of d is generally set by the thermal budget and the diffusivity of the dopant species. Further, it should be noted that very small values of d require non-standard process [9] such as solid phase epitaxy or laser thermal annealing in order to be practically feasible.

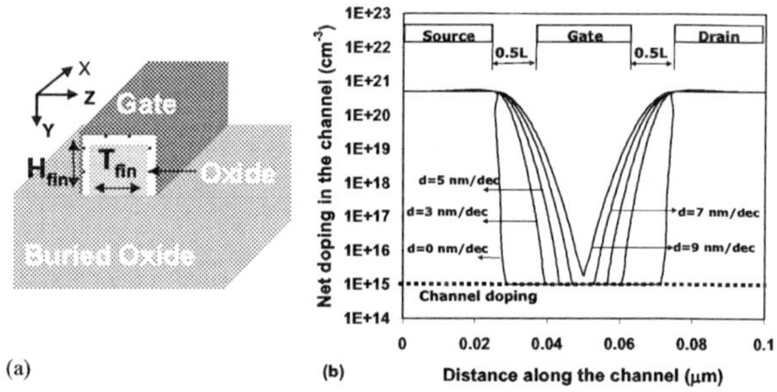

Fig. 1: (a) Schematic diagram of a TG FinFET (b) Variation of d at s = (0.5) L in the middle of the fin along the x direction.

Table 1 attempts to summarize the complexity of Fig. 2-4 for the reader, showing that with an increase in AR the possible set of s, d and φ_m that meet the ITRS criterion is much reduced. Moreover, as AR increases, the difference in I_{off} values between TG and DG devices becomes insignificant. Although AR values upto 4-5 are technology feasible, AR >2 would certainly not be useful in achieving the ITRS targets because of larger I_{off}. Also, an increase in φ_m from 4.52 eV to 4.72 eV reduces the possible set of s, d, φ_m and AR values for HP and LOP logic technologies whereas higher φ_m is more suitable for LSTP logic technology for both TG and DG devices. On comparing the possible set of s, d, φ_m and AR values for respective logic technologies, we find that a DG device offers greater design flexibility for HP and LOP logic technologies, in selecting device parameters as it achieves a higher I_{on} as compared to TG FinFET, along with an acceptable higher I_{off} (Table 1). For HP ITRS targets, an increase in φ_m from 4.52 eV to 4.72 eV shifts the possible set of s and d values in the I_{on}

0-7803-9212-4/05/$20.00 ©2005 IEEE

limiting region for TG devices whereas DG FinFETs could still be designed without limiting I_{on}. Also, TG FinFETs cannot achieve LOP ITRS target with φ_m of 4.72 eV, whereas DG devices show a larger set of s and d values to design from.

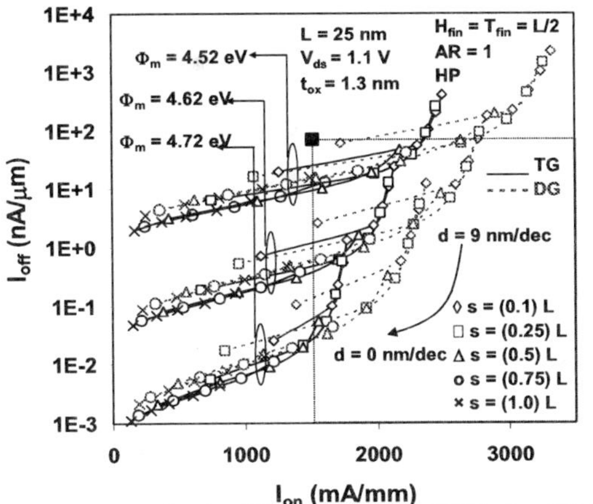

Fig. 2: I_{on}-I_{off} graphs for TG and DG FinFETs for HP logic technology.

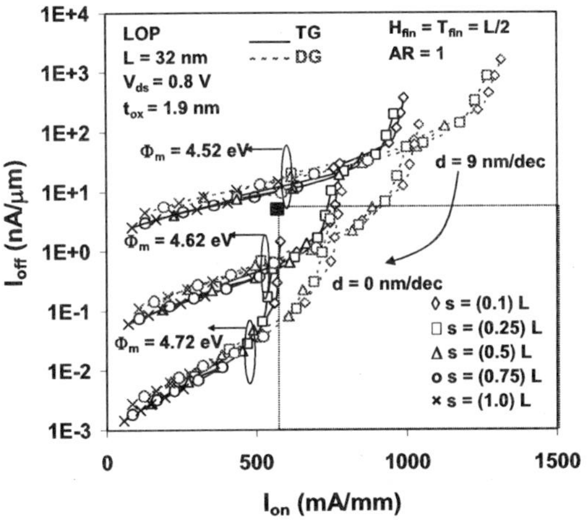

Fig. 3: I_{on}-I_{off} graphs for TG and DG FinFETs for LOP logic technology.

It is clear from Table 1, that AR and φ_m are not critical for DG devices if designed with s=(0.5)L and d in the range of 5–9 nm/dec to meet HP ITRS targets. However for optimal TG design the spacer must be reduced from (0.5)L to (0.25)L to achieve the same design flexibility on AR, φ_m and d. However, for LOP ITRS targets, FinFETs should be designed with smaller values of spacer (~ (0.25)L) along with AR=1 and φ_m=4.72 eV for DG devices and φ_m=4.62 eV for TG FinFETs. The ITRS LSTP targets offer minimum flexibility in terms of all device parameters for both TG and DG devices with φ_m in the range of 4.52-4.72 eV. In order to achieve LSTP ITRS targets, (i) higher φ_m value (> 4.72 eV) resulting from Hf-Mo based binary alloys that yield a continuous and linear gate workfunction from 3.93 - 4.93 eV [11] or (ii) a thinner EOT (< 2.3 nm) [8] with φ_m = 4.72 eV could be used. DG FinFETs should be designed with higher s values (≥ (0.5)L) to offer maximum flexibility in the choice of d. Also, in nanoscale regime, current conduction with an extra top gate does not increase the current drive in proportion to the increase in channel width due to a higher parasitic series resistance [10]. TG devices with a top gate contributing effectively to the current will only be beneficial at larger values of T_{fin} that result in larger widths and even with lower AR would yield a higher I_{off}, and thus not be useful for achieving ITRS targets.

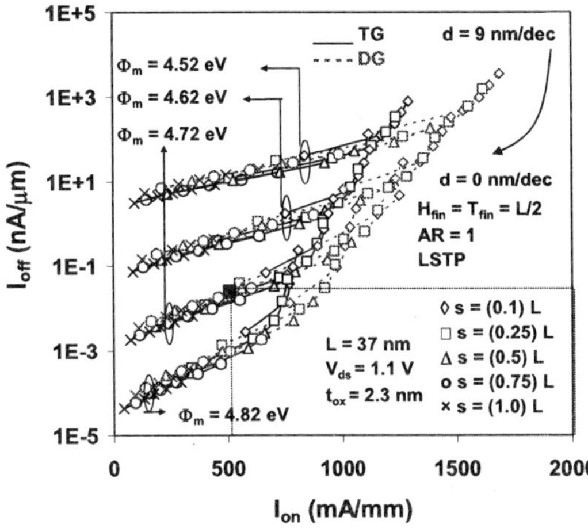

Fig. 4: I_{on}-I_{off} curves for TG and DG devices for LSTP logic technology.

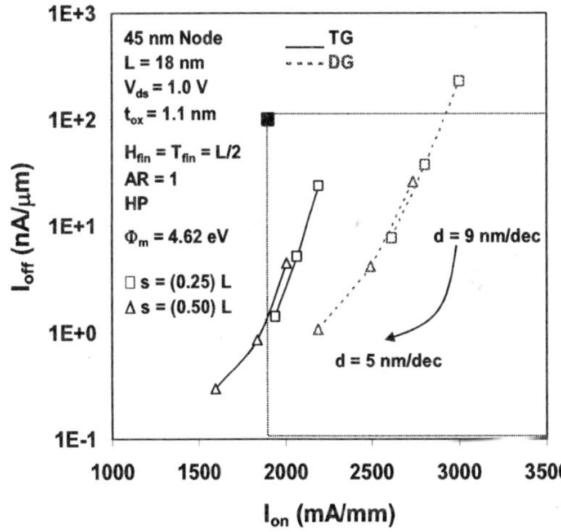

Fig. 5: I_{on}-I_{off} curves for TG and DG devices for 45 nm HP logic technology.

Based on the scaling guidelines for the 65 nm technology node presented above, we evaluate the performance of TG and DG FinFETs for 45 nm node. Fig. 5 shows the I_{on}-I_{off} graphs for HP technology for two different spacer values i.e. (0.25)L and (0.50)L with doping gradient varying from 9 nm/dec to 5 nm/dec. DG FinFETs show greater design flexibility as compared to TG devices in order to meet the targets for 45 nm node. DG FinFETs designed with s = (0.5)L and (0.25)L achieve ITRS values with d=7 nm/dec whereas TG FinFETs need to be designed with a spacer of (0.25)L and doping gradient of 7 nm/dec. Also, new materials like strained silicon could be used improve the drive current and thus enhance the design space for TG and DG FinFETs. If a typical ITRS specification for contact resistivity of 4 Ω μm^2 is included in the simulation then I_{on} reduces by 6% for both TG and DG FinFETs. However, even in that case DG FinFETs would show a greater design space as compared to TG devices.

Conclusions

Based on 3D device simulations, we report performance assessment of TG and DG FinFETs for HP, LOP and LSTP logic technologies for the 65 nm technology node. An investigation of s, d, φ_m and AR shows that DG devices offer higher I_{on} (mA/mm) while achieving acceptable values of I_{off} (nA/μm) and thus offer greater design flexibility in selecting device parameters as compared to TG FinFETs to meet ITRS targets for all the three logic technologies. TG and DG FinFETs should be designed with lower aspect ratios (1 ~ 2), along with lower values of fin height and thickness to achieve ITRS projections. If a target value for spacer and doping gradient is set to (0.5)L and 7 nm/dec for both TG and DG FinFETs then EOT and gate workfunction need to be adjusted to achieve the ITRS targets. TG and DG FinFETs with gate workfunction values in the range of 4.62 – 4.72 eV are more likely to meet ITRS targets for HP and LOP technologies whereas higher workfunctions ~ 4.82 eV would be more suitable for LSTP applications. A spacer width of (0.25)L with a doping gradient of 7 nm/dec would be more suitable for TG FinFETs for achieving ITRS targets for 65 nm and 45 nm nodes, whereas a spacer width of (0.5)L with the same doping gradient would be more appropriate for DG FinFETs. The present work provides valuable design insights in the performance of TG and DG FinFETs and serves as a tool for optimizing device and technological parameters for 65 nm node and below.

Acknowledgement

A. Kranti is grateful to Invest N.I., UK for the financial assistance to carry out the research work.

References

[1]. http://public.itrs.net
[2]. B.S Doyle et al., IEEE EDL, 24, 263-265, 2003.
[3]. J.-T. Park et al. IEEE TED, 49, 2222-2228, 2002.
[4]. V. Trivedi et al., IEEE TED, 52, 56-62, 2005.
[5]. S.H Kim et al., IEEE SOI Conference, 145-147, 2004.
[6]. ATLAS Users manual: SILVACO International 2004.
[7]. K. Banoo et al., Solid-State Electronics 44, 1689-1695, 2000.
[8]. T.C. Lim et al., Solid-State Electronics, 49, 1034-1043.
[9]. R.S. Shenoy et al., IEEE Nanotechnology, 2, 265-270, 2003.
[10]. E. Leobandung et al., J. Vac. Sci. Technol. B, 15, 2791–2794, 1997.
[11]. T-L Li et al., IEEE TED, 52, 1172-1179.

	AR	φ_m (eV)	TG FinFETs [s ; d]		DG FinFETs [s ; d]	
			Less flexibility	Enhanced flexibility	Less flexibility	Enhanced flexibility
HP	1	4.52	[(0.75)L; 9]	[(0.25)L; 3, 5], [(0.5)L; 5, 7, 9],	[(0.25)L; 3]	[(0.5)L; 3, 5, 7], [(0.75)L; 7, 9]
		4.62	-	[(0.25)L; 3, 5, 7, 9], [(0.5)L; 7, 9]	-	[(0.25)L; 3, 5, 7], [(0.5)L; 5, 7, 9], [(0.75)L; 7, 9]
		4.72	[(0.5)L; 9]	[(0.25)L; 5, 7, 9]	[(0.75)L; 9]	[(0.25)L; 3, 5, 7, 9], [(0.5)L; 5, 7, 9]
	2	4.62	[(0.5)L; 7]	[(0.25)L; 3, 5, 7]	[(0.75)L; 9]	[(0.25)L; 3, 5], [(0.5)L; 5, 7, 9],
		4.72	-	[(0.25)L; 7, 9]	[(0.5)L; 9]	[(0.25)L; 5, 7, 9]
LOP	1	4.62	-	[(0.25)L; 3, 5, 7], [(0.5)L; 7, 9]	[(0.25)L; 3], [(0.75)L; 9]	[(0.5)L; 5, 7]
		4.72	-	-	-	[(0.25)L; 3, 5, 7, 9], [(0.5)L; 7, 9]
	2	4.62	[(0.25)L; 3]	-	-	-
		4.72	-	-	[(0.5)L; 9]	[(0.25)L; 5, 7, 9]
LSTP	1	4.72	[(0.5)L; 5], [(0.75)L; 9]	-	-	-
		4.72	[(0.75)L; 9] (t_{ox} = 2 nm)	[(0.5)L; 5,7] (t_{ox} = 2 nm)	[(0.5)L; 3], [(0.75)L; 7], [(1.0)L; 9] (t_{ox} = 2 nm)	-
		4.82	-	[(0.25)L; 3, 5, 7], [(0.5)L; 7, 9]	[(0.25)L; 3]	[(0.5)L; 5, 7, 9], [(0.75)L; 7, 9]
	2	4.82	[(0.25)L; 3], [(0.5)L; 7]	-	[(0.75)L; 9]	[(0.50)L; 5,7]

Table 1: Possible set of s, d, φ_m and AR values for HP, LOP and LSTP logic technologies for TG and DG FinFETs that achieve ITRS targets. Some φ_m values have not been shown because neither TG nor DG devices can be designed with those to meet ITRS requirements. Also, device design with spacer=(0.1)L and d=0 nm/dec are practically infeasible and have thus not been included in Table 1.

Fabrication of Self-aligned Drain and Source on Insulator MOSFET with Dielectric Pocket by Local SIMOX Technology

Zhichao Lv[1*], Hao Zhang[1], Jian Wang[1], Lilin Tian[1], Zhijian Li[1] Jiaying Sun[2], Jing Chen[2], Xi Wang[2]
[*]Email: lzc02@mails.tsinghua.edu.cn, Tel: (86-10) 62784684-124
[1]Institute of Microelectronics, Tsinghua University, Beijing, 100084, P. R. China
[2]Ion Beam Laboratory, Shanghai Institute of Microsystem and Information Technology, Chinese Academy of Science, 865 Changning Road, Shanghai, 200050, P. R. China

Introduction

Drain and Source on Insutlator (DSOI) MOSFET has been proposed [1,2,4]. Non-self-aligned DSOI (NSA-DSOI) is fabricated by local SIMOX technology, it shows superior characteristics in solving Floating-Body Effect and Self-Heating-Effect (SHE), increasing device's speed [3,4]. However, non-self-aligned process results in the mismatch between buried oxide (BOX) and gate (Fig.1), which deteriorates device's performance, such as immunity to SHE and SCE, speed perfermance, thus it limits DSOI to scaling down. In this work, combined local SIMOX technology with CMOS technology, we realize self-aligned DSOI (SA-DSOI) with dielectric pocket. The experimental results show that it not only maintains the NSA-DSOI's advantages on alleviating SHE and increasing speed, but also shows its further superiority in suppressing SCE and decreasing leakage current due to its dielectric pocket.

Device Fabrication

Fig.2 is the process flow. First, a p-type, (100), $1\times10^{15}cm^{-3}$, wafer is prepared. After gate oxide (20nm) growth, 300nm undoped poly is deposited (Fig.1a). TEOS and Nitride, which serve as the oxygen implantation mask to prevent oxygen ion implanting into the gate, are deposited on the poly (Fig.2b). Next, TEOS, Nitride and poly are patterned (Fig.2c). Then, 50nm TEOS is deposited, round corners are formed at the two sides of gate. TEOS thickness is different at the corner, the difference results in the different oxygen ion implantation depth. After implantation, the profile of oxygen ion is shown in Fig. 1d, which will realize the pocket and BOX at the same time after anneal process. Then, 300nm TEOS is deposited to prevent poly and top silicon from oxidation during the succedent anneal process. Then the wafer is annealed in Ar/0.5% O_2 at 1300°C for 5 hours to form the dielectric pocket and the BOX under S/D (Fig.2e). Then, the protection TEOS layer is removed. LDD, spacer, S/D implantation and other 0.8μm conventional CMOS process are used to fabricate this device. The final structure is illustrated in Fig.2f.

Results and Discussion

Fig. 3 is the SEM photo. In Fig. 3a, good formation of BOX under S/D and dielectric pocket can be clearly seen. In higher resolution SEM photo Fig.3b, details of dielectric pocket can be seen. ***h*** (~20nm) is the distance between the surface of the channel and pocket. ***d*** (~60nm) is the distance between gate and pocket. The corners of the poly seems a little round. This is caused by the poly oxidation during anneal process. Due to long time and high temperature in anneal process, 300nm TEOS cannot prevent poly from oxidation thoroughly.

SA-DSOI with dielectric pocket and conventional bulk device are fabricated on the same wafer in this work. From their I-V characteristics (Fig. 4). SA-DSOI's drive current (I_{on}) and suppression of SCE are all superior than bulk device. Further comparison of this work with former results has been made in Tab.1. The Drain-Induced Barrier Lowering (DIBL) of SA-DSOI with dielectric pocket is 88mV, leakage current is 35fA/μm, and Sub-threshold Slope (SS) is 95mV/Dec. All these values are superior to the other devices in Tab.1. This is due to the dielectric pocket and BOX[2,5], which cut off the path of DIBL, restrain charge sharing and elminate p-n junction leakage. On the other hand, this structure offers the flexibility to control P-sub and S/D doping concentration independly due to dielectric isolation between them. The device also maintains the advantage of NSA-DSOI in suppression of SHE and speed performance due to BOX window under gate.

Though the device in this work has shown advantage on suppression of SCE and leakage current, it seems that the improvement is not signifcant. This is due to the channel doping in this work ($1\times10^{15}cm^{-3}$) is lower than former work ($1\times10^{17}cm^{-3}$). Poly oxidation causes the gate oxide becoms thicker at the fringe of the gate, it makes I_{on} not as high as simulation.

Conclusion

In this work, a method to fabricate SA-DSOI MOSFET with dielectric pocket has been presented. Dielectric pocket and BOX are realized by Local SIMOX technology This novel structure results in good SCE and SHE suppression and higher speed performance, which is very important in nanoscale device design. With this novel self-aligned process, DSOI MOSFET can be scaled down to nano-scale and becomes a promising device.

0-7803-9212-4/05/$20.00 ©2005 IEEE

Acknowldgement

This work is supported by National NSF of China under 90207004 grant, and the Special Funds for Major State Basic Research Projects under G2000036501 grant.

References

[1]T.I.Kamins, *et al, J.Appl.Phys*, pp.423, 1988. [2]Y.Tian,*et al, TED*, pp.561, 2005. [3]P.He, *et al, Int'l SOI conference* , pp.55, 2002 [4]Z.Zhang, *et al, TED*, pp.1002, 2004 [5]M. Jurczak, *et al, TED*, pp.1770, 2001 [6]P. He, *PhD Thesis*, Tsinghua University., 2003

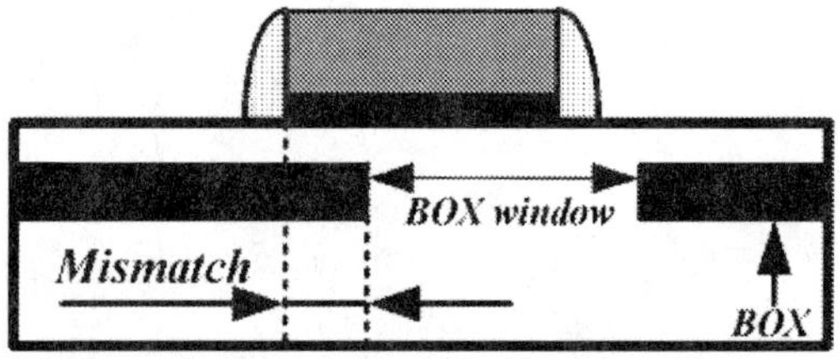

Fig.1 Mismatch due to non-self-aligned process

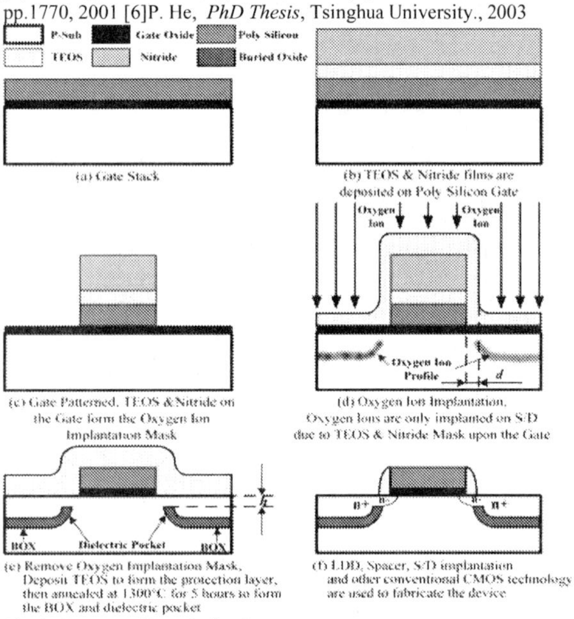

Fig.2 Process flow & final structure

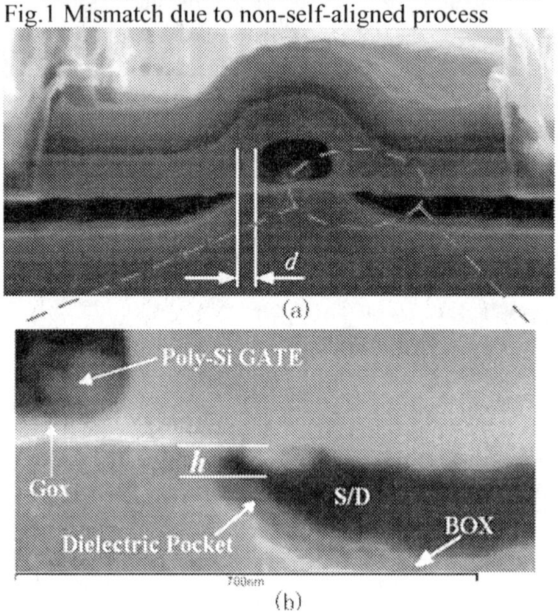

Fig.3 (a)The whole structure (b)Dielectric pocket & BOX

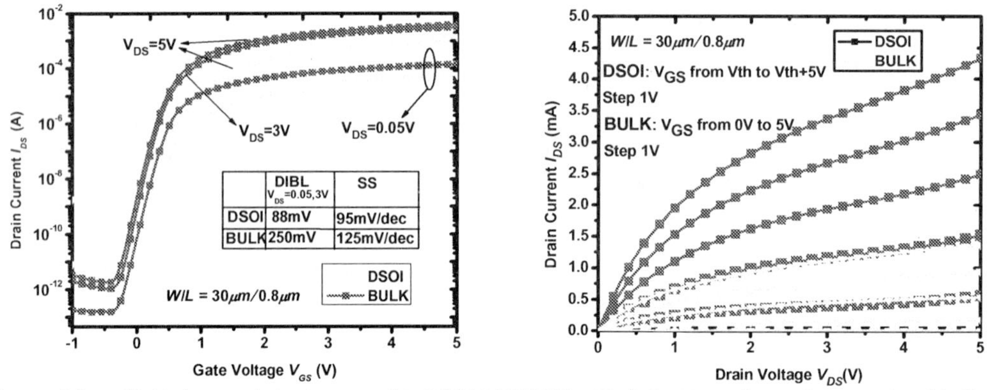

Fig.4. Comparision of I-V characteristics between SA-DSOI MOSFET with dielectric pocket and conventional bulk MOSFET. They are fabricated in the same wafer. (a) I_d-V_g (b) I_d-V_d

	Experimental results in this work (P-Sub doping concentration is $1\times10^{15}cm^{-3}$)		Former experimental results (published on 2002 IEEE Int'l SOI conference ref[3][6]) (P-Sub doping concentration is $1\times10^{17}cm^{-3}$)		
Device* (30μm/0.8μm)	SA-DSOI with dielectric pocket	Conventional Bulk	NSA-DSOI	Conventional Bulk	Conventional SOI
Drive Current(I_{on})	1.66mA (V_{DS}=V_{GS}@3V) 4.01mA (V_{DS}=V_{GS}@5V)	1.40mA (V_{DS}= V_{GS}@5V)	1.52mA (V_{DS}=V_{GS}@3V) 6mA (V_{DS}=V_{GS}@5V)	1.67mA (V_{DS}=V_{GS}@3V)	2.03mA (V_{DS}=V_{GS}@3V)
Leakage Current(I_{off})	1.07×10^{-12}A	1.8×10^{-9}A	2.5×10^{-12}A	2.06×10^{-12}A	1.1×10^{-12}A
$DIBL_{(V_{DS}=0.05V,3V)}$	88mV	250mV	100mV	125mV	609mV
SS	95mV/Dec	125mV/Dec	136mV/Dec	140mV/Dec	107mV/Dec

Table.1 Electrical Characteristics Comparision between the former work and this work

V_{th} Control of t_{pd}-Degradation-Free FD SOI MOSFET with Extremely Thin BOX Using Variable Body-Factor Scheme

Tetsu Ohtou[1], Kouki Yokoyama[1,2], Toshiharu Nagumo[1] and Toshiro Hiramoto[1]

[1]*Institute of Industrial Science, University of Tokyo, 4-6-1 Komaba, Meguro, Tokyo, 153-8505, Japan*, [2]*Chuo University*

Email: ohtou@nano.iis.u-tokyo.ac.jp, Tel: +81-3-5452-6264, Fax +81-3-5452-6265

1. Introduction

The threshold-voltage (V_{th}) control by a back-bias [1-3] is considered to be a key low-power technology for the next generation. Fully-Depleted (FD) SOI MOSFET, including FinFET [4], is also promising for a low-power and high-speed device. However, the combination of the two technologies [5-7] has a serious problem: too small body-factor (γ) due to a thick buried-oxide (BOX). Thin BOX, on the other hand, results in performance degradation because of a large parasitic capacitance. To solve the problem, we have proposed variable-γ FD SOI MOSFET [8], where γ is varied by substrate depletion layer and excellent I_{on}-I_{off} characteristics can be obtained.

In this paper, the bias scheme for the variable-γ device is re-examined. It is found that the bias scheme in Ref. [8] (*Type-I*) severely degrades the inverter delay (t_{pd}) especially when BOX is extremely thin. We propose a new bias scheme (*Type-II*) to solve the problem.

2. Issues of Back-bias for FD SOI MOSFETs

The sub-threshold leakage is suppressed by raising V_{th} using back-bias. V_{th} shift is given by [3]

$$|\Delta V_{th}| = \gamma |\Delta V_{sub}| \cdot \qquad (1)$$

where V_{sub} is the back-bias. Thus, a proper γ is needed in a given V_{sub} range. The γ is expressed as [3]

$$\gamma = C_B / C_G, \qquad (2)$$

where C_B is the channel-to-back-gate capacitance and C_G is the channel-to-front-gate capacitance. C_B should be increased for a sufficient γ with a constant t_{ox}. However, for FD SOI MOSFETs, thin BOX for obtaining a sufficient γ leads to the serious increase in the drain capacitance and the degradation in S swing and gm. Hence, there are tradeoffs between the device performance and γ (Table I).

3. Variable-γ FD SOI MOSFET (Type-I)

i) Operation Principle: Fig. 1 shows schematics of a (inversion-type) variable-γ FD SOI MOSFET. The features are: (1) an extremely thin BOX (~10nm), (2) a low impurity concentration in the substrate (N_{sub}) (~10^{15}-10^{16}cm^{-3}), and (3) a side-gate (n$^+$ region) for minority carrier injection into the BOX/substrate interface.

In Type-I, the substrate (V_{sub}) and the side-gate (V_{side}) are biased to change V_{th} with V_{sub} = V_{side} (See Table II). In the standby-state, negative bias is applied and the BOX/substrate interface is inverted. Then, γ becomes large and a low I_{off} is obtained by a large V_{th} shift. In the active-state, slightly positive bias is applied and the BOX/substrate interface is depleted. Then, γ becomes small due to the depletion layer and a high I_{on} is obtained. t_{pd} is also small due to small drain capacitance [9].

ii) Disadvantages: If the effective gate oxide thickness (t_{ox}) is less than 1 nm in the future, BOX should be very thin (< 10 nm) to obtain a sufficient γ. Moreover, in a 3D channel device like a FinFET, γ becomes even smaller [3,10] and thinner BOX is required. However, with an extremely thin BOX, the BOX/substrate interface is easily inverted or accumulated even in the active-state, as shown in Fig. 2 and Fig. 3(a). The inversion below the drain causes the t_{pd} degradation [11] and the accumulation causes further decrease in V_{th} resulting in the dramatic increase in I_{off}. Furthermore, the range of V_{th} control in the active state is narrow with V_{sub} connected to V_{side}. Therefore, the stable depletion in the BOX/substrate interface and wide V_{th} control are strongly required in the active-state, and Type-I with an extremely thin t_{BOX} does not have the substantial solution for the degradation.

4. New Bias Mode of the Variable-γ (Type-II)

We propose a new bias mode (Type-II) where V_{sub} is fixed and V_{side} is varied to change V_{th}. In the active-state, a positive V_{side} is applied, resulting in the reverse-bias between the side-gate and back-gate. Then, the minority carrier (component of the inversion layer) and majority carrier are absorbed into the side-gate and back-gate, respectively. Thus, the depletion is obtained even with an extremely thin t_{BOX} (Fig. 2 and Fig. 3(b)). In the standby-state, on the other hand, the inversion in the BOX/substrate interface is obtained by a negative V_{side}, and large V_{th} shift is obtained by a large γ.

5. Measured/Simulated Results and Discussion

Figs. 4 and 5 show simulated and measured V_{th} shift by a back-bias in Type-I and Type-II. Both V_{side} and V_{sub} are varied in Type-I and only V_{side} is varied in Type-II. Please note that the measured data in Fig. 5 is for a p-MOSFET. In Type-II, a wide-range of substrate depletion for the active-state is obtained by a forward V_{side}. Furthermore, the V_{th} control in the active-state is realized by V_{sub} modulation, so the V_{th} adjustment for reducing inter-die V_{th} fluctuations using back-bias is possible [2,3]. Fig. 6 shows simulated t_{pd} dependence on V_{sub} and V_{side} in the Types-I and II. Although low I_{off} in the standby-state and small t_{pd} in the active-state are realized by both types, Type-II has better I_{off} in the active-state than Type-I because the inversion or accumulation in the BOX/substrate interface is avoided. Fig. 7 shows simulated t_{BOX} dependence of t_{pd} at the constant I_{off} ($5\text{x}10^{-8}$Aμm^{-1}). FD SOI MOSFET with a high N_{sub} in the active-state is also shown for comparison. Type-II shows only a slight degradation of t_{pd} in the range of t_{BOX} less than 10 nm. Fig. 8 shows simulated I_d-

0-7803-9212-4/05/$20.00 ©2005 IEEE

V_g of a FinFET with a low aspect-ratio [3,10] using Type-II. Even with a FinFET, a low I_{off} in the standby-state and small gate/drain parasitic capacitance due to the depletion in the BOX/substrate interface in the active-state are realized by Type-II.

6. Conclusion

A new bias scheme of variable-γ FD SOI MOSFET is proposed. Using the scheme, almost no degradation of t_{pd} in the active-state is achieved even in the t_{BOX} of a sub-10 nm while I_{off} is sufficiently suppressed in the standby-state. Reducing the inter-die V_{th} fluctuation on a wide V_{sub} range in the active-state is realized. This device scheme is also well applicable to 3D channel MOSFETs including a FinFET.

Acknowledgements: This work was supported in part by the program for the "Promotion of Leading Researches" in the Special Coordination Funds for Promoting Science and Technology from the MEXT, Japan. The device simulators (Medici™ and Davinci™) have been supplied through VLSI Design and Education Center (VDEC), the University of Tokyo in collaboration with Synopsys Inc.

References: [1] T. Kuroda et al., IEEE JSSC **31** (1996) 1770. [2] J. W. Tschanz et al., IEEE JSSC **37** (2002) 1396. [3] T. Hiramoto et al., JJAP **42** (2003) 1975. [4] X. Huang et al., IEDM (1999) 67. [5] T. Kachi et al., Symp. VLSI Tech. Dig. (1996) 124. [6] I.Y. Yang et al., IEEE TED **44** (1997) 822. [7] R. Tsuchiya et al., IEDM (2004) 631. [8] T. Ohtou et al., JJAP **43** (2004) 3311. [9] Y. Omura et al., IEICE Trans. Elec. **75-C** (1992) 1491. [10] T. Nagumo et al., JJAP **44** (2005) 50. [11] T. Ohtou et al., JJAP **44** (2005) 3885.

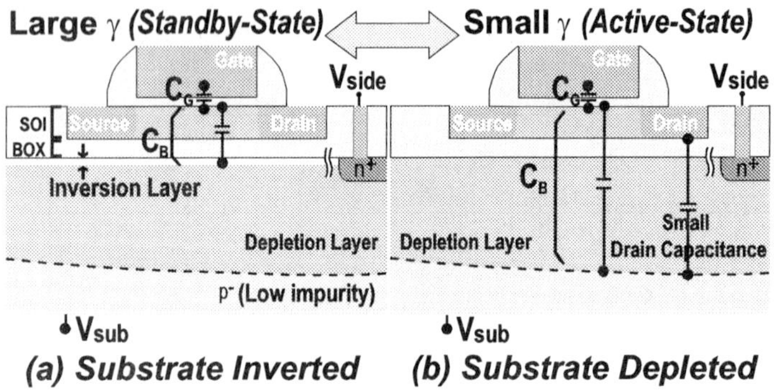

Fig. 1: Schematics of the inversion-type variable-γ FD SOI MOSFET [8]. γ is expressed as C_G/C_B [3]. Substrate condition is changed by a back-bias (V_{sub} and V_{side}). The side-gate (n+ region) is for the supply of minority carriers (inversion charges) into the BOX/substrate interface during the standby-state.

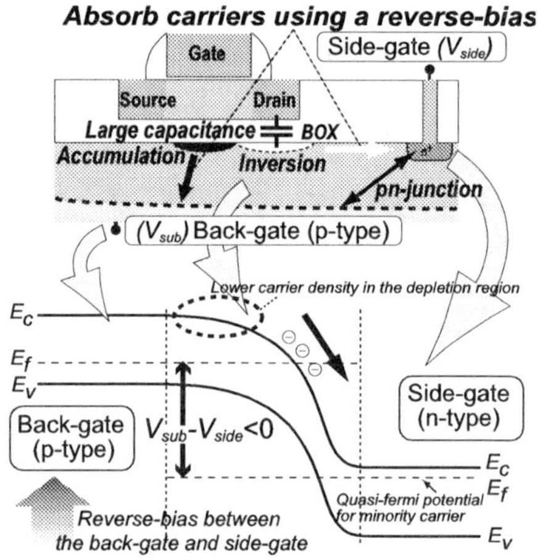

Fig. 2: Mechanism of the variable-γ device of Type-II in the active-state. The depletion in the BOX/substrate interface is obtained by a reverse-bias between the side-gate and back-gate because of deeper quasi-Fermi levels of carriers in the substrate.

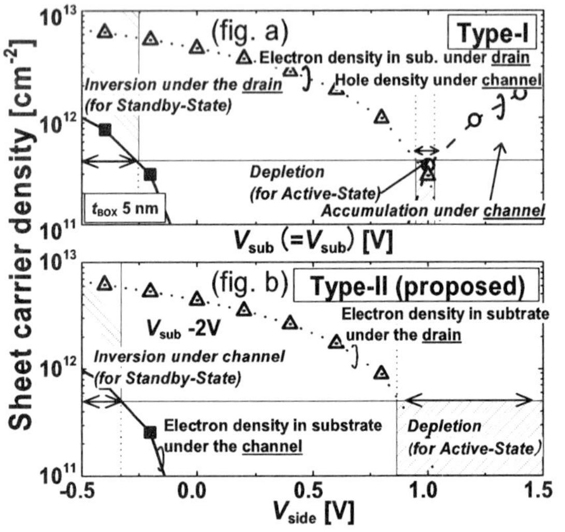

Fig. 3: Simulated sheet carrier density in the BOX/substrate interface. (a) Type-I and (b) Type-II with t_{BOX} of 5 nm and N_{sub} of $1x10^{16}$ cm^{-3}. A large depletion window for active-state is obtained in Type-II while very small in Type-I. The inversion and accumulation are defined as the condition that the density exceeds $5x10^{11}cm^{-2}$.

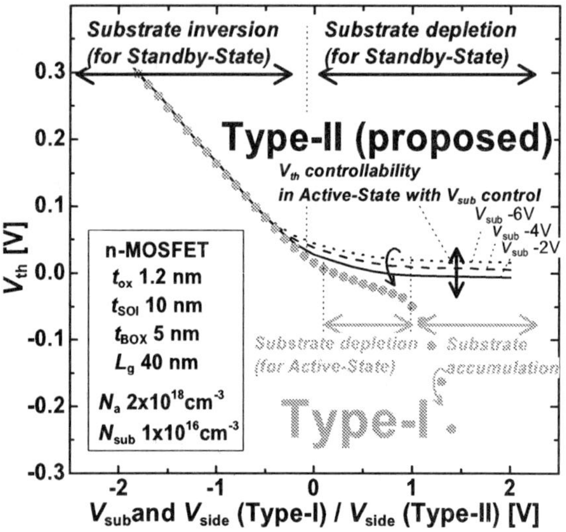

Fig. 4: Simulated dependence of V_{th} on V_{side} in the Type-II and on V_{sub} (=V_{side}) in the Type-I of a variable-γ n-MOSFET in the short-channel regime. A large depletion window is obtained in Type-II while small in Type-I. Note that V_{th} control in the active-state on Type-II is realized by controlling V_{sub}.

Table I: Device characteristics tradeoffs of thin BOX for back-bias scheme.

Advantages	Disadvantages
• Large V_{th} shift thanks to large γ	• Large parasitic capacitance • Degraded transconductance • Degraded S swing

Table II: Bias voltages of the two modes for example.

Bias Scheme	Type-I	Type-II (proposed)	
	$V_{side}=V_{sub}$	V_{side}	V_{sub}
Active-State	+0.5V	+1V	-3V
Standby-State	-3V	-2V	

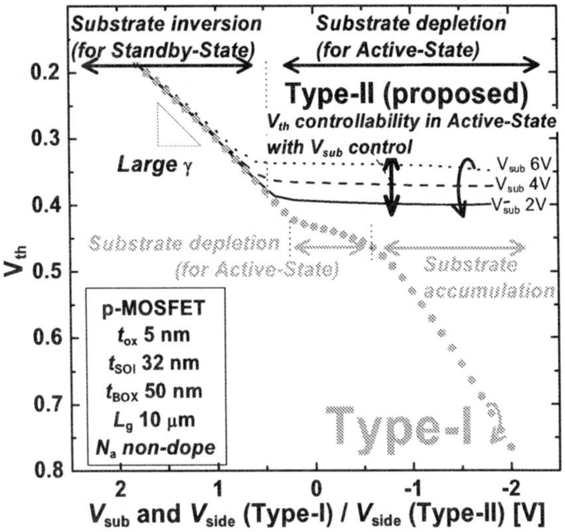

Fig. 5: Measured dependence of V_{th} on V_{side} in the Type-II and on V_{sub} (=V_{side}) in the Type-I of a variable-γ p-MOSFET. Note that the vertical and horizontal axes are inverted. The similar tendencies to Fig. 4 are obtained.

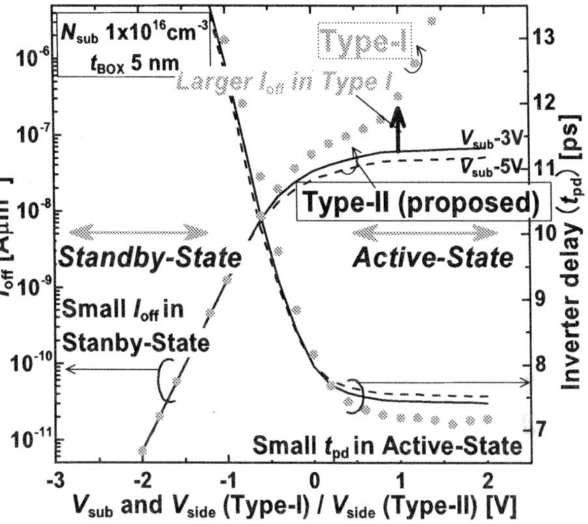

Fig. 6: Simulated dependence of t_{pd} and I_{off} on V_{side} in Type-II and on V_{sub} (=V_{side}) in Type-I. Extremely low I_{off} in the standby-state and high-speed in the active-state are realized. Type-I has larger I_{off} in the active-state than Type-II.

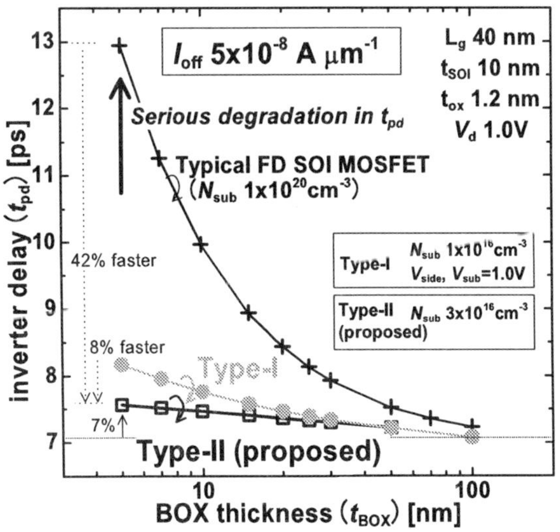

Fig. 7: Simulated dependence of t_{pd} on t_{BOX} in Type-I, Type-II, and a typical FD SOI MOSFET at the active-state. I_{off} is set to $5x10^{-8}$A μm^{-1}. Only 7% degradation on t_{pd} compared with t_{BOX}=100nm is obtained in Type-II with t_{BOX}=5nm.

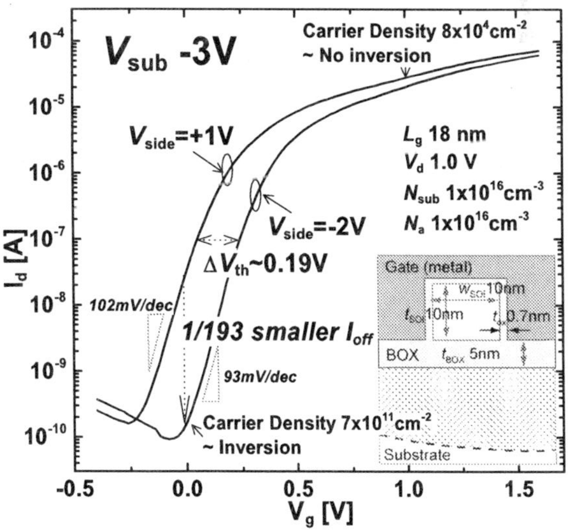

Fig. 8: Simulated I_d-V_g of a FinFET with a low aspect-ratio [3,10] using the variable-γ scheme of Type-II. No inversion in the substrate in the active-state, good SCE immunity and 1/193 smaller I_{off} in the standby-state is achieved at the ITRS target in 2010. Gate leakage is ignored.

3D Via Etch Development for 3D Circuit Integration in FDSOI*

J. M. Knecht, D.R.W. Yost, J. A. Burns, C. K. Chen, C.L. Keast, K. Warner
Lincoln Laboratory, Massachusetts Institute of Technology, Lexington, MA 02420
Email: jmk@ll.mit.edu , Tel. (781) 981-2636

A dense, high-yielding 3D via etch process has been developed and has become an integral part of Lincoln Laboratory's 3D IC technology. The 3D vias form vertical electrical connections between two or more stacked wafers, called tiers. Since Lincoln's process technology is SOI-based these vias are etched through oxide, rather than silicon, as would be the case in a bulk technology. Details of Lincoln Laboratory's 3D IC technology have been discussed elsewhere [1]. The vertical interconnect structure, shown in Fig. 1, is designed to use a concentric, self aligned process where the connection is made through a via etched through a metal annulus on the top tier down to a metal land on the tier below. It is unique from a traditional oxide via etch in that the metal annulus forms the top electrical contact and its presence directly impacts the etch process. Vertical connections constructed this way minimize the pitch of 3D vias, increasing the packing density. Current 3D via design rule is 1.75 μm square with a submicron version in development. 3D via depth is about 8 μm. This abstract will describe the development of the 3D via etch process.

The oxide via etch was developed in a Trikon Technologies low pressure, high density, helicon-based cluster tool. A response surface design-of-experiments (DOE) was performed varying etch pressure and wafer bias to examine their effect on etch profile and etch rates. An anisotropic etch is essential for high packing density. There was an excellent fit between the data and the model. Low pressure and high bias were required to give vertical profiles. Higher etch pressure caused excessive polymer deposition resulting in etch stop. Low wafer bias could not remove the deposited polymer fast enough, also resulting in etch stop. Cross-sectional SEM's from two extreme points of the DOE are shown in Fig. 2. The etch conditions chosen are shown in Table 1. It was expected that the metal annulus exposed during the etch (ledge) would form a hard mask for the lower part of the etch. In fact, the presence of the metal ledge played a crucial role in the etch development and will be described below.

Electrically testable 3D via kelvins and via chains were designed to quantify via resistance and yield. Early test results showed high yield and low resistance to the annulus metal but poor contact yield to the landing pad on the lower tier. Cross-sectional SEM photos of 3D vias showed that the metal ledge had been completely removed and that the vias had pinched off below the annulus metal. Test samples etched without the annulus metal displayed no via pinch-off (Fig. 3). We believe that the removal of the TiN/Ti/Al/Ti metal ledge is accomplished both by chemical and physical etching. The Ti/TiN is volatile in F_2-based chemistries and while AlF is not, the high wafer bias is sufficient to remove it. It has been reported that via etching down to Ti-containing metals produces excessive polymer [2]. It is hypothesized that when the oxide etch hits the annulus, excessive metal organic polymer is formed which is then deposited on the via sidewalls. This can lead to via pinch-off. A yield model, shown in Fig 4, was constructed from test devices with varying annulus and 3D via sizes. The model indicated that the major yield driver is the annulus opening size. In order to increase 3D via yield the ledge width needed to be minimized and/or the process modified. In practice, both options were selected. The current three step process includes adding a small amount of O_2 to the oxide etches for polymer removal with an intermediate in-situ Cl_2 metal etch to remove the ledge. Fig. 5 shows a plot of via chain yield that compares two sizes of ledge and two different processes. The reduction in ledge width increased via chain yield from approximately 70% to 100%. Also, the process changes decreased the 3D via etch sensitivity to the ledge. With these process changes we routinely yield chains of 10,000 3D vias at 100% and with a 3D via resistance of 0.68 Ω. A plot of 3D via resistance versus 3D via count is shown in Fig. 6 for both the current 3D via and the submicron 3D via. A SEM image of a completed W-filled 3D via is shown in Fig. 7.

A number of fully functional 3D IC circuits have been fabricated at Lincoln Laboratory using this vertical interconnection technology. They include a 99-stage 3-tier ring oscillator and a 2-tier 1024 x 1024 imager. An image acquired from the latter is shown in Fig. 8 [3]. Each pixel in the imager contains one 3D via that must function for the pixel to work. This imager had >99% 3D via yield and is the most advanced 3D circuit built in Lincoln Laboratory's 3D process technology.

References

[1] J. Burns, et al, *GOMACTech-05,* pp. 407-409, April, 2005.
[2] G. Colovos, et al., *SPIE Conference on Multilevel Interconnect Technology III,* vol. 3883, pp. 52-59, 1999.
[3] V. Suntharalingam, et al, *ISSCC Dig. Papers,* pp. 356-357, Feb. 2005.

* This work was sponsored by the United States Air Force under contract FA8721-05-C-0002. Opinions, interpretations, conclusions, and recommendations are those of the author and are not necessarily endorsed by the United States Government.

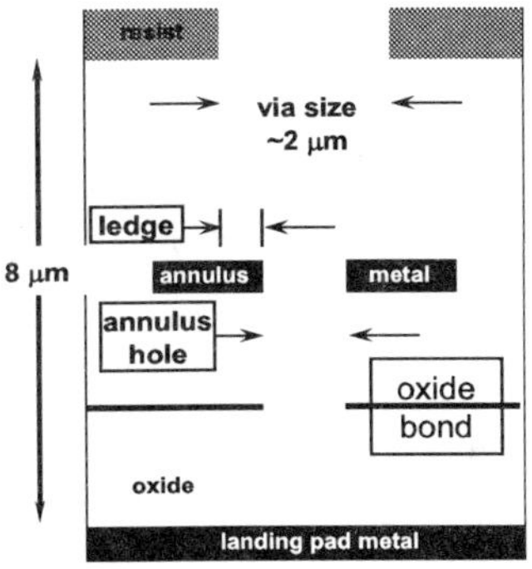

Fig. 1: Cross-section of 3D Via

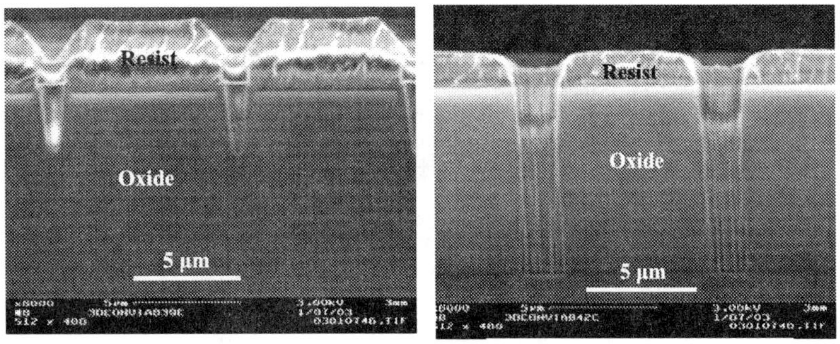

a) High pressure, Low Bias

b) Low Pressure, High Bias

Fig. 2: SEM Results from Etch DOE

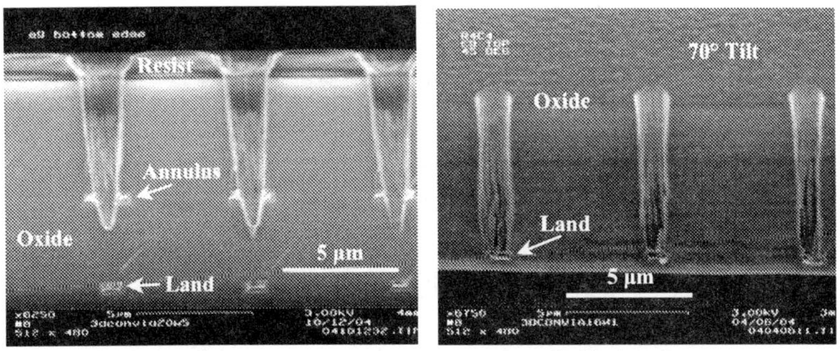

a) Annulus

b) No Annulus

Fig. 3: SEM Photos with and without Annulus Metal

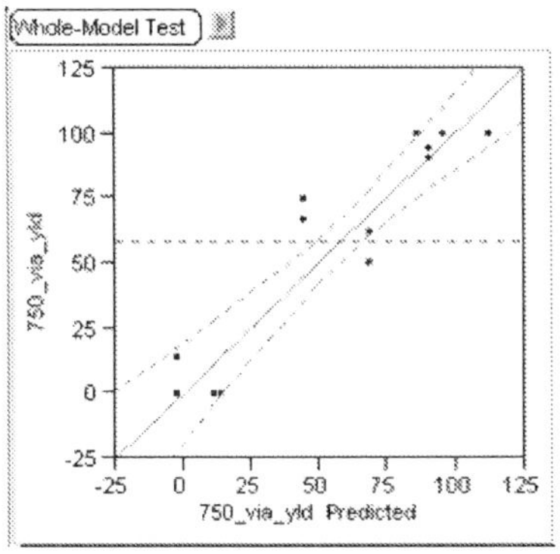

Fig. 4: Yield Model (Actual Yield vs. Predicted Yield)

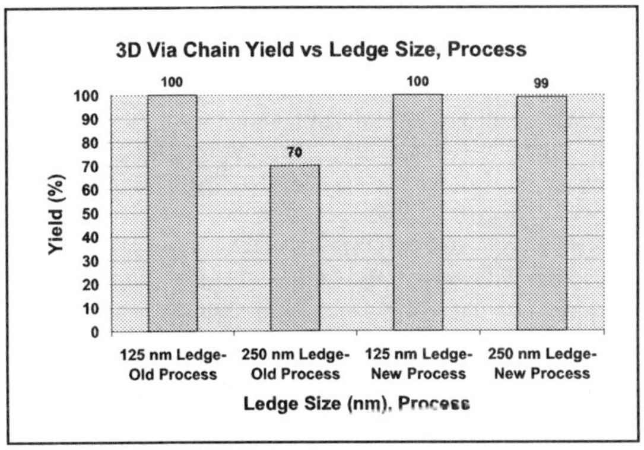

Fig. 5: 3D Via Chain Yield vs. Ledge Sizes, Process

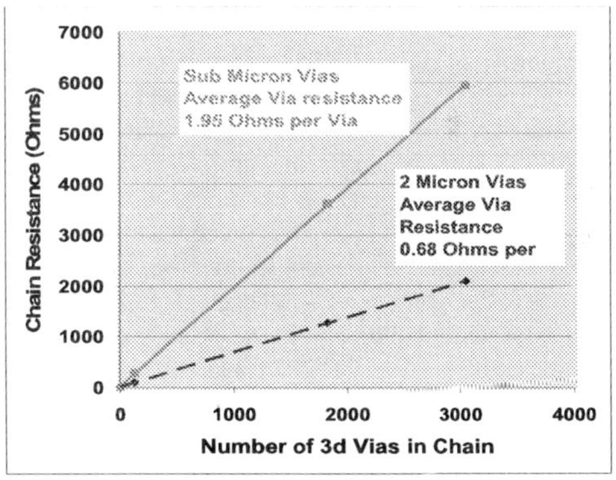

Fig. 6: 3D Via Chain Resistance vs Number of Vias

3D Via Etch		
	Oxide	Metal
Source Power (W)	1750	2750
Bias Power (W)	1300	250
Press (mTorr)	3.5	6
C4F8 (cc)	30	—
CH2F2 (cc)	40	—
CO (cc)	24	—
Ar (cc)	80	—
Cl2 (cc)	—	75
BCl3 (cc)	—	10
N2 (cc)	—	15

Table 1: 3D Via Etch

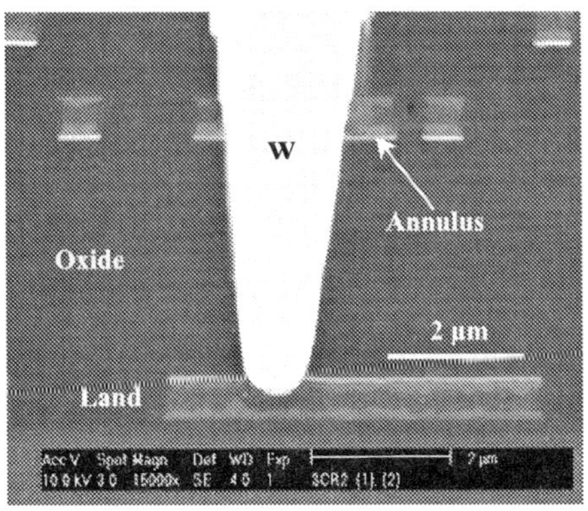

Fig. 7: SEM Cross-section of W-filled 3D Via

Fig. 8: Image from $(8\ mm)^2$ 1024x1024 Imager

Stress Technology Impact on Device Performance and Reliability for <100> sub-90nm SOI CMOSFETs

Wen-Kuan Yeh[1], Chieh-Ming Lai[2], Chien-Ting Lin [2], Yean-Kuen Fang [2], W. T. Shiau[3]
1. Department of Electrical Engineering, National University of Kaohsiung, Taiwan.
2. Institute of Microelectronics,National Cheng Kung University, Taiwan. 3. United Microelectronics Corporation, Central R&D Division
No.700, Kaohsiung University Rd., Nan-Tzu Dist., Kaohsiung, Taiwan *Emial: wkyeh@nuk.edu.tw/Phone:886-7-5919234/FAX: 886-7-5919374*

1. Introduction

<100> channel orientation substrate and high tensile stress gate capping layer (GC liner-SiN) have been adapted for the mobility improvement of CMOSFETs [1-4]. There are few reports studying the effects of GC liner-SiN film thickness on device's characteristic and reliability. In this work, the impacts of high tensile stress GC liner-SiN thicknesses on device performance and hot-carrier induced degradations for 90nm PD-SOI CMOSFETs were investigated. We also inspect the stress come from GC liner-SiN and STI edge impact on device characteristic using the length of diffusion (LOD) check.

2. Experiments

PD-SOI CMOSFETs were fabricated on <100> channel orientation SOI substrate based on advanced 90nm CMOSFETs process with 200nm buried oxide layer (BOX) and 40nm Si layer using STI isolation technology. Sub-90nm devices were fabricated with different thicknesses of high tensile stress GC liner-SiN layer (700A, 1100A) and LOD (0.4 μm~4.5μm). Device measurements and hot-carrier stressing were performed on probe station using various drain voltages (V_D=0~1.4V), and various gate voltages (V_G=0~1.4V) with 100 minutes hot-carrier stress time.

3. Results

Figure 1 shows the control methods of stress orientation. High stress liner-SiN induces a tensile stress in channel region, and STI induces a compressive one. From previous studies [1], it is found that <100> channel orientation substrate will enhance the mobility of pMOSFET and channel tensile stress induced by GC liner-SiN layer will enhance the mobility of nMOSFET; besides, a thicker (140A) liner-SiN layer will induce higher stress and improve the device performance[2]. However, in this work we found that thick GC liner-SiN layer (1100A) have no apparent improvement on device performance. The threshold voltage (V_{TH}) roll-off of 90nm BC-SOI nMOSFETs was shown in Fig. 2. It is obvious that devices with 700A GC liner-SiN layer have a lower V_{TH} and a smaller V_{TH} roll-off than devices with 1100A GC liner-SiN layer. And devices with 700A GC liner-SiN also possess smaller subthreshold swing and better I_{on}-I_{off} characteristics than 1100A ones (Fig.3 and 4). The I_D-V_D characteristics of 90nm BC and FB-SOI nMOSFETs are shown in Fig. 5. For body-tied device (BC-SOI), it is apparent that devices with GC liner-SiN possess larger drain current density (I_D) than conventional devices (SiN 380) does, and I_D in 700A device is much larger than that in 1100A one. Compared with BC-SOI, I_D of FB-SOI is high because of the floating body effect, and 700A devices possess the highest I_D. There are no apparent different between conventional (SiN 380) and 1100A GC liner-SiN devices. Same tendency was found in G_m characteristic (Fig. 6). In this work, larger mobility can be found on device with 700A GC liner-SiN. Figure 7 shows the I_D-V_D characteristics of 90nm BC-SOI nMOSFETs after 100min hot-carrier stressing. Because device's V_{TH} shifted after hot-carrier stressing, as shown in Fig. 8; thus, I_D rises apparently after hot-carrier stressing first and then degrades finally especially in 1100A device. Similar tendency was found on device's G_m characteristic. It is apparent that devices with 700A GC liner-SiN are more stable (I_D and G_m) after hot-carrier stress. It is presumably that thicker GC liner-SiN layer (1100A) will induce larger tensile stress and cause more serious damage to lattice structure, resulting in more serious hot-carrier induced device degradations (HCIDD). Figure 9 shows that lower V_{TH} was found on 90nm BC-SOI nMOSFETs with 1100A GC liner-SiN layer. We believed that more holes were trapped in the Si-SiO_2 interface during the hot-carrier stressing, thus lowering the V_{TH} and raising the I_D. Besides, hot-carrier induced V_{TH} variation in devices with 700A GC liner-SiN is more stable than that in devices with 1100A GC liner-SiN. In order to inspect the stress come from GC liner-SiN film and STI edge impact on device performance, various length of diffusion (LOD 0.4 μm~4.5μm) was used for comparison. The insert shows the test structure of LOD. Figure 10 shows the I_D-V_D of 700A GC liner-SiN SOI nMOSFETs with various LOD. Increasing the LOD will decrease the compressive stress caused by shallow trench isolation (STI), and increase tensile stress caused by high stress GC liner-SiN. Thus I_D increases apparently with LOD increases. Same tendency was found in the G_m characteristic for mobility check, as shown in Fig. 11. No V_{TH} shift was found while LOD changed. For pMOSFETs, It had been known that the major factor of mobility improvement is <100> channel orientation substrate [1]. In this work, it is found that for 90nm BC-SOI and FB-SOI pMOSFETs, even with various GC liner-SiN thicknesses, there is no apparent change on device characteristic (I_D and G_m), as shown in Fig. 12 and 13. Therefore, the effect of GC liner-SiN for SOI pMOSFETs is not apparent. Figure 14 shows the G_m characteristics for 90nm BC-SOI pMOSFETs with different stress GC liner-SiN thicknesses before and after 100min hot-carrier stressing; the inserts show the hot carrier induced I_D degradation. Similar tendency of hot-carrier induced device degradations was found for both stress GC liner-SiN device (700A, 1100A). The G_m characteristics degraded apparently just after 20min hot-carrier stressing and had a V_{TH} shift, then very serious device breakdown was found. Compared with nMOSFETs, the variations of pMOSFETs characteristics were sensitive; thus pMOSFETs is easy to be degraded and lost the MOSFET characteristics especially after the hot-carrier stressing. Table I summarizes the mobility modulations induced by various stress items. It is found that the stress technology impacts of device performance for mobility improvement are different for n and pMOSFETs in respectively. For nMOSFETs, thin GC liner-SiN layer (700A) has apparent improvement on device performance than thick one (1100A) does. For pMOSFETs, the GC liner-SiN doesn't show apparent impact on device's performance.

4. Summary

In this work, for 90nm PD-SOI CMOSFETs on <100> Si substrate, the impacts of high tensile stress GC liner-SiN thicknesses on device performance and hot-carrier induced degradations were investigated. For nMOSFETs, devices with 700A GC liner-SiN possess apparent mobility enhancement and hot-carrier reliability immunity than devices with 1100A GC liner-SiN do. We believed that thicker GC liner-SiN (1100A) will induce large stress defects and makes damage to the device's channel lattice structure, thus degrading device characteristics. For pMOSFETs, the effects of high tensile stress GC liner-SiN thicknesses on device performance are not apparent. The major factor of mobility improvement is <100> channel orientation Si substrate. It is necessary to optimum high tensile stress GC liner-SiN technology to enhance pMOSFETs reliability.

Acknowledgement

The National Science Council of Taiwan, R.O.C., under Contract NSC 93-2215-E-390-001, supported this work.

References

[1] T. Komoda, *et. al*, *in IEDM Tech. Dig.*, 2004, p. 217
[2] K. Goto, *et. al*, *in IEDM Tech. Dig.*, 2004, p. 209
[3] S. Pidin, *et. al*, *in IEDM Tech. Dig.*, 2004, p. 213
[4] T. Ghani, *et. al*, *in IEDM Tech. Dig.*, 2003, p. 11.6.1

0-7803-9212-4/05/$20.00 ©2005 IEEE

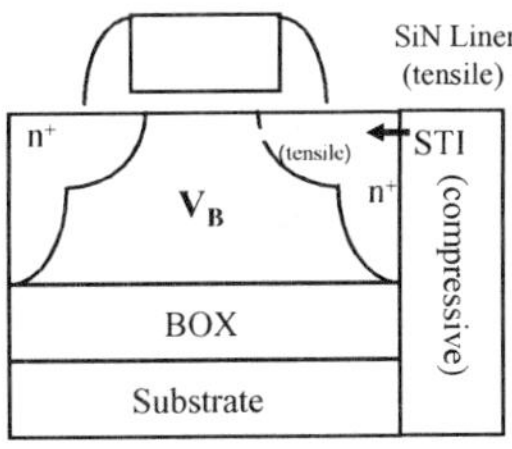

Fig. 1. The stress item in MOSFET channel region. High stress liner-SiN induces a tensile stress, and STI induces a compressive one.

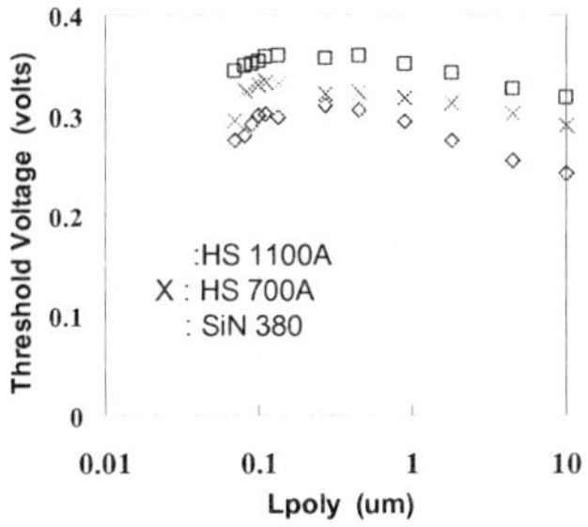

Fig. 2. V_{TH} of BC-SOI nMOSFETs with various stress GC liner-SiN layer.

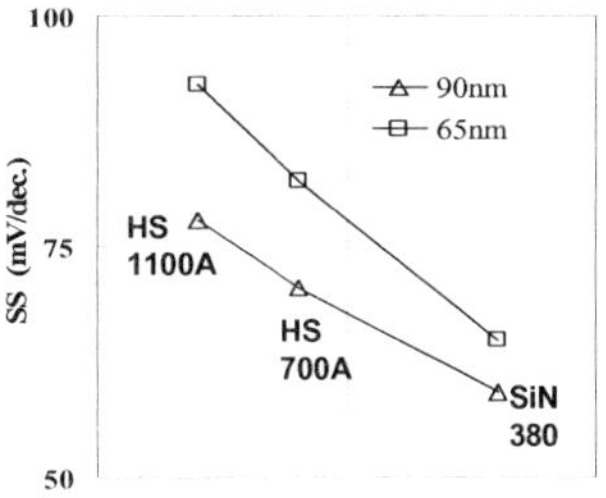

Fig. 3. Subthreshold Swing of BC-SOI nMOSFETs with various stress GC liner-SiN layers.

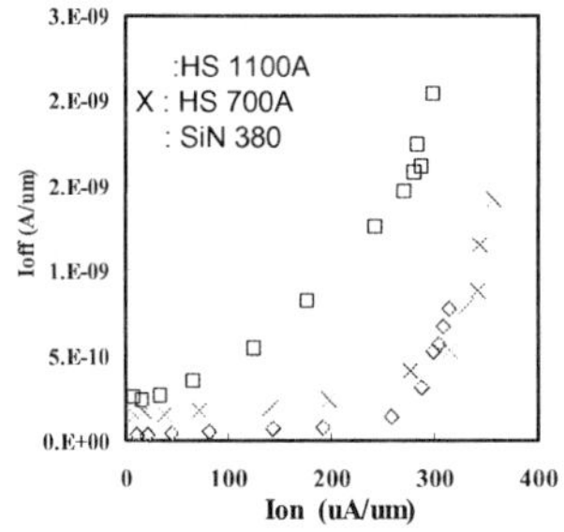

Fig. 4. I_{ON}-I_{OFF} characteristics of BC-SOI nMOSFETs with various stress GC liner-SiN layer.

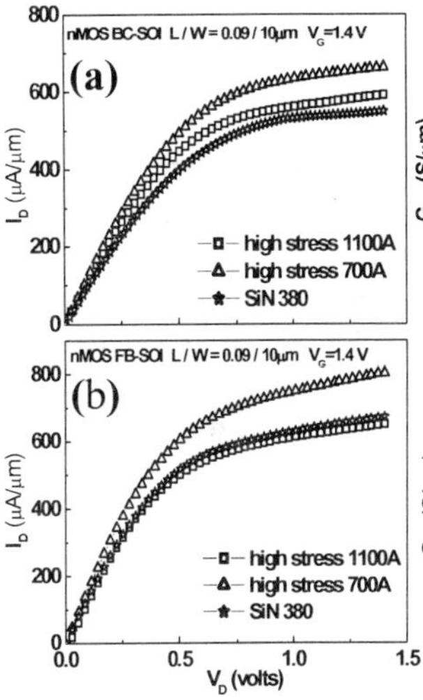

Fig. 5. I_D-V_D of 90nm (a)BC-SOI (b)FB-SOI nMOSFETs with various stress GC liner- SiN thicknesses.

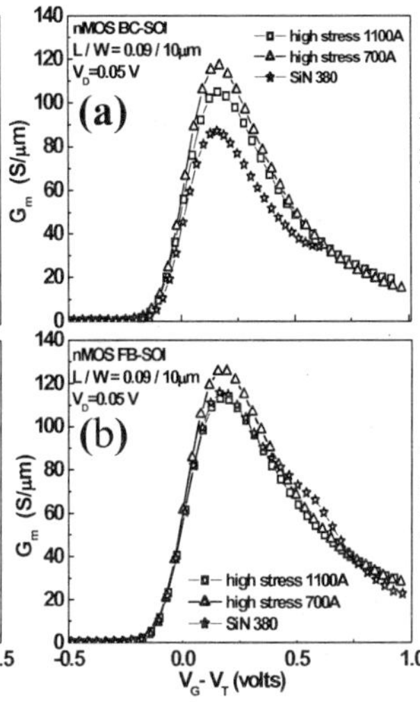

Fig. 6. G_M of 90nm (a)BC-SOI (b)FB-SOI nMOSFETs with various stress GC liner-SiN thicknesses.

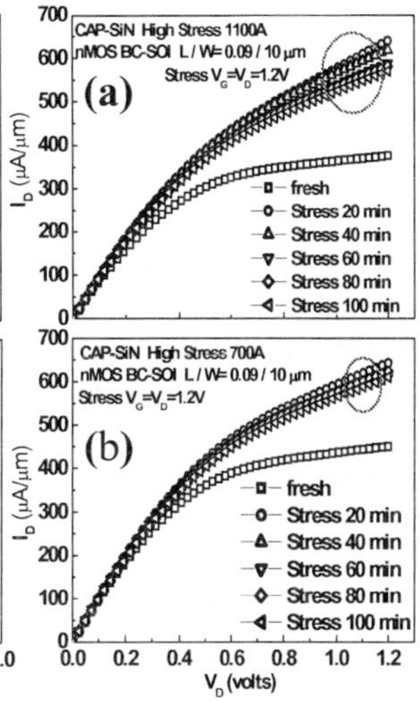

Fig. 7. I_D-V_D of 90nm BC-SOI nMOSFET with (a) 1100A (b) 700A high stress GC liner-SiN after 100 min hot carrier stressing.

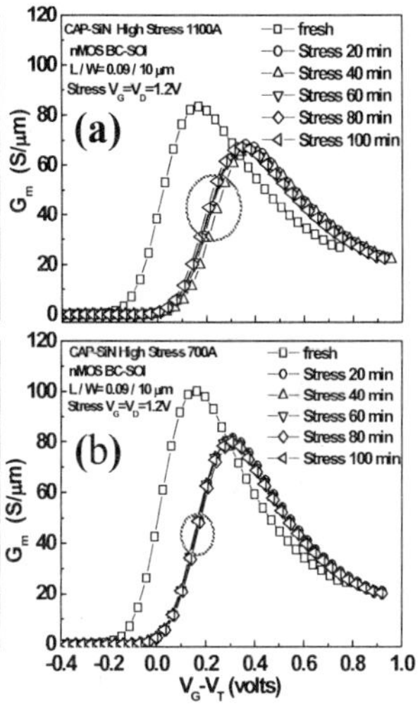

Fig. 8. G_M of 90nm BC-SOI nMOSFET with (a) 1100A (b) 700A high stress GC liner-SiN after 100 min hot carrier stressing.

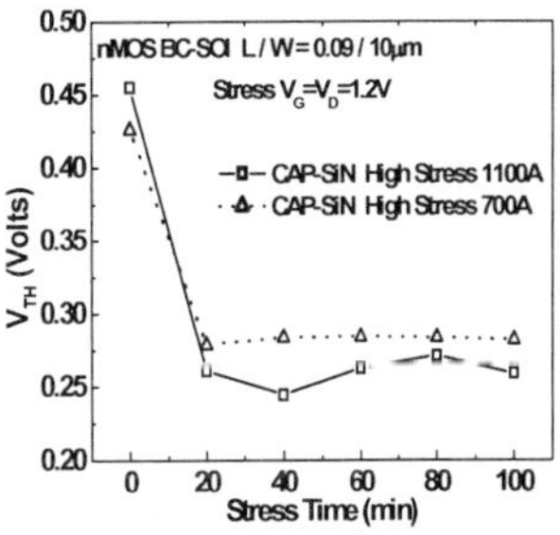

Fig. 9. V_{TH} variation of 90nm BC-SOI nMOSFET with different high stress GC liner-SiN after 100 min hot carrier stressing.

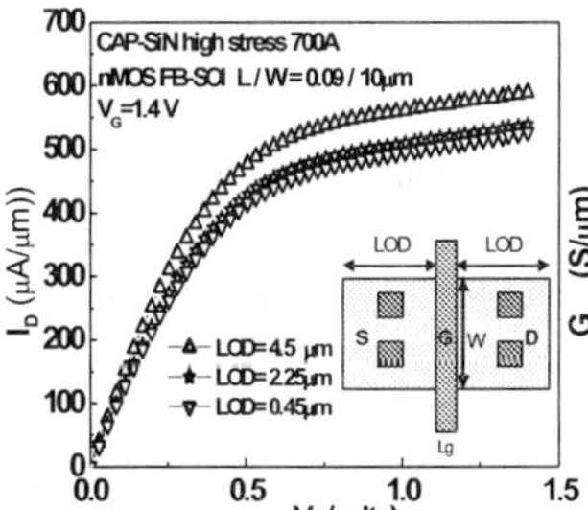

Fig. 10. I_D-V_D of 90nm FB-SOI nMOSFETs with various LOD. The insert shows the design of LOD.

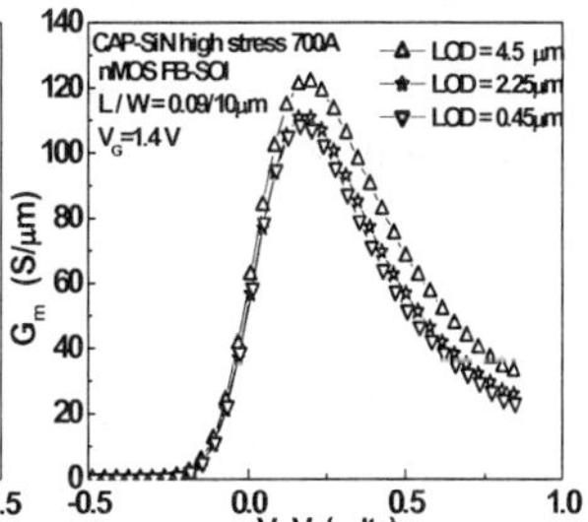

Fig. 11. G_M of 90nm FB-SOI nMOSFETs with various LOD.

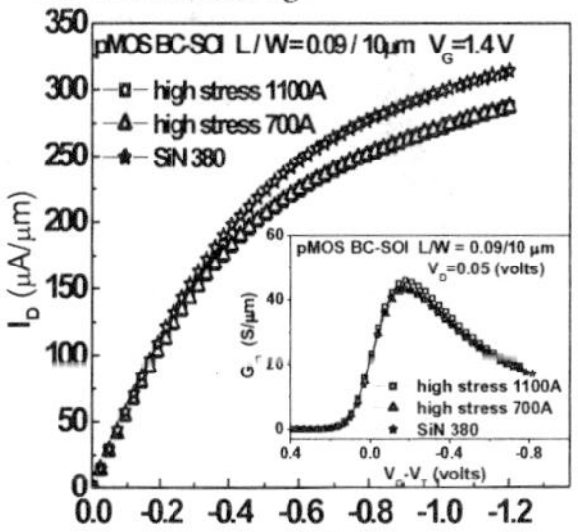

Fig. 12. I_D-V_D of 90nm BC-SOI pMOSFETs with various stress GC liner-SiN thicknesses. The insert shows the G_M characteristics.

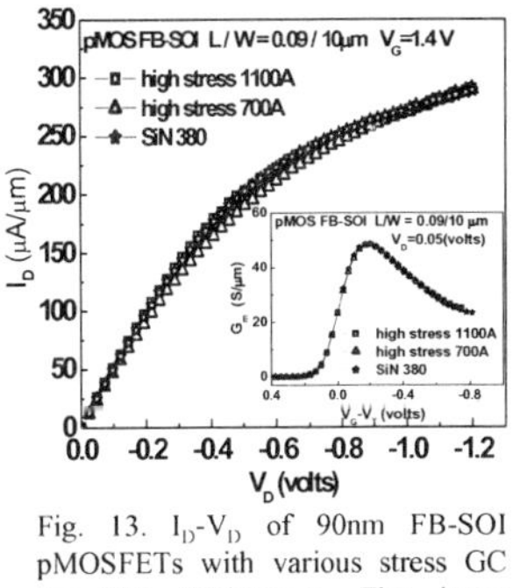

Fig. 13. I_D-V_D of 90nm FB-SOI pMOSFETs with various stress GC liner-SiN thicknesses. The insert shows the G_M characteristics.

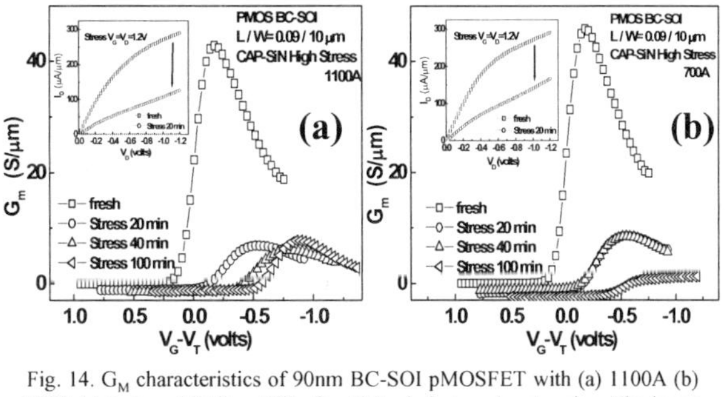

Fig. 14. G_M characteristics of 90nm BC-SOI pMOSFET with (a) 1100A (b) 700A high stress GC liner-SiN after 100 min hot carrier stressing. The insert shows the I_D-V_D characteristics. Both devices show unstable characteristics during hot-carrier stressing process.

Stress item	nMOS		pMOS
<100> channel	→		↗
GC Liner-SiN	700A ↗ +	1100A ↗	→
LOD	↗		↗ -
HCIDD	700A stable	1100A unstable	All unstable

Table I. The summary of mobility modulations induced by various stress item.

Scaling of GeOI NFET in sub-50nm regime using a Monte-Carlo method

S. Barraud, L. Clavelier
CEA-DRT/LETI 17 avenue des Martyrs 38059 Grenoble Cedex-9, France
Email: sylvain.barraud@cea.fr, Tel: (33)-4-38-78-98-45

Introduction

The recent fabrication of fully-depleted pMOSFETs with high-K dielectric and metal gate on 200mm GeOI [1] is an interesting way to reach the 22nm ITRS node. In the present work, we propose a theoretical investigation on the scaling of Germanium-On-Insulator (GeOI) using Monte-Carlo device simulations. A preliminary study is dedicated to electron transport in bulk Germanium. The improvement of low field electron mobility and energy relaxation time in comparison to silicon is highlighted. Then, the scaling of GeOI NFET is studied for a channel length L varying from 600nm to 40nm. An understanding of physical mechanisms underlying the improvement of linear and saturation current compared to Strained-Silicon-Directly-On-Insulator (SSDOI) is reported for both diffusive to ballistic transport. The results clearly show the interest of Ge for future CMOS technology.

Bulk transport in Germanium

First, the Monte-Carlo simulator MONACO has been suited in order to efficiently describe the electron transport in Ge [2]. Figure 1 displays the calculated velocity-field curves for different lattice temperature $T_L \in [77K\text{-}300K]$ in the <100> direction. Considering the experimental uncertainty, a good agreement is achieved between simulation results and experimental data [3]. The Ge band structure consists of four <111> minima at the edge of the Brillouin zone (named 'L valleys'), a single minimum at the center of the zone (named 'Γ valley'), and six <100> minima lying near the zone edge (named 'Δ valleys'). The electron repartition between the valleys as a function of the latteral electric field is shown in Figure 2 for three lattice temperature (300K, 190K and 77K). At low electric field, all electrons are located in the 'L valleys'. When the electric field increases, electrons are transferred from 'L valleys' to 'Δ valleys'. This new electron repartition between the Ge valleys may induce a conductivity effective mass modulated by the latteral electric field. The energy relaxation time calculated in Ge (Figure 3) is higher than its Si counterpart ($\sim 4\times10^{-13}$ s). It is mainly induced by a conductivity mass lower in Ge than in Si and results in an increase of electron-velocity overshoot. This effect, important in the nanometer regime, will be dealt in the following section.

Scaling of GeOI and SSDOI NFET

Figure 4 illustrates the simulated structure. The film thickness is equal to 15 nm with a doping of 10^{16} cm^{-3}. The shift of threshold voltage V_{th} exhibits in Figure 5 (~0.5V in GeOI against ~0.14V in SSDOI devices) shows a similar V_{th} roll-off. Figure 6 shows the channel length dependent enhancement of drain current induced by sSi and Ge on the linear current. For long channel, a 60% increase is reached in GeOI device against 80% in SSDOI device. The on-current improvement versus L is shown in Figure 7. The results clearly show two opposite behavior between GeOI and SSDOI NFET. As experimentally observed [4], the I_{on} improvement decreases with L in strained-Si based devices. In long channel length devices (L = 600nm), the transport is mainly diffusive and the electron velocity in GeOI device is lower than in SSDOI device. Figure 8 shows the average electron velocity along the device. When the channel length is shrinked, out-equilibrium and ballistic transport become more and more important. Thus, the electron velocity in GeOI device become higher than in SSDOI and SOI devices as observed in Figure 9. This is due to the more important energy relaxation time (cf. Figure 3). The decrease of the I_{on} improvement observed in SSDOI device (Figure 7) when L is reduced is explained by the modulation of the conductivity mass induced by the electron transfert from the Δ_2 to Δ_4 valleys (Figure 10). Around 30% of current enhancement is obtained in short SSDOI structures. In GeOI device, the I_{on} improvement increases when L decreases. The electron transfert from 'L valleys' to 'Δ valleys' observed in Figure 11 is not a critical point in Ge. An I_{on} improvement of ~55% is shown in 40nm-GeOI device. Moreover, an analysis of ballistic electrons along the channel (Figure 11) shows a ballisticity two time more important in GeOI device than in SSDOI device at the end of channel.

Conclusion

Electrical properties in Germanium and performances of GeOI devices have been studied using MC simulations. We have shown an I_{on} improvement increasing when the channel length is shrinked that is not the case in strained-Si based NFET. The results clearly demonstrate that the Germanium may be a promising material for the future CMOS technology.

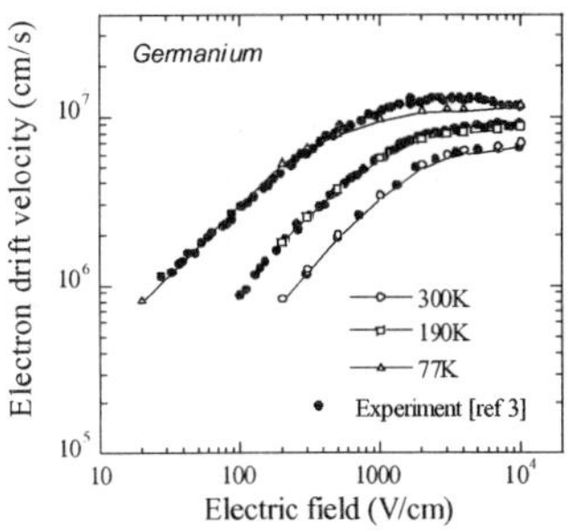

Fig. 1. v-E characteristics calculated for different temperatures in Germanium.

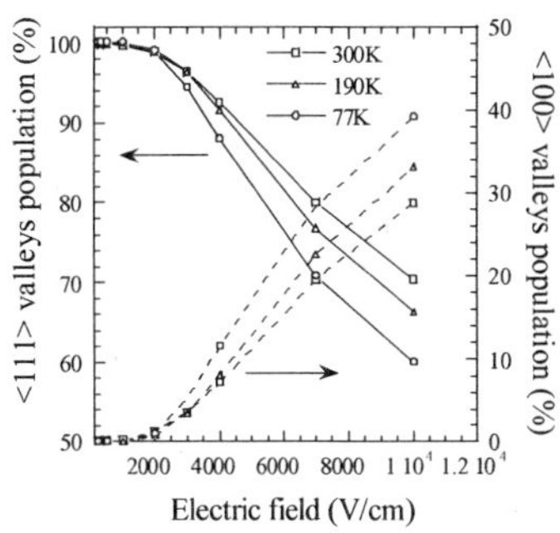

Fig. 2. Electron repartition between the valleys of Germanium.

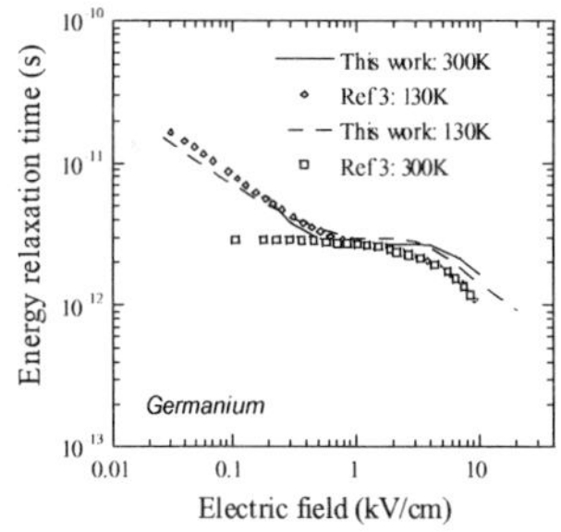

Fig. 3. Energy relaxation time as a function of the electric field in Germanium.

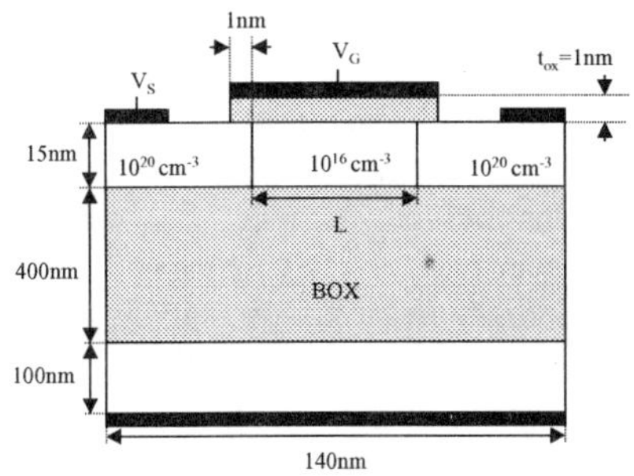

Fig. 4. Schematic representation of simulated device.

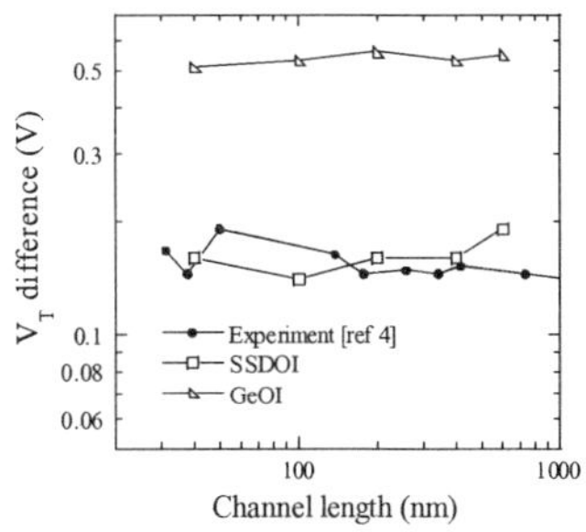

Fig. 5. $V_{th}^{SSDOI} - V_{th}^{SOI}$ and $V_{th}^{GeOI} - V_{th}^{SOI}$ as a function of L.

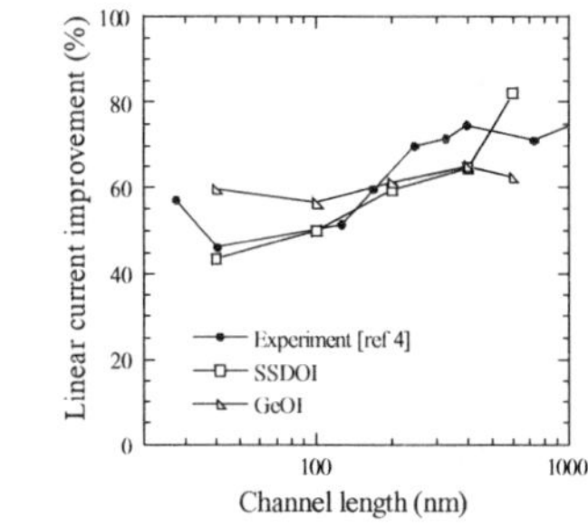

Fig. 6. I_{lin} improvement in SSDOI and GeOI NFET. (V_{GS}-V_T=1V and V_{DS}=0.05V)

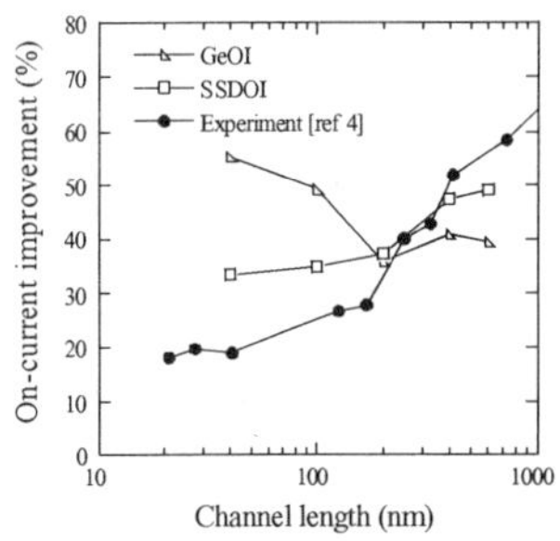

Fig. 7. I_{on} improvement in SSDOI and GeOI NFET. (V_{GS}-V_T=V_{DS}=1V)

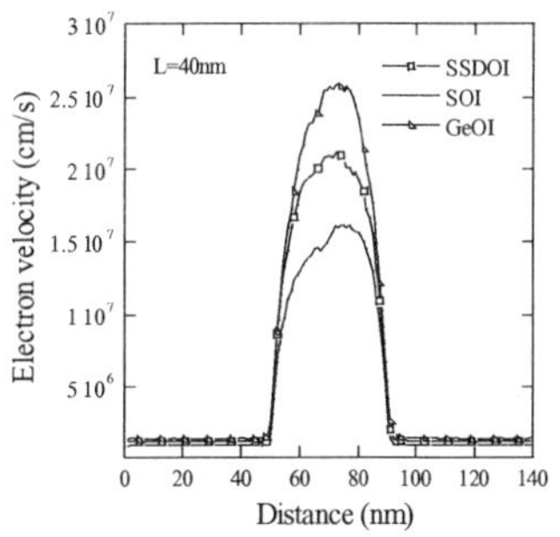

Fig. 8. Electron velocity along the device. (V_{GS}-V_T=V_{DS}=1V)

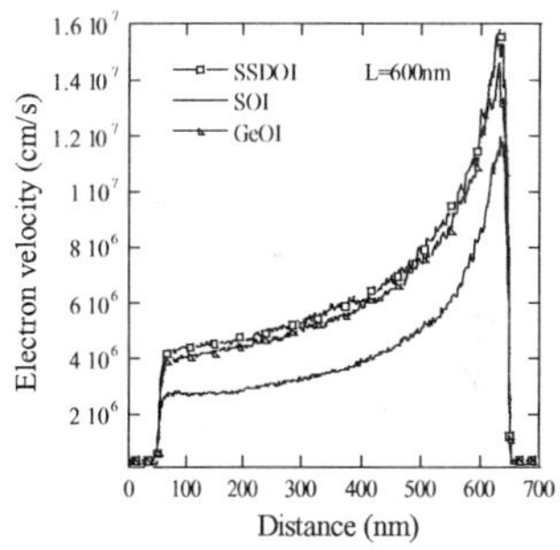

Fig. 9. Electron velocity along the device. (V_{GS}-V_T=V_{DS}=1V)

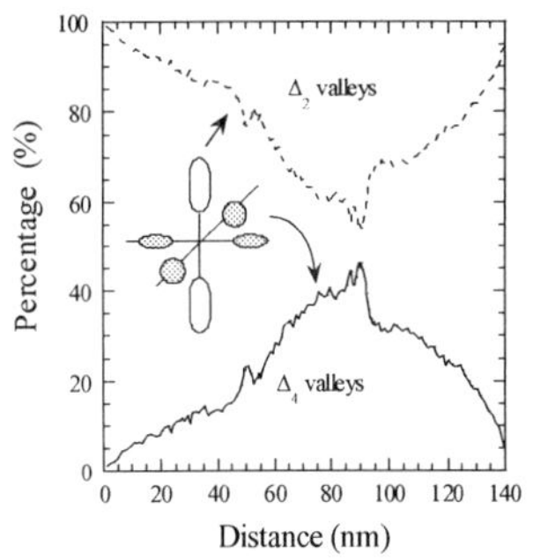

Fig. 10. Percentage of electron in the valleys of sSi: SSDOI (L=40nm and V_{GS}-V_T=V_{DS}=1V)

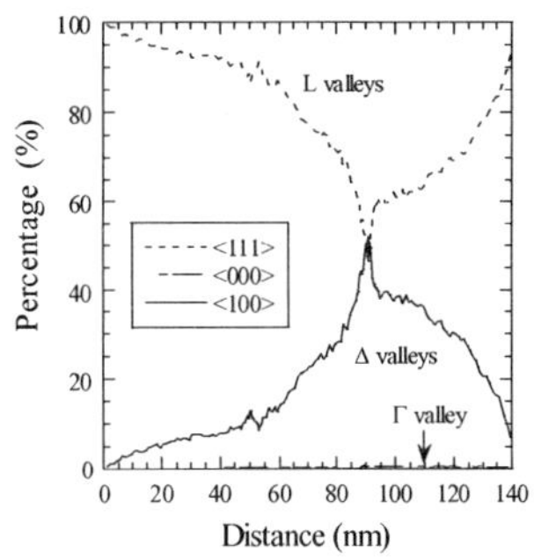

Fig. 11. Percentage of electron in the valleys of Ge: GeOI (L=40nm and V_{GS}-V_T=V_{DS}=1V)

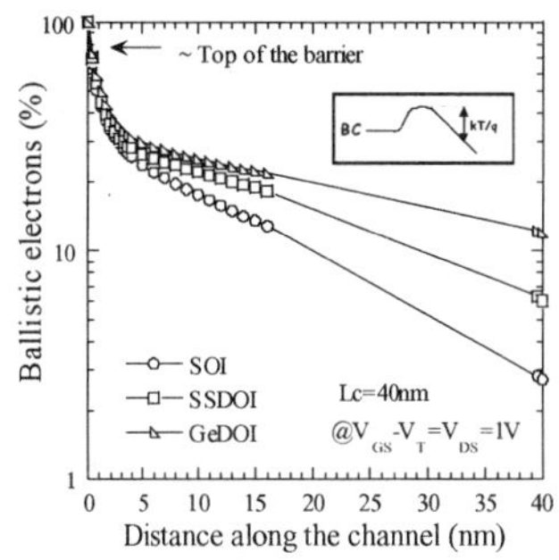

Fig. 12. Intrinsic balisticity in SSDOI, GeOI and SOI NFET.

Ref: [1] L. Clavelier, *et al.*, accepted for publication in SNW2005
[2] S. Barraud, *et al.*, accepted for publication in Solid State Electronics
[3] C. Jacoboni, *et al.*, Phys. Rev. B, p.189, 1981
[4] Q. Xiang, *et al.*, VLSI Tech., p.52, 2003

Revised split C-V technique for mobility investigation in advanced devices

V. Kilchytska[(1)], D. Lederer[(2)], P. Simon[(2)], N. Collaert[(3)], J.-P. Raskin[(2)], and D. Flandre[(1)]
[(1)] Microelectronics and [(2)] Microwave Laboratories, UCL, 3 Place du Levant, 1348 Louvain-la-Neuve, Belgium
[(3)]IMEC, Kapeldreef 75, B-3001 Leuven, Belgium.
Email: kilchytska@dice.ucl.ac.be, Tel.: 32(0)10 472564

In this paper we revise the split C-V technique widely used for mobility extraction. To extend its applicability we propose to use an integral of transconductance measured at high frequencies (HF) instead of the DC drain current. For the first time it is shown that such procedure allows not only to suppress parasitic gate-induced floating body effect (GIFBE), but also to improve the accuracy of mobility extraction in weak inversion (WI) regime. In this paper we apply our technique to partially-depleted (PD) SOI MOSFETs from a FinFET process and demonstrate its advantages in comparison to the standard method.

Introduction

Split C-V [1] is a well known powerful technique for mobility extraction, thanks to its robustness when other more simple methods [2] fail. Many improvements have been introduced in the initial method e.g. to overcome limitations of high leakage currents through thin gate oxide [3, 4] and to extend its applicability to short-channel devices [5]. Recently, some specific advantages of this technique have been demonstrated for mobility extraction in advanced structures like FinFETs [6, 7] and sub-0.1μm devices with HALOs [5]. This is especially promising for multi-gate FETs, since it allows to extract effective mobility without knowledge neither of gate oxide capacitance nor of channel width, which are not reliably known in such devices. However, in the same time, [6] showed that the split C-V procedure suffers from gate-induced floating body effect, which is unavoidably present in advanced SOI devices due to direct tunneling through the thin gate oxide.

In this work we propose a radical modification in the split C-V procedure, which suppresses the influence of floating body and transient effects on extracted mobility values. We demonstrate the applicability of this improved method on wide-fin FinFETs, which we used just as example. We believe that our revised method will be advantageous for any other advanced, especially PD SOI, MOS devices.

Device processing

Devices used in this study were microwave 2-gate fingers n-FinFETs with a 100 nm thick poly-Si gate deposited on top of 2nm nitrided oxide. High angle extension implantations, 70nm RTCVD nitride spacers, As+P and B deep S/D implants and NiSi silicide were used [8]. For this study we used large Fin widths, W_{Fin}, of 10-20 μm, so that for the high doping ($\sim 1\cdot 10^{18}$ and $6\cdot 10^{18}$ cm^{-3}) and Fin height, H_{Fin}, of 75nm, the devices operate in PD mode, which was desirable to test GIFBE. Furthermore, we used only 1-2μm long devices to avoid problems related to the effective channel length definition. However, despite of PD operation, only lower doped devices exhibited GIFBE, while higher doped devices do not show it, since their V_T coincide with V_g of GIFBE appearance and so the effect is masked.

Experimental procedure

Standard split C-V technique combines measurements of gate-to-channel capacitance, C_{gc}, and DC drain current, I_d, in linear regime versus gate voltage, V_g, to extract effective mobility in the following way [5]:

$$\mu_{eff} = \frac{L^2}{V_d}\cdot\frac{I_d}{\int_{V_0}^{V_g} C_{gc}(V_g)dV_g} \qquad \text{Eq. 1}$$

with L the device length, V_d the low drain voltage, V_g the gate voltage, and V_0 the gate voltage at which Q_{inv}~0.

However, in [6] it was demonstrated that this extraction procedure fails for devices with pronounced GIFBE (present in any floating body PD devices with thin gate oxide), giving an error in extracted effective mobility which can reach ~15-25% depending on device geometry. This occurs due to increased DC drain current values, while HF capacitance measurements are insensitive to this effect [6]. Indeed, [9] showed that GIFBE is suppressed for frequencies above 100kHz.

Hence, to exclude the influence of GIFBE on the extracted mobility values, in this work we propose to replace the drain current in Eq. 1 by the integral of the gate transconductance, G_m, measured at high frequency:

$$I_d = \int_{V_0}^{V_g} G_m(V_g)dV_g + I_{d_0}, \text{ where } I_{d0}=I_d \text{ at } V_g=V_0 \qquad \text{Eq. 2}$$

After substitution of Eq. 2 in Eq. 1, effective mobility can be expressed as follows:

$$\mu = \frac{L^2}{V_d}\cdot\frac{\int_{V_0}^{V_g} G_m(V_g)dV_g + I_{d_0}}{\int_{V_0}^{V_g} C_{gc}(V_g)dV_g} \qquad \text{Eq. 3}$$

In the next section we compare the mobility extracted by revised and standard split C-V methods.

Results and discussion

Firstly, comparing measurements at different frequencies, 1MHz has been chosen to assure the suppression of GIFBE and at the same time to not distort CV curves through series resistance or gate leakage. Fig. 1 presents the G_mV_g curves extracted from DC and 1MHz measurements. DC curve clearly exhibits strong GIFBE, which appears as a 2nd peak on G_mV_g curve, slow DC measurements only attenuate it, while at 1MHz this effect is, indeed, completely suppressed.

Comparison of the effective mobility extracted from I_d measurements (Eq. 1), and from HF G_m measurements (Eq. 3) is presented in Fig. 2. Mobility extracted in a standard way (i.e. using I_d measurements) suffers from GIFBE, which appears as a rather big hump at V_g~1V and leads to overestimation of the maximum mobility value by ~20% in this case. The mobility curve, extracted using the integral of HF G_m is free from this hump. This really confirms our expectations that the proposed improvement of split C-V method allows to suppress the negative influence of the gate-induced kink effect on the extracted mobility values.

Moreover, suppression of GIFBE is not the only advantage which our improved split C-V technique can yield. A first benefit is related to the transient (e.g. thermal) effects which do not affect HF measurements. The slight differences observed in the WI regime for devices subject to GIFBE (Figs. 1, 2), as well as on higher-doped devices which did not exhibit GIFBE (Fig. 3), can be related to transient effects and floating body in PD SOI MOSFETs without body contact. These devices need a certain time to reach the stationary operation. This statement was supported by I_dV_g measurements with different delay times. In the WI regime, with delay time increase, the curve obtained from I_dV_g measurements approaches the one calculated from HF G_m measurements, which is directly reflected on the mobility curves (Fig. 3). Another reason for an improvement of the mobility extraction from HF G_m can be the better accuracy, resulting from the integration of G_m and C_{gc} in the same voltage range. Therefore, better correlation between the current and the inversion charge can be expected as compared to the fixed drain current value used in Eq. 1 at a given V_g whereas the C_{gc} is integrated over a range of different V_g values. This improvement should be significant in the range of steep current changes, i.e. in the WI regime.

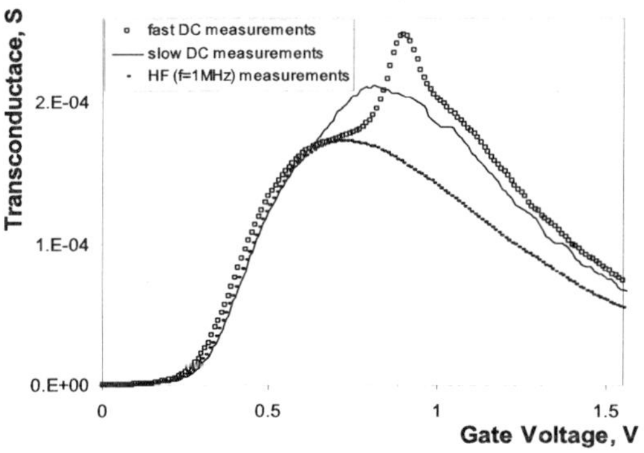

Figure 1. G_mV_g curves of wide nFinFET measured at DC (with short and long delay times: 10^{-4} and 1 sec, respectively) and HF (1MHz). L=2μm, W_{Fin}=2x10μm. V_d=50mV.

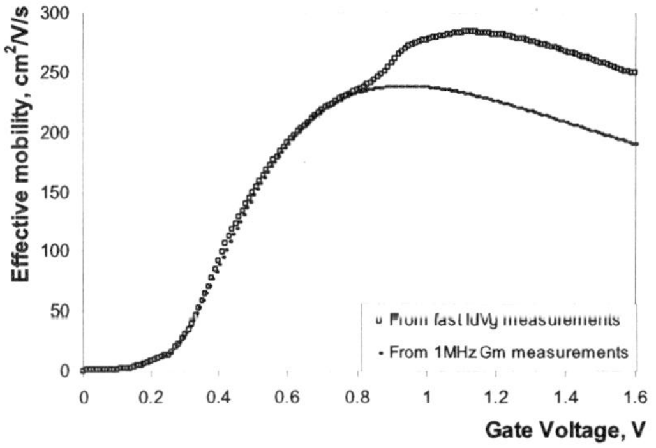

Figure 2. μ_{eff} as a function of V_g, extracted by standard split C-V method (Eq. 1) and by improved one (Eq. 3) for the device of Fig. 1.

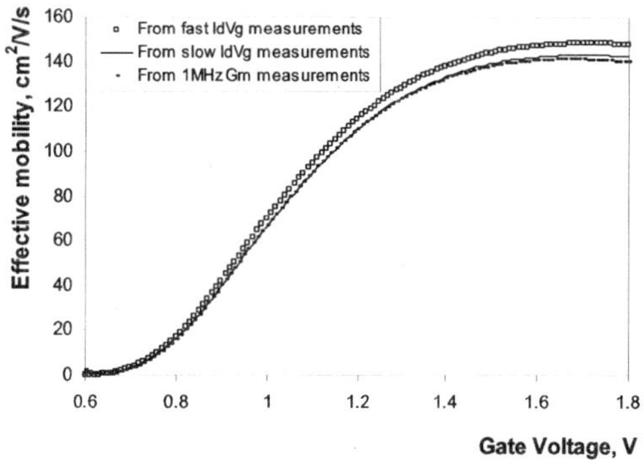

Figure 3. Effective mobility as a function of gate voltage, extracted by standard split C-V method (Eq. 1) from fast and slow (~1s delay) I_dV_g measurements and by improved method (Eq. 3) for a device which did not exhibit GIFBE. L=2μm, W_{Fin}=2x10μm. V_d=50mV.

Conclusions

A modification of the split C-V technique has been proposed to suppress the influence of floating body effects on the mobility extraction. The advantages of our revised extraction procedure, especially for PD SOI advanced devices, have been demonstrated based on measurements performed on wide-fin FinFETs. It was shown that our revised split C-V technique, in addition to avoiding overestimation of mobility due to GIFBE, also improves the extraction accuracy in the WI regime.

Acknowledgements

This work has been partly funded by the European Commission under the frame of the Network of Excellence "SINANO" (Silicon-based Nanodevices, IST-506844).

References:

[1] J. Koomen, *Sol. St. El.* **16** (1973) 801-810.
[2] S. Cristoloveanu, S.S. Li, "*Electrical characterization of SOI materials and devices*", 1995.
[3] F. Lime et al. *Sol. St. El.* **47** (2003) 1147-1153.
[4] P. Zeitzoff et al. *IEEE El. Dev. Lett.* **24** (2003) 275-277.
[5] K. Romanjek et al. *IEEE El. Dev. Lett.* **25** (2004) 583-585.
[6] V. Kilchytska, et al., *ULIS 2005*, pp.117-120.
[7] T. Rudenko, et.al. *INFOS 2005*.
[8] N. Collaert, et al. IEEE El. Dev. Lett., **25** (2004) 568-570.
[9] D. Lederer et al., *IEEE El. Dev. Lett.* **25** (2004) 104-106.

A New Dose Rate Model for SOI MOSFETs and Its Implementation in SPICE

H. Y. Liu, K. W. Golke, S. T. Liu
Honeywell Defense and Space Electronics System, Plymouth, MN 55441

Introduction

Unlike localized single event interactions, a dose rate event simultaneously perturbs every device. The basic equations describing the photocurrents generated in p-n junctions under ionizing radiation were proposed by Wirth and Rogers [1]. The steady state value of the photocurrent in a Si p-n junction can be estimated as

$$I_{ph} = q\, g_o \times (\dot{\gamma} \times V_{col}) \quad (1)$$

where q is the elementary charge ($q=1.6\times10^{-19}$C), g_o is the generation rate in silicon (4×10^{13} electron-hole pairs $rad^{-1}cm^{-3}$), $\dot{\gamma}$ is the dose rate (rad(si)/s) and V_{col} is the collection volume (cm^3). V_{col} can be estimated as the depletion region of the p-n junction plus a diffusion length on each side.

Photocurrent modeling becomes more complicated in an SOI MOSFET, where the collection volume includes any Si region that is lightly doped. The more heavily doped source and drain regions may have to be included as well, depending on the technology and how layout is done. Many parasitic devices are present in an SOI MOSFET and they interact with each other to determine the magnitude of the photocurrent at each terminal. These parasitic devices include source/body junction diode, drain/body junction diode, as well as a source/body/drain bipolar transistor.

The challenge for dose rate simulation has been to concurrently model localized junction behavior and their interactions with each other at a global circuit level. Most previous dose rate studies for CMOS circuits have been done in a simplified manner. Some studies were conducted at single device level, where hand calculation was used to estimate the dose rate at which the parasitic bipolar transistor would turn on. Some other studies were carried out at circuit level [2], where the most common approach is to place current sources between manually constructed V_{dd} and V_{ss} resistor networks, then run SPICE simulations to evaluate rail-span collapse.

These approaches provide useful insight to certain circuits, but are not adequate when more detailed information at a discrete node is needed to guide a dose rate sensitive design.

Photocurrent Generation

Consider an SOI NMOS device under no bias, where all terminals are discrete nodes but at identical voltage. During a dose rate event the built-in electric fields in the source and drain junctions will pull excess holes to the body and electrons to the source and drain. The source current I_s and the drain current I_d are primarily electron current and flow into the NMOS, whereas the body current I_b is primarily hole current and flows out of the NMOS. In an ideal case where the layout of the device is perfectly symmetric and every electron-hole pair is collected, the source and drain junctions will each collect 50% of the excess carriers in the MOSFET, which can be described by

$$I_b = I_s + I_d \quad (2)$$

$$I_s = I_d = 50\%\ q\, g_o \times (\dot{\gamma} \times V_{col}) \quad (3)$$

where V_{col} is the total collection volume in the SOI MOSFET. If the drain junction is more reverse biased, the depletion region of the drain junction will be larger and more efficient at collecting charges. Accordingly, the drain current will be somewhat greater than the source current. The difference will depend on layout, process parameters as well as bias. In the case where all terminals are at identical voltage but the body and source terminals are physically connected together, the source junction will still participate in the collection of photo generated excess carriers. However, the holes collected by the source junction will flow back to the source/body ohmic contact through the body tie recombining with the electrons collected. Therefore the net external source current is essentially the photocurrent collected by the drain junction, which is still $50\%\times I_{ph}$, but flows out of the NMOS. The situation in an SOI PMOS can be described in a similar manner, except the current polarities are reversed.

Another interesting phenomenon, which has not been reported before, is that the drain junction can get forward biased during a dose rate event under certain conditions. One example is illustrated in Fig. 1 where a current source and a diode are used to model the drain junction during a dose rate event. The current continuity equation dictates $(V_{dd}-V_d)/R = 50\%\times I_{ph}-I_{diode}$, where I_{ph} is the ideal photocurrent due to dose rate, I_{diode} is the "dark" current caused by the bias across the drain junction, and V_d is the voltage at drain. Under high dose rate conditions when $V_{dd}/R < 50\%\times I_{ph}$, the resistor will not be able to keep up with $50\%\times I_{ph}$ even when V_d is at V_{ss}. Since the anode of the drain diode is at V_{ss}, V_d will have to be pulled to below V_{ss} in order to satisfy the current continuity equation. As a result, the drain junction is forward biased and the drain current is greater than V_{dd}/R. In this case, the maximum current from V_{dd} to V_{ss} under dose rate is roughly $(V_{dd}+V_{for\_n})/R$, where V_{for_n} is the turn-on voltage of the NMOS drain diode. Similarly, in the case of an inverter, the maximum current can be as great as $(V_{dd}+V_{for\_n}+V_{for\_p})/R$, where V_{for_p} is the turn-on voltage of the PMOS drain diode.

These ideas were verified by 2D Medici [3] simulations based on process and layout parameters of Honeywell's 0.15μm SOI technology. In addition, it was found that even for maximum width devices (both nch and pch SOI MOSFETs), negligible parasitic bipolar effect is present until 5×10^{13} rad/sec. The parasitic bipolar transistor has a greater tendency to turn on in maximum width devices during a dose rate event. However, the simulated result of 5×10^{13} rad/sec is greater than the hand calculated result by almost two orders of magnitude. In the hand calculation of I×R drop across the source or emitter junction, the body resistance R is not a function of dose rate. However, the body resistance is modulated by

0-7803-9212-4/05/$20.00 ©2005 IEEE

the excess carriers created by dose rate, in which case higher dose rate would lead to greater I but smaller body R. Therefore I×R increases much slower than it would have with a fixed R as dose rate increases. Since in most designs the survivability specification is 1×10^{12} rad/sec, it is reasonable to completely ignore bipolar effect without sacrificing any accuracy in dose rate analyses.

Results and Discussions

Based on the understandings that have been established in the sections above, it is proposed that each junction in an SOI MOSFET be modeled using a constant current source in parallel with a diode for dose rate simulation. The polarities of the current sources and diodes are illustrated in Figs. 2(a) and 2(b). The magnitude of each current source is 50% $\times I_{ph}$. V_{col} is obviously dependent on different layout styles. However, V_{col} can be approximated as a function f(L, W) of transistor width and length according to the worst case layout allowed by rules. The strength of the diode depends on process and layout as well. The value of parameter "IS" in the diode model is proportional to the junction perimeter to the first order, but the accuracy of this parameter is not of primary concern. This is because the diode current under forward bias would vary by a factor of 10 when the bias changes only 60mV at room temperature. Any inaccuracy in "IS" can easily be absorbed by a small change in bias. Therefore, it is reasonable to assume that IS is proportional to transistor width, with the scaling factor being dependent on the technology.

To conduct a dose rate simulation at the circuit level, each SOI MOSFET in the circuit needs to be replaced by a sub-circuit whose components include current source(s), diode(s) and the MOSFET itself. Connections of these components should be done according to Figs 2(a) and (b). Parameters used in the current sources and diodes are transistor size, dose rate and some scaling factors dependent on technology. Standard parasitic RC extraction tools can produce an extracted netlist including the V_{dd}/V_{ss} networks. Once all SOI MOSFETs in the netlists are replaced with the appropriate sub-circuits, a dose rate simulation becomes an "average" SPICE based simulation.

In order to evaluate the validity of this methodology, nch and pch arrays fabricated with Honeywell's 0.15µm technology were tested at a variety of dose rate levels and V_{dd}'s. We chose to use these simple structures because dose rate tests are notoriously known for being noisy and hard to interpret. Reasonable correlation was seen between simulation and test results. An example is given in Fig 3, where for an nch array of 80000 devices with L=0.15µm and W=0.40µm, the measured dose rate currents at the body node are plotted as solid symbols. The solid line in the plot represents the simulation results, which are in good agreement with measured results.

Summary

A new SPICE based dose rate model is proposed for SOI MOSFETs, which accounts for collections of excess carriers by both source/body and drain/body junctions. It is also identified that the parasitic bipolar transistor does not play a significant role within the range of dose rate of interest. An implementation method for circuit level simulation is described. The validity of this model has been verified by test results.

References

[1] J. L. Wirth and S. C. Rogers, "The Transient Response of Transistors and Diodes to Ionizing Radiation," *IEEE Trans. On Nuclear Science*, NS-11, pp. 24-38, Nov. 1964.

[2] L. W. Massengill and S. E. Diehl-Nagle, "Transient Radiation Upset Simulations of CMOS memory Circuits," *IEEE Trans. On Nuclear Science*, NS-31, pp 1337-1343, Dec. 1984.

[3] D. E. Fulkerson and H. Liu, "A charge-control SPICE engineering model for the parasitic bipolar transistor action in SOI CMOS SEU," *IEEE Trans. Nuclear Science*, Vol. 51, No. 1, Feb. 2004, pp. 275-287.

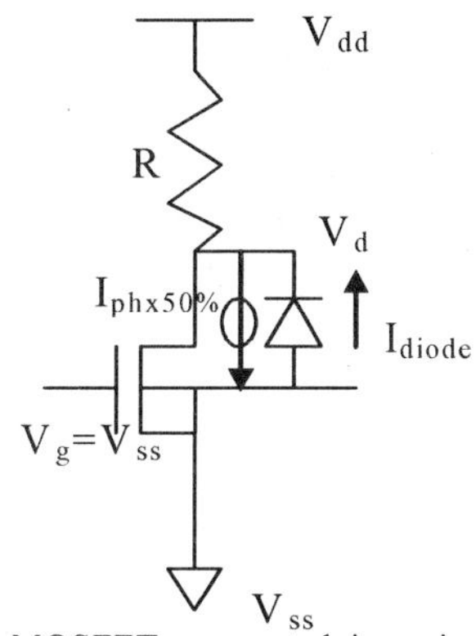

Fig. 1 A MOSFET connected in series with a resistor between V_{dd} and V_{ss}. V_d can be pushed to below V_{ss} at high dose rate.

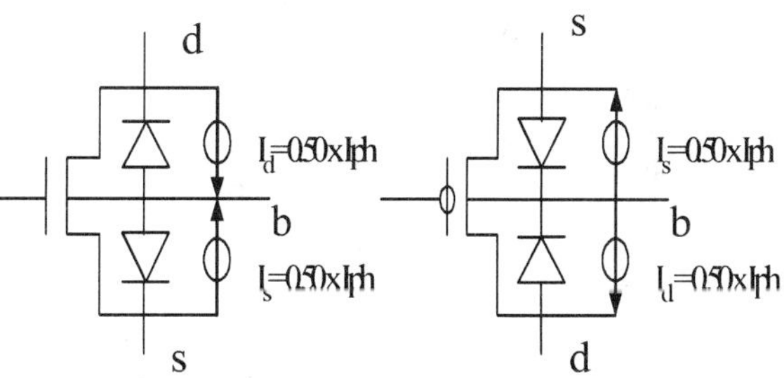

Fig. 2(a) nch under dose rate 2(b) pch under dose rate

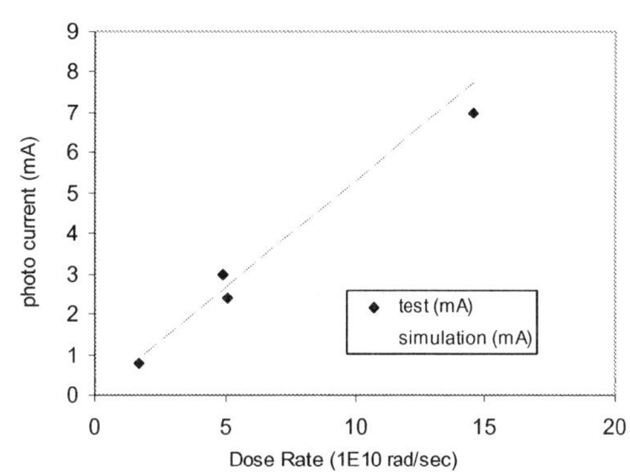

Fig.3 Tested and simulated photo currents at the body node for a large nch array as a function of dose rate.

Design and Fabrication of Temperature-Independent Arrayed-Waveguide Grating Demultiplexers on SOI Substrates

M. Uenuma, Y. Ikoma, Y. Kato, T. Motooka
Department of Materials Science and Engineering, Kyushu University
6-10-1 Hakozaki, Fukuoka 812-8581, Japan
Email: uenuma@zaiko8.zaiko.kyushu-u.ac.jp, Tel: +81-92- 642-3677

Introduction

Wavelength division multiplexing (WDM) optical communication systems are expected to become an important part of the infrastructure in the future society. One of the key devices for WDM optical communication is arrayed waveguide grating (AWG) multi/demultiplexers [1]. Although the silica-based AWG has already been playing a key role in the backbone WDM systems,• there exists the temperature sensitive problem due to the effective index change and the thermal expansion [2.3]. In addition, it is desired to integrate the AWG into the existing Si technology. In this paper, we propose a new temperature independent AWG using Silicon-on-Insulator (SOI). Figure 1 shows our proposed configuration for temperature independent AWG. Since the length of each waveguide is equal and the second star coupler consists of the hollow slab, the temperature dependence can be eliminated. Wavelength-multiplexed light is divided by the first diffraction order in the output slab region. We fabricated the AWG composed of equal-length waveguides on SOI substrates and the demultiplexing characteristic was measured.

Experimental procedure

We utilized SOI (100) substrates with a 2 μm surface Si layer on a buried oxide layer of 0.3 μm. Figure 2 shows a prototype SOI AWG composed of 21 equal-length waveguides. The shape of each waveguide is illustrated in Fig.3. The device was fabricated by UV-Lithography and isotropic wet etching. The device characteristics were examined by using an IR vidicon camera and a multichannel infrared laser.

Results and discussion

Figure 4 shows the diffraction spots which were obtained by three incident wavelengths 1483, 1548, and 1610 nm. The clear diffraction spots as well as the zero-th order spots were observed at the end of the slab. The first-order spots are linearly shifted with the wavelength. Figures 5 (a) and 5 (b) show the position dependence of the calculated and observed the first-order diffraction spot, respectively. The observed peak positions are in good agreement with the calculated ones.

Conclusion

We fabricated a prototype equal-length AWG on SOI substrates and observed the demultiplexing characteristics for the wavelength 1483, 1548, and 1610 nm. These results indicate that the equal-length SOI waveguides are promising for realization of a temperature independent AWG in Dense-WDM systems.

References

[1] Hisato. Uetsuka, "AWG Technologies for Dense WDM Apprications," *IEEE J. Select. Topics Quantum Electron.*, vol. 10, pp. 393–402, 2004.

[2] Hidenobu Hirota, Mikitaka Itoh, Manabu Oguma, and Yoshinori Hibino, "Athermal Arrayed-Waveguide Grating Multi/Demultiplexers Composed of TiO_2–SiO_2 Waveguides on Si," *IEEE Photon. Technol. Lett.*, vol. 17, pp.375-377, 2005.

[3] L. Grave de Peralta, A. A. Bernussi, V. Gorbounov, and H. Temkin, "Temperature-Insensitive Reflective Arrayed-Waveguide Grating Multiplexers," *IEEE Photon. Technol. Lett.*, vol. 16, pp.831-833, 2004.

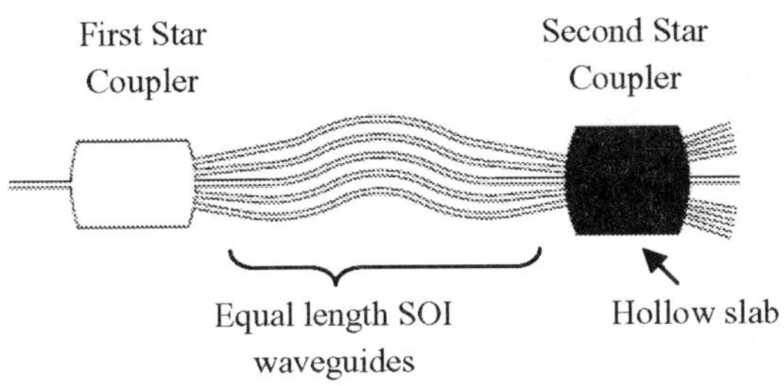

Fig.1 Design of temperature-independent AWG.

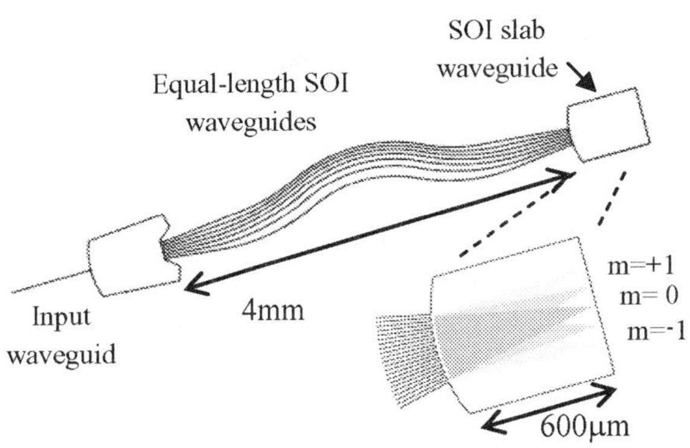

Fig.2 Schematic diagram of a prototype device of equal-length AWG. The diffraction image of second star coupler is also shown. the letter m shows the diffraction order.

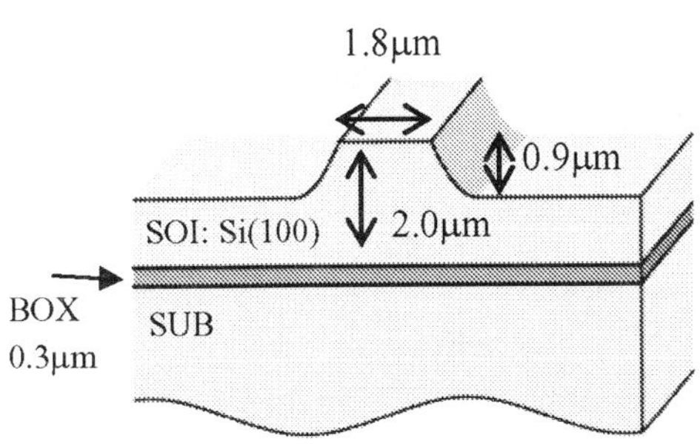

Fig.3 Cross-sectional view of the SOI rib waveguide.

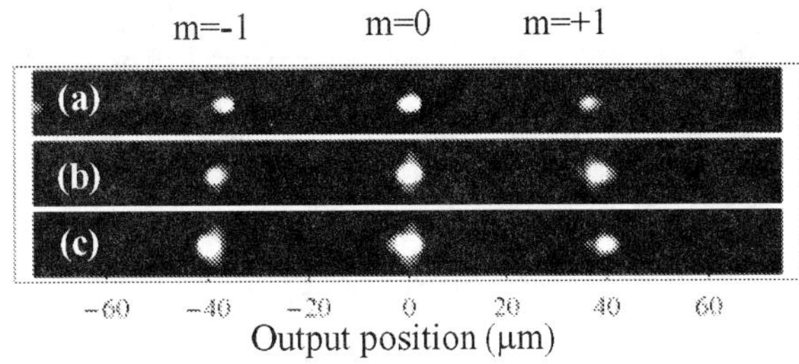

Fig.4 Diffraction patterns at the end of the SOI slab waveguide in Fig.2: (a) 1483 nm, (b) 1548 nm, and (c) 1610 nm.

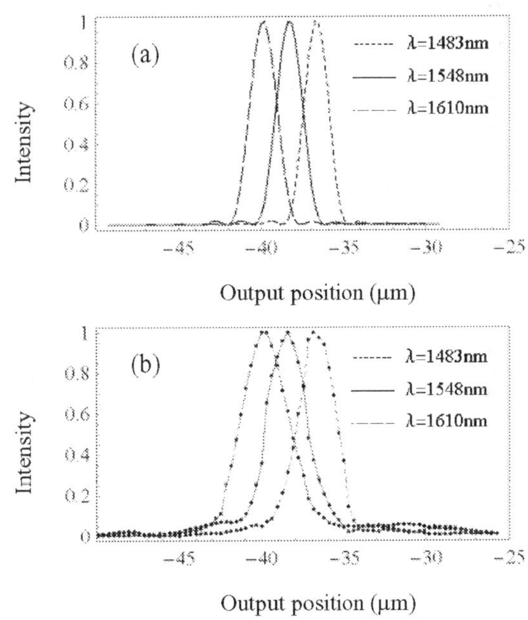

Fig.5 (a) Calculated and (b) measured intensities of the first-order diffraction spots for λ=1483, 1548, and 1610 nm.

New Design Concept of ESD Protection Circuits in SOI BiCDMOS LSI for Automotive Applications

Hiromasa Fuwa[1] and Kiyoharu Hayakawa[2]
[1]Toyota Motor Corp. [2]Toyota Central R&D Labs.,Inc.
543, Kirigahora, Nishihirose-cho, Toyota, Aichi, 470-0309, Japan
Email: fuwa@hiromasa.tec.toyota.co.jp, Tel: +81-565-46-6372, Fax: +81-565-46-3382

Introduction

Recently electronics systems are increasing in a car. And automotive electronic control units (ECUs) that include analog circuits, digital circuits and power devices are required to be small size and low cost. SOI BiCDMOS technology is one of solutions for these requirements[1][2]. Also electrostatic discharge(ESD) protection of LSIs for automotive applications is a very important consideration for high reliability. Fig.1 shows a general ESD protection circuit. During an ESD event on the I/O pad, ESD current flows through the diode and the protection circuit to the ground. The protection circuit clamps the voltage not to damage the internal circuits. In this work, we considered suitable devices and design parameters for ESD protection circuits to achieve high ESD robustness.

Results of measurement

Devices for ESD protection circuits are required to have low trigger voltage(Vt1), low holding voltage(Vh), low on-resistance(Ron) and high failure current(It2). We considered Nch LDMOS and NPN that are high voltage devices of 35V SOI BiCDMOS technology. And the I-V characteristics of these devices were measured using transmission line pulse(TLP) tester. Fig.2 and Fig.3 show schematics and cross sections of a Nch LDMOS and a NPN respectively.

Fig.4 shows the I-V characteristic of Nch LDMOS with gate width=2000um. Since snapback is not observed, the clamp voltage increases as the current increases. Non-snapback devices like the Nch LDMOS are not suitable for ESD protection circuits because the high clamp voltage will cause damage to the internal circuits. In addition, the high power dissipation of the protection circuit will cause damage to itself.

Fig.5 shows the I-V characteristics of NPN with emitter area=1920um^2. The NPN with collector to base(CB) space=5um is not suitable because It2 is very low(=1.4A). The NPN with CB space=15um has low Vh(=31V), low Ron(=2.1ohm) and high It2 (=7.6A). These characteristics are very suitable for ESD protection circuits. Also these results show that CB space improves It2 extremely. Vt1 is too high(=76V) to protect the internal circuits and it is necessary to be lower Vt1. Vt1 is controlled easily by adding a low trigger voltage device between the collector and the base[3].

We summarize the measurement results in Table1.

Results of device simulation

As a result of the characteristics measurements, we found that CB space improves failure current. The behavior of the NPN were investigated in detail by device simulation.

Fig.6 shows the current density and the electric field of the NPN with CB space=5um. In high current region, the lateral NPN turns on and the current mainly flows through the silicon surface. This behavior will cause high power dissipation close to the silicon surface and thermal damage.

Fig.7 shows the current density and the electric field of the NPN with CB space=15um. The vertical NPN turns on and the current flows through Deep N+ and N+ buried layer to the emitter. The power is dissipated at wide area under collector-base junction. This behavior results in high failure current.

Conclusion

Using TLP tester and device simulation, we investigated the I-V characteristics of two devices and the behavior of the NPN in high current region. The NPN is suitable device for ESD protection circuits. Especially, collector to base space is an important design parameter. We achieved low holding voltage, low on-resistance and high failure current by optimizing collector to base space of the NPN. Therefore we propose the ESD protection circuits that consist of the NPN for automotive applications.

Reference

[1] M.Hattori et al, "A Very Low On-resistance SOI BiCDMOS LSI for Automotive Actuator Control " Proceedings 2004 IEEE SOI Conference, PP.171-173, Oct 2004

[2] F.Kawai et al, "Multi-voltage SOI-BiCDMOS for 14V&42V automotive applications" Proceedings 2004 ISPSD 2004 International Symposium, PP.165-168, May 2004

[3] A.Amerasekera and C.Duvvury, "ESD in Silicon Integrated Circuits" John Wiley & Sons, 2002

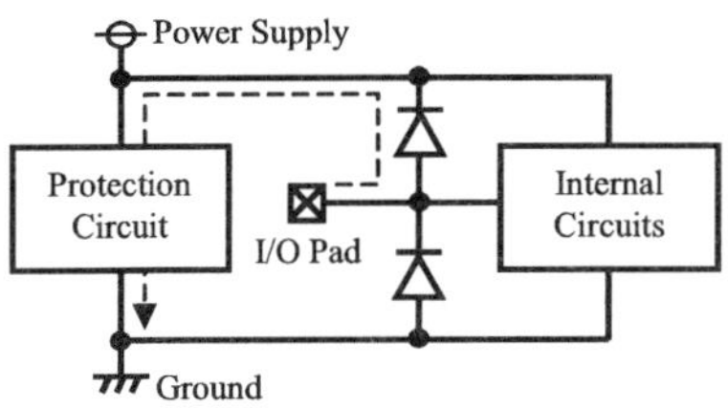

Fig.1 General ESD protection circuit

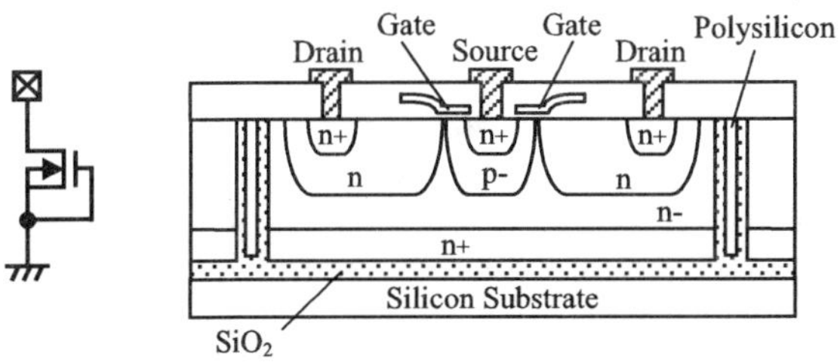

Fig.2 Schematic and cross section of a Nch LDMOS

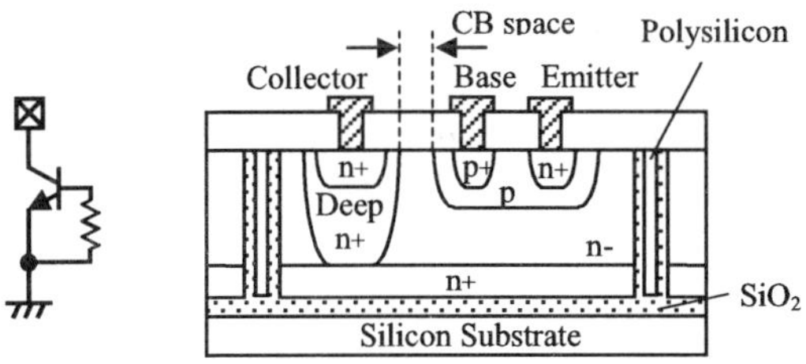

Fig.3 Schematic and cross section of a NPN

Table1 Summary of measurement results

	Nch LDMOS	NPN with CB=5um	NPN with CB=15um
Vt1(V)	48	76	76
Vh(V)	Non snapback	-	31
Ron(ohm)	16.8	-	2.1
It2(V)	1.4	1.4	7.6

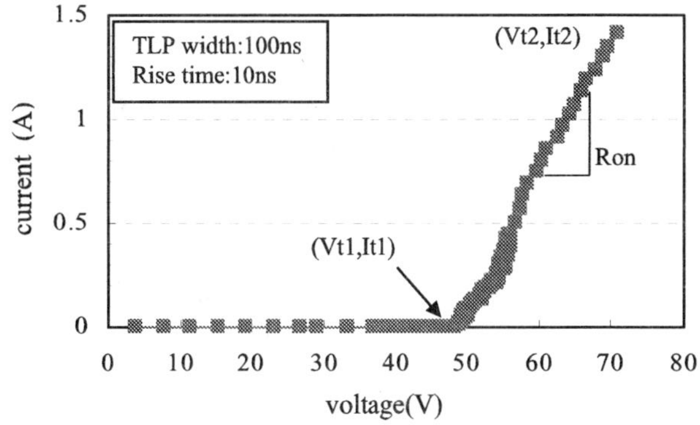

Fig.4 I-V characteristic of Nch LDMOS

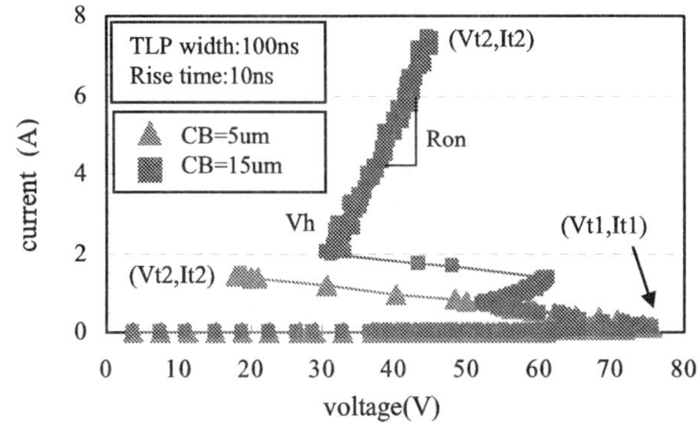

Fig.5 I-V characteristics of NPN

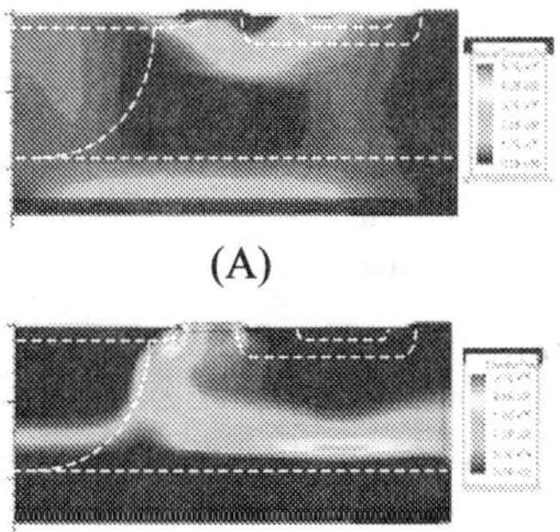

(A)

(B)

Fig.6 (A)The current density and (B)the electric field of the NPN with CB space=5um

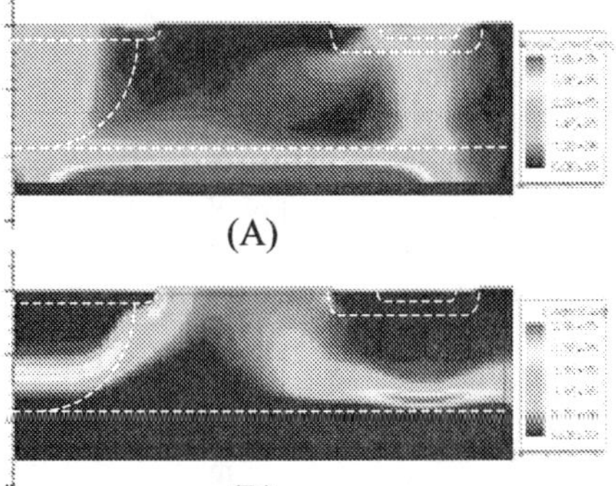

(A)

(B)

Fig.7 (A)The current density and (B)the electric field of the NPN with CB space=15um

Sense amp design in SOI

M. Golden, J. Tran, B. McGee, B. Kuo
Advanced Micro Devices, Sunnyvale, CA
Email: michael.golden@amd.com, Tel: (800) 548-8450

Abstract

Microprocessors require cutting edge technology to deliver competitive performance. AMD manufactures microprocessors in a 90nm, triple-Vt, SOI process. In this process, static transistor mismatch, caused by process variation, and dynamic transistor mismatch, caused by SOI effects, combine to increase the input referred offset of sense amplifier circuits used in SRAMs. A silicon experiment comparing different sense amp topologies reveals that body-tied transistors provide significant improvement in input referred offset without performance degradation. **Keywords: VLSI, CMOS, SOI, Sense amplifier, SRAM.**

Introduction

Line edge roughness, random dopant density fluctuation, and other static process variations play a greater role as feature sizes shrink. Partially-depleted SOI transistors experience floating body effects. SOI transistors are thermally as well as electrically insulated from the silicon substrate, so they suffer from self-heating effects. Both floating body potential and junction temperature change depending on the dynamic usage pattern of a device. These static and dynamic effects mean that two "matched" transistors may have different electrical behavior. Sense amplifiers in SRAM arrays require well-matched transistors for correct functionality.

For a given sense amp topology, careful layout can mitigate some static mismatch between transistors. The choice of topology can reduce both dynamic and static mismatch. Line width variation is usually constant, so increasing transistor width and length reduces relative variation between two transistors. Connecting the body of an SOI transistor to a known voltage source, or to the body of its matched pair, reduces dynamic mismatch. This work compares circuit design choices to produce a sense amp with lower input referred offset, the minimum input differential at which an amp will evaluate to the correct value. A 1MB SRAM test chip has been built in a 90nm triple-Vt SOI process [1] with a variety of sense amp designs. Failure rates give a metric for the merits of each sense amp circuit topology.

Circuit topologies

The test chip contains a 1MB SRAM, in which there are 20,224 instances of the sense amp, and a BIST engine. The read circuitry for a bit column is shown in Figure 1. Although only one bit cell is shown, each column contains 128 bit cells.

PMOS prechargers M6 and M7 are enabled during a read and disabled during a write. The ratio of the PMOS precharger and the bit cell pull-down transistors sets the bit-line slew rate and the maximum differential. NMOS clamps M8 and M9 are disabled during a write. They provide enough current to precharge the bitlines after a write, but don't interfere with a read as they don't conduct until the bitline voltage drops below the NMOS threshold voltage.

Sense amp designs

The experimental SRAM uses a current-controlled latch sense amplifier as shown in Figure 2 as a baseline [2]. One of eight alternatives was instantiated in each sense amp location. All devices in the baseline sense amp are floating-body, regular-Vt (RVT) devices. The experimental sense amps vary from the baseline as follows:

1. Double the width of all transistors.

Doubling the width reduces relative error caused by static differences in width, length, and threshold voltage between matched transistors.

2. Optimize width, length, and threshold voltage.

Transistor sizes for the baseline sense amp were chosen as they might be for any CMOS gate, based primarily on drive strength considerations. For this variation, extensive spice simulations with silicon-based spice models were performed to find transistor sizes and threshold voltages (see Table 1) that would produce an amp with the smallest possible input referred offset.

3. Double the width of transistors in option 2.

4. Tie the bodies of M2 and M3 together. Tie the bodies of M4 and M5 together as shown in Figure 3. Make these transistors high-VT (HVT).

The bodies of the matched transistors float together, reducing dynamic threshold voltage variation.

5. Tie the body of M2, M3, M4, and M5 to VSS.

The body-to-body connection is made in active silicon. The high resistance of silicon means that the body tie may have a poor frequency response. Tying the body to ground provides a better frequency response at the cost of greater area.

6. Repeat option 4, use RVT body-tied transistors.

7. Repeat option 5, use RVT body-tied transistors.

SOI issues in SRAM testing

The on-chip BIST engine was then used to test the SRAM. Because the test chip is built in an SOI process, it does not suffice to rely on standard march tests [3], which do not capture the dynamic changes in electrical characteristics of an SOI transistor. None of the parts used in the experiment show any failures with normal march tests. Although the dynamic mismatch between transistors is necessary to cause a failure, it is not sufficient. In all experiments, a given amp would fail only a single logic state, demonstrating the role played by static mismatch.

To test a sense amp, set every inactive cell in the column to the opposite value of the final read. Skew the sense amp threshold voltages by pre-reading the active cell many times with the opposite value of the final read. Finally, write the active cell to the final value and, in the following cycle, read the bit cell and compare the sense amp output to the expected result.

Experimental results

The SRAM array was tested using patterns containing varying pre-read loop counts at increasing clock frequencies until a single failure appeared. Programmable pulse shapers on the test chip were then used to force additional failures by delaying the assertion of the wordline to reduce the amount of time allowed for the generation of bit line differential.

Figure 4 shows the total number of sense amp failures in a single part, which had the most failures of all parts tested. Optimizing and increasing the size of the floating-body amps both improve the offset of the amp. This effect is cumulative. The body-tied amps have an order of magnitude fewer failures than the floating-body amps. As the number of pre-reads increases, the number of failures increases, demonstrating the importance of taking dynamic mismatch into account in SOI designs.

Further evidence of this importance can be seen in Figure 5, which shows failure rates for a population of parts with varying gate oxide thickness. The parts with thinner gate oxide are more affected by floating-body hysteresis and show more failures.

Raising supply voltage also increases the effect of the floating body in the sense amp, as seen in Figure 6. The increase in supply voltage increases the number of failures on this part for the floating body amps, even though bitline differential also increases with increased supply voltage. The body-tied amps show an opposite trend, because they are not susceptible to floating-body effects. The effect is less pronounced for a part fabricated with thicker gate oxide, as shown in Figure 7.

Modulating the idle time between the pre-reads and the final read produces the results shown in Figure 8. Increasing idle time from 0 read cycles to 768 read cycles produces fewer failures in a population of parts. As idle time increases, the floating body potential of the sense amp transistors relax from a dynamic state, resulting in more mismatch, to an equilibrium state, resulting in less mismatch.

Conclusions

Even carefully crafted sense amplifiers will have static differences in the electrical characteristics of matched transistors. Dynamic changes in transistor properties in SOI exacerbate static differences. Carefully chosen transistor sizes and body-tied devices reduce the input referred offset of a sense amp. BIST algorithms must take SOI effects into consideration. With careful circuit design and well-chosen test algorithms, reliable, high-performance, high-yielding SRAMs can be built with sense amps in advanced SOI processes.

References

[1] D. Greenlaw, et al., "Taking SOI substrates and low-k dielectrics into high-volume microprocessor production", Electron Devices Meeting, 2003. IEDM '03 Technical Digest, IEEE International, Dec. 2003, pp. 11.1.1 - 11.1.4B.

[2] B. Wicht, T. Nirschl, and D. Schmitt-Landsiedel, "Yield and speed optimization of a latch-type sense voltage sense amplifier", IEEE Journal of Solid-State Circuits, vol. 39, no. 7, July 2004

[3] A.J. Van De Goor, "Using march tests to test SRAMs", Design & Test of Computers, IEEE, vol.10 , no. 1 , March 1993, pp. 8 - 14

Table 1 Transistor sizes and threshold voltages for each sense amp variant

	M0/M1	M2/M3	M4/M5	M6	M7	M8/M9	M10/M11
0. Baseline	7:1	10:1	10:1.08	2.5:1	7:1.08	20:1	10:1
1. 2x Baseline	14:1	20:1	20:1.08	5:1	14:1.08	40:1	20:1
2. Optimized	7:1	10:1 (lvt)	10:1.25	2.5:1	7:1 (lvt)	20:2.13	10:2.13
3. 2x Optimized	14:1	20:1 (lvt)	20:1.25	5:1	14:1 (lvt)	40:2.13	20:2.13
4. Body-tied	7:1	15:1.25 (hvt)	15:1.25 (hvt)	2.5:1	7:1.08	20:1	10:1
5. Body-tied to VSS	7:1	15:1.25 (hvt)	15:1.25 (hvt)	2.5:1	7:1.08	20:1	10:1
6. Body-tied, RVT	7:1	15:1.25	15:1.25	2.5:1	7:1.08	20:1	10:1
7. Body-tied to VSS, RVT	7:1	15:1.25	15:1.25	2.5:1	7:1.08	20:1	10:1

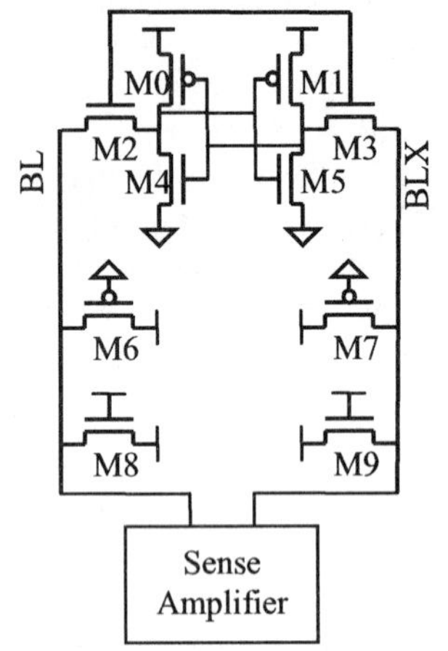

Figure 1 Read circuitry

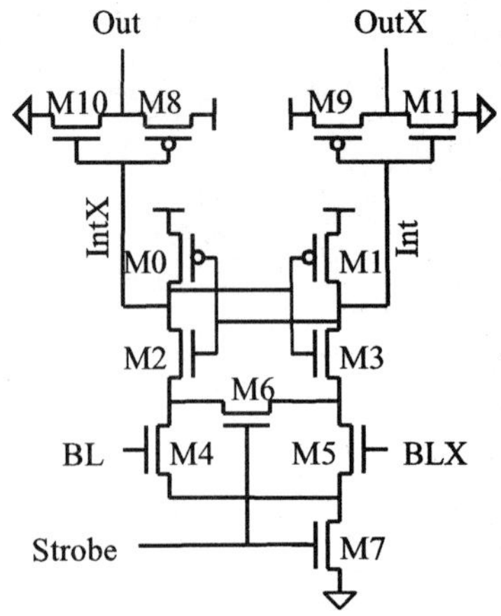

Figure 2 Basic amp topology

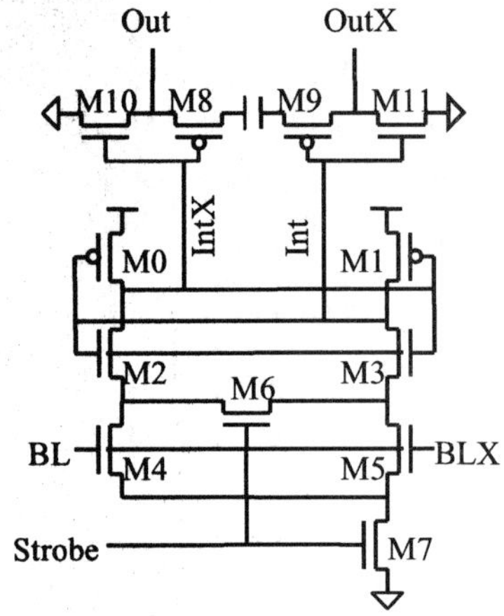

Figure 3 Body-tied amp topology

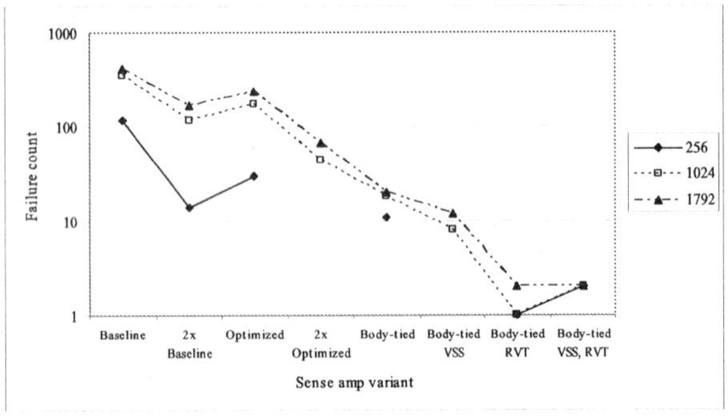

Figure 4 Vary pre-read loop, part with most failures

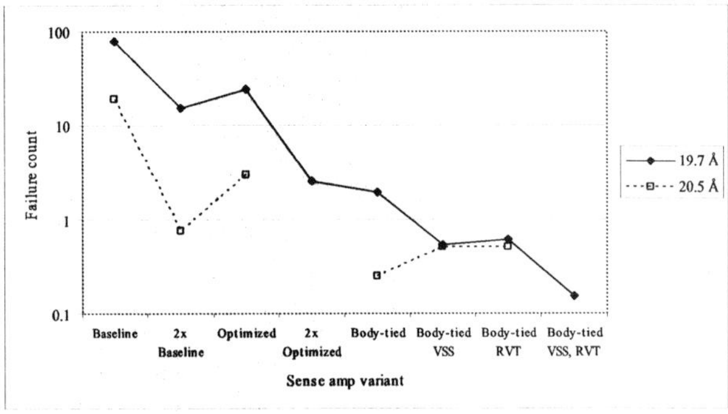

Figure 5 Failures per part, varying GOX

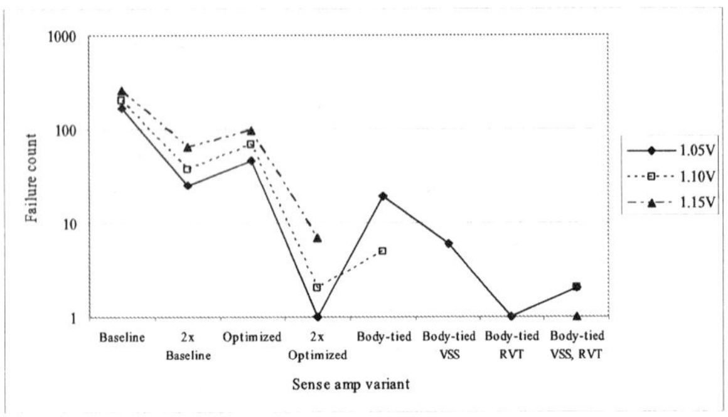

Figure 6 Vary VDD, thin GOX part

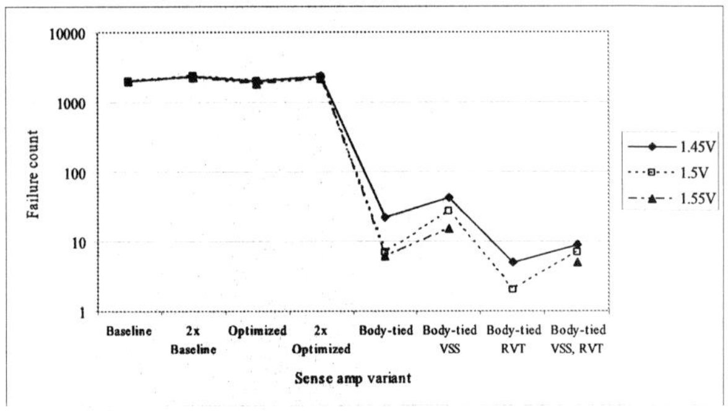

Figure 7 Vary VDD, thick GOX part

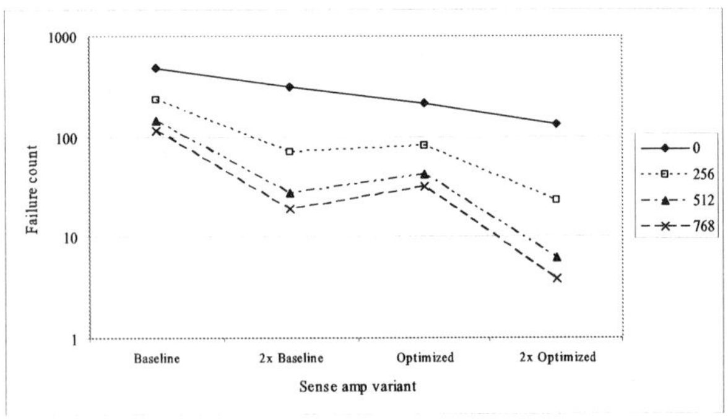

Figure 8 Failures per part, varying idle time

Layer-Transfer Process Modifications for Fabricating Hybrid Crystal Orientation Engineered Substrates

Jim Sullivan, Harry R. Kirk, Sien Kang,
Philip J. Ong, and Francois J. Henley

Silicon Genesis Corporation, 61 Daggett Drive, San Jose, CA 95134, USA
Email: mail@sigen.com, Tel: (408)-228-5858

INTRODUCTION

Numerous alternative device and substrate materials have been proposed to optimize sub-65nm device performance [1]. One of the most promising is a silicon substrate system having different crystal orientations to independently tailor NMOS and PMOS device performance enhancement. Generally referred to as a hybrid-orientation substrate, these engineered substrates necessarily use layer-transfer processes to laminate silicon crystal layers of different orientation [2-5]. The approach relies on the well-known dependence of effective MOS device carrier mobility to silicon surface crystal orientation. Generally, the (100) silicon crystal surface orientation is advantageous for fabricating NMOS devices while the effective PMOS mobility is nearly doubled for PMOS devices processed on a (110) crystal surface orientation [5].

Hybrid-orientation substrates have other integration advantages over alternative wafer-level solutions. Since these substrates rely on fundamental carrier mobility effects, the technology is expected to be less susceptible to scaling or process induced mobility degradations. In addition, they are also compatible with strained silicon mobility enhancement approaches such as wafer-level (global) and process-level (local) strain to further improve device performance. For example, the use of global uniaxial or biaxial strained layers, S/D SiGe compressive PMOS or Double-Stress Liners (DSL) technology can be readily integrated onto the substrate process [6,7].

Hybrid-orientation substrates can be fabricated using either a dual film laminate or having one device type built on the base (handle) substrate. The films can also be laminated using an oxide interface layer or directly. Table I shows the two basic types of hybrid-orientation substrates: single-film or dual-film. More than 24 major alternative substrate variations are possible depending on choosing specific interface layers (including none) and silicon layer orientation and twist. Adding biaxial or uniaxial global strain on the transferred films further multiply the substrate design space.

Successful introduction of hybrid-orientation therefore necessitate that the layer-transfer technology be robust and flexible to allow transfer of films either directly or using an oxide interface. The (110) orientation surfaces should also be smoothed using non-contact surface technologies previously developed for (100) SOI surfaces [8].

RESULTS AND DISCUSSION

Modifications of SiGen's SOI layer-transfer process were undertaken to confirm its applicability to fabricate hybrid-orientation substrates. Four major process areas were researched:

- (110) silicon film surfaces were layer-transferred to confirm that the layer-transfer processes could be applied to this film orientation.
- Direct crystal film transfer without oxide or interface layers were tested.
- Multiple film transfer was reviewed to allow (110) and (100) orientation films to be present on a handle substrate.
- Gas-phase epitaxial smoothing was tested on (110) surfaces.

Direct silicon to silicon bonding was demonstrated showing the possibility of a sub-20 Å void-free interface. Figure 1 shows a (110) film layer-transferred onto an oxide layer in a SOI-like process as a control. Modifications of the layer-transfer parameters were not needed to successfully transfer (110) films. Figure 2 shows a silicon to silicon bond process with a residual 10-20 Å interface layer. It is expected that (100) or (110) direct film transfer can be successfully accomplished in high-volume manufacturing.

Multiple film transfer of up to three oxide-silicon stacks has been previously developed for photonic applications [9]. Figure 3 shows a (110) on (100) film stack fabricated without an interface layer.

Finally, gas-phase smoothing tests were undertaken since it was not clear if there would be a crystal orientation dependence on etch performance or rates. Figure 4 shows a 2x2um AFM of a (110) layer-transfer surface after gas-phase smoothing. The sub-Angstrom RMS roughness confirms that the process can be applied to (110) orientation films.

0-7803-9212-4/05/$20.00 ©2005 IEEE

CONCLUSIONS

Successful integration of hybrid-orientation substrates into high-volume production requires a layer-transfer process to prepare the substrates. Modifications of SiGen's NanoCleave™ layer-transfer process were tested for compatibility to cleave and smooth (110) films, produce multi-stack composites and prepare direct bonded film structures. The technology has the promise to be a viable manufacturing process for strained and unstrained hybrid-orientation substrates used for deep submicron mobility enhancement.

REFERENCES

[1] V. Moroz et al., Solid State Tech., 77, July 2005.
[2] F. J. Henley et al, European Semiconductor, 25, Feb. 2000.
[3] Y.C. Liu, paper 9.4, VLSI Tech. Dig., 2004.
[4] M. Yang et al., paper 16.1, VLSI Tech. Dig., 2004.
[5] M. Yang et al., 453, IEDM 2003.
[6] Q. Ouyang, paper 3A-4, VLSI Tech. Dig., 2005.
[7] C.D. Sheraw, paper 2.1, VLSI Tech. Dig., 2005.
[8] A. Thilderkvist et al, 12, IEEE Int'l SOI Conf., 2000.
[9] I. J. Malik et al., in Proc. SiGe: Materials, Processes & Devices Symp., Fall 2004 ECS Mtg.

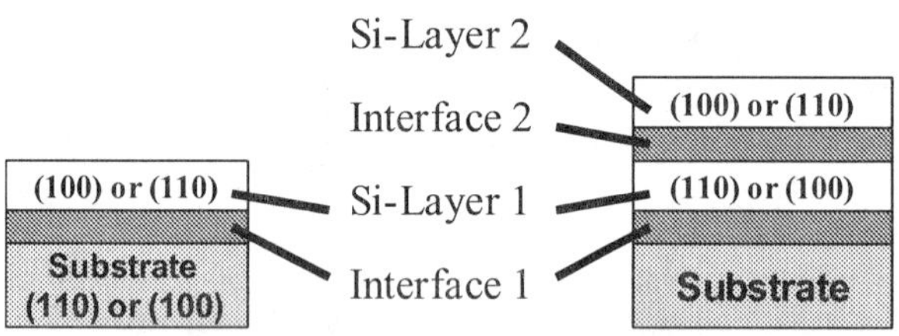

Table I: Partial variants of hybrid-orientation substrates.

Fig. 1: TEM of a 260nm (110) film layer-transferred onto a (100) handle substrate (240 nm oxide interface).

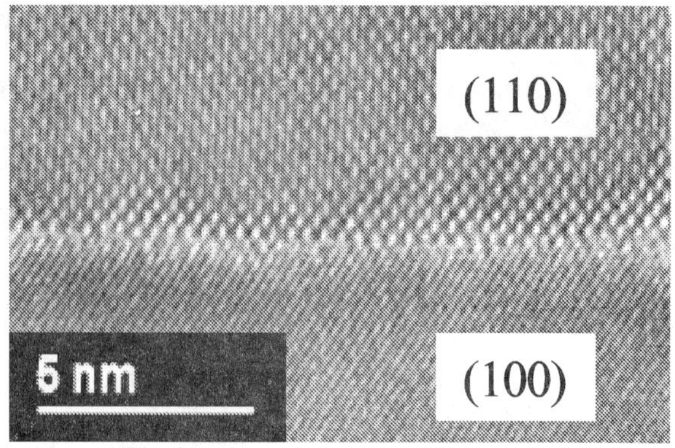

Fig. 2: TEM of a (110) on (100) silicon to silicon interface using a modified SiGen bond process.

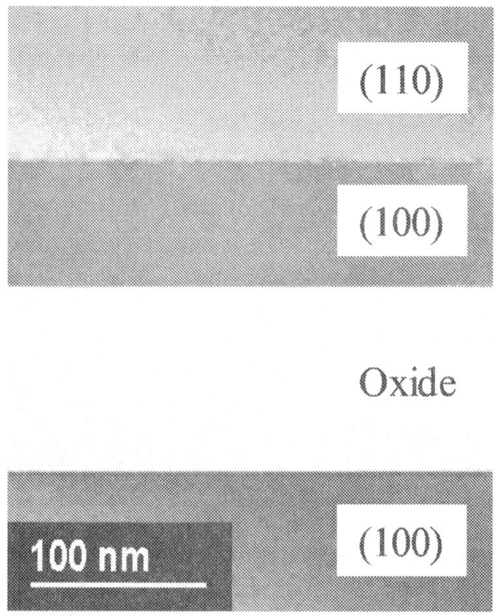

Fig. 3: TEM of a (110), (100) film stack on a (100) handle wafer.

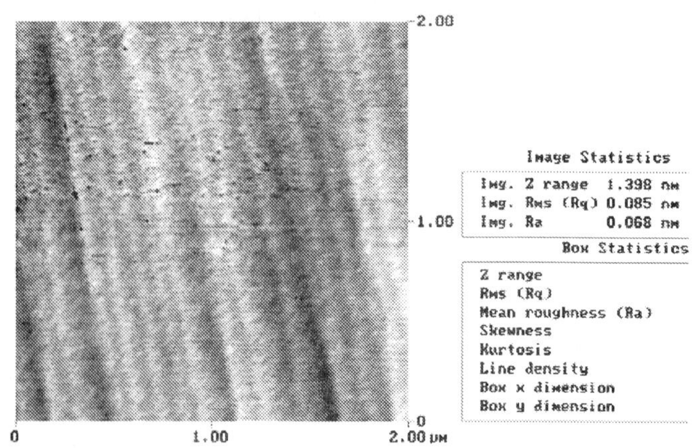

Fig. 4: 2x2um AFM of cleaved (110) surfaces after epitaxial gas-phase smoothing. RMS roughness is 0.85 Å, identical to the (100) control.

Novel ultra-low power RF Lateral BJT on SOI-CMOS compatible substrate for SoC Applications

I.-S. M. Sun, W. T. Ng*, H. Mochizuki, K. Kanekiyo, T. Kobayashi, M. Toita, H. Imai, A. Ishikawa, S. Tamura, K. Takasuka**
**Asahi Kasei Microsystems Co.Ltd. Kanagawa 243-0021, Japan
*University of Toronto, Electrical and Compter Engineerig, Toronto, Ontario
Email: {suni,ngwt}@vrg.utoronto.ca, Tel: (416) 978-6249

Abstract

This work presents a novel ultra-low power RF LBJT on SOI. The Johnson's product ($f_\tau \times BV_{CEO}$) of the fabricated LBJTs ranges between 190-300 GHz·V. The f_{max} of the optimal device reaches 46 GHz at collector current density of only 0.15mA/μm^2. This LBJT is compatible with SOI-CMOS for SOI-BiCMOS integration, an ideal technology for RF and mixed-signal SoC.

Introduction

SOI-BiCMOS technology is emerging as the next promising technology platform for RF system-on-chip (SoC). The SOI CMOS can implement faster digital functions, while bipolars handle all RF and analog blocks [1]. The SOI substrate also enables higher packing density and superior dc/ac isolation, a substantial advantage over silicon for large-scale SoC. However, the integration of vertical bipolar structure with thin-film SOI is problematic. Conventional vertical BJT requires thick-film SOI (>1μm) for the collector region, which is incompatible with thin-film SOI-CMOS [3]. Thin-film SOI vertical HBTs was demonstrated [2]. Although this technology promises low power consumption, the RF and breakdown characteristics is limited by the horizontal collector design. To alleviate this problem, the trivial yet ingenious solution is to build lateral BJT horizontally on CMOS compatible thin-film SOI substrate. Several works were proposed and extremely high device f_{max} at very low current consumption were demonstrated [4-6].

Novel LBJT Device Concept

The proposed LBJT device is based on the novel Poly-Side-Wall-Spacer (PSWS) LBJT structure presented in our previous work [5]. The cross-sectional 3D schematic of the PSWS-LBJT is as shown in Fig. 1. The structure significantly reduces the extrinsic R_B and parasitic C_{BC}, allowing higher f_τ and f_{max} to be achieved without sacrificing the breakdown voltage. In this design, Selective-Implanted Collector (SIC), commonly used in vertical BJTs, is added to optimize speed-breakdown tradeoff. Also, high dose boron at 1×10^{16}cm^{-2} is implanted into the PSWS to reduce R_B and enhance f_{max}.

Device Fabrication

The device is fabricated using a 0.35μm generation technology. Only five masks were used to realize both High-Voltage (HV) and High Gain (HG) versions of the LBJT. The SEM micrograph of the fabricated LBJT (Fig. 2) highlights the novel self-aligned PSWS base structure. The measured PSWS width is only 100nm wide, comparable to the minimal contact size in an advanced 90nm CMOS process. Without relying on high resolution photolithography, the smallest self-aligned base contact to date has been realized. This allows the fabrication of the LBJT base width in the 100nm range.

Experimental Results

The performances of both HV/HG LBJTs have shown considerable improvement over our previous work, as shown in Table 1. Notably, the HG LBJT exhibits 50% higher f_{max} and 2 times higher $f_\tau \times BV_{CEO}$ product. The measured Gummel plot, IV curves and the f_τ/f_{max} vs. I_C curves for both HV/HG LBJTs are shown in Fig. 3, 4 and 5, respectively. The peak f_{max} for the HG LBJT, extrapolated from measured U and MAG power gains (Fig. 6), reaches 46 GHz. The current density at peak f_τ and f_{max}, is only two-tenth compared to previously published low power SOI SiGe HBTs [2] data (Fig. 7). The ultra low current density required to operate the device at high frequency is due to the minimized parasitic capacitances that is inherent to this structure. Furthermore, the achieved Johnson's products of 190/300GHzV are competitive with state-of-the-art 0.25μm SiGe process (Fig. 8). The comparisons to the references are summarized in Table 2 [2-7].

Conclusion

A simple process for realizing a low power, RF LBJT on CMOS compatible SOI substrate is demonstrated. The ability to deliver high breakdown (BV_{CEO} >10V) and speed (f_τ/f_{max} = 16/46GHz) at ultra low current density (~0.15mA/μm^2) would benefit a wide range of circuit applications for wireless ICs. Based on this work, it is possible and feasible to implement a SOI BiCMOS process with complementary NPN and PNP lateral BJTs as shown in Fig. 9. This process is simple and

0-7803-9212-4/05/$20.00 ©2005 IEEE

inexpensive, and offers SOI CMOS and LBJTs on the same chip for today's RF and mixed-signals SoC.

References

[1] T. H. Ning, IBM J. R&D. pp.181-186,03/05, 2002.
[2] J. Cai et. Al., , BCTM, pp.215-218. 2003.
[3] K. Washio et. al, *IEEE T-ED*, pp. 271-278, 2002.
[4] H. Nii et al., *IEEE T-ED*, pp. 1536-1541, 2000.
[5] I-S. MSun et al., *IEEE T-ED*, pp.1376-1383, 2000.
[6] A. Suligoj, et al., *IEEE EDL*, pp.200-203, 2005.
[7] A. Joseph, et al., in Digest of SMIC, pp.1-4, 2004.

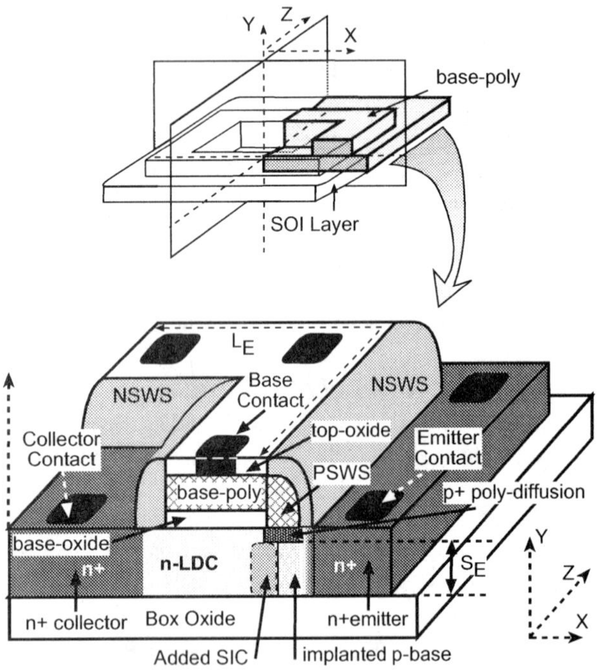

Fig. 1 Cross-sectional schematic of the 3-D PSWS-LBJT structure.

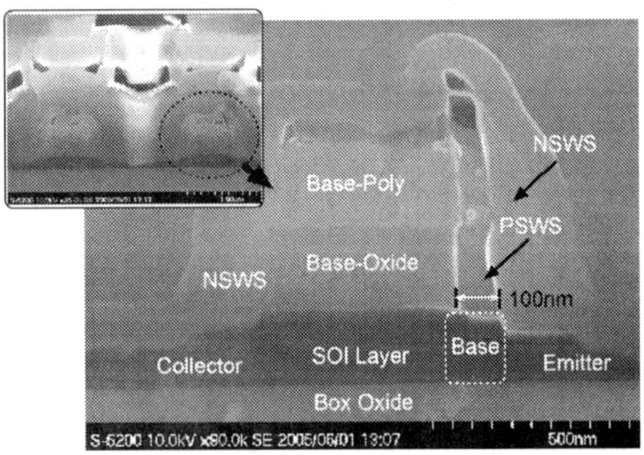

Fig. 2 SEM Image of the LBJT. Picture-in image shows the complete cross-section of device.

Table 1 Summary of LBJT device characteristics

	HG	HV	HG [5]	HV [5]	Note
A_E	2.85	2.85	2.28	2.85	in μm^2
h_{FE}	66	48	77	36	V_{CE}=2V, V_{BE}=0.8V
Peak f_T	16GHz	12GHz	17GHz	12GHz	V_{CE}=3V
Peak f_{max}	45GHz	36GHz	28GHz	30GHz	V_{CE}=3V
V_A	7.1V	11.8V	4.0V	12.6V	V_{CE}=3V, I_B=1μA
R_B	97	68	190	102	V_{BE}=1V
BV_{CEO}	11.9V	25.7V	4.3V	27.5V	I_C=1μA
BV_{CBO}	24.7V	31.5V	21.9V	33.6V	I_C=1μA
BV_{EBO}	8.4V	6.4V	8.6V	8V	I_E=1μA

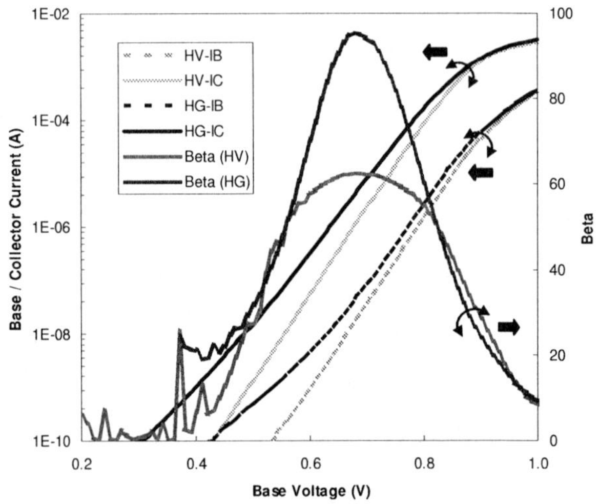

Fig. 3 Gummel and Beta plots of HG/HV LBJTs.

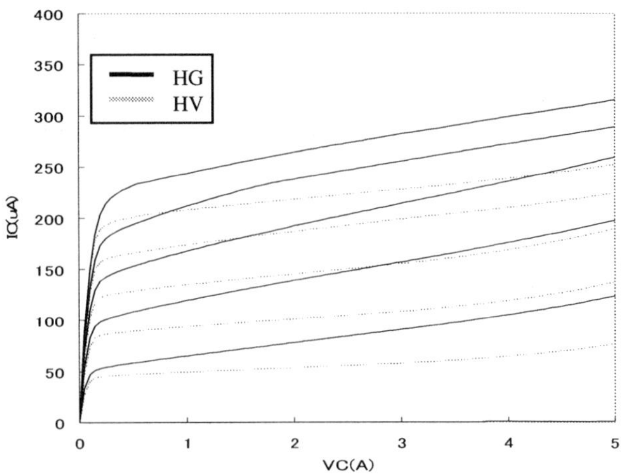

Fig. 4 I-V characteristics of HG/HV LBJTs. I_B are 1, 2, 3, 4, 5 μA, respectively.

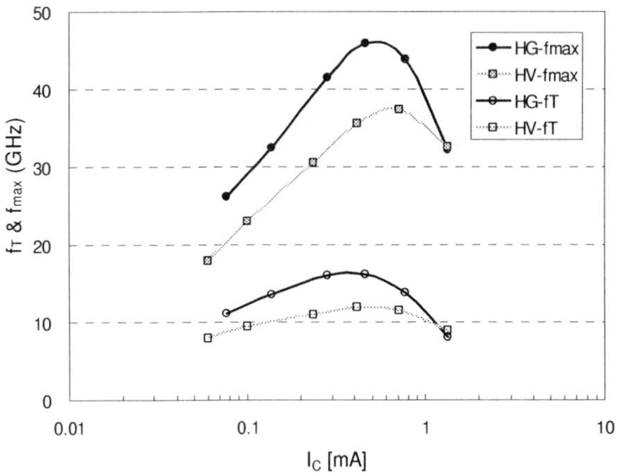

Fig. 5 Measured f_T/f_{max} vs. I_C characteristics of HG/HV LBJTs.

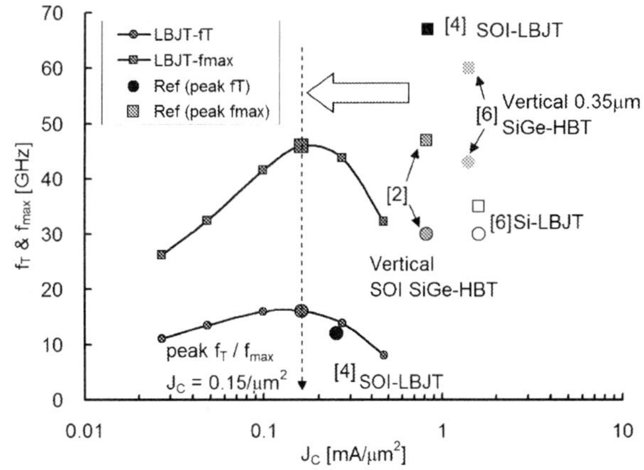

Fig. 7 Peak f_T/f_{max} vs. I_C comparison with published data. 2 to 10 times reduction in J_C at peak f_T/f_{max}.

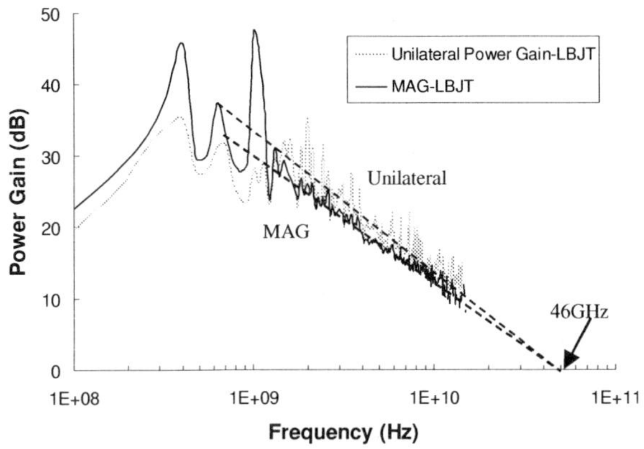

Fig. 6 Dependence of U and MAG vs. frequency for HG LBJTs at V_{CE}=3V.

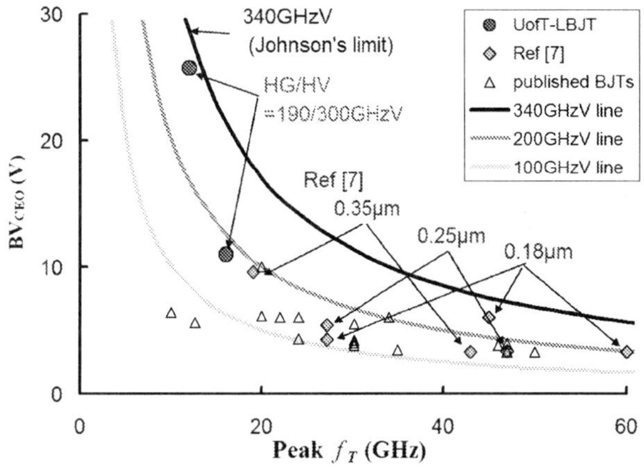

Fig. 8 Johnson's product comparison to advanced Bipolar processes from literatures.

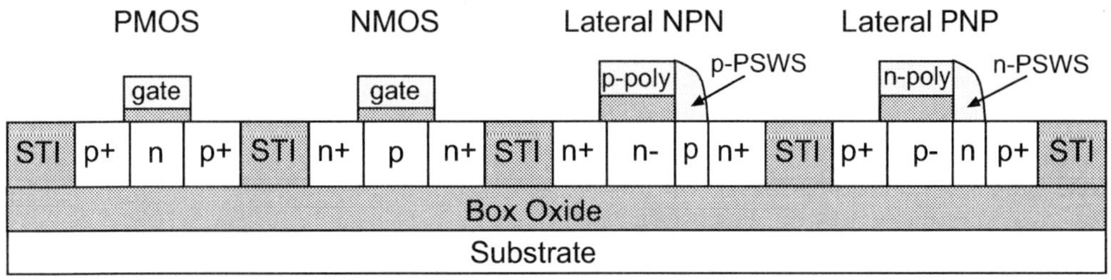

Fig. 9 Schematic illustration of a full complimentary BiCMOS technology with Lateral BJTs.

Table 2 Technology comparison between various processes.

	This work	Nii [4]	Suligoj [6]	Cai [2]	Washio [3]	Joseph [7]		
Device Type	LBJT	LBJT	LBJT	SiGe-HBT	SiGe-HBT	SiGe-HBT		
Substrate	SOI	SOI	Si	SOI	SOI	Si		
Ext. Base	Poly	CoSix	Poly	Poly	Poly	Poly		
Lithography	0.35	0.12	0.5	0.13	0.2	0.35	0.25	0.18
f_τ (GHz)	12 / 16	12.6	30	30 / 56	76 / 47	19 /43	27 / 47	45 / 60
f_{max} (GHz)	36 / 46	67	35	48 / 47	180 / 125	35 / 45	55 /65	73 / 85
BV_{CEO}	25.7 / 11.9	5.6	4.2	5.5 / 3.2	2.5 / 3.9	9.6 / 3.3	5.4 / 3.35	6 / 3.3
$f_\tau \times$ BV (GHzV)	308 / 190	70.5	128	165 / 179	190 / 183	182 / 142	146 / 157	270/ 198

Planar Double Gate Transistors with Asymmetric Independent Gates

G. Ilicali[1,2], W. Weber[1,2], W. Rösner[1], L. Dreeskornfeld[1], J. Hartwich[1], J. Kretz[1], Tarek Lutz[1], J.-P. Mazellier[1], Martin Städele[1], Michael Specht[1], J.R. Luyken[1], E. Landgraf[1], F. Hofmann[1], L. Risch[1], R. Käsmaier[1], and W. Hansch[2]

[1] *Infineon Technologies AG, Corporate Research, D-81730 Munich, Germany*
[2] *Institute for Technical Electronics, Technical University Munich, D-80333 Munich, Germany*
Email: gurkan.ilicali@infineon.com, Tel: +49 (89) 23457570

Abstract

Planar double-gate field effect transistors with asymmetric (p++/n++) independent gates down to 55nm physical gate lengths are successfully fabricated. A novel fabrication concept, Epi-Before-Bonding, is introduced and demonstrated to be highly successful in achieving ultra-thin and planar Si bodies. Various modes of operations are extensively analyzed and compared to 2D simulations. It is experimentally shown that specific off-current requirements can be fulfilled with conventional poly-Si gates.

Introduction

Multi-gate transistors are the most promising candidates to surmount the scaling roadblock beyond 32nm node [1]. Among them, the planar double gate transistors (PDG) offers an attractive option, because an excellent body thickness control without lithography, a fully planar construction promising for strain engineering, and an easier integration of independent gates are possible. Separate control of the gates puts forward new application and optimization possibilities [2,3]. A DG with symmetric gate work functions (SDG) necessitates an new gate material with a near midgap work function so that desired off-current specifications can be realized. On the other hand, extension of conventional poly-Si gates to DG era could be possible in a p^{++}/n^{++} asymmetric form, which yields normally-off transistors [1]. However, compared to existing theoretical studies and debates, experimental works on asymmetric DG transistors (ADG) are still lacking in the literature [4,5]. In this work, we report functional (p^{++}/n^{++}) ADGs fabricated with a novel epi-before-bonding concept (EBB), which makes an ultra-thin body (UTB) PDG easier to realize.

Device Fabrication

The basic process flow is briefly sketched in Fig.1. Upon definition of alignment marks [6], an (n^{++}) bottom gate (BG) was patterned on a 3nm gate oxide (GOX) using an e-beam lithography system. After TEOS gate spacers, epitaxial Si was selectively grown for raised source/drain (S/D) regions, which takes advantage of a thick starting SOI material. It is known that an UTB layer is a serious process concern in epi-growth [7]. Deposition of a 30nm nitride layer was then followed by a thick TEOS layer, which is polished (CMP) for a reliable layer transfer [8]. Buried bottom gate structures are next transferred onto a handling wafer covered with a thick oxide. Subsequent back-etching of the donor wafer, and the buried oxide (BOX), exposes the opposite surface for the top gate (TG) fabrication. The transferred silicon layer can now be thinned down by oxidation and etch-back to obtain an UTB Si. The EBB concept, therefore, eliminates the need for a raised S/D on an UTB after the top gate formation. After definition of active areas, patterning of TGs (in-situ p^{++} doped) on a 3nm GOX was performed with e-beam litho. S/D implantations were aligned to the TG. BGs were designed 40nm (nominal) oversized to account for a ± 20nm alignment accuracy.

Results and Discussion

An AFM image and a TEM cross-section of an ADG with the EBB concept were shown in Fig. 2& 3. The body was thinned down to 20-24 nm. Various operational modes such as Double Gate(DG), Ground Plane(GP) were analyzed. The DG performance is clearly better than both GP configurations, except the I_{off} since all render the same I_{off} current regardless of L and V_{ds} as expected (Fig.4a). Subthreshold slope (Ss) and DIBL read 89mV/dec (at V_{ds}=1V) and 96mV/V, respectively. Near ideal values of 68mV/dec with DIBL=30mV/V was achieved with 80nm $L_{physical}$ (not shown). Smaller work function of the BG leads to a smaller V_T at the bottom surface. It is revealed that Ss degradation in GP-mode with V_{TG}=0 is significantly less than the contrary case,i.e., V_{BG}=0 (S_{BG}=113 mV/dec vs. S_{TG}=220 mV/dec at V_{ds}=0.5V). These results suggest that, in GP-mode, BG-mode of the ASD with realistic Vt and Ss values could be a better alternative to a SDG with midgap gates [9]. Observed experimental tendencies were also justified by 2D ATLAS simulations (Fig.4b). Transconductance (g_m) of the same device is shown in Fig.5. Integration of the curves results in 11% higher g_m in DG-mode than the sum of the individual modes of GP. Modulation of the conduction with independent gates and associated V_T dynamics are also presented in (Fig.6). As the front channel starts to invert weakly for a fixes TG voltage (V_{TG}>0), BG starts to lose the control over the channel and SS degrades significantly whereas an optimum SS for an active BG-mode is achieved at V_{TG}=-0.5V for the given Vds.

0-7803-9212-4/05/$20.00 ©2005 IEEE

Conclusions

The novel EBB concept is proven to be favorable to carry out UTB bonded DGFETs. Conventional poly-Si gates with asymmetric work functions deliver very promising results to satisfy application specific I_{on}-I_{off} requirements. ASD yields a superior performance in DG-mode compared to the other modes of operations. Near ideal SS, low DIBL and high on current can be realized with ASD having an optimized gate stack and an UTB. It was shown that an n-poly active gate with a grounded p-poly gate is a better option for a GP application. This operation provides a realistic V_T with a less Ss degradation compared to a GP operation of an SDG with a midgap.

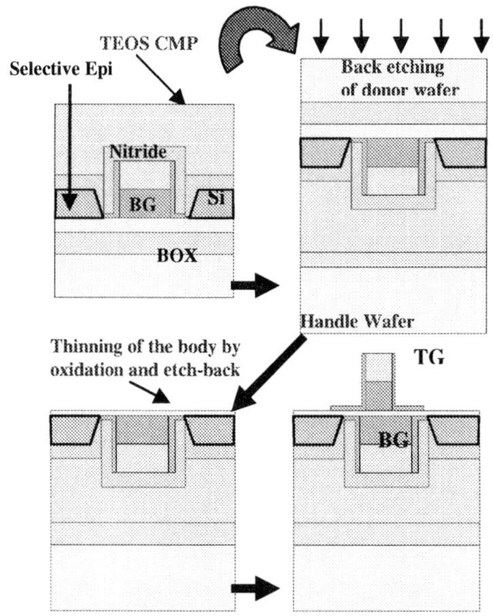

Fig.1: Brief description of the process flow. (100) SOI wafers with 50nm undoped Si body on top of a 100nm-buried oxide (BOX) were used.

References

1. H. -S.Wong, IBM J.Res.&Dev. Vol. 46, No2/3.p133 (2002)
2. S. Balasubramanian,et.al., Proc. Int. SOI Conf., pp.27 (2004)
3. L.Mathew, et.al., Proc. Int. SOI Conf., pp.187 (2004)
4. Tanaka, et.al., VLSI Symp. Tech.Dig., pp.11 (1994)
5. K.W. Guarini, IEDM Tech. Dig.,pp, 425 (2001)
6. W.Weber,et.al., Micro.Elec.Eng., 78-79, 206 (2005)
7. N.Nuryadi., et.al, J. Vac. Sci. Technol. B, Vol.20, 167 (2002)
8. G. Ilicali, et.al, Proc. Int. SOI Conf., pp.44 (2004)
9. Lolivier, J, et.al., Pro. 34th ESSDERC, pp. 77 (2004)

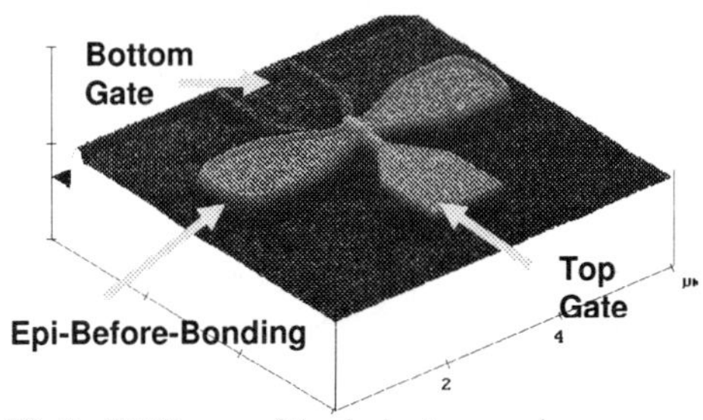

Fig.2: AFM image of the device topography. Raised active area is clearly shown. The mesa is isolated with a raised nitride spacer to avoid sidewall parasitic devices.

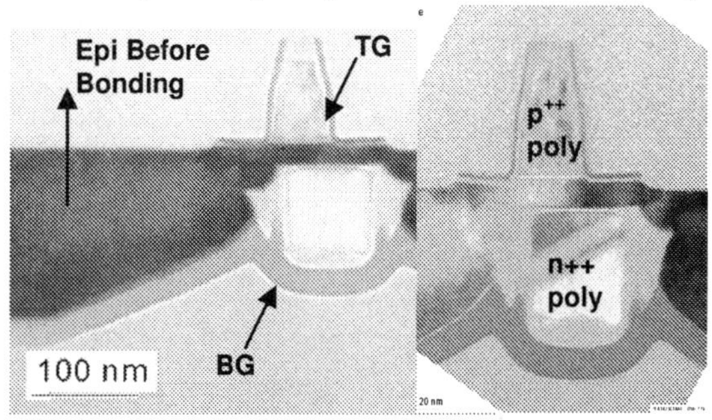

Fig.3: TEM cross-section of a DGFET with Epi-Before-Bonding concept.

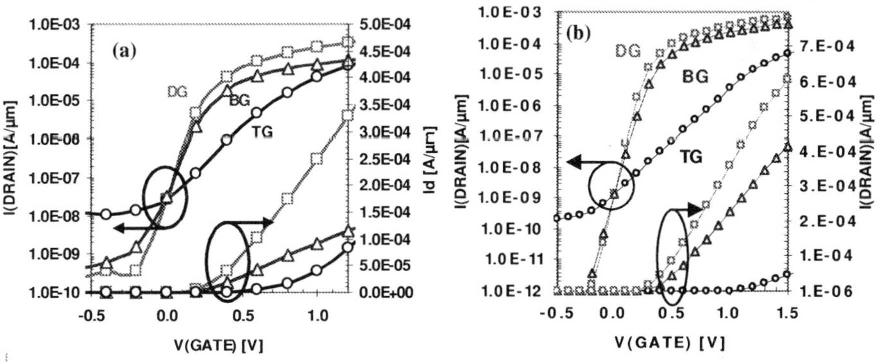

Fig.4: **(a)** ADG with L=55nm (physical) & t_{Si}=24nm. BG and TG represent the two active gates with corresponding GP-modes, for which the opposite gate is grounded. Vds=1V **(b)** 2D simulation of an ADG with L=50nm for comparison. Vds=1V with nominally similar device parameters

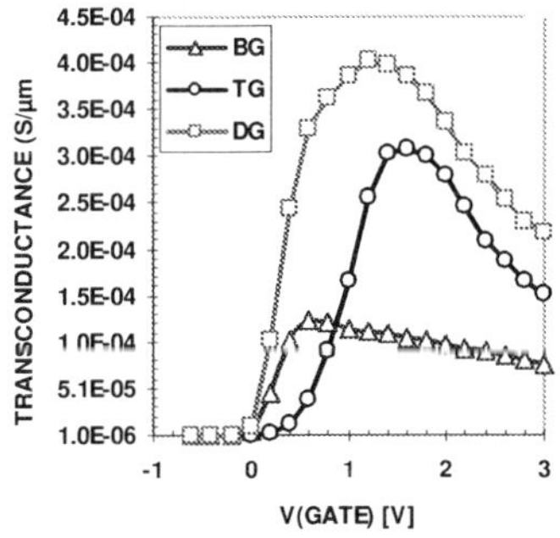

Fig.5: Transconductance of ADG device in various modes of operations. V_{ds}=1 V.

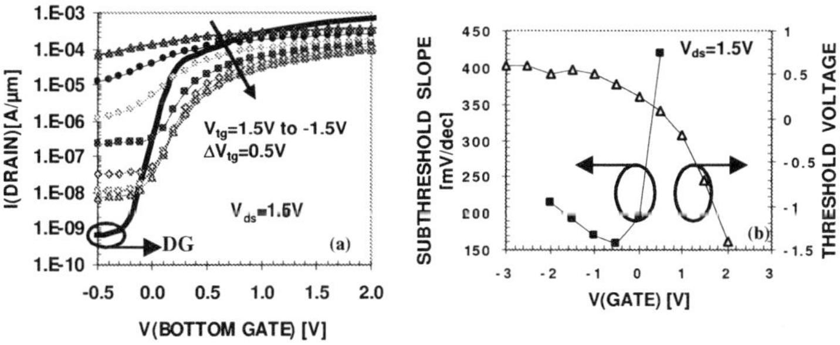

Fig.6: **(a)** Conducting channel modulation with a fixed V_{TG}. **(b)** Dynamic V_T control and corresponding variation of the subthreshold slope.

The Effect of Integration of Strontium-Bismuth-Tantalate Capacitors onto SOI Wafers.*

Vikram Joshi (1), Morifumo Ohno (2), Jiro Ida (2), Yoshiki Nagatomo (2), Karl Strauss (3)
(1) Symetrix Semiconductor Corporation, Palo Alto, California.; (2) Oki Electric Industry Co. Ltd., Hachiouji, Japan; (3) NASA Jet Propulsion Laboratory, Pasadena, California.;
Email: vikjos@msn.com, Tel: (650) 861-2341

We report for the first time the successful integration of Strontium-Bismuth-Tantalate ferroelectric capacitors on an SOI Substrate. We have verified that the unique processing requirements of SBT capacitors does not affect the properties of the surrounding FD-SOI transistors, and, conversely, we have verified that the SOI processing does not affect the qualitry of the SBT capacitors.

Motiviation

The integration of SBT capacitors on SOI substrate was borne out of a request by NASA's Jet Propulsion Laboratory to develop a suitable replacement for a non-volatile memory element used for start-up/boot-up code for a processor selected for use on the Prometheus Deep Space mission.

The use of SOI was considered a natural selection due to its well known capabilities in terms of operation through and after exposure to intense ionizing radiation fields.[1] The use of SBT capacitors as data storage elements was also a natural selection due to its well-known improved performance and processability when compared to other materials, such as PZT (Lead-Zirconium-Tantalate).[2]

Figure 1: Fully integrated SBT capacitors on FD-SOI

For this development, Metal Organic Decomposition (MOD) SBT solution commercially available from Kojundo Chemicals, Japan was applied to Unibond SOI wafer by Oki Electric Industry at their Hachiouji fab. The technology chosen was Oki's 0.2 μm scale FD-SOI transistors.

Novelty

We have, for the first time, succesfully integrated ferro-electric capacitors of any type (in this case, SBT) onto SOI substrate by overcoming several factors, namely the etching of very-deep tungsten plugs with an aspect ratio of 5.5 on ultra-thin S/D region (50 nm) , the development of two-step plugs under ferroelectric capacitor without the use of an interstitial pad in order to mitigate the stress to ultra thin SOI, and the modification of the SBT capacitor formation temperature profile resulting in elimination of an increase in back-channel leakage.

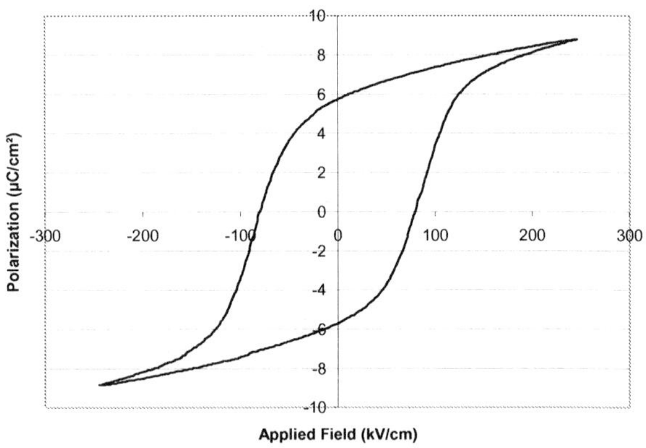

Figure2: Hysteresis of an integrated SBT capacitor(2 μm^2) on FD-SOI

* This development funded, in part, by NASA's Jet Propulsion Laboratory, Pasadena, California, under contract 1265898, in support of NASA Task Order 7107141 and the NASA Prometheus Program.

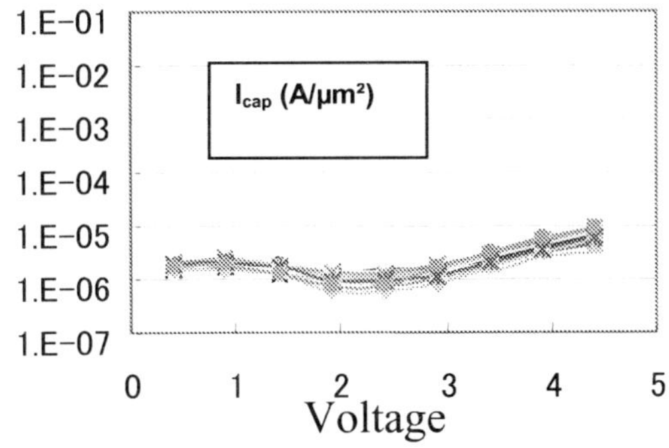

Figure 3: Leakage of an integrated SBT capacitor(2 μm²) on FD-SOI. Multiple point test.

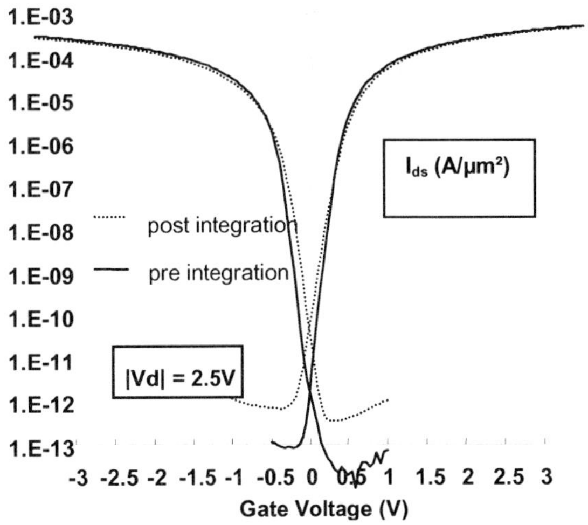

Figure 4: Comparison of Ids-Vgs curves of FD-SOI transistors (0.35 x 10 μm) with and without SBT capacitors.

Device Peformance

Figure 2 and 3 show the characteristics of SBT ferroelectric capacitors fully integrated into 0.2 μm FD-SOI with three layers of metallization, nitride passivation, and $CoSi_2$ contact process. The SBT film thickness is 120 nm. As can be seen, good hysteresis loops with excellent leakage characteristics are obtained on these integrated capacitors thus qualifying them for memory operation. Figure 4 shows the Ids-Vg characteristics of SOI transistors. The two plots compare the transistor characteristics with and without SBT capacitor integration. The data confirms that integration process of SBT capacitor has negligible effect on the SOI transistor characteristics and thus FD-SOI transistors can be integrated with SBT ferroelectric capacitors for ultra high speed and ultra low power non-volatile memory fabricaton.

Total Ionization Dose

The TID testing of FD-SOI transistors, post integration, is currently being carried out. We will present that information and compare that to data obtained earlier from TID testing of 0.2 μm FD-SOI transistors. The devices were subject a total ionization (gamma) radiation to a level of 300 kRads (Si).

For this purpose a test chip was designed with numerous transistor varying in the width-to-length ratio. Strutures with the use of body ties and enclosed body architecture to control cell leakage were also designed.

Conclusion

We have shown that SBT ferroelectric capacitors can be successfully integrated onto FD-SOI wafer with no degradation in either leakage or capacitor performance. We have shown that the process challenges of etching of deep tungsten plugs with large aspect ratios on ultra-thin Si is possible and finally we have shown processes specific to ferroeletric capacitor do affect the FD-SOI transistor characteristics. The radiation effects on FD-SOI transistors, post integration, are under investigation.

Future Developments

With the proof-positive that one can succesfully integrate SBT ferroelectric capacitors onto SOI with no deletrerious effect on either, opens the door to a new generation of memory devices for almost any practical use: from wristwatches and cell phones to deep space missions, the Holy Grail of non-volatility, low power, high speed and high radiation immunity is, at once, achieved.

References

[1] Mitsubishi Heavy Industry and JAXA, "SEE characteristics of 0.2 μm FD-SOI SRAM," Presentation IRPS, 2004

[2] C. A. Araujo, J. D. Cuchiaro, M. Scott, L.D. Mcmlllan, "Fatigue free properties of SBT on Pt electrodes," Nature, vol. 374, pp. 627, 1995

2005 IEEE INTERNATIONAL
SOI CONFERENCE

Session #5

SOI MATERIALS TECHNOLOGY AND CHARACTERIZATION

CO-CHAIRS

Carlos Mazure
SOITEC
Samuel Fung
Taiwan Semiconductor Mfg. Co.

SOI MATERIALS TECHNOLOGY AND CHARACTERIZATION Session 5

The Joys of Equipment Manufacturing and the Agony of Making Materials

Andrew Wittkower
Soitec USA, Inc.
Email: wittkower@verizon.net, Tel: (978) 546-7895

Abstract

This paper describes two different aspects of endeavor in the supply chain of the semiconductor industry. (1) Semiconductor equipment manufacturers have great problems dealing with the cyclical nature of the business as periods of overactive sales activity are followed by substantial periods of inactivity. Some management solutions are suggested. By comparison, sales of advanced materials tend to follow more smoothly the ups and downs of the market. However, the never-ending quest for higher quality material, defined by tighter specifications, generates its own challenges including improved capabilities for next generation characterization equipment. An example is given.

The Joys of Equipment Manufacturing

From its earliest days, equipment manufacturing has required great skill and dedication in the quest to achieve a functional and reliable product. Greater than any technical challenge, the managing of an equipment company through the cyclical upward surges, and subsequent declines of the business cycle, has been an intransigent problem. Since IC's are the end product of the manufacturing process, intuition would suggest that the appropriate business model one would use to predict the number of machines needed, would be "More or Less" (Fig. 1), following closely an increasing and decreasing IC demand. This theory may be applicable to many industries, including the IC industry itself, but for the equipment supplier the correct model is actually close to "All or Nothing." (Fig. 2).

When the IC manufacturing plants (or even a single fab) are in growth, they need more equipment and they need it immediately. For the equipment manufacturer, the scramble starts for an increase in everything, including some or all of the following:

- The physical plant may be too small to be able to accept the new demand. A long-term plan, which would have anticipated future needs, often incurs unacceptable financial exposure and may not have been implemented.

- Qualified people are needed immediately. There are few in view, and those few already have jobs; so inexperienced people must be found and trained. Since many high tech jobs are difficult and require substantial learned skills, it may take 9-12 months to have a fully functional employee. This training activity occupies the time and effort of those already skilled, and they can no longer now spend their entire workday on the activity they were hired to do. This puts additional stress on the organization during growth.

- Greatly increased deliveries are urgently needed from OEM and other vendors (large and small) while they are going through exactly the same growth problems. Their lead times increase and quality may be compromised in the rush for speedy deliveries.

When our industry is in recession, it does not need less equipment; it needs <u>no</u> new equipment for existing technologies. During these times, fab lines are operating at less, often much less, than full capacity. Maybe an emerging technology – a larger wafer diameter, perhaps – can soften the blow, but a severe blow it is to the equipment builder in any case. The scramble is on for cost savings and as a result, a decrease – in everything.

- Since at the beginning of a downturn everything is still turned "full-on", internal inventory levels and orders placed with vendors are too high. Purchasing departments spend their days damaging relationships with their vendors, while wincing as they are forced to accept million-dollar cancellations from their customers. Meanwhile, the stock price plummets.

People, the very same people just trained at great expense – are laid-off. Particularly pernicious are serial layoffs; these misjudged efforts by management to be "kind" result in a dispirited workforce waiting around for the other shoe to drop. Employees are told that they are wanted back just as soon as the downturn is over; however, by then – with luck – they have found employment elsewhere, or they choose not to re-enter a business which has let them down once and can be expected to dump them again when the next downturn occurs. Maintaining a percentage of contract workers who can be removed and rehired easily is a partial solution, but these workers have no loyalty to the company and cannot be expected to be fully involved.

The most successful solutions to these problems involve building more stable, less cyclical business units which may be expected to bring in revenue even when equipment sales head to zero. These units include replacement parts and service sales. Other availble solutions do not work very well, although, by now, many have been tried, for example:

- A short plant shutdown is a stopgap measure at best, which can have unexpected lingering effects as people and equipment start up again after a week or two.

- Price reductions have little short-term effect since the demand is not there, but cause great long-term harm to the equipment supplier when demand increases again.

The bright side of this story is that however bad it gets in the bad times – the good times will surely return. But when they do, who has the time to enjoy them?

The Agony of Making Materials

Many advanced materials are often treated as commodities even if they are not. As examples, here we look at the increasingly demanding technology requirements for bulk silicon and silicon-on-insulator (SOI) wafers over time, and the effect this has had on design and manufacturing of characterization equipment. Fortunately, unlike equipment manufacturing, the demand for materials does tend to follow the demand for IC end product, so the "More or Less" model fits well. (Fig. 3). Here, rather than company management, the critical issue is technology advancement. As devices evolve, following, at least for now, Moore's Law, line-widths decrease. This means that defects which previously were irrelevant and often unnoticed become critical.

As defects get smaller, the signal-to-noise performance of many characterization tools becomes unacceptable, and the issues of detection and measurement lead to the need for better tools. First attempts to solve this problem usually involve a series of improvements to the existing tool, but all to often the solution requires a new operating principle. Look ahead to the year 2016, the ITRS [2,3,4] roadmap lists specification after specification changing in ways that make life more and more difficult. The wafer diameter gets larger and becomes more difficult to manufacture, while almost every other of nearly 40 specifications gets smaller and more difficult to measure. Fig. 4 illustrates for bulk silicon, the desired particle reduction evolution – both their number and size – from 1997 to 2016. Similarly, Fig. 5 illustrates the evolution to thinner layers of the device layer thickness of silicon-on-insulator (SOI) over a similar time period.

Although some specifications may be less critical than others, all must be measured. The general rule is that measurements become increasingly more sophisticated and expensive as time goes by.

As an example, material quality is often determined by the measurement of its defectivity, which is defined as the number of defects of a specific size or type per unit area. The defectivity detection threshold is predicted to decrease with time in the manner shown in Table I. The 90nm node initiates a major cost increase transition for final inspection, caused by the need for improved detection capabilities now applied to a 300mm diameter wafer.

However, the equipment needed to measure defectivity is only one of a number of characterization tools that a wafer manufacturer must have in place before shipping a single wafer. For SOI, with its 40 specifications, the measurement tool situation is even worse than for bulk. The tool set requirement for SOI in 2004 is listed in Table II; the substantial increase in content and cost from 1997 to 2005 is apparent and somewhat frightening. It takes a great deal of imagination to foresee what tool set will be required in 2016 and even more imagination to guess what the cost might be.

The race for perfection never ends and this is the essence of the agony of making advanced materials. Every time that a material maker or characterizer believes that he/she has finally developed the "perfect product," the technology has moved on and it's back to the (metaphorical) drawing board again. It's enough to make Sisyphus [5] weep.

Acknowledgements

The author acknowledges with gratitude the contributions of colleagues who, over the years, have chosen to suffer alongside him.

References

[1] Part of this section was first published in Semiconductor Manufacturing, November, 2003, p.176.
[2] The National Roadmap for Semiconductors Technology Needs, Semiconductor Industry Associationi (1997).
[3] International Technology Roadmap for Semiconductors (ITRS), Semiconductor Industry Association (2001).
[4] International Technology Roadmap for Semiconductors (ITRS). (http://public.itrs.net 2003.
[5] Homer, The Odyssey, xi, 593 (c.850 B.C.).

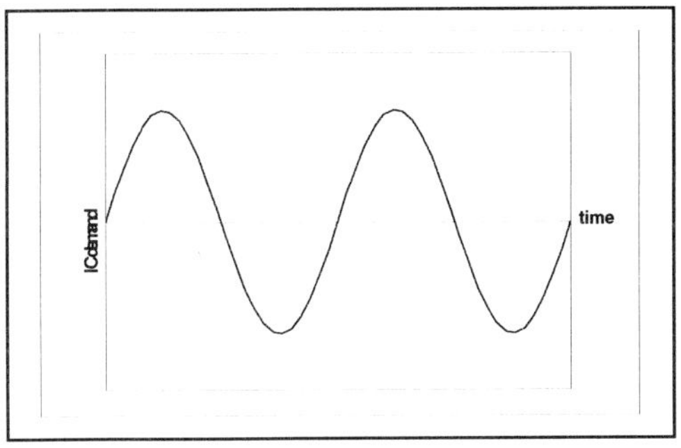

Fig.1. A schematic representation of the IC demand cycle as a function of time

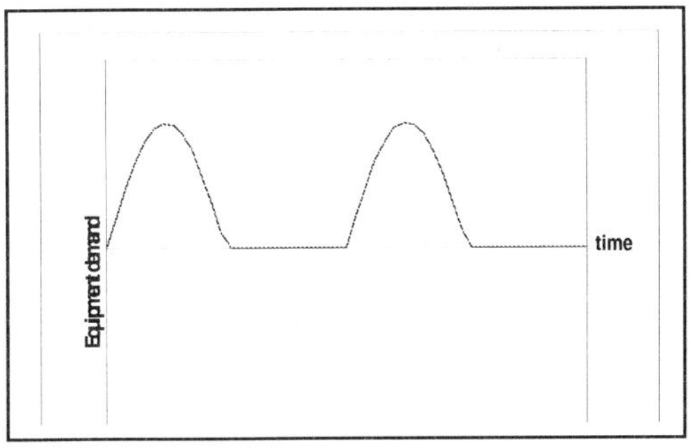

Fig.2. A schematic representaiton of the equipment demand cycle as a function of time: The "All or Nothing" model.

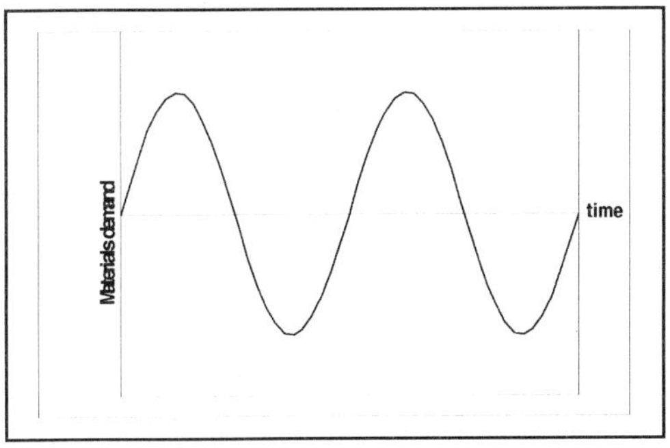

Fig.3. A schematic representation of the materials demand cycle as a function of time: The "More or Less" model.

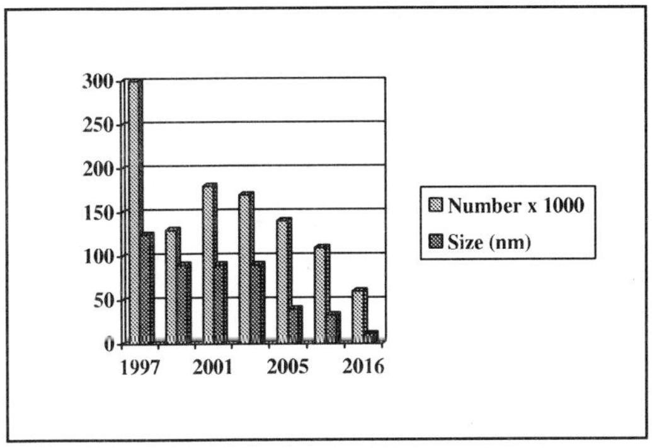

Fig.4. Particle reduction evolution for bulk silicon as a function of time [2,3,4].

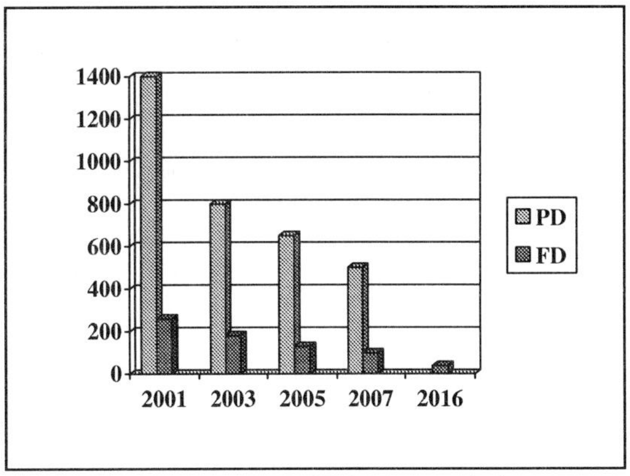

Fig.5. SOI device layer thickness evolution as a function of time [2,3,4].

Source ITRS	'97	'01	'04	'07	'10	'13	'16
Technology Node (nm)	230	180	90	65	45	32	22
Required Sensitivity (nm)	125	90	90	90	90	65	45

Table I. Defectivity Measurement Evolution

• Silicon Thickness	Spectroscopic Ellipsometer Spectroscopic Reflectometer
• SOI Etch Pit	Selective Chemical Etch
• HF Defect	HF Etch
• Voids	SEM
• Roughness	AFM
• Metal Contamination	ICP-MS
• Particles/Defects	Laser Scanner
• Nanotopography	Interferometer
• Visual Inspection	Optical Microscope
• Bulk Micro-Defects	Light Scattering Tomography
• Strain	Raman Microprobe
• Electrical	Various
• Approx Total Cost	1997 (for 230nm) \$3.6M 2005 (for 90nm) \$10.9M

Table II. Characterization Equipment Required for SOI.

This document was created with Win2PDF available at http://www.daneprairie.com.
The unregistered version of Win2PDF is for evaluation or non-commercial use only.

200mm Germanium-On-Insulator (GeOI) by Smart Cut^{TM} technology and recent GeOI pMOSFETs achievements.

T. Akatsu[1], C. Deguet[2], L. Sanchez[2], C. Richtarch[1], F. Allibert[1], F. Letertre[1], C. Mazure[1], N. Kernevez[2], L. Clavelier[2], C. Le Royer[2], J.M. Hartmann[2], V. Loup[2], M. Meuris[3], B. De Jaeger[3], and G. Raskin[4].
[1]Soitec S.A., Parc Technologiques des Fontaines, 38190 Bernin, France
[2]CEA-DRT-LETI-CEA/Grenoble, 17, rue des Martyrs, 38054 Grenoble Cedex 9, France
[3]IMEC, Silicon Process Technology Division, Kapeldreef 75, B-300 Leuven, Belgium
[4]Umicore EOM, Watertorenstraat 33, B-2250 Olen, Belgium
Email: takeshi.akatsu@soitec.fr Tel: +33 476 92 88 28

Introduction Germanium with its better transport properties than Silicon and the high-K dielectrics increases the interest for the 32 nm node solution or beyond. Recently, GeOI has opened the freedom in Ge device design in the partially and fully depleted regimes. The Smart-Cut^{TM} technology has demonstrated the path to the high quality front end GeOI fabrication [1]. The wafer diameter up to 200mm using epitaxial Ge on Si substrate was shown[2]. We present our recent achievements on 200mm GeOI formation from bulk Ge wafers and the resulting device charactereistics. Pseudo-MOS measurements were done at Soitec, and Ge MOSFET fabrication was done at LETI and IMEC from epitaxial and bulk Ge starting materials, respectively.

Characterics of GeOI substrates Ge epitaxial films on Si (100) were grown using an industrial RPCVD tool [2], with the threading dislocation (TD) density around $6x10^6$ cm^{-2} for annealed 2.5 μm-thick Ge layers. The bulk substrates used were supplied by Umicore for n and p types with no measurable TD.
The GeOI substrates by the Smart Cut^{TM} technology was described in [1, 2]. Recently, the 200mm GeOI from bulk Ge substrates have been demonstrated. The Ge film thickness (T_{Ge}) in the range between 40 and 200nm with the BOX thickness (T_{BOx}) between 150 and 400 nm can be obtained. These thickness ranges cover the platform for partially, fully depleted and multi-gate devices, which is comparable to the state-of-the-art SOI technology. Good homogeneity of the Ge film thickness has been obtained with a standard deviation of 10% for 40 nm thick Ge. AFM measurements confirmed the final surface roughness down to 3Å rms for a 5x5 μm² scan. TEM analyses with detect limit around 10^5/cm² found no observable crystalline defects. The surface inspection using a Surfscan SP1 from KLA-Tencor has shown the defectivity level down to 0.2 μm, with our calibration, even below 20 def/cm², independent of the nature of the starting material (Fig. 1).

Pseudo(Ψ)-MOSFET Fig. 2 shows a Ψ-MOSFET characterization [3] of an un-processed 2000/4500A p-type "bulk" GeOI wafer. Typical results obtained are around 5^E12 $eV^{-1}.cm^{-2}$ for Dit at Ge/SiO2 interface, around 300 $cm^2.V^{-1}.s^{-1}$ for electron mobility, μ_e, and over 350 $cm^2.V^{-1}.s^{-1}$ for hole mobility, μ_h. The lower μ_e value is probably due to the stronger impact of Dit on electrons than holes. For epitaxial GeOI (550/3000A), Dit in the order of 1 $^E13$ $eV^{-1}.cm^{-2}$, μh around 250-350 $cm^2.V^{-1}.s^{-1}$ were obtained, but μe could not be extracted.

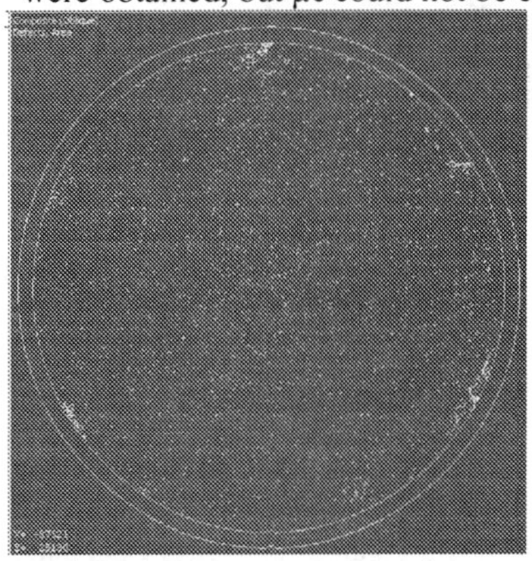

Fig. 1: SP1 inspection on a GeOI substrate shows a defectivity lower than 20 def/cm² @ 0.2 μm.

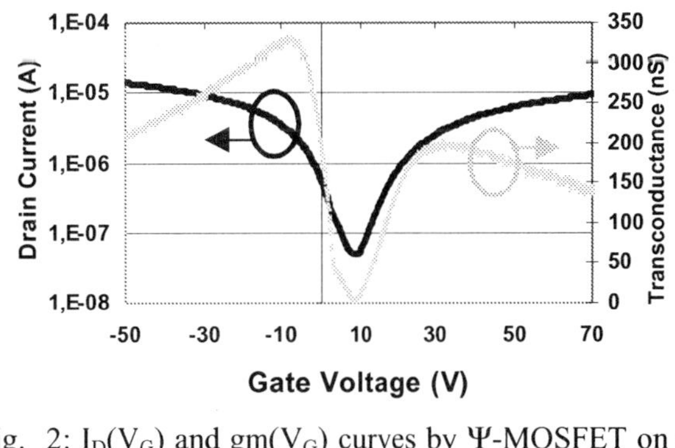

Fig. 2: $I_D(V_G)$ and $gm(V_G)$ curves by Ψ-MOSFET on a GeOI wafer at low V_D (200mV). Dit =5E12 $eV^{-1}.cm^{-2}$, μe=270 $cm^2.V^{-1}.s^{-1}$, μh=340 $cm^2.V^{-1}.s^{-1}$ were extracted.

MOSFET on epitaxial GeOI Fig. 3 shows Id-Vg characteristics of FD GeOI p-MOSFETs [4]. Ge surface passivation was done by an NH_3 thermal treatment[5], the gate stack consists of 4.5nm HfO_2, 10nm TiN and 50nm doped polysilicon. Instead of using metal alloys for contacts, Kelvin MOSFETs were made to avoid contact resistances and reduce series resistances. Due to the Ge large thickness (60nm) and the measured doping (P type / N_A=4E17At/cm^3), a back gate voltage V_{BG} is applied to ensure the full depletion regime. A I_{ON}/I_{OFF} ratio above 10^3 and a 300mV/decade sub-threshold slope were measured. These results suggest that both the Ge layer quality and the gate stack have to be improved. Nevertheless I_{ON} vs. L_G shows state-of-the-art values [4]. With the HfO_2/Ge gate stack of 1.5nm EOT, the leakage current was 6 decades lower than SiO_2/Si [4]

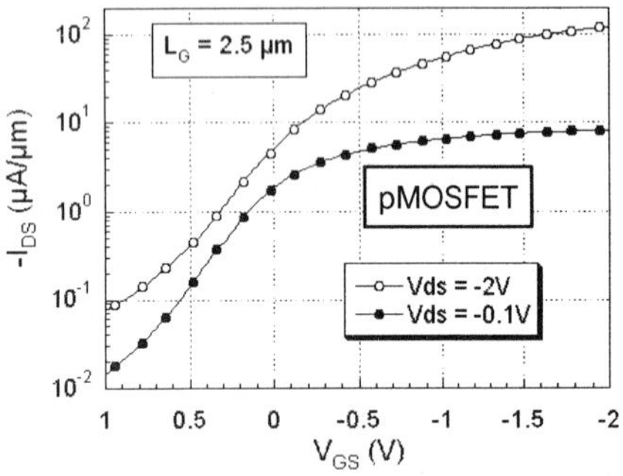

Fig. 3. $I_{DS}(V_{GS})$ characteristics (for V_{DS} = -0.1V and -2V, with V_{BG} = 50V) of a Kelvin P-MOSFET (L = 2.5µm, W = 0.25µm) on a 200mm epitaxial GeOI (T_{Ge} ~ 60nm - T_{BOx} = 400nm). The I_{ON}/I_{OFF} ratio equals ~10^3.

MOSFET on bulk GeOI Prior to gate oxide deposition, an epitaxial Si layer was grown for 4.6 Å. A precise thickness control was crucial to achieve optimal CV characteristics [7]. In this work, the Si layer is partially oxidised in a N_2O plasma at room temperature. Finally, a 300°C ALD HfO_2 gate dielectric is deposited. 200mm <100> n- and p-type 10 Ωcm GeOI wafers are used (T_{Ge} = 200nm) [8]. The silicon-like process flow for Ge device fabrication is outlined in [9]. The gate dielectric is 10nm HfO_2 and the metal gate is a 10nm PVD TaN / 70nm PVD TiN stack. Fig. 4 shows I-V characteristics of pMOSFETs from 0.15 up to 10 µm. A channel mobility of 170 cm^2.V^{-1}.s^{-1} could be extracted.

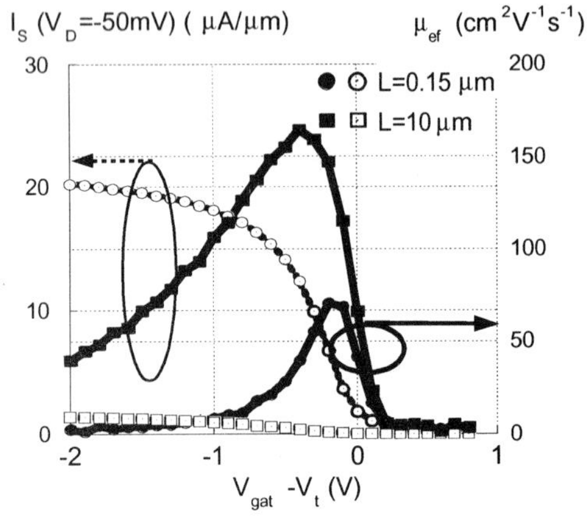

Fig. 4: low field effective mobility μ_{eff} and corresponding I_S-V_G characteristics of 10 µm and 0.15µm p-FETs

Discussions The higher mobility by Ψ-MOS on bulk GeOI may origine from its current that flows near the bottom interface of GeOI, while the current flows near the upper surface of Ge in FETs (Fig. 5). Moreover, many FET process steps may degrade the Ge layer, such as implantation and heat treatment, the latter one especially is known to be very delicate for Ge. The difference in Ψ-MOS between the bulk and epitaxial GeOI may be attributed to the existance of TD in the epi layer (10^6/cm²), whereas the bulk material is dislocation free, even though the passivation method is not the same in this comparison.

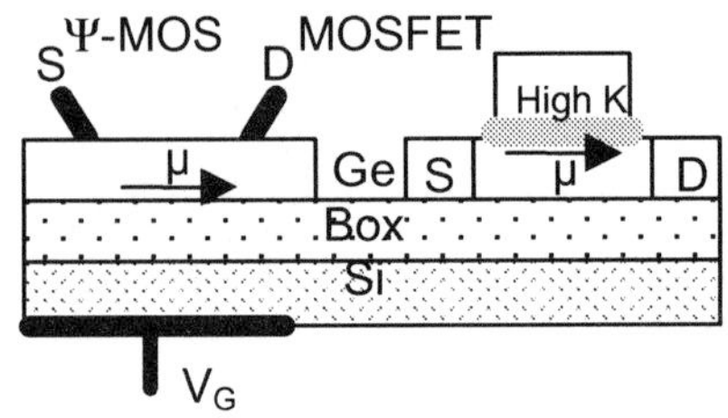

Fig. 5. schematic of structure difference between pseudo-MOS and MOSFET.

Conclusions We have demonstrated a recent 200mm GeOI formation from Ge bulk wafers by the Smart Cut^{TM} technology. T_{Ge} could be from 200 down to 40nm, T_{BOx} 400 to 150 nm, which is compatible with state-of-the-art MOS processing. Ψ-MOS measurements on bulk GeOI showed 300 and 350 cm^2.V^{-1}.s^{-1} for μ_e and μ_h, respectively, and Dit at Ge/BOX interface of 5E12 eV^{-1}.cm^{-2}. On epitaxial GeOI, μ_h was 250-350 cm^2.V^{-1}.s^{-1} with Dit in the order of 1 ^{E}13 eV^{-1}.cm^{-2}. A full integration process of 0.25µm FD p-MOSFETs was shown with high-K dielectric and metal gate on 200mm epitaxial GeOI. An Ion/Ioff ratio of 3 decades was measured. Also, p-MOSFETs with a mobility up to 170 cm^2.V^{-1}.s^{-1} were obtained on bulk GeOI using the Si epitaxially grown layer as passivation technique. The impact of starting material and the measurement on the mobility has been discussed. Thus, the GeOI substrates and MOSFETs can be a solution beyond 32nm node, with some improvements to do in the coming years.

References

[1] F. Letertre et al., Mater. Res. Soc. Proc. 809, B 4.4, 153 (2004).

[2] C. Deguet et al., to be presented at ECS (Québec), May 2005

[3] S Cristoloveanu et al, IEEE T.E.D. 47, p.1018, 2000.

[4] L. Clavelier et al., to be presented at SNW (Kyoto), June 2005.

[5] B. Guillaumot et al., IEDM 2002, p.355.

[6] N. Wu *et al.*, Appl. Phys. Lett., vol.85, n.18, p. 4127.

[7] B. De Jaeger et al., to be presented at the INFOS Conference (Leuven), June 2005.

[8] C. Deguet et al., proc. First EUROSOI Workshop, 2005, p. 31.

[9] B. De Jaeger *et al.*, proc. ESSDERC, 2004, p. 189.

A Novel Technique To Calculate The Critical Temperature of Thermal Agglomerations on Patterned SOI Wafers

Lisa Widodo, Daniel Pham[a], Barry Sassman, Larry Larson
SEMATECH, [a]Freescale Semiconductor Assignee, Austin, Texas
Email: Lisa.Widodo@sematech.org, Tel: (512) 356-3292

Introduction: Thermal agglomeration of ultra-thin SOI (<20nm) is an undesirable issue for device fabrication. In this article for the first time, a methodology to calculate the critical temperature and agglomeration rate on patterned SOI wafers with different thicknesses, ranging from 5.9nm to 12.1nm has been established. This study also presents the agglomeration effect on patterned SOI wafers for silicon structure shape, undoped, n-type doped, and p-type doped. The developed method has been tested and compared in context of previous papers published on non-patterned films [1]. In addition, detailed analysis on various surface morphologies at different temperatures has been included.

Fabrication: Commercial bonded SOI wafers with initial Si thickness of 60nm were thermally thinned down the thickness is ranging from 5.9 to 12.1 nm and patterned with the production isolation mask set. Initially, the SOI wafers were chemically cleaned using dilute hydrofluoric acid (HF/50A) solution to take away native oxide on the surface. The experiment was done under a vacuum condition at a low pressure of 20 torr using single wafer CVD tool. Each different Si thickness wafer was treated with H_2 baked recipe, using H_2 chemistry flowing at the top of the wafer at a rate of 18 slm, for 300 seconds. The H_2 flow was used because it is known as the source of Si migration. The process was repeated at an increment of 5°C.

To identify any Si agglomeration sites on each thickness for every increment of temperature, top view SEM and optical microscope have been used. The images were taken at the smallest silicon active structures.

Experimental Results and Discussion: The preliminary results from the micrographs is shown in table 1, where the first agglomeration site was spotted in the inspection stage using SEM and optical microscope.

Table 1. Data shows preliminary results from micrographs inspection of agglomeration sites.

t_{Si} (nm)	$T_{H2\text{-baked}}$ (°C) increment of 5°C	T_i (°C) First agglomeration sited
5.9	770 – 810	810
6.7	770 – 815	815
8.3	770 – 830	830
9.9	770 – 840	840
10.3	770 – 840	840
11.0	850 – 855	855
12.1	770 - 860	860

Figure 1 shows the small silicon active and non active areas. The lighter area indicates the active region whereas the darker area represents the non active region (i.e. exposed buried oxide). The analyses were done by calculating the area fractions for the agglomeration sites transformed in both the specific square and rectangular active-regions of the smallest capacitor. The area fractions of agglomeration sites formed were determined using ImageJ analysis software, which can be downloaded from the website of NIST.

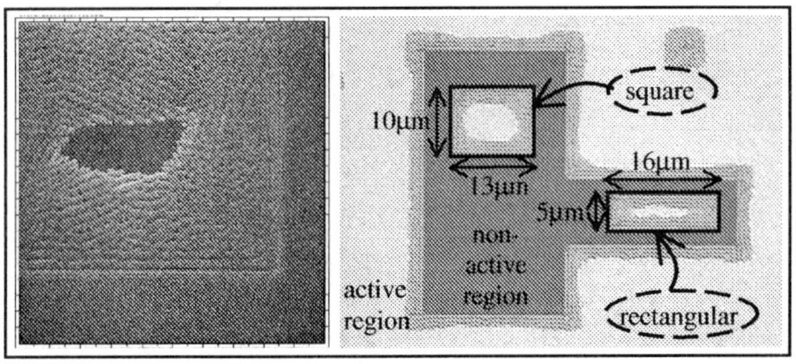

Figure 1. SEM and optical micrograph image of AK6, the smallest capacitor, indicating the active region and the non-active region that are prone to Si agglomeration.

Figure 2 shows the progress of the growth of agglomeration sites in the active region. For each of Si thickness and incremental temperature, the average of three area fractions data were taken across the center of the wafer. The temperature incremental is needed to determine the rate of agglomeration sites for each of the Si thickness. The data and plots for determining the rate of agglomeration are presented in figure 3. From plots, a linear correlation between the H_2 baked temperature and area of agglomeration sites is generated, in order to get the rate of agglomeration per °C. An approximation of 10% difference in the rate of agglomeration for square and rectangular active-regions is observed. The agglomeration rate difference of different shapes could be explained by the imbalance surface tension of rectangular structure vs. more balance surface tension of square silicon structure.

The critical temperature for the first agglomeration site is determined by the following formula:

$$T_{crit} = T_i \quad dT \qquad (1)$$

$$dT = \frac{A_f}{aggl} \qquad (2)$$

where,
T_{crit} is the critical temperature for agglomeration sites

T_i is the corresponding initiation H_2 bake temperature during experimentation when the first agglomeration site is sited in an optical microscope.

dT is the difference in H_2 bake temperature before and after the agglomeration starts. It is defined as the area fraction of agglomeration sites divided by the agglomeration rate.

A_f is the area fraction of agglomeration sites (represented by the darker area in figure 1).

$\overline{aggl.}$ or the rate of agglomeration is how fast the agglomeration sites transformed per °C.

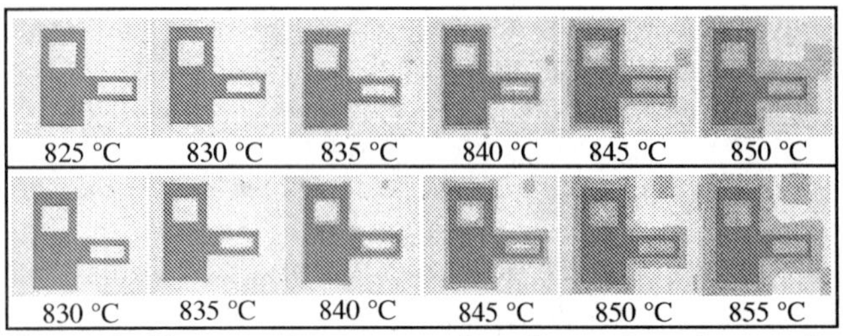

Figure 2. The growth progress of agglomeration sites as the temperature increases is shown, for Si thickness of 6.7nm (top) and 8.3nm (bottom).

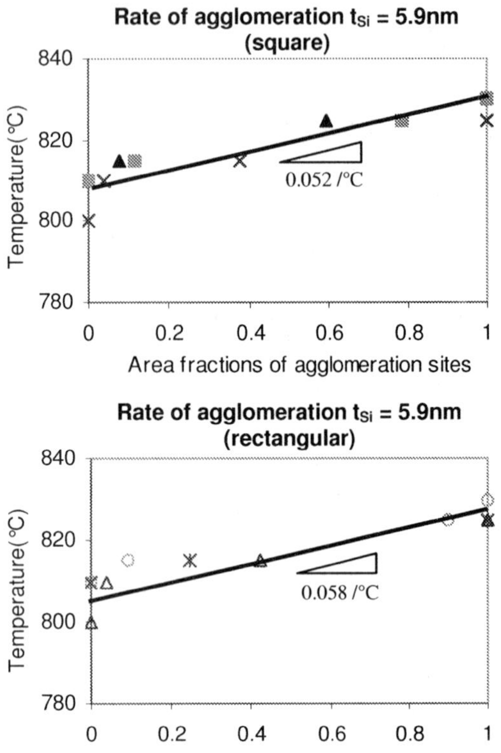

Figure 3. H_2 baked temperature is plotted against area fractions of agglomeration sites, to determine the rate of agglomeration.

Table 2 summarizes data for each Si thickness and its critical agglomeration temperature. Even though the rectangular active region have greater rate of agglomeration than the square active region, the critical agglomeration temperature between the square and rectangular active region is not significantly different, with less than 1°C difference. It concludes that the critical agglomeration temperature does not depend on the shape of active areas but it depends on the Si layer thickness.

Table 2. Summary of experimental and calculated data to determine the critical agglomeration temperature.

t_{Si} (nm)	Square active region			Rectangular active region		
	T_i(°C)	$\overline{agglo}$	T_{crit} (°C)	T_i(°C)	$\overline{agglo}$	T_{crit} (°C)
5.9	810	0.052	809.74	810	0.058	809.79
6.7	815	0.043	814.69	815	0.048	814.71
8.3	830	0.040	829.40	830	0.046	828.67
9.9	840	0.038	839.71	840	0.043	839.69
10.3	840	0.055	837.16	840	0.060	837.07
11.0	855	0.037	854.86	855	0.042	854.80
12.1	860	0.041	858.38	860	0.050	858.37

Figure 4 shows a linear correlation between the critical temperature for agglomeration sites and the Si layer thickness of this work on patterned wafer. In the same plot, the linear relationship is also claimed by Ishikawa et al. [1] for the un-patterned wafers.

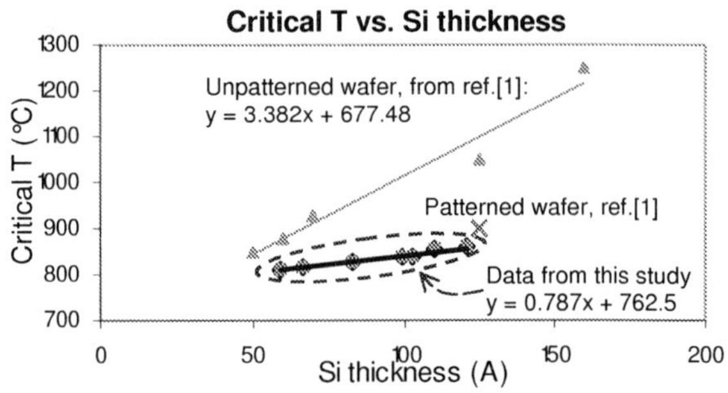

Figure 4. The critical temperature for agglomerations is plotted against Si layer thickness, for t_{Si} of 5.9 to 12.1nm.

Since critical temperature for agglomeration sites linearly increases with increasing Si layer thickness, therefore a general formula for calculating the critical temperature for each Si layer thickness of patterned-UTB-SOI wafer has been developed. The linear equation for the square active region and the rectangular active region is as follow:

$$T_{crit} = 0.787\ t_{Si} + 762.5\ °C \qquad (3)$$

Conclusion: A methodology to calculate the critical temperature and rate of thermal agglomerations for patterned-UTB-SOI wafer has been developed. The critical agglomeration temperature has a strong dependency with Si layer thickness. Equation predicting the critical agglomeration temperature per silicon thickness is obtained for patterned wafer. The agglomeration rate, on the other hand, depends on the shape of the active silicon structure. Concentric shapes such as square or circle could reduce the agglomeration rate significantly. Understanding the rate and critical temperature of the agglomeration could help circuit designer and process engineers to minimize the issue for the device fabrication.

References:

[1] Y. Ishikawa et al., Appl. Surf. Sci., Vol. 190, pp. 11-15, 2002.

Ultra-thin strained Si on insulator substrate using laser annealing

Yasuyoshi Mishima, Atsushi Mimura, Masatoshi Fukuda, and Hirosato Ochimizu
Fujitsu Laboratories Ltd., 10-1 Morinosato-Wakamiya, Atsugi 243-0197, Japan
Email: yasu.mishima@jp.fujitsu.com, Tel: (046) 250-8237

ABSTRACT

We describe a low-temperature process for fabricating strained silicon on insulator (SOI) substrates using excimer laser annealing technology. We used an excimer laser to relax the SiGe layer on the SOI substrate. We can fabricate ultra-thin strained SOI by removing the relaxed SiGe layer on the SOI after laser annealing. Strains are investigated using Ultraviolet (UV)-Raman spectroscopy and Transmission electron microscopy (TEM).

INTRODUCTION

SOI MOSFETs are attractive materials for low-power and small-sized devices. Ultra-thin SOI MOSFETs are effective for suppressing short channel effects[1]. Ultra-thin strained SOI MOSFETs are particularly effective substrates for fully depleted SOI devices and a promising option for aggressively scaled devices[2]. These substrates are usually fabricated using the layer transfer technique[3]. However, this technology needs other more complex processes. We previously described strained-silicon formation on relaxed SiGe/Si on insulator substrates using laser annealing[4,5]. In this paper, we describe a new ultra-thin strained silicon on insulator substrate fabrication process that uses excimer laser annealing technology.

EXPERIMENT

Figure 1 shows the fabrication process for strained SOI structures. First, we prepared 6-inch Si (100) 14-Ωcm p-type bonded SOI wafers, containing 10-20 nm of Si and 140-nm of buried oxide. Second, strained 30-40 nm $Si_{1-X}Ge_X$ (X = 0.2 - 0.25) layers were grown on the SOI substrate using ultra-high-vacuum chemical-vapor deposition (UHV-CVD). Third, an XeCl excimer laser beam was irradiated on the SiGe surface. We used a single exposure in the same place. The Raman spectra were almost independent of the number of pulses between 1 and 3. The temperature of the SiGe layer at the irradiated position rapidly rose to over 1,000℃ because of the short pulse-duration times[4,5]. Finally, we removed a 30-40 nm relaxed SiGe layer from the SOI substrate using a chemical etching technique after laser annealing. Figure 2 shows the experimental set-up for laser annealing. We set the SOI wafer in the vacuum chamber and used a XeCl excimer laser with a pulse duration of 160-ns and λ=308 nm. The laser irradiation was done at room temperature under N_2 atmospheric pressure. We used a beam size of 20 x 20 mm^2.

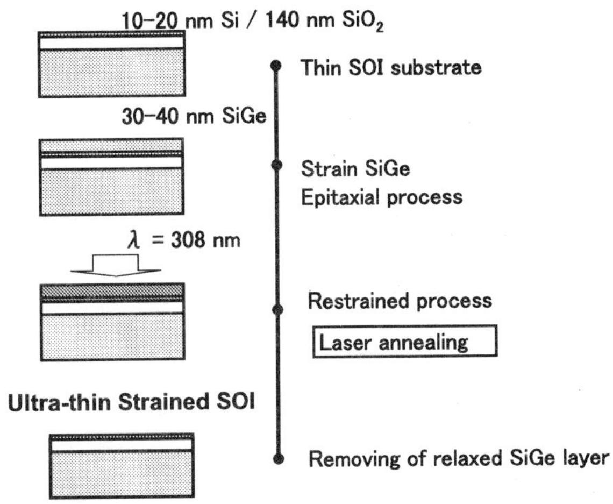

Fig. 1: Schematic process flow of ultra-thin strained SOI substrate formation using laser annealing technology.

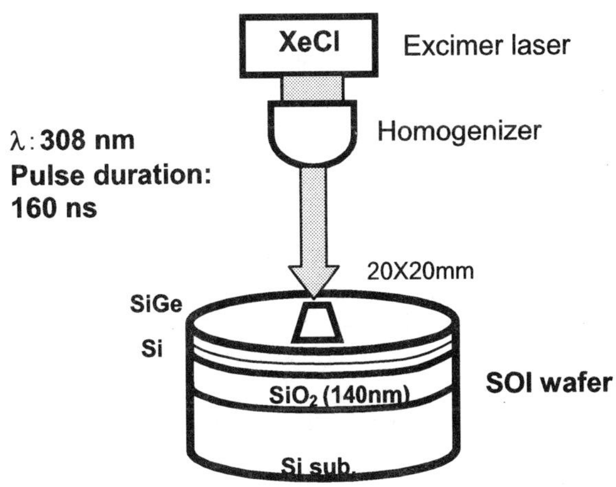

Fig. 2: Experimental set-up for laser annealing. We can change the beam size by controlling the optics and the irradiated place by moving the substrate stage..

RESULTS AND DISCUSSION

Figure 3 shows the Si-Si Raman spectra of the $Si_{0.75}Ge_{0.25}$ layer on SOI. We used 325-nm UV-Raman spectroscopy to measure the strained Si layer on the insulator. The peak of the Si-Si Raman spectrum for the as-grown $Si_{0.75}Ge_{0.25}$ was nearly 511 cm^{-1}, depending on the Ge content[6]. This result confirms that the $Si_{0.75}Ge_{0.25}$ layer is strained[7]. The peak position was changed to a lower Raman shift side from 511 to 507 cm^{-1} using laser annealing at 1,200 mJ/cm^2. That of $Si_{0.8}Ge_{0.2}$ moved from 515 to 510 cm^{-1}. These results were almost the same as those described in Ref. 5, where we used a laser with a 40-ns pulse duration at 240 mJ/cm^2 for strained-Si formation on relaxed SiGe/SOI substrates. This proves that the SiGe layer stress was relaxed by laser annealing. The ultra-thin Si layer under the SiGe layer was stressed by releasing the SiGe layer stress. Because laser irradiation of SiGe on SOI increases the temperature of the SiGe, the ultra-thin Si and buried oxide layers, the ultra-thin Si layer can be strained by extending the SiGe lattice.

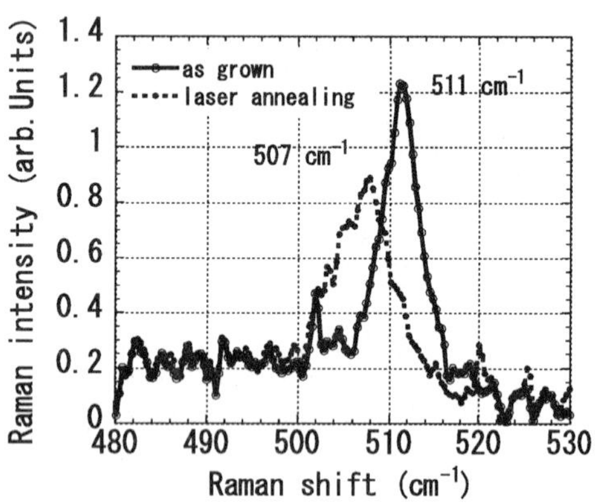

Fig. 3: UV-Raman spectra of $Si_{0.75}Ge_{0.25}$ on SOI. The solid line represents SiGe/SOI without laser annealing. The dashed line represents SiGe/SOI with laser annealing.

Figures 4(a) and (b) show the Si-Si UV-Raman spectra from 480 to 540 cm^{-1} for 17-nm thick SOI after removing the $Si_{0.8}Ge_{0.2}$ layer on the SOI. (a) is the result of 30-nm $Si_{0.8}Ge_{0.2}$. and (b) is that of 40-nm $Si_{0.8}Ge_{0.2}$. The strained SOI were prepared using laser annealing at 1,300 mJ/cm^2. The 516 cm^{-1} strained Si spectra showed that the Si on insulator was strained about 1 GPa by the relaxed $Si_{0.8}Ge_{0.2}$ layer[8]. The $Si_{0.8}Ge_{0.2}$ layer thickness did not influence the laser annealing effect. On the other hand, the $Si_{0.75}Ge_{0.25}$ layer could not enough strain the ultra thin Si layer by using these structures. The reason may be due to the quality of the thick $Si_{0.75}Ge_{0.25}$ layer.

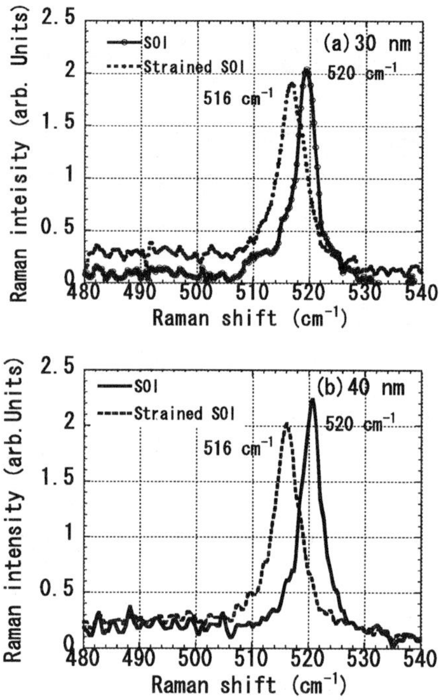

Fig. 4(a), (b): UV-Raman spectra of 17-nm SOI and 17-nm strained SOI. We used 30-nm SiGe in (a) and 40-nm SiGe in (b).

Figure 5 shows the depth profiles of Ge for 17-nm SOI by backside-SIMS. We compared the two samples. One was the strained SOI using laser annealing and other was SOI without laser annealing. We measured the SIMS profiles after removing the $Si_{0.8}Ge_{0.2}$ layer. The Ge depth profile of the strained SOI was almost the same as that in SOI without laser annealing. This shows that the Ge depth profile is not influenced by laser annealing because of the short pulse duration of 160 nm.

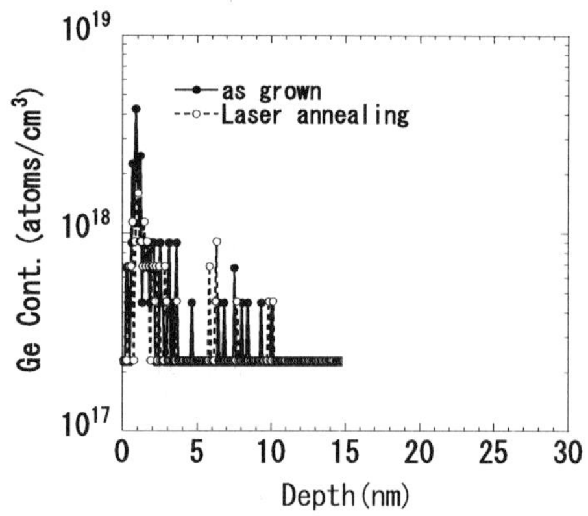

Fig. 5: Depth profile of Ge for ultra-thin SOI and strained SOI. Solid line is for SOI and dashed line is for strained SOI by laser annealing.

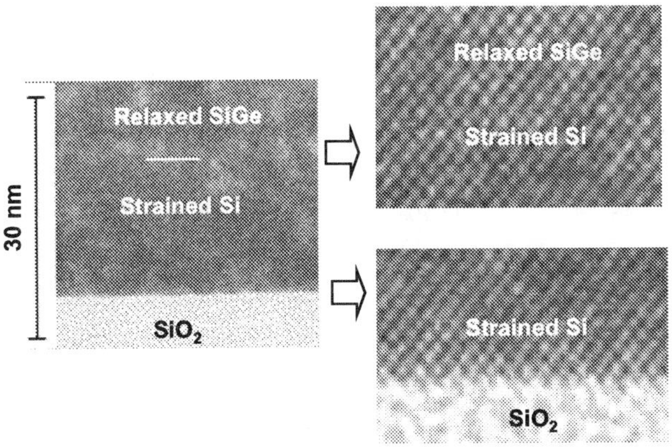

Fig. 6: Cross-sectional TEM photograph of strained SOI. SiGe layer is left to compare the lattice image of strained Si layer with that of SiGe layer.

The crosssectional TEM micrograph in Fig. 6 shows the good crystalline quality of the strained Si SOI substrate. The $Si_{0.8}Ge_{0.2}$ layer on the SOI substrate was left to compare the lattice constant of SiGe and strained Si layers. These micrographs show that the lattice constant of ultra thin Si layer on insulator was the same as that of SiGe layer horizontally. The interfaces of SiGe/Si and Si/buried oxide layers are very smooth.

CONCLUSION

We proposed a new fabrication technology for ultra-thin strained SOI structures. We can fabricate ultra-thin strained SOI by removing the relaxed SiGe layer on the SOI after laser annealing. We can produce an ultra-thin strained SOI structure exactly where we want it. Using this technology, obtaining ultra-thin strained SOI substrates is a very simple process.

REFERENCES

[1] B.Doris, M.Leong, T.Kanarsky, Y,Ahang, R.A.Roy, O.Dokumaci, Z.Ren, F.F.Jamin, L.Shi, W,Natzle, H.J.Huang, J.Mezzapelie, A.Mocuta, S.Wromack, M.Grivelyuk, E.C.Jones, R.J.Miller, H.S.P. Wong and W.Haensch, "Extreme scaling with Ultra-Thin Si channel MOSFETs," in IEDM Tech. Dig., 2002. pp. 267-270.

[2] K. Rim, K.Chan, L.Shi, D.Boyd, J.Ott, N.Kiymko, F.Cardone, L.Tai. S.Koester, M.Cobb. D.Canaperi, B.To, E.Duch, I.Babich, R.Carruthers, P.Saunders, G.Waler, Y.Zhang, M.Steen and M.Ieong, "Fabrication and Mobility characteristics of ultra-thin strained Si directly on insulator (SSDOI) MOSFETs," in IEDM Tech.Dig., 2003. pp. 48-51.

[3] G. K. Celler and S. Cristoloveanu, " Frontiers of silicon-on-insulator," J. Appl. Phys. vol. **93**, pp. 4955-4978, 2003.

[4] Y. Mishima, H. Ocimizu nad A. Mimura, "Strained-silicon formation on relaxed silicon-germanium/siliocn-on-insulator substrate using laser annealing," Appl. Phys. Lett. vol. **86**, pp. 72106-72107, 2005.

[5] Y. Mishima, H. Ochimizu and A. Mimura. "New strained silicon-on-insulator fabricated by laser annealing technology," Jpn. J. Appl. Phys. vol. **44**, pp. 2336-2339, 2005.

[6] M. I. Alonso and K. Winer, "Raman spectra of c-$Si_{1-X}Ge_X$ alloys," Phys. Rev. vol. **B 39**, pp. 10056-10062, 1989.

[7] J. C. Tsang, P. M. Mooney, F. Dacol and J. O. Chu, "Measurement of alloy composition and strain in thin Ge_xSi_{1-x} layers," J. Appl. Phys. vol.**78**, pp. 8098-8108, 1994.

[8] Th. Englert, G. Abstreiter and J. Pontcharra, "Determination of existing stress in silicon films on sapphire substrate using Raman spectroscopy," Solid-State Electronics. vol. **23**, pp. 31-33, 1980.

Development of Stacking Faults in Strained Silicon Layers

S. W. Bedell[1], A. Reznicek[1], B. Yang[3], H. J. Hovel[1], J. A. Ott[1], K. Fogel[1], A. G. Domenicucci[2] and D. K. Sadana[1]
[1]IBM T. J. Watson Research Center, 1101 Kitchawan Road, Yorktown Heights, NY 10598
[2]IBM Semiconductor R&D Center, Hopewell Junction, NY 12533
[3]Depatment of Chemistry, University of Wisconsin, Madison, WI 53706
Email: bedells@us.ibm.com, Tel: (914) 945-2232

Strained silicon layers are currently being used to extend the performance of CMOS IC technology. All of the methods currently used [1,2] rely on locally straining the channel region of the transistor using various process-level techniques. An alternative method of providing a strained silicon channel layer is to have strain present in the starting substrate. Starting wafers that have a continuous strained Si layer are referred to as "globally strained". Strained silicon layers formed directly on an insulating layer (SSOI) allow the advantages of strained Si to be combined with those of SOI technology. Modern SOI-based CMOS devices are built using partially-depleted SOI technology (PD-SOI). This technology requires that the top Si layer has a thickness greater than about 300 Å. For sufficient performance gain to be achieved from strained silicon, at least 0.6% tensile strain is required. SSOI wafers are formed by growing the strained-Si layer on a relaxed SiGe virtual substrate, and transferring the Si onto a handle wafer using layer transfer techniques. Because the critical thickness of a 0.6% strained Si layer is less than 300 Å, these layers will be unstable against dislocation formation and growth. Therefore, low-defect PD-SSOI can only be realized if thick (> 300 Å) strained Si layers can be grown on high-quality relaxed SiGe layers without creating additional defects.

Stacking fault (SF) defects have been shown to form in Si layers under tensile strain [3,4]. In this work, we investigate the development of SF defects in strained silicon layers grown on low-defect SiGe graded buffer layers. Si layers were grown to various thicknesses at different temperatures and the resulting SF densities are measured using a specialized etching technique [5].

In this study, compositionally graded SiGe buffer layers were grown up to a surface concentration of 20 atomic percent Ge. Silicon layers were then deposited at various thicknesses without breaking vacuum at a temperature of 600

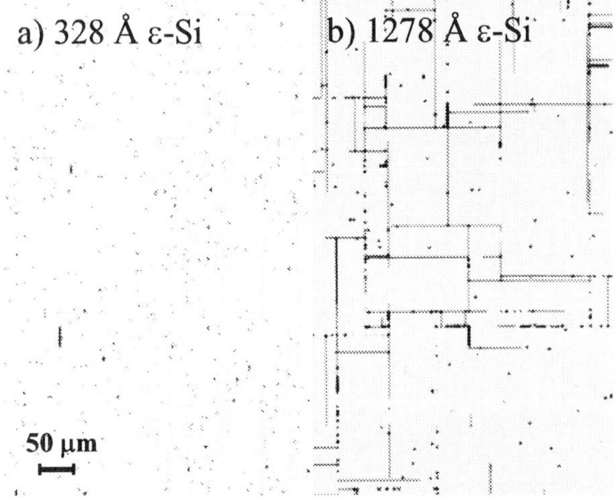

Figure 1. Defect micrographs of 328 Å (a) and 1278 Å (b) strained Si layers grown on a 20% SiGe graded buffer layer. The strain was measured to be 0.64% in (a) and 0.62% in (b) by Raman spectroscopy.

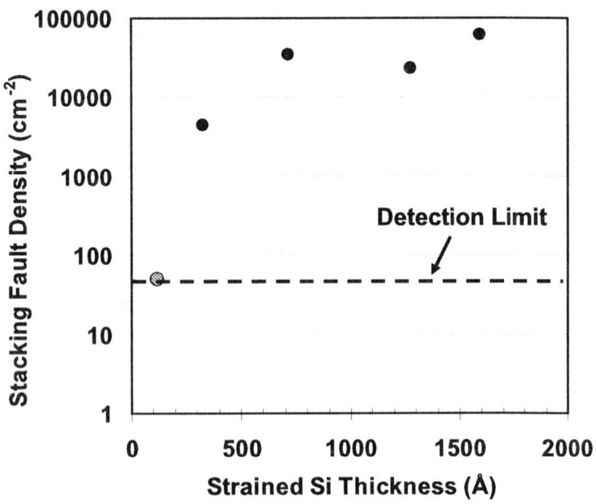

Figure 2. Measured areal density of stacking faults as a function of strained Si thickness. No SFs were observed in the thinnest layer and the data point represents the detection limit (50 cm^{-2}).

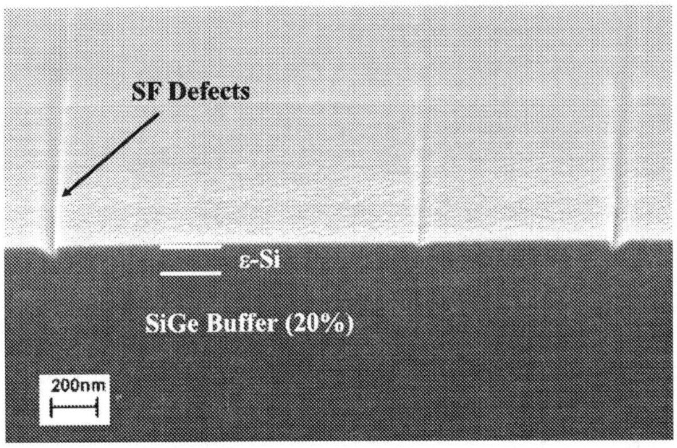

Figure 3. Tilt-view SEM of defect-etched 1278 Å strained Si layer shown in Fig. 2.

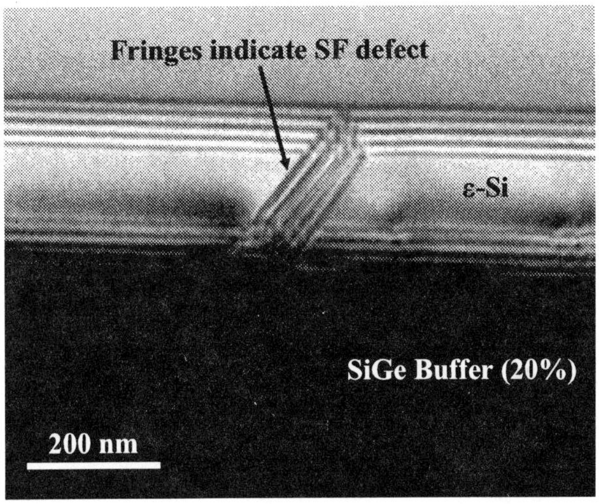

Figure 4. Cross-section TEM of strained Si layer shown in Figs. 2 and 3 indicating that the planar defects observed after etching are stacking faults.

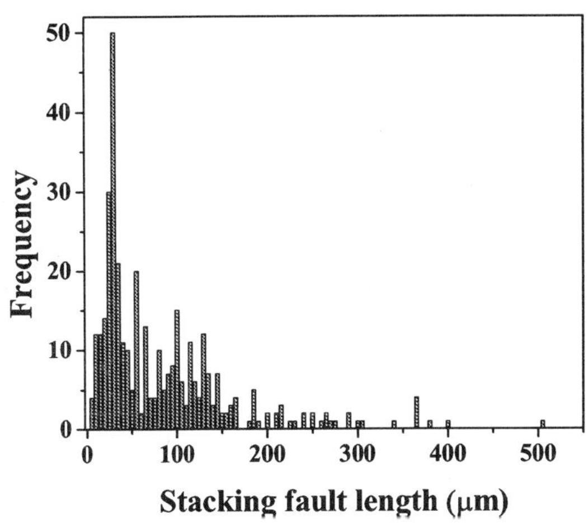

Figure 5. Histogram showing the distribution of SF lengths in a 1595 Å strained Si layer.

°C. The thinnest Si layer (measured by spectroscopic ellipsometry) was 115 Å and defect etching gave an etch pit density (threading dislocations) of about 1.5×10^5 cm^{-2} and the SF density was below the limit of detection (< 50 cm^{-2}). Figure 1 shows optical micrographs of samples with (a) 328 Å and (b) 1278 Å strained Si layers. Although the etch pit densities are similar for all the samples (varying between 1 and 2.5×10^5 cm^{-2}), the SF defects increase by several orders of magnitude with increasing Si thickness. Figure 2 shows the measured relationship between SF density (areal) and strained Si thickness. Figures 3 and 4 are tilt-view SEM and cross-section TEM images, respectively, showing the line defects are indeed stacking faults. The sample shown in Fig. 3 has been defect etched prior to imaging.

A proper analysis of stacking faults should include some information about the distribution of defect lengths. Figure 5 shows the distribution of stacking fault lengths in a sample with a 1595 Å strained Si layer (the area of the microscope image was 5.9×10^{-3} cm^2). All Si thicknesses show a primary peak; for the 330 Å Si layer it is centered at ~ 5 μm. The peak of this distribution shifts to ~ 30 μm for the 1595 Å thick Si sample. Also, as the strained Si layer thickness increases, a significant number of long stacking faults appear. This data shows that even at relatively low growth temperatures, SF defects can grow appreciably long.

The investigation into the formation of SF defects in strained Si layers grown on high-quality relaxed SiGe layers suggests that thick strained Si layers with low defectivity may be difficult to produce. This poses a significant technical challenge for the production of high quality SSOI substrates for use with partially-depleted CMOS technology.

References

1. S. Yang *et al., IEEE IEDM Conference Proceedings*, **28.8**, 1075 (2004).

2. T. Ghani *et al., Tech. Dig. International Electron Devices Meeting*, Dec. (2003).

3. P. M. J. Marée *et al., J. Appl. Phys.*, **62**, 4413 (1987).

4. S. W. Bedell *et al., Appl. Phys. Lett.* **85**, 2493 (2004).

5. S. W. Bedell *et al., Electrochemical and Solid State Letters*, **7**, G105 (2004).

UV scanning for SOI defectivity monitoring

C. Moulin, D. Delprat and C. Maleville
SOITEC SA; Parc Technologique des Fontaines ; 38926 Crolles Cedex ; France
W. McMillan, J. Payne, K. Birdwell, R. Brun and R. Moirin
KLA-Tencor Corp; Milpitas; CA
Email: Christophe.Maleville@Soitec.fr, Tel: +33 (4) 76 92 75 84

Device speed improvement has been ensured by shrinking dimensions as illustrated by Moore's law. New starting substrate have been introduced to move beyond 130nm, including SOI or strained silicon. SOI is used in volume production today for high performance microprocessors. Device performance has to be combined with high manufacturing yield that is closely related to defect monitoring. This paper shows how a new generation of inspection system enables us to monitor SOI as requested in the ITRS roadmaps [1]; it shows also the large amount of key information we can collect, giving us important feedback on the SOI process.

SP1 SOI defect inspection

SOI inspection with visible light is limited to 0.15µm threshold because of reflectivity effect [2]: light is transmitted and reflected through the silicon and oxide layers leading to constructive and destructive interferences (Fig.1). This results in different defect sizing and different background (haze) level depending on silicon and oxide thickness combination. With such wavelength, thresholds are limited for two reasons:

- High reflectivity products generate a magnification effect on small defects - defects smaller than 0.15µm can be detected using 0.30µm threshold [3]. This effect tends to enhance scattering coming from "surface" defects limiting the use of low threshold unless corrected by using customized SOI calibration curve.
- Low reflectivity products generate a very high background (haze) level. This effect can be corrected by using specific optical configurations, reducing the haze level. Best achievable threshold with such optical configuration is around 0.15µm.

SP2 SOI defect inspection

Limitations described in the previous paragraph disappear with UV wavelength. Incoming light is absorbed in the first nm of the silicon layer, no more reflectivity variations is seen, leading to three major advantages :

- SOI can be measured at very low threshold: XUT Unibond™ products can be measured at 85nm, showing similar defectivity as silicon wafers (Fig. 2).
- No more dedicated calibration curve is needed, and defectivity comparison becomes possible between different SOI products.
- Low thresholds achieved on SOI, enabled in one pass only, to get high sensitivity for all defect types.

SP2 for yield improvement

In SOI characterization flow, threading defects are discriminated by using a combination of HF revelation (to increase their size), SP1 measurement and SEM review. SP2 inspection at 85nm allows isolating impacted wafers without the help of any destructive characterization, saving significantly time and cost. Developments are now on their way to track and classify voids with SP2.

The SURFimage, a high resolution surface background, available on SP2, is also a very helpful feature: the SURFimage maps give access to defectivity below the threshold. One example is shallow scratches visible on high haze resolution maps of bare silicon wafers (Fig. 3). Particles, cops and shallow scratches can be detected at the same time, avoiding the use of an additional recipe for scratch detection. This surface analysis feature can be helpful for fast process improvement, as demonstrated for polishing in figures 4 and 5. Such surface features were not detectable either with visual inspection or SP1 measurements. As no more interface artifact is contributing to surface scattering on SOI, specific features are easily detected and can be fixed.

0-7803-9212-4/05/$20.00 ©2005 IEEE

Conclusion

SP2 is the metrology tool of choice for SOI inspection: SOI calibration curves are no longer needed, very low threshold can be achieved, and SOI can now be measured as bulk Silicon, helping with its adoption. Superior haze resolution analysis gives additional information below the threshold. SURFimage analysis gives access to surface defects that are invisible with most of other inspection tools. Low thresholds combined with the high haze resolution found in the SURFimage leads to a fast and accurate feedback on SOI process quality. SP2 maps obtained on XUT Unibond product show that SOI has similar defects levels than silicon substrate. Direct analysis of the surface allows surface roughness analysis: very good correlation with AFM monitoring is obtained and will be shown.

References

[1] http://public.itrs.net
[2] C.Maleville and al., 2002 SOI conference
[3] C.Maleville and al., 2004 STS Semicon West

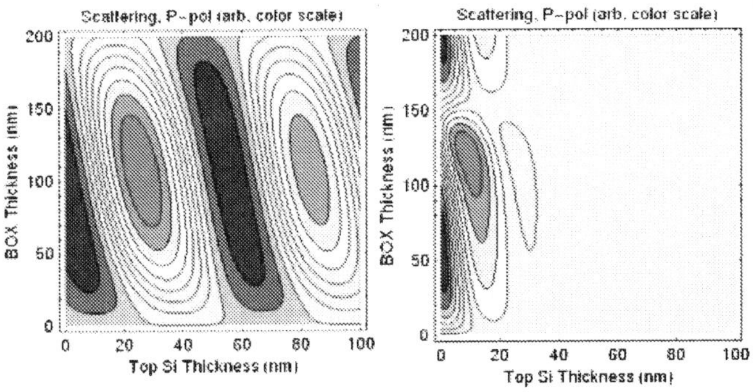

Fig. 1: Scattering intensity simulation depending on Silicon and Oxide thickness on SP1 (left) and SP2 (right)

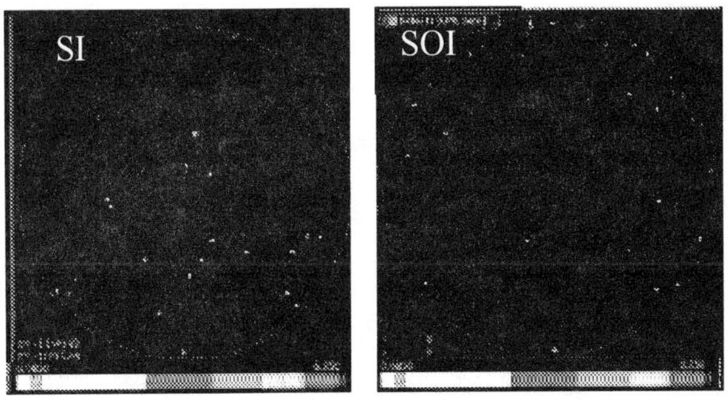

Fig. 2: Bulk silicon and XUT SOI wafers scanned at 85nm both showing a low defectivity level.

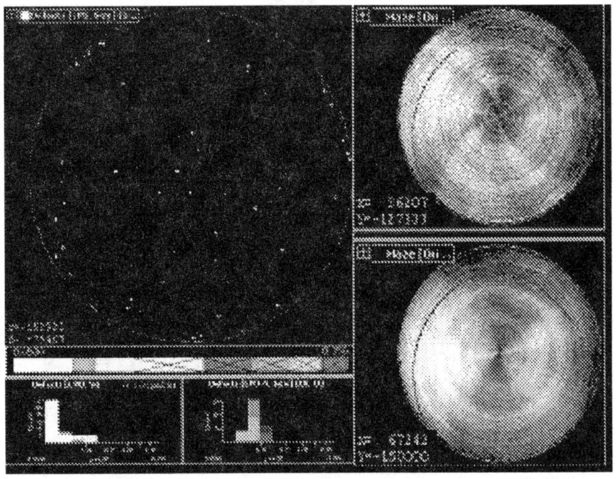

Fig. 3: Silicon wafer -shallow scratches visible on SURFimage maps only – Invisible on composite map.

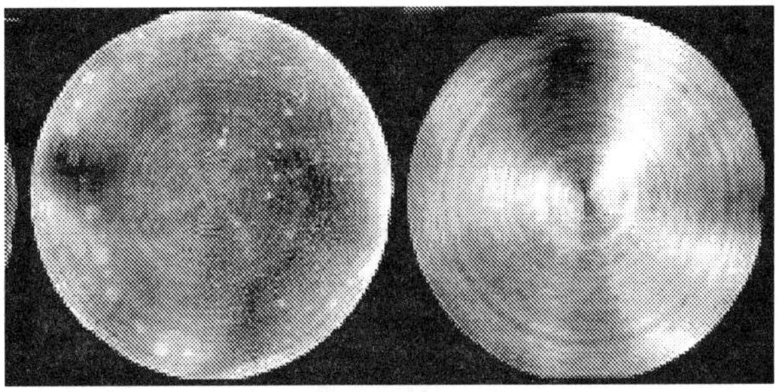

Fig. 4: Surface features visible on SP2 SURFimage maps – Polishing issue removal from left image to right image

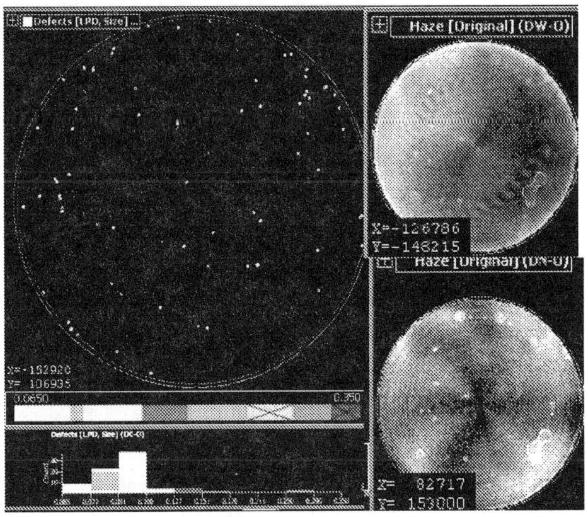

Fig. 5: Polishing features are visible on SURFimage maps only – Invisible on composite map whatever the threshold level.

2005 IEEE INTERNATIONAL
SOI CONFERENCE

Session #6

MULTIPLE GATE DEVICES

CO-CHAIRS

Toshiro Hiramoto

University of Tokyo

Wade Xiong

Texas Instruments

A Hybrid FinFET/SOI MOSFET

W. Zhang, J. G. Fossum, and L. Mathew*
University of Florida, Gainesville, FL 32611-6130, U.S.A.
**Freescale Semiconductor, Austin, TX 78721, U.S.A.*

Introduction

With CMOS technology being rapidly scaled, the double-gate (DG) FinFET [1] has attracted interest because of its excellent suppression of short-channel effects (SCEs) [2]. The fin height (h_{Si}) is limited, subject to its thin width (w_{Si}) needed for the SCE control, and only one fin per pitch is allowed. Hence, the current per pitch attainable is limited, the device width is "digital," and layout-area efficiency can be compromised. To improve the layout efficiency, or current per pitch, we have suggested that a single-gate (SG) SOI/UTB MOSFET be fabricated in the unused portion of the pitch area, and merged with the DG FinFET to form a single device [3]. In this paper, we use our process/physics-based generic compact model for DG MOSFETs, UFDG [4], in Spice3, combined with simulations done with the 3-D numerical simulator Davinci [5], to gain insights regarding the design and performance of this hybrid device, which we call the Inverted-T FET (ITFET) because of its cross-sectional shape (Fig. 1). Based on UFDG/Spice3 simulations, we also note the potential benefit of the ITFET in effecting a good design of the FinFET-CMOS 6T-SRAM cell with regard to the area-performance tradeoff.

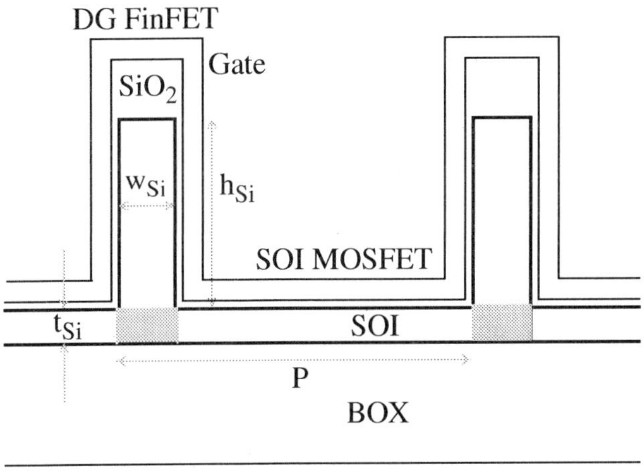

(a)

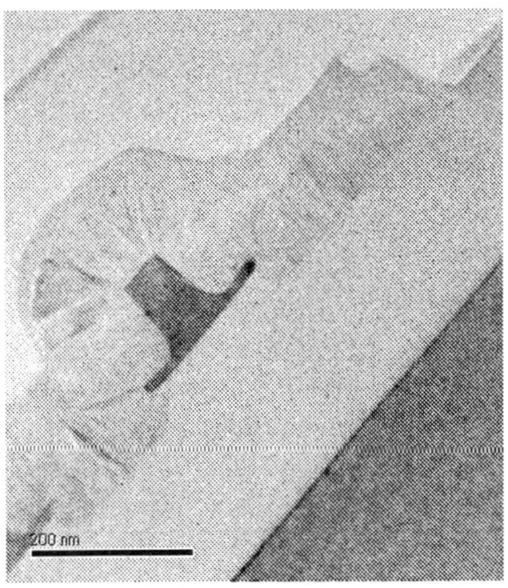

(b)

Fig. 1. (a) Cross-section of the ITFET structure, between the source and drain, with the ungated fin-base shown shaded. **(b)** TEM picture of a fabricated ITFET.

The ITFET

A cross-sectional illustration of the ITFET, with one fin per pitch (P) and SOI covering the rest of the pitch area, is shown in Fig. 1(a); Fig. 1(b) shows a fabricated ITFET. We first use a gate length L_{gate}=28nm, and assume that it equals the metallurgical, or effective channel length (L_{eff}). We further assume a gate-oxide thickness (t_{ox}, or EOT) of 1nm for both the FinFET and the FD/SOI MOSFET (FDFET) that constitute the ITFET. And, since nanoscale devices should be designed with undoped bodies [6], we let the common gate be midgap metal. For good SCE control [2], we let $w_{Si}=L_{eff}/2$=14nm, and the SOI thickness $t_{Si}=L_{eff}/5$=5.6nm.

The ITFET design goal is to achieve good current-voltage characteristics, i.e., high (and variable) I_{on}/pitch and high I_{on}/I_{off} with acceptable threshold voltage (V_t). Relative to the FinFET, we expect substantially more current drive from the ITFET, as implied by the increased per-pitch $W_{eff} \cong h_{Si}+P-w_{Si}$ and the ratio (f_{DG}) of the per-width on-state currents in the FinFET and FDFET. For f_{DG}=2, the relative I_{on} increase is $(P-w_{Si})/(2h_{Si})$, which is 113% for our device with an $R_f=h_{Si}/w_{Si}$=4 fin. However, I_{off} can be much higher due to leakage in the ungated fin-base indicated in Fig. 1(a).

To check the fin-base leakage, we use Davinci. We plot in Fig. 2 the predicted per-pitch current-gate voltage characteristic, at V_{DS}=1.0V, of the n-channel ITFET with R_f=4, contrasted with those of the constituent FinFET and FDFET. As expected, the Davinci-predicted I_{on} of the ITFET is

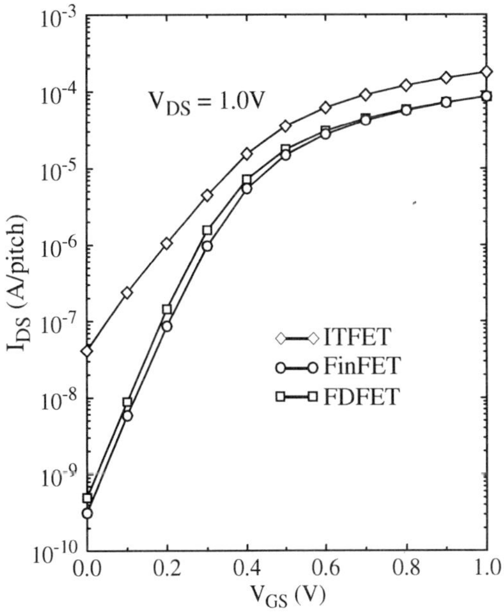

Fig. 2. Davinci-predicted I_{DS}-V_{GS} characteristics per pitch of the L_{gate}=28nm nITFET, along with the separate characteristics of the constituent FinFET and FDFET; R_f=4.

0-7803-9212-4/05/$20.00 ©2005 IEEE

approximately the sum of those of the FinFET and FDFET, but the ITFET I_{off} is about two-orders of magnitude higher than that of the FinFET, a difference that is much more than the contribution of the constituent FDFET. Clearly, the ungated fin-base supports prodigious weak-inversion current. Davinci-predicted electron density profiles down the fin of the ITFET confirm this conclusion. This nanoscale ITFET, with high drive current but also very high off-state current, could be useful in particular applications, but not generally.

To make the ITFET more viable, the fin-base leakage must be eliminated. One way of doing this might be to heavily dope the fin-base to suppress the diffusion of gate-induced electrons into it. Davinci simulations show that this works, but the doping density required is $\sim10^{19}cm^{-3}$, and such local doping seems undoable. Alternatively, we have developed processing that enables extension of a thinner version of the FinFET gate down to the underlying BOX, thereby gating the fin-base, and in fact effecting use of the entire height of the fin ($h_{Si} \rightarrow h_{Si}+t_{Si}$). Measurements and Davinci simulations of this modified ITFET indicate that it is virtually equivalent to a parallel combination of the constituent FinFET and FDFET, yielding composite current characteristics that can be optimized with proper choice of w_{Si} and t_{Si}.

Presuming such an ITFET, we now use UFDG to model it as a composite DG FinFET and SG FDFET, and examine its design and potential performance. We plot, in Fig. 3, UFDG-predicted per-pitch current-voltage characteristics of our $L_{gate}(=L_{eff})$=28nm nITFET, but with t_{Si} varying over a viable range of 5nm ($<L_{eff}/5$) up to 9nm ($\sim L_{eff}/3$), along with those of the constituent FinFET and FDFET. Note that substantive increase in I_{on} is yielded by the ITFET, relative to the FinFET, with very little dependence on t_{Si}. With t_{Si}=5nm, good subthreshold characteristics are obtained; I_{off} is virtually unchanged from that of the FinFET due to the higher V_t of the constituent FDFET, which results from the combined effects of

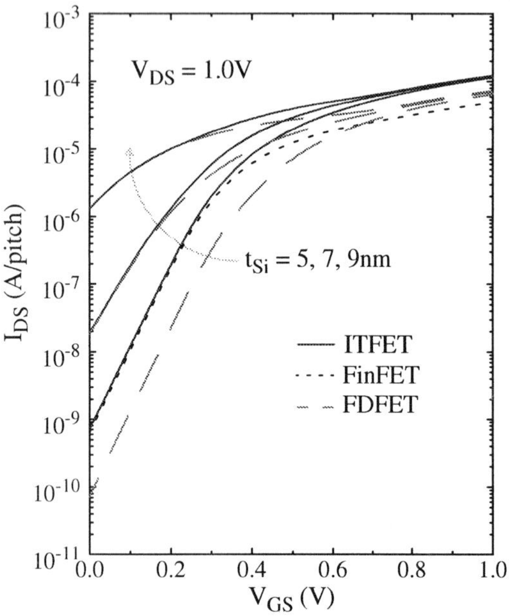

Fig. 3. UFDG-predicted per-pitch I_{DS}-V_{GS} characteristics of the L_{gate}=28nm nITFET, for a viable range of SOI thickness, compared with those of the constituent FinFET and FDFET; R_f=4.

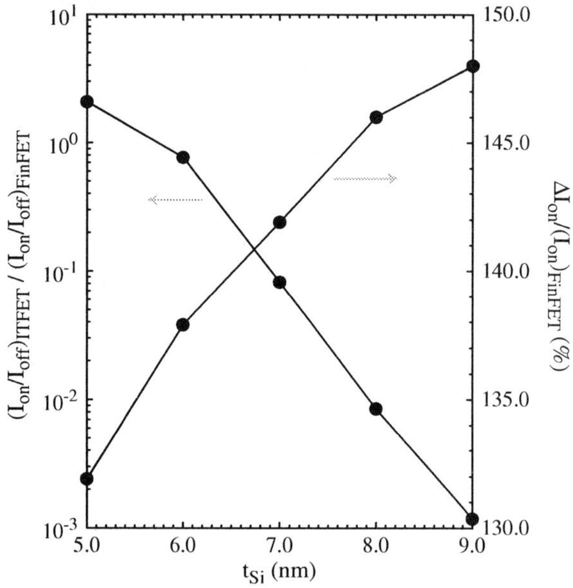

Fig. 4. UFDG-predicted ratio of I_{on}/I_{off}, along with the percentage I_{on} enhancement, of the L_{gate}=28nm nITFET relative to the constituent FinFET versus SOI thickness; R_f=4.

carrier-energy quantization and SCE suppression [6]. However, with $t_{Si}>L_{eff}/5$, the ITFET subthreshold characteristics are degraded and are virtually identical to those of the constituent FDFET, which are undermined by SCEs and decreased V_t due to less quantization.

Further insight is gained from the plots, versus t_{Si}, in Fig. 4 of the predicted ITFET I_{on}/I_{off} and I_{on} enhancement (ΔI_{on}) relative to the FinFET. Note that the relative I_{on}/I_{off} degrades dramatically with increasing t_{Si}, despite the increasing relative ΔI_{on}. Only with the well-controlled SCEs and quantization yielded by the thinnest achievable t_{Si}=5nm does the enhanced I_{on} render a much higher I_{on}/I_{off} ratio (by >100%) for the ITFET compared to the FinFET.

Hence, the scalability of the ITFET beyond 28nm is questionable since t_{Si} cannot be thinned to <~5nm [6]. The UFDG-predicted per-pitch current-voltage characteristic of an $L_{gate}(=L_{eff})$=18nm nITFET, with t_{Si}=5nm, is compared with those of the constituent FinFET and the FDFET in Fig. 5. Here, t_{ox}=0.7nm, and w_{Si}=8.5nm$\cong L_{eff}/2$ gives reasonable SCE control in the FinFET. However, the predicted characteristics in Fig. 5 show that even though I_{on} is dramatically enhanced in the ITFET (by about 160%) with the thinnest t_{Si}, the I_{on}/I_{off} ratio is substantively degraded due to the severe SCEs in the constituent FDFET. So, future-generation applications of the undoped ITFET will probably be limited to those in which drive current is a premium and off-state current is less significant, e.g., I/O devices.

ITFET SRAM Cell

A properly designed ITFET could have widespread applications because of the added current drive afforded, with good control of the SCEs via a viable thin t_{Si}. For example, it could benefit optimal design of the FinFET-CMOS 6T-SRAM cell [7] by providing the higher drive (which could be tuned by varying the width of the constituent FDFET) needed in the pull-down nMOSFETs for better read stability and speed, without

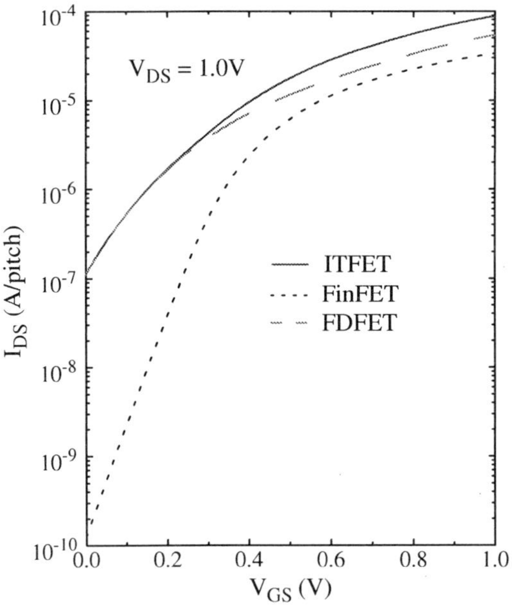

Fig. 5. UFDG-predicted per-pitch I_{DS}-V_{GS} characteristic of an $L_{gate}(=L_{eff})$=18nm nITFET, compared with those of the constituent FinFET and the FDFET; R_f=4, w_{Si}=8.5nm, t_{Si}=5nm, t_{ox}=0.7nm.

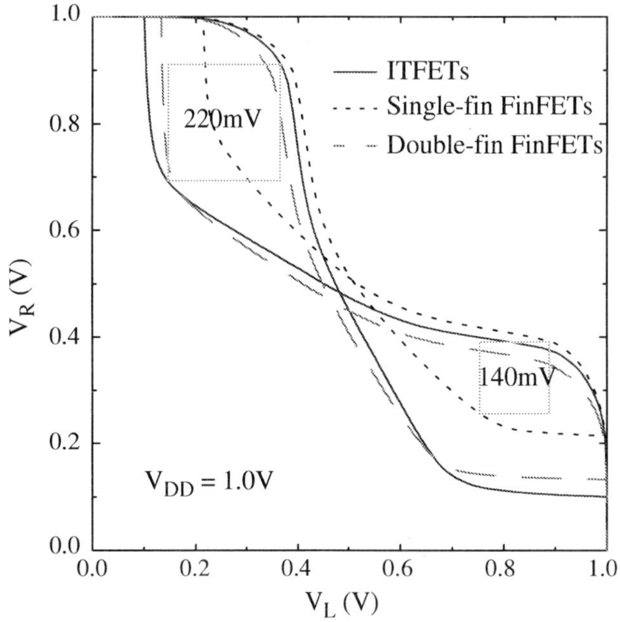

Fig. 6. UFDG/Spice3-predicted butterfly curves for the L_{gate}=28nm SRAM cell with ITFETs for pull-down, compared with those of SRAM cells with single-fin FinFETs and double-fin FinFETs for pull-down, respectively.

increasing the leakage current of the cell, and with no area penalty at all. To illustrate this application, we simulate the SRAM cell with UFDG/Spice3, using the properly designed L_{gate}=28nm ITFETs for the pull-down nMOSFETs, and just the FinFETs, with their p-type counterparts, for the access and load transistors, respectively. Minimum-size, i.e., single-fin, devices are used for all transistors for maximum SRAM density, and R_f=4 (i.e., h_{Si}=56nm) is assumed.

The UFDG/Spice3-predicted butterfly plot of the 28nm ITFET/FinFET SRAM cell with V_{DD}=1.0V is shown in Fig. 6. Note that excellent static noise margin (SNM ≅ 220mV) is obtained, due to the added strength afforded the pull-down transistors by the ITFETs. For comparison, we also show in the same figure the butterfly plots predicted when single-fin FinFETs are used for the pull-down nMOSFETs, instead of the ITFETs, and when larger-area double-fin FinFETs are used. In the former case, the SNM (≅140mV) is significantly smaller. In the latter case, the SNM (≅200mV) is improved by the added strength of the pull-down transistors, but is still smaller than that of the cell with the ITFETs even though the cell area is enlarged considerably. With 45nm design rules, an SRAM cell layout-area penalty of about 17% would result from using two fins instead of one [7].

As we have noted, the SOI thickness is a crucial parameter in the ITFET design, and we have thinned it to near the limit [6]. We use UFDG/Spice3 to check the sensitivity of the ITFET-SRAM performance to variations in t_{Si}. Remember that increasing t_{Si} does not strongly affect I_{on} of the ITFET, but does increase I_{off} and lower V_t due to the deteriorating SCEs with less quantization in the constituent FDFET. This deterioration is clearly reflected by predicted degradation of the SNM with increasing t_{Si}. This loss of SRAM performance, due to increased I_{off}, further demonstrates the importance of the control of the fin-base leakage in the ITFET. For the same reason, process-induced increases in L_{gate} and w_{Si} of the ITFET also decrease the SRAM SNM, as additional UFDG/Spice3 simulations show. And, because it controls I_{off}, t_{Si} also will govern the static power consumption (P_{static}) of the cell. UFDG/Spice3 SRAM simulations at different supply voltages (V_{DD}) show how P_{static} increases with t_{Si}, as well as how it and the SNM depend on V_{DD}. Indeed, t_{Si} must be properly scaled to limit I_{off} in order to exploit the benefits of the undoped ITFET.

References

1. X. Huang, et al., *IEDM Tech. Dig.*, pp. 67-70, Dec. 1999.
2. J-W. Yang and J. G. Fossum, *IEEE Trans. Electron Devices*, vol. 52, pp. 1159-1164, June 2005.
3. L. Mathew and J. G. Fossum, Freescale U.S. Patent Application, Feb. 2005.
4. J. G. Fossum, "UFDG User's Guide (Ver. 3.0)," Univ. of Florida, Gainesville, Aug. 2004.
5. *Davinci-2003.06 User Guide*. Durham, NC: Synopsis, Inc., 2003.
6. V. P. Trivedi and J. G. Fossum, *IEEE Trans. Electron Devices*, vol. 50, pp. 2095-2103, Oct. 2003.
7. E. J. Nowak, et al., *IEDM Tech. Dig.*, pp. 411-414, Dec. 2002.

Fin Thickness Asymmetry Effects in Multiple-Gate SOI FETs (MuGFETs)

T. Schulz[1], W. Xiong[2], C. R. Cleavelin[2], K. Schruefer[1], M. Gostkowski[3], K. Matthews[3], G. Gebara[3], R. J. Zaman[3], P. Patruno[4], A. Chaudhry[5], A. Woo[5], and J.P. Colinge[5]

1. Infineon Technologies, 2706 Montopolis Drive, 78741 Austin, TX
2. SiTD, Texas Instruments Incorporated, 13560 North Central Expressway, 75265 Dallas, TX
3. ATDF Incorporated, 2706 Montopolis Drive, 78741 Austin, TX
4. SOITEC S.A., Parc Technologique des Fontaines – 38190 Bernin, France
5. Dept. of Electrical and Computer Eng., University of California. Davis, CA
Email: Thomas.Schulz@infineon.com, Tel: 1-512-356-7094, Fax: 1-512-356-7118

Abstract

Fin thickness non-uniformity is a potential shortcoming of vertical multiple-gate devices such as FinFETs and Tri-Gate FETs. In this paper a test structure with intentionally misaligned gates is used to investigate the sensitivity of electrical characteristics on fin thickness variations.

Introduction

To achieve desired current drive Multiple-Gate SOI MOSFETs (MuGFETs) are usually laid out as a succession of parallel fins or fingers (Fig. 1a). Due to the inherent imperfections of lithography the width of the fins may vary from source to drain (Fig.1b-3). Device characteristics degradation has been shown to occur if the width of the fins varies from top to bottom [1]. Here we show that similar degradation occurs if the fin width varies from source to drain (Fig. 4).

Device fabrication

Both FinFETs and Tri-Gate SOI MOSFETs were fabricated and show similar results. Standard Unibond® SOI wafers were used as starting material. The silicon film and BOX thicknesses are 88 nm and 145 nm, respectively. Lithography and RIE were used to etch the SOI layer to define 50 nm-wide silicon fins. A 2nm gate oxide was grown by wet oxidation. Amorphous silicon was deposited and partially pre-doped by phosphorus ion implantation. Lithography and RIE trim-etch were used to define gates length down to 80 nm. Source and drain extensions were formed by ion implantation. Impurity activation was achieved by a 10 second, 1000°C RTA step. Conventional Al/Si metallization and Ni silicidation was used to complete the process.

Device characteristics

Figure 5 compares two devices with gates intentionally misaligned by 15nm (=triangle symbol) and 60nm (=square symbol). The nMOS device with the larger misalignment has a 200mV higher threshold voltage mainly because of the average increase in fin thickness towards the drain. To investigate the impact of the fin thickness difference under the gate ($\Delta t_{Si} = t_{Si,D} - t_{Si,S}$) the transfer characteristics of the same devices were measured again but with the source and drain electrode exchanged. When the drain voltage is 50 mV little difference is seen when S&D are swapped. At V_{DS}=1.2V, however, larger DIBL is observed when the drain is wider than the source (Fig.6). The same observation is made on curves simulated using DESSIS [2] (Fig. 7): at high drain bias the penetration of the drain field lines in the channel region is larger if the fin is wider at the drain ($W_{D,D} > W_{D,S}$). In practical devices the threshold voltage shift and the DIBL variation due to swapping S&D is on the order of 10 mV, which is comparable to variations due to dopant fluctuations. To further study the effects of gate misalignment a set of FinFETs with intentionally misaligned gates was fabricated. The intentional gate misalignment is 0,±15nm, ±30nm and ±60nm (Figure 4).

Figure 8 and 9 shows CMOS devices with a drawn gate length of 85nm and 100nm, respectively. The PMOS devices are less sensitive to the gate misalignment because of a longer effective channel length due to a skipped LDD implant. The larger the misalignment, the wider the fin width and the larger the threshold voltage (Figure 10). This is explained by the larger depletion charge controlled by the gate in the wider devices ($\Delta V_{TH} \propto 0.5qN_D \Delta t_{Si} / C_{ox}$ where t_{Si} is the average fin width). The V_{TH} shift is now observed for V_{DS} = 50 mV and 1.2V because is not caused only by DIBL effect, but rather by a variation of depletion charge under the gate due to a different fin thickness. DIBL and subthreshold slope degradation caused by the increase of fin width is observed in the misaligned devices (Figs. 11-12).

Conclusion

Misalignment of the gate electrode in MuGFETs causes a variation of fin width that gives rise to variations of threshold voltage and DIBL. Short-channel effects are more pronounced if the fin width is larger at the drain than at the source. Large misalignments of the gate yield large threshold voltage variations.

References

[1] X. Wu, P.C.H. Chan, M. Chan, Proceedings of the IEEE International SOI Conference, p. 151, 2003
[2] http://www.synopsys.com/

0-7803-9212-4/05/$20.00 ©2005 IEEE

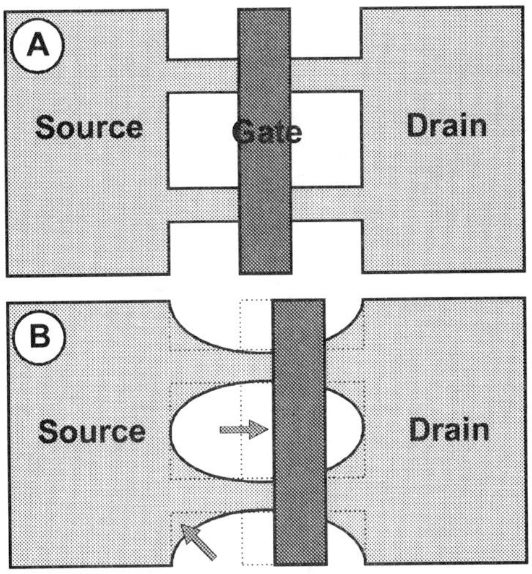

Fig. 1: A: Original layout. B: Actual pattern after device processing.

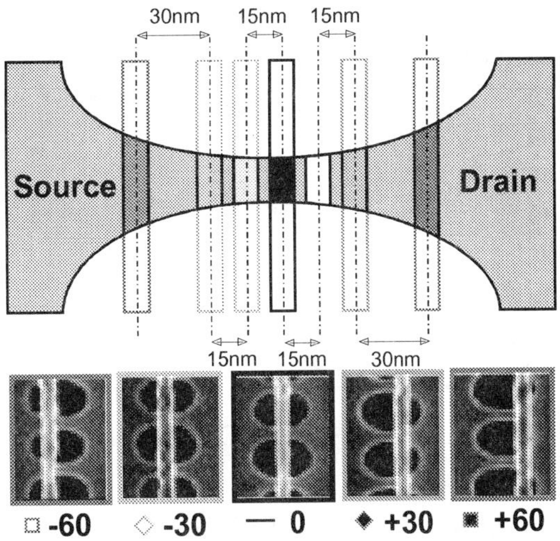

Fig. 4: Devices with intentionally misaligned gates of 0, (15), 30, and 60nm.

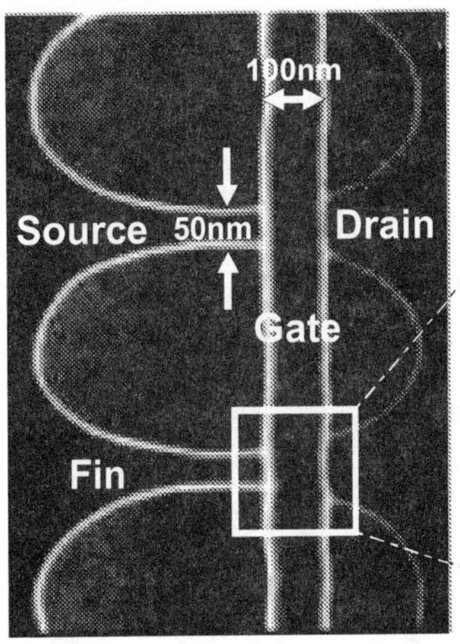

Fig. 2: SEM of a MuGFET with misaligned gate.

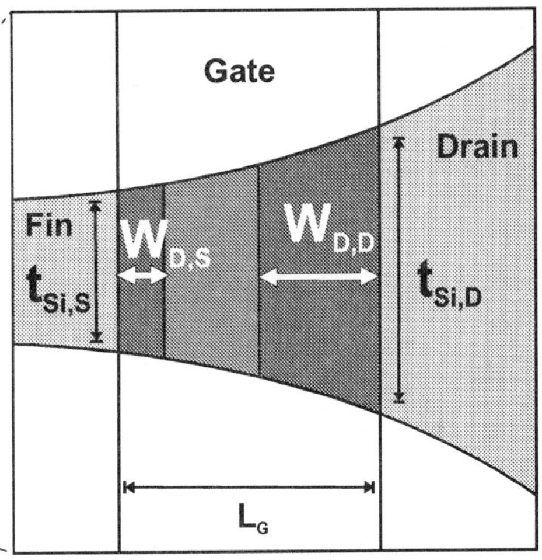

Fig. 3: Schematic of fin width variation from source to drain and under the gate ($t_{Si,D} > t_{Si,S}$).

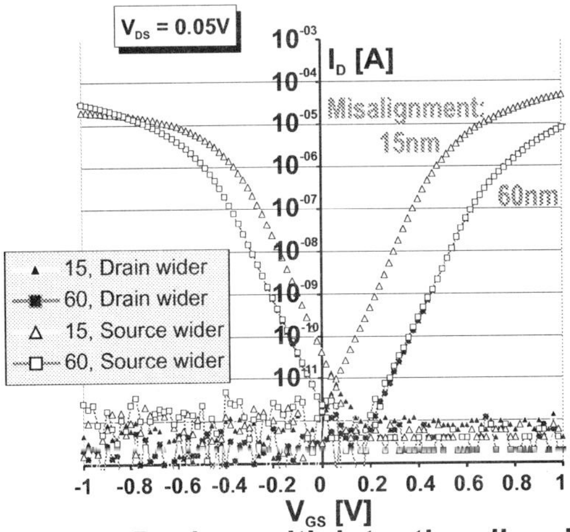

Fig. 5: Devices with intentionally misaligned gates and swapped source and drain electrodes (with white symbols).

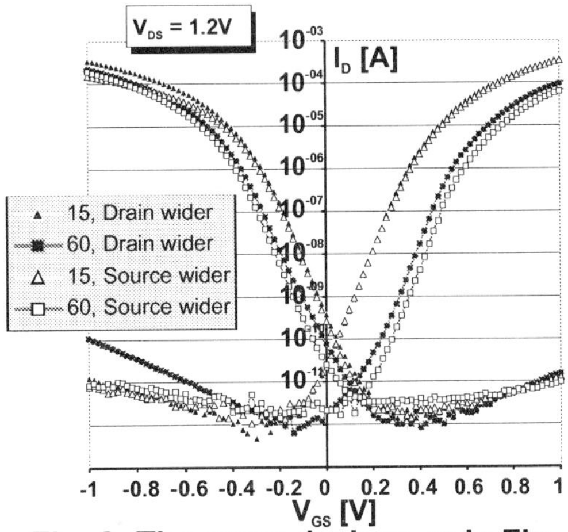

Fig. 6: The same devices as in Fig. 5 but now at V_{DS}=1.2V. The short channel effects are stronger for wider Drain.

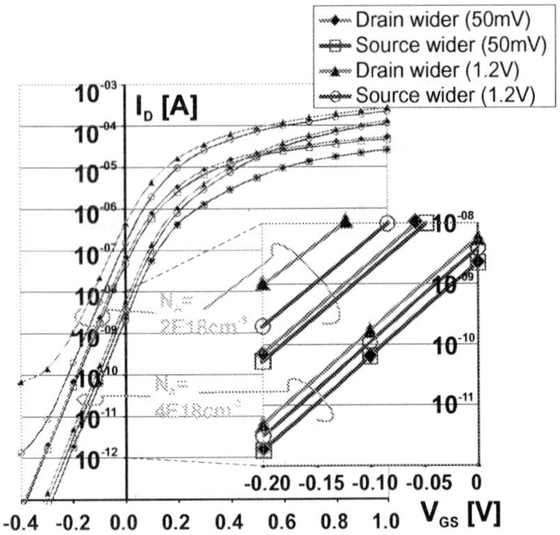

Fig. 7: L_G=50nm Pi-Gate simulations, $t_{Si,S}$=150nm; $t_{Si,D}$=250nm; taper angle 45°

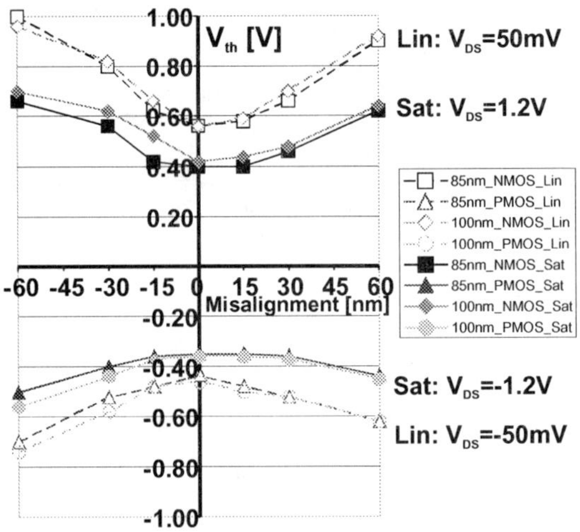

Fig. 10: V_{th} versus Gate misalignment.

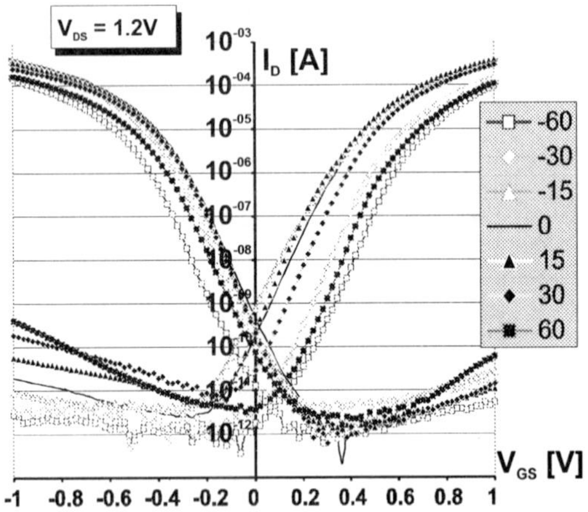

Fig. 8: Devices with $L_{G,drawn}$=85nm and Gate misalignment as shown in Fig. 4.

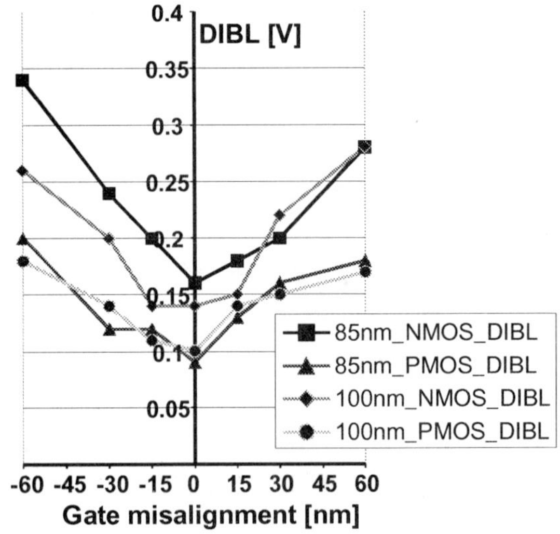

Fig. 11: DIBL versus Gate misalignment.

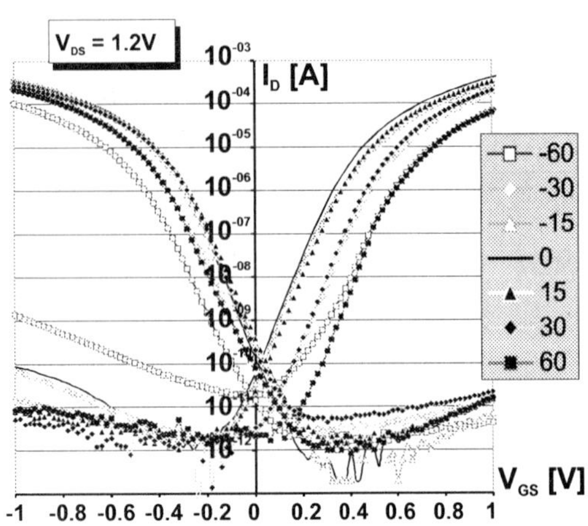

Fig. 9: Devices with $L_{G,drawn}$=100nm and Gate misalignment as shown in Fig. 4.

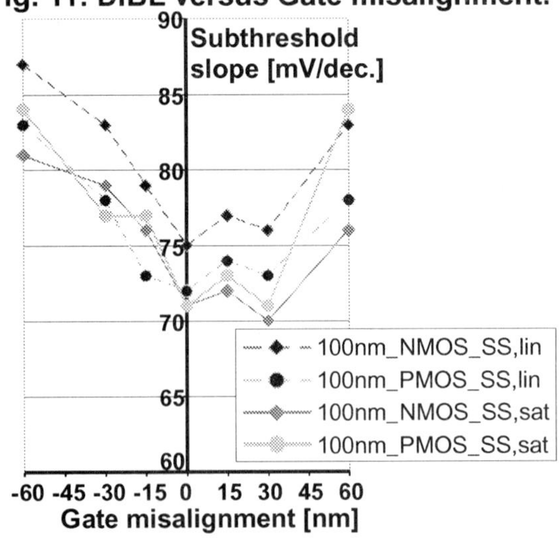

Fig. 12: SS versus Gate misalignment.

An Integratable Dual Metal Gate/High-k CMOS Solution for FD-SOI and MuGFET Technologies

Z.B. Zhang[1], S.C. Song, K. Choi, J.H. Sim, P. Majhi[2], and B.H. Lee[3]
SEMATECH, [1]Texas Instruments Assignee, [2]Intel Assignee, and [3]IBM Assignee
2706 Montopolis Drive, Austin, TX 78741
Email: zhibo.zhang@sematech.org, Tel: (512) 356-3864

Introduction

To continue improving MOSFET performance and reducing gate leakage current, there is great interest in introducing high-k dielectrics and metal gate electrodes into CMOS technology. For fully-depleted MOSFETs (FD-FET) such as FD-SOI and multi-gate FET (MuGFET), dual metal gates with work functions close to the quarter-gap of Si (~4.4eV for NMOS and ~4.8eV for PMOS where electron affinity of Si equals to 4.05eV) together with un-doped or lightly-doped channels are desired [1,2]. Integrating dual metal gate CMOS with two completely different metals is challenging because of the complicated processes involved [3,4]. This paper describes a simple process that can tune the work function of ALD TaCN gate electrode on HfO_2 from 4.47eV to 4.77eV by adding a CVD TiN overlayer. It also discusses the device characteristics of TaCN and TiN/TaCN (TaCN with a TiN overlayer) metal gate/high-k MOSFETs and presents a manufacturable process for integrating dual metal gate/high-k CMOS in a FD-FET technology.

Device Fabrication

To fabricate TaCN metal gate/high-k MOSFETs, after the ALD deposition of 20Å HfO_2 gate dielectrics, 100Å TaCN gate electrode is deposited by ALD technique. For TiN/TaCN devices, the HfO_2 is 30Å thick and a 100Å MOCVD TiN overlayer is deposited on 100Å ALD TaCN. Next, a 1000Å amorphos Si capping layer is deposited, followed by gate patterning and etching. Other process steps follow a conventional CMOS process flow.

Reults and Discussion

Fig. 1 show the SIMS profiles of the TaCN/HfO_2 and TiN/TaCN/HfO_2 gate stacks after the full CMOS process. For the TiN/TaN/HfO_2 stack, there are evidences of Ta and Ti inter-diffusion, suggesting that the tuning of TaCN work function by the TiN overlayer is likely due to the intermixing of TaCN and TiN after high-temperature processes. Fig. 2 shows the long-channel V_t distributions of TaCN NMOSFET and TiN/TaCN PMOSFET. Using the V_t's of the baseline poly/SiO_2 transistors (0.35V for NMOS and 0.45V for PMOS) as a reference and accounting for the EOT and Q_i differences, the effective work functions of TaCN and TiN/TaCN on HfO_2 are calculated to be 4.47eV and 4.77eV, respectively, which match well with the work functions determined by the V_{fb} vs. EOT technique [5]. The symmetric work functions around the mid-gap suggest that TaCN and TiN/TaCN are promising candidates for dual metal gate CMOS integration in a FD-FET technology (Fig. 3). The ALD and CVD depositions of metal gate electrodes are especially suitable for MuGFET application.

The CV curves measured on the 20μmX20μm TaCN NMOSFET and TiN/TaCN PMOSFET are shown in Fig. 4. The inversion capacitance matches well with the accumulation capacitance for both devices, confirming the metallic nature of the ALD TaCN. The NMOSFET T_{inv} is only 15.5Å. Fig. 5 illustrates well-behaved I_d-V_d characteristics of the 10μmX0.2μm devices. Fig. 6 presents their I_d-V_g curves. The sub-threshold slope (SS) is 70mV/dec and 73mV/dec for TaCN NMOSFET and TiN/TaCN PMOSFET, respectively. The gate leakage current densities for both devices are ~3 orders of magnitude lower than those of poly/SiO_2 devices with the same EOT's (data not shown). The N_{it} is 2.2×10^{10} cm^{-2} for the TaCN/HfO_2 gate stack and 7.7×10^{10} cm^{-2} for the TiN/TaCN/HfO_2 stack; both are low for HfO_2 gate dielectrics. The electron and hole mobility is presented in Fig. 7. At E_{eff} = 1MV/cm, the TaCN/HfO_2 NMOSFET achieves 83% and the TiN/TaCN/HfO_2 PMOSFET 89% of universal mobility.

Conclusions

The effective work function of ALD TaCN on HfO_2 gate dielectrics can be tuned from 4.47eV to 4.77eV by adding a TiN overlayer. Well behaved TaCN/HfO_2 NMOSFET and TiN/TaCN/HfO_2 PMOSFET have been demonstrated, and high electron mobility and hole mobility have been achieved, making them promising candidates for providing the dual metal gate/high-k CMOS solution in FD-SOI and MuGFET technologies.

Reference

[1] L. Chang et al., *IEDM Tech. Digest 2000*, pp. 719.
[2] J. Kedzierski et al., *IEEE TED*, **51**, 2115 (2004).
[3] S.B. Samavedam et al., *IEDM Tech. Digest 2002*, pp. 433.
[4] Z.B. Zhang et al., *VLSI Tech. Digest 2005*, in press.
[5] K. Choi et al., submitted to *APL*.

0-7803-9212-4/05/$20.00 ©2005 IEEE

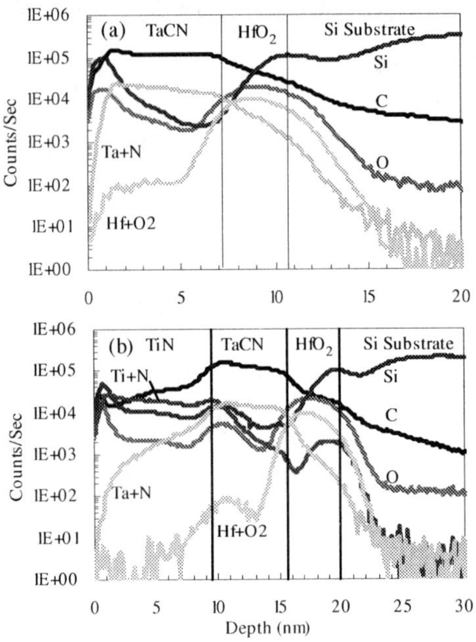

Fig. 1. SIMS profiles of (a) the TaCN/HfO_2 gate stack taken from an NMOSFET and (b) TiN/TaCN/HfO_2 taken from a PMOSFET, after the full CMOS process.

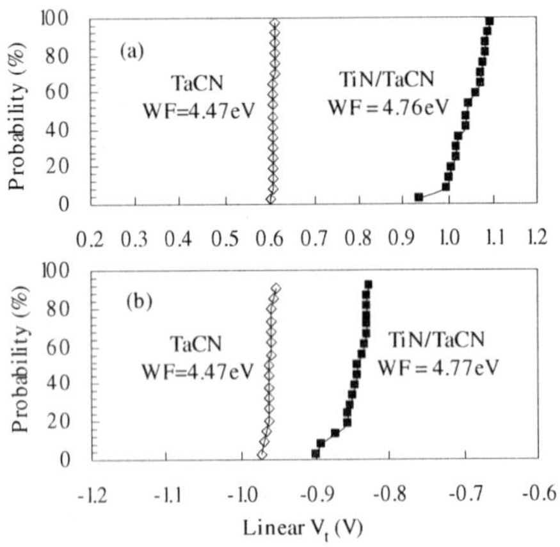

Fig. 2. V_t distribution of 10μmX1μm (a) NMOSFETs and (b) PMOSFETs. The work functions are calculated using the V_t's of poly/SiO_2 NMOSFET and PMOSFET as references.

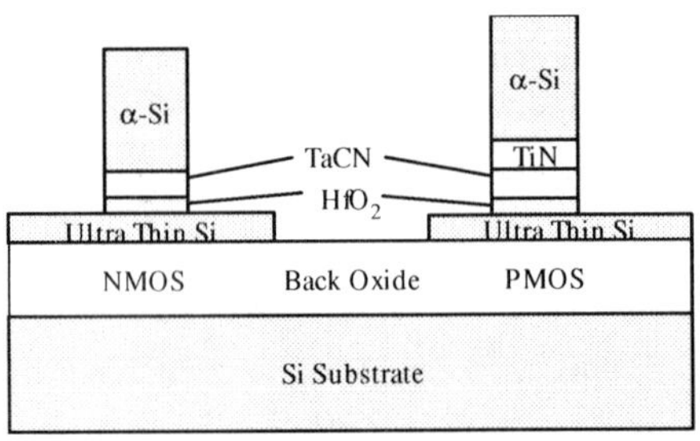

Fig. 3. Schematic of the dual metal gate integration for a fully-depleted CMOS technology, using FD-SOI as an example.

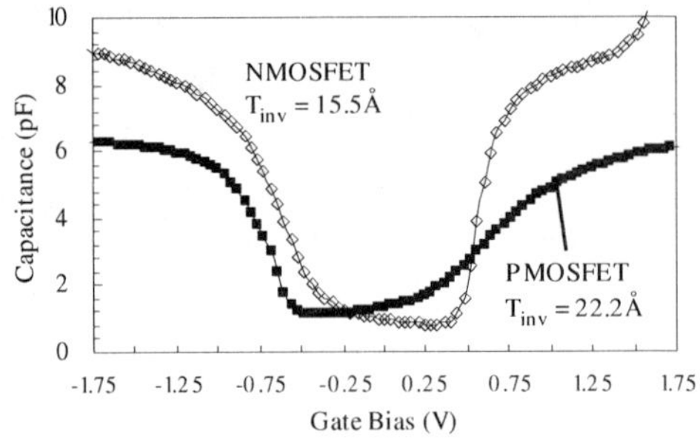

Fig. 4. CV characteristics of TaCN/20Å HfO_2 NMOSFET and TiN/TaCN/30Å HfO_2 PMOSFET measured on 20μmX20μm devices.

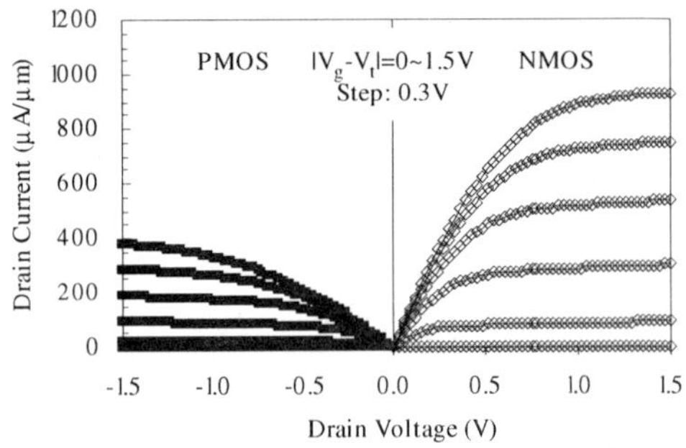

Fig. 5. I_d-V_d characteristics of a 10μmX0.2μm TaCN/HfO_2 NMOSFET and TiN/TaCN/HfO_2 PMOSFET.

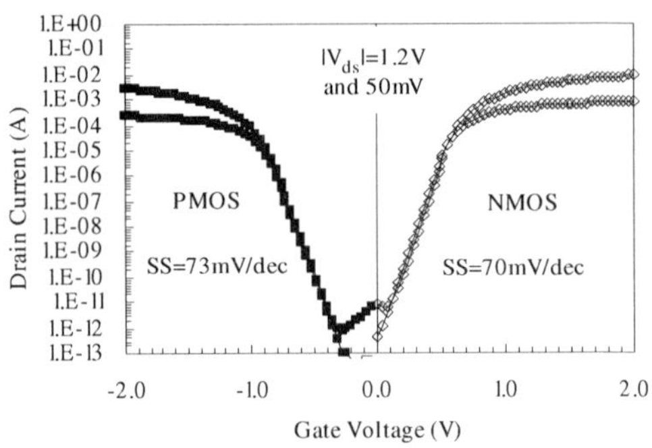

Fig. 6. I_d-V_g characteristics of a 10μmX0.2μm TaCN NMOSFET and TiN/TaCN PMOSFET.

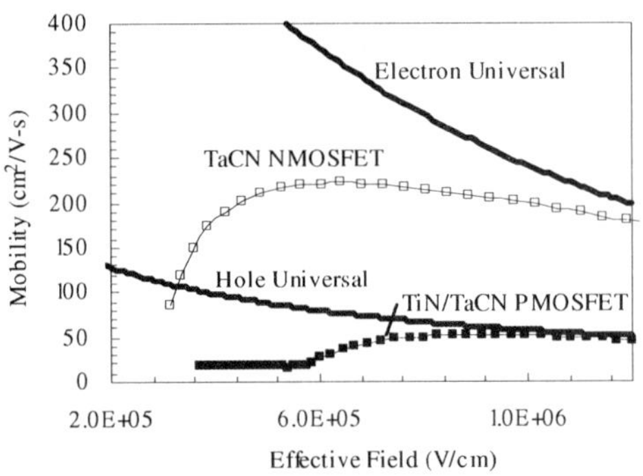

Fig. 7. DC electron and hole mobility.

Analysis of Drain-to-Body Band-to-Band Tunneling in Double Gate MOSFET

Hari Ananthan, Aditya Bansal and Kaushik Roy
Dept. of ECE, Purdue University, West Lafayette, IN, USA
Email: hari@purdue.edu, Tel: (765) 494-0750

Abstract – An analytical model is proposed for drain-to-body band-to-band tunneling leakage in nanoscale symmetric and asymmetric double-gate MOS devices. The model is used to analyze the impact of technology and circuit parameters, and suggest ways of minimizing this leakage.

Introduction

Band-to-band tunneling (BTBT) of electrons across a reverse-biased p-n junction from the p-side valence band to the n-side conduction band, is becoming an increasingly important component of leakage in nanoscale bulk and SOI CMOS technologies. It is manifested both as gate-induced drain leakage (GIDL) in the drain-gate overlap region and reverse-biased junction leakage in the halo-implant region.

With double-gate (DG) CMOS likely to become the mainstream technology around the 45nm or the 32nm node, it is important to study the BTBT phenomenon in these devices. These devices are likely to be designed with intrinsic or lightly doped channels ($N_a < 5e16$ cm^{-3}). The potential in the fully-depleted (FD) body, and hence the potential across the drain-body junction (V_{db}), is thus set predominantly by gate workfunction (WF). For symmetric devices,

$$V_{db}= (V_{DS}\pm(kT/q)\log(N_{sd}/n_i)) - (V_{GS}+WF-4.72).$$

Unlike bulk NMOS, which uses n^+-poly gate (p^+ for PMOS), DG-NMOS must be designed with either close-to-midgap symmetric or n^+/p^+ asymmetric workfunctions for the two gates to get the appropriate threshold voltage. At positive drain bias and/or negative gate bias, V_{db} can exceed the bandgap voltage (E_g), especially at the p^+ surface in asymmetric devices, causing BTBT between the body and the drain. Secondly, the parasitic lateral bipolar gain of the floating-body device amplifies the BTBT current (I_{btbt}) [1]. Finally, while supply voltage scales below 1.0V in Low-Operating-Power systems (which reduces V_{db}) at around the 45nm node, it is projected to remain constant at around 1.0V in High Performance and Low-STandby-Power systems. LSTP systems such as memory arrays demand strict I_{off} control of the order of 10-100 pA [2]. I_{btbt} can thus become comparable to source-drain sub-threshold (I_{sub}) and gate tunneling (I_g) currents in nanoscale DG circuits. Hence, it is important to model this leakage component and understand the impact of technology and circuit choices on its magnitude.

In digital CMOS circuits, the bias regions of interest are the (V_{ds}=V_{dd}, Vgs=0) state and the (V_{ds}=V_{dd}, V_{gs}=slightly negative) state. The latter situation arises when the negative wordline-biasing technique is used to save I_{sub} in memory circuits. Our analysis is restricted to these bias regions.

Analytical Model for BTBT estimation

We have developed analytical models for the band structure in the silicon body for the generic DG-MOS structure shown in Fig. 1. The parameters are L_g, L_{un} (underlap), t_{oxf}, t_{oxb}, t_{si}, WF_f, WF_b, N_{sd}, Lateral straggle (σ), V_{DS}, V_{GSF}, and V_{GSB}. TAURUS device simulator with drift-diffusion transport and self-consistent local field-dependent BTBT is used to verify our model. We solve the 2D Poisson's equation [3] in regions (I) the drain-side underlap region, (II) under the gate, and (III) the source-side underlap region, to get the lateral and vertical band profiles (see inset at bottom of next page). The potential is assumed to be nearly constant in the transverse direction in the underlap regions. I_{btbt} is a function of the local electron-hole pair generation rate $(A_{btbt}*E/E_g^{0.5})*\exp(-B_{btbt}*E_g^{1.5}/E)$, where E is the magnitude of the local electric field. This is integrated over the tunneling volume, that is, the region where band-bending is greater than E_g. Since transverse band-bending is much smaller than lateral band-bending in underlapped devices, tunneling volume is calculated based on the lateral band profile independently at different depths in the body. While the exact value of the current from our model does not match the simulation results well (due to the extreme sensitivity of Ibtbt to the band structure near the gate edge), it predicts the dependence of I_{btbt} on various technological and circuit parameters quite accurately. Figs. 2 through 6 show a good fit between the normalized model (dotted lines – normalized to the nominal point of the simulation results) and the simulation results (discrete points) for various technological and circuit parameters. The model deviates from simulation results for devices with larger σ, that is, non-abrupt doping profiles (Fig. 4). This is expected, because our model assumes that region (I) is free of n-type dopants. A separate model must be developed for this case by solving Poisson's equation in the n-type region.

Discussion

The dependence of I_{btbt} on WF, t_{ox} and V_{BGS} (Fig. 2, 3, and 6) can be easily explained based on the magnitude of the drain-to-body potential and the electric field (lateral and vertical) near the gate edge. Fig. 4 suggests that while underlapping decreases I_{btbt}, increasing σ increases it because of higher drain-to-gate transverse field. However, increasing σ implies smaller lateral field (gentler potential gradient); thus Ibtbt starts decreasing beyond a point. L_{spacer} and σ must thus be carefully co-designed to minimize I_{btbt} while achieving desired short-channel effect, on-current, etc. Fig. 5 suggests that I_{btbt} is independent of t_{si} since it does not affect lateral or transverse field near the gate edge signifcantly. Asymmetric devices follow the same trends as symmetric devices, but they suffer from larger I_{btbt} because V_{db} is very large at the back (p^+) interface.

One of the interesting findings of this study is that the most common techniques, both technological and circuit, used to reduce I_{sub} tend to worsen I_{btbt}. For example, higher workfunction and negative back-gate bias increase I_{btbt} exponentially. Using a thicker t_{ox} reduces both I_{btbt} and I_g while worsening I_{sub} through DIBL. An optimal strategy must be devised to minimize total leakage. This is especially important for circuits such as memory that operate at relatively lower temperatures, because I_g and I_{btbt} are likely to be comparable to I_{sub} only at low temperatures.

References

[1] J.G. Fossum et al., *IEEE TED*, 1999, pp. 2195-2200
[2] ITRS 2004 Update : http://public.itrs.net
[3] K. Suzuki et al., *IEEE TED*, 1996, pp. 732-738

0-7803-9212-4/05/$20.00 ©2005 IEEE

Fig. 1 Double-gate Structure

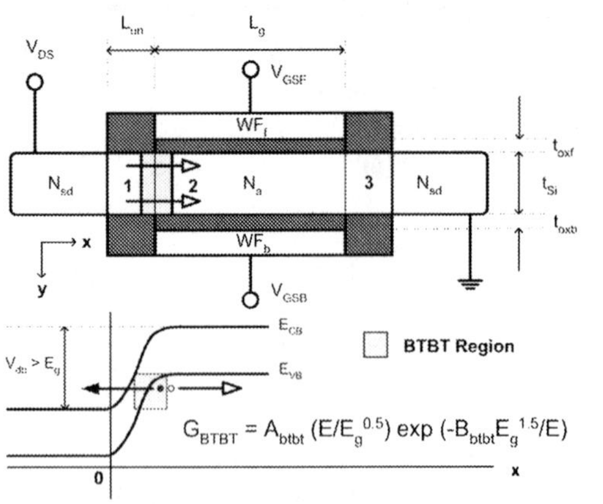

Fig. 2 : Dependence on Gate Workfunction (Symmetric)

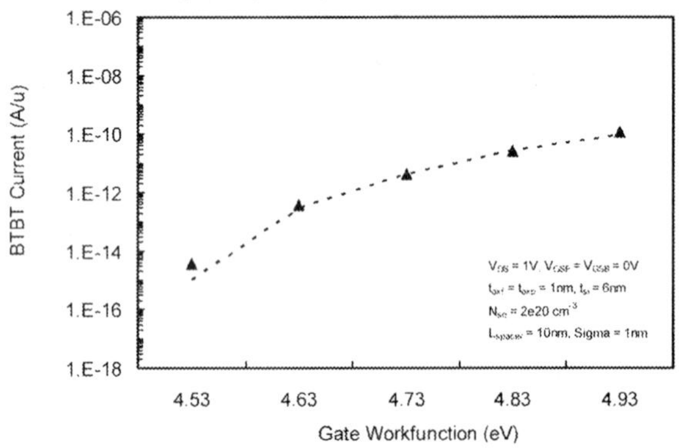

Fig. 3 : Dependence on Oxide Thickness

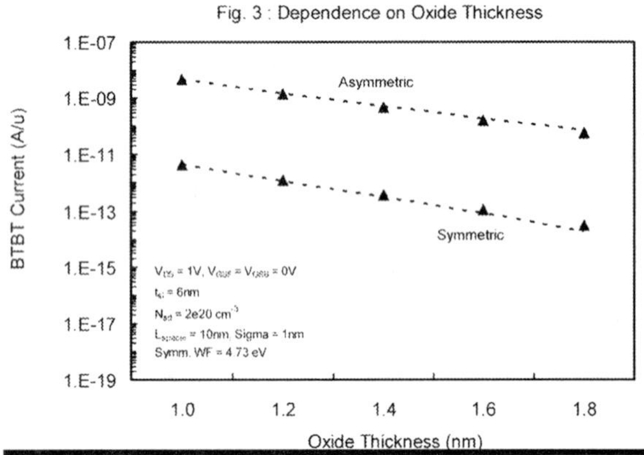

Fig. 4 : Dependence on Lateral Diffusion Length (Sigma)

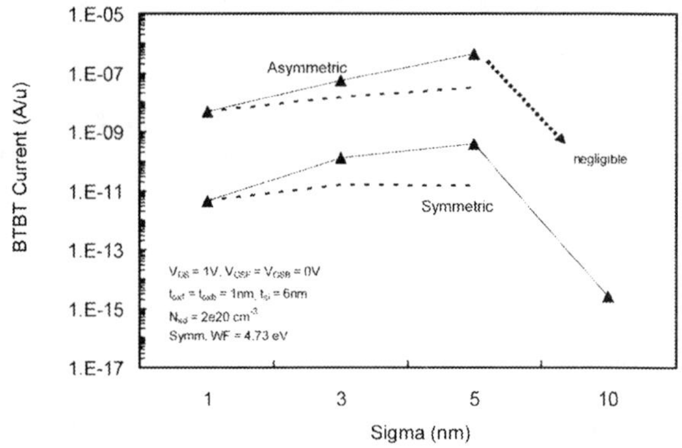

Fig. 5 : Dependence on Body Thickness

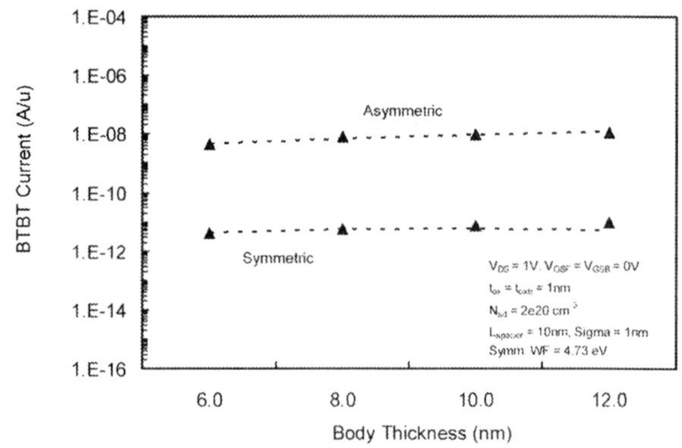

Fig 6 : Dependence on Back Gate Voltage (Asymmetric 4-T)

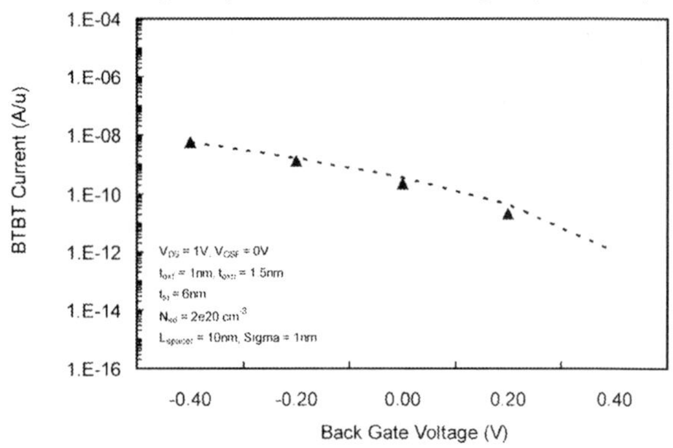

MODEL EQUATIONS

$$\psi_1(x,0) = \frac{qN_a t_{ox}^2}{2\epsilon_{si}}\left(\frac{L_{un}-x}{t_{ox}}\right)^2 + C_1\left(\frac{L_{un}-x}{t_{ox}}\right) + C_2,$$

$$\psi_2(x,y) = \psi_{s2}(x) + a_{12}(x)y + a_{22}(x)y^2,$$

$$\psi_3(x,0) = \frac{qN_a t_{ox}^2}{2\epsilon_{si}}\left(\frac{x-L_g-L_{un}}{t_{ox}}\right)^2 + C_5\left(\frac{x-L_g-L_{un}}{t_{ox}}\right) + C_6,$$

$$\psi_{s2}(x) = C_3 e^{\alpha(x-L_{un})} + C_4 e^{-\alpha(x-L_{un})} - \beta/\alpha^2,$$

$$\beta = \frac{qN_a}{\epsilon_{si}} + 2\frac{\left((V_{GSF}-\phi_{MSF})\left(\frac{1}{t_{oxf}} + \frac{\epsilon_{ox}t_{si}}{\epsilon_{si}t_{oxf}t_{oxb}}\right) + (V_{GSB}-\phi_{MSB})\frac{1}{t_{oxb}}\right)}{\frac{t_{si}^2}{t_{oxb}} + \frac{2t_{si}\epsilon_{si}}{\epsilon_{ox}}},$$

$$\alpha^2 = 2\frac{\left(\frac{1}{t_{oxf}} + \frac{1}{t_{oxb}} + \frac{\epsilon_{ox}t_{si}}{\epsilon_{si}t_{oxf}t_{oxb}}\right)}{\frac{t_{si}^2}{t_{oxb}} + \frac{2t_{si}\epsilon_{si}}{\epsilon_{ox}}},$$

$$a_{12}(x) = -\frac{\epsilon_{ox}}{\epsilon_{si}}\left(\frac{V_{GSF}-\phi_{MSF}-\psi_{s2}(x)}{t_{oxf}}\right),$$

$$a_{22}(x) = \frac{qN_a}{\epsilon_{si}} - \beta - \alpha^2\psi_{s2}(x).$$

$$\psi_1(0,0) = V_{bi} + V_{ds},$$

$$\psi_1(L_{un},0) = \psi_{s2}(L_{un}),$$

$$\frac{d\psi_1(x,0)}{dx}\Big|_{x=L_{un}} = \frac{d\psi_{s2}(x)}{dx}\Big|_{x=L_{un}},$$

$$\psi_{s2}(L_g+L_{un}) = \psi_3(L_g+L_{un},0),$$

$$\frac{d\psi_{s2}(x)}{dx}\Big|_{x=L_g+L_{un}} = \frac{d\psi_3(x,0)}{dx}\Big|_{x=L_g+L_{un}},$$

$$\psi_3(L_g+2L_{un},0) = V_{bi}.$$

$$G_{btbt} = A_{btbt}\frac{E}{E_g^{0.5}}\exp\left(\frac{-B_{btbt}E_g^{1.5}}{E}\right),$$

$$E(x,y) = \sqrt{\left(\frac{d\psi}{dx}\right)^2 + \left(\frac{d\psi}{dy}\right)^2},$$

$$A_{btbt} = 3.5e21\ \mathrm{eV^{0.5}/cm/s/V^2},$$

$$B_{btbt} = 22.5e6\ \mathrm{V/cm/eV^{1.5}}.$$

Surface vs. Bulk Noise in SOI Four-Gate Transistors

K. Akarvardar[1], B. Dufrene[2], S. Cristoloveanu[1], J. A. Chroboczek[1], P. Gentil[1], B. J. Blalock[3], M. Mojarradi[4]
[1] IMEP, ENSERG, BP257, 38016 Grenoble Cedex 1, France
[2] IBM Microelectronics, Burlington, VT 05452, USA
[3] ECE Department, University of Tennessee, Knoxville, TN 37966, USA
[4] Avionic Equipment Section, Jet Propulsion Laboratory Pasadena, CA 91109, USA
akarvard@enserg.fr, +33 476 85 60 45

Abstract

Low-frequency noise characteristics of four-gate transistors (G^4-FETs) are presented distinguishing the surface conduction (MOSFET mode) and volume conduction (JFET mode). As the conducting channel moves from the surface to the bulk we observe that: (i) the noise level dramatically decreases and (ii) the nature of the noise changes. The validity of the existing noise models for different conduction modes is discussed.

Introduction

The G^4-FET [1,2] is defined as an accumulation-mode partially-depleted SOI MOSFET provided with two junction-gates (JG1 and JG2, Fig. 1) exercising an extra JFET control over the channel. In the G^4-FET, along with the polysilicon top gate (G1) and the substrate emulating a back-gate (G2), there are four gates which can be independently biased. The operation of the G^4-FET is analyzed by distinguishing two different operation modes (in both cases the junction gates are tied together for simplicity, $V_{JG1} = V_{JG2} = V_{JG}$):

(i) MOSFET mode: the G^4-FET is driven by G1 while JG and G2 are biased to a constant voltage and fully deplete the body (Fig. 2a). The drain current flows at the surface. The operation and the resulting static characteristics are the same as that of fully-depleted accumulation-mode MOSFET (Fig. 3a).

(ii) JFET mode: the G^4-FET is driven by JG while G1 and G2 are biased to a constant voltage inducing depletion or inversion at the interfaces (Fig. 2b). Drain current flows in the volume far from the surfaces. The device functioning and the static characteristics are the same as that of a JFET (Fig. 3b).

The drain current noise power spectral densities (PSD) in these two modes are shown in Fig. 4 for the same bias values as in Fig. 3. The following observations are made: (i) The noise is of the 1/f type above threshold and either Lorentzian or 1/f in the subthreshold region, and (ii) the frequency exponent above the threshold (ν of $1/f^{\nu}$) is systematically lower in JFET mode (ν = 0.7) than in MOSFET mode (ν = 1).

The Quasi-Variance of Drain Current Fluctuations

The bias dependence of the noise is most commonly studied by monitoring the variation of the noise PSD *at a single frequency* (10 Hz, for example), as a function of the bias voltage or current. However this approach may lead to different trends depending on the choice of the spot frequency, when the type of noise (Lorentzian, 1/f) does not remain the same for the whole bias range (as in Fig. 4). Moreover the inherent fluctuations of the spectras give rise to a further dispersion of the data points and makes the interpretation of the results difficult. Therefore we have introduced a new parameter, the "quasi-variance of drain current fluctuations", $Var(I)^*$, which takes into account the noise PSD not at a single frequency but in a *frequency range*. $Var(I)^*$ depends on multiple PSD values and thus enables us to obtain smoother variations and reproducible trends. It is defined as:

$$Var(I)^* = \int_{f_{min}}^{f_{max}} S_I(f)df \cong \sum_{i=0}^{n} S_I(f_i)\Delta f \quad (1)$$

where $\Delta f = f_{i+1}-f_i$ is the bandwidth, $f_{min} = f_0$ and $f_{max} = f_n$ are the minimum and maximum frequencies, respectively. The quasi-variance differs from the variance by the finite and non-zero limits of the integral in (1). Throughout this paper $\Delta f = 1$ Hz, $f_{min} = 1$ Hz and $f_{max} = 500$ Hz are used in $Var(I)^*$ calculations. The variation of $Var(I)^*$ and noise power S_I at 10 Hz as a function of I_D are shown in Fig. 5. Note that the tendencies are similar for $Var(I)^*$ and S_I, but $Var(I)^*$ curves are much smoother.

Noise Reduction

For a given drain current, the noise is maximum in MOSFET mode (uppermost curve, Fig. 5) where the dominating fluctuation mechanism is the dynamic capture and release of the channel electrons by oxide traps [3]. Noise PSD decreases in JFET mode (bottom two curves, Fig. 5) by 1-2 orders of magnitude as the conducting channel moves away from the surface. $Var(I)^*$ is lower for strong inversion (bottom curve) than for depletion at the front interface (middle curve) because the minority carriers shield the front interface and the exchange between the channel majority carriers and oxide traps is minimized. There are important implications associated with noise reduction, particularly in the design of analog circuits.

Noise Transformation

"Normalized quasi-variance of drain current fluctuations", $Var(I)^*/I_D^2$, is shown in Fig. 6 as a function of the drain current. In MOSFET mode (upper curve) the behavior is similar to that observed in inversion-mode MOSFETs, *i.e.*, a plateau in the subthreshold region followed by a $1/I_D^2$ roll-off above threshold. This variation is basically explained by McWhorter's carrier number fluctuation model [3]. When the current flows sufficiently far from the interfaces and the interaction between the oxide traps and the conducting channel is suppressed (JFET-mode with front interface in strong inversion, bottom curve), the plateau observed in surface conduction disappears and $Var(I)^*$ decreases as $1/I_D$ over about three decades. Such a bias dependence is predicted by the Hooge's empirical relation [4] that can be adapted to $Var(I)^*$ via an integration from f_{min} to f_{max}:

$$\frac{Var(I)^*}{I_D^2} = \frac{\alpha'_H}{N} \quad (2)$$

where α'_H is a mobility dependent constant and N is the carrier number, proportional to I_D (assuming a bias-independent mobility when the conduction is far from the surfaces). When the conduction is in the volume but not completely isolated from the interface (JFET mode with front-interface depleted, middle curve), the behavior of $Var(I)^*/I_D^2$ is a mixture of McWhorter and Hooge models. These remarks suggest a transition from the carrier number to mobility fluctuations as the conduction is moved from the front-surface to the volume of the G^4-FET.

Conclusion

Bulk and surface low-frequency noise characteristics of the SOI four-gate transistors are compared. It is shown that for the same drain current, as the conducting channel moves away from the surface towards the volume: (i) the noise PSD decreases by one to two orders of magnitude and, (ii) the bias dependency and nature of the noise change, indicating the presence of different fluctuation mechanisms at the surface and in the bulk.

References

[1] B. Dufrene et al., *IEEE Int. SOI Conference*, pp. 55 56, 2003.
[2] K. Akarvardar et al., *IEEE Int. SOI Conference*, pp. 89-90, 2004.
[3] J. Brini, *Microelectronic Engineering*, 40, pp. 167-179, 1998.
[4] F. N. Hooge, *IEEE Trans. Electron Devices*, 41, pp. 1926-1935, 1994.

0-7803-9212-4/05/$20.00 ©2005 IEEE

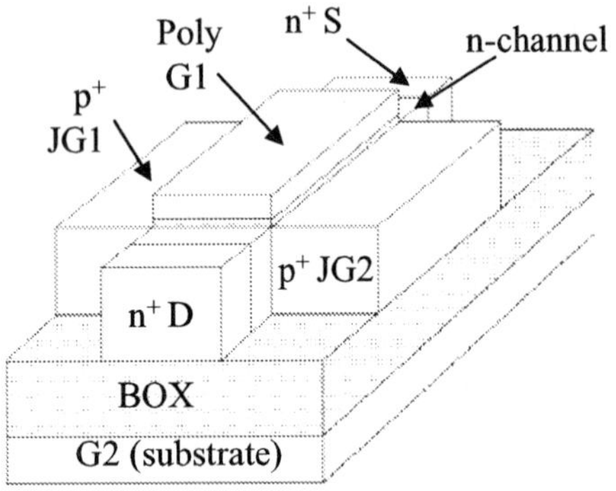

Fig. 1. Configuration of the n-channel G^4-FET.

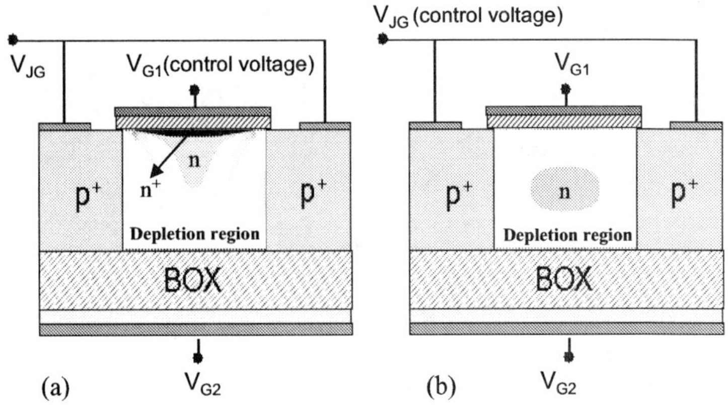

Fig. 2. N-channel G^4-FET cross-sections: (a) MOSFET mode, (b) JFET mode (drain current is perpendicular to figure).

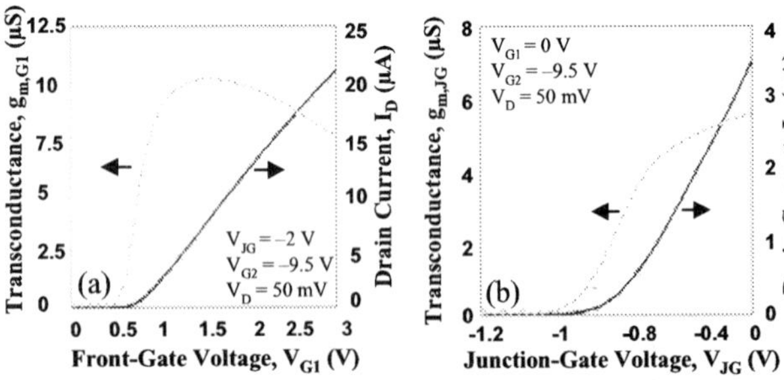

Fig. 3. N-channel G^4-FET static characteristics: (a) MOSFET mode, (b) JFET mode (standard 0.35 µm PD SOI, W = 0.35 µm, L = 3.4 µm).

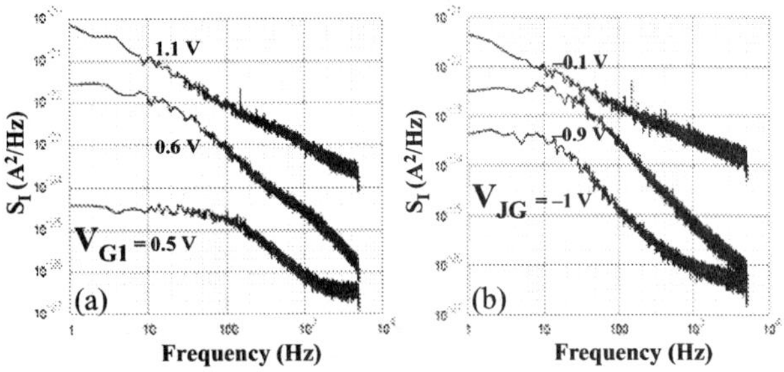

Fig. 4. N-channel G^4-FET drain current noise power spectral density: (a) MOSFET mode, (b) JFET mode (bias and transistor parameters correspond to those of Fig. 3).

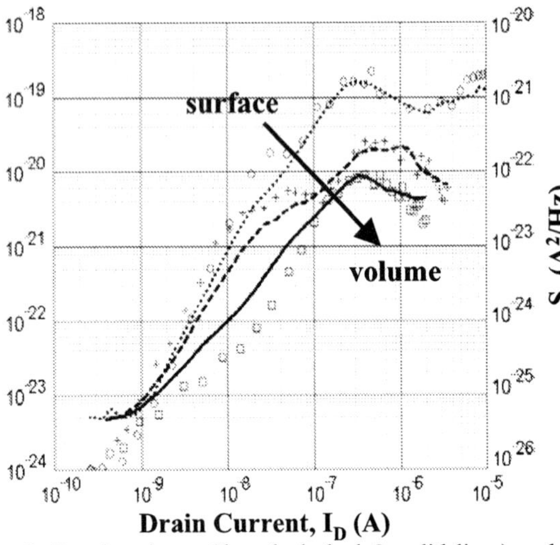

Fig. 5. Quasi-variance (dotted, dashed & solid lines) and drain current noise power at 10 Hz (circles, hatches and boxes) as a function of the drain current (V_{G2} = – 10 V). Dotted line & circles: MOSFET mode (V_{JG} = –2 V), dashed line & hatches: JFET mode (V_{G1} = 0 V, front interface depleted), solid line & boxes: JFET mode (V_{G1} = –2 V, front interface in strong inversion).

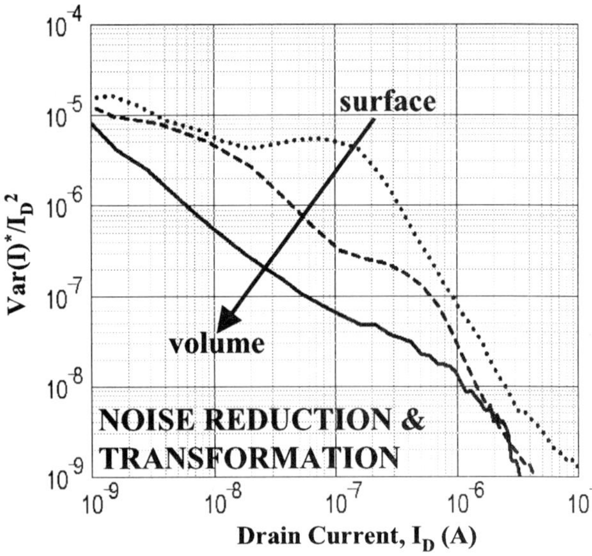

Fig. 6. Normalized quasi-variance as a function of the drain current (V_{G2} = 0 V). Dotted line: MOSFET mode (V_{JG} = –4 V), dashed line: JFET mode (V_{G1} = 0 V, front interface depleted), solid line: JFET mode (V_{G1} = –3 V, front interface in strong inversion).

Oxidation Induced Stress Effects on Hole Mobility as a Function of Transistor Geometry in a 0.15 μm Dual Gate Oxide CMOS SOI Process

P. S. Fechner and E. E. Vogt, Honeywell Solid State Electronics Center, Plymouth, MN 55441

Introduction

Trench liner oxidations used in Shallow Trench Isolation (STI) SOI fabrication processes impact local stress in silicon islands which reduce yield [1-4] and change transistor transconductance [5,6]. With the continued reduction in power supply voltages it has become standard to incorporate a dual gate oxide thickness option to permit I/O interfaces to operate at higher voltages than the low voltage, high-speed core.

It is common in dual thickness gate oxide STI processes to grow the high voltage oxide after STI formation, etch it off in the core area and then grow the gate oxide in the core area to the required thickness. Just prior to growing the thin oxide the high voltage gates are thinned so as to achieve the desired thickness after thin oxide oxidation. For this work the liner oxidation thickness was 11nm followed by a 8.0nm high voltage oxide growth which was subsequently thinned to achieve 5.9nm after the 3.0nm nitridized low voltage gate oxidation. Final active area silicon thickness is 145nm.

An examination of the p-channel transistor hole mobility measured at maximum gm (gmmax) on a wide variety of lengths(L) and widths(W) revealed a complex dependence of mobility on L and W. (Figure 1) L and W effects on mobility are observed to vary by up to 2X and 3X on the 1.8V(LV) and 3.3V(HV) transistors respectively. A similar complex relationship also was observed on the n-channel transistors but the variation was 5-10X smaller.

To understand this complex relationship a study using SUPREM4 simulations and cross-sectional TEM analysis of the buried oxide thickness was undertaken.

Experimental Investigation

Because all devices measured were essentially "isolated" from other active areas by at least 10 μm it was assumed that these relationships were primarily related to stress induced effects on mobility via the piezoresistive effects associated with the oxidations after silicon island definition [5]. However, the observation that the mobility vs W goes through a minimum at ~1μm and then increases again below 1μm was not obvious.

Cross-sectional TEM of a large island shown in Figure 2a revealed that the increase in buried oxide thickness due to the oxidations after island definition extends approximately 1μm underneath the silicon island. As indicated by En et al.[5], this causes an upward bending of the silicon due to volume expansion which creates a compressive stress in the W direction of the channel. For p-channels this stress causes a reduction in hole mobility proportional to this stress over some region near the edges of the island. As geometries decrease below 2 μm then the oxidation profiles from both sides will begin to overlap causing the thickness in the center of the island to increase as well. (Figure 2b) It is reasoned that in the limit as the W approaches 0 then the center thickness will be the same as the edge and there will be no curvature to create the stress and the mobility should return to the unstressed value at W=0.

To verify this theory and quantify the value of this stress, SUPREM4 simulations were performed. However, first attempts using default parameters only increased the oxide thickness at the island edge from 145nm to ~160nm rather than the 164nm measured by TEM on the device in Figure 2b. Experiments demonstrated conclusively that the thickness of the deposited oxide has no measurable impact on the oxidation of the underlying silicon in this process. Simulations of the structure without depositing the field oxide accurately matched the TEM results. Since this is consistent with the "isolated" islands assumption mentioned previously, all further simulations were carried out without depositing the STI oxide.

Figure 3a demonstrates that the horizontal stress components near the surface of a large island reach a peak about 0.35μm from the edge and decays to <1E9 dynes/cm^2 after about 1.3μm. When the device island is about 1 μm, the stress from both edges is focused in the middle of the island and reaches in excess of 6E9 dynes/cm^2. (Figures 3b) Piezoresistive coefficients predict approximately 7%/1E9 dynes/cm^2[7] indicating the stress levels present may be sufficient to explain the mobility effects observed in the data.

An experiment was also run eliminating the 8.0nm HV oxidation from the process sequence to reduce the oxidation induced stress post STI formation.

Results and Discussion

Figure 4a plots the simulated average island stress near the surface and the measured mobility as a function of island W for the two different processes. The simulation correlates with the measured data. Figure 4b is a similar plot as a function of device length L. The simulation island size has been adjusted to reflect that the source and drain length (~.8μm) needs to be added to L to reflect the actual size of the island. The stress was averaged only over the gate and spacer region for the L. A plot of the 2D stress profile for a L=.15μm gate device is shown in Figure 3b. The average stress results explain the increase in the W=10μm mobility as a function of L observed on the 3.3V transistor. The peak at L=1μm for the 1.8V transistor is believed to be related to the threshold voltage roll up. This process actually uses the same Nwell for both 1.8V and 3.3V transistors. To achieve the desired threshold voltage on the 1.8V 3.0nm gate oxide transistors an aggressive Arsenic halo implant is used to roll the device threshold up from .35V at L=10μm to 0.55V at 0.15μm. This effectively increases the vertical electric field at gmmax counteracting the stress induced improvements in mobility. No halo implant is used on the 3.3V transistor which demonstrates about .1V Vt roll off between L=10 and L=0.35μm. Thus, while some of the apparent increase in mobility may be vertical electric field related in the 3.3V transistor, it appears to be primarily related to the stress effects on mobility.

Figure 4 also demonstrates that elimination of the HV oxidation reduces the overall mobility variation. At minimum transistor L=.15 μm and W=1-2μm geometries commonly used in ASIC core logic this stress reduction results in a 5% improvement in drive current.

Conclusion

Additional process complexity required for multiple gate oxide thicknesses has been demonstrated to cause significant impact on p-channel mobility through oxidation induced stress effects. This can occur even after the STI oxide planarization step. Therefore, process flow options to minimize all oxidations after island definition should be considered.

The data presented here also underscores the need to understand and model the oxidation induced stress associated with the specific device of interest. Since these stresses have a major impact on device mobility and current drive, care must be taken to make sure variation in layout options such as stacking transistors in a single device island rather than individual islands do not lead to circuit design modeling errors sufficient to compromise circuit performance margins. SUPREM4 appears to be able to model this behavior but most circuit level transistor models are not capable of handling this level of detail or complexity.

Acknowledgement

This material is based upon work supported by the Defense Threat Reduction Agency under Contract No. DTRA01-03-D-0018-0001. Any opinions, findings and conclusions or recommendations expressed in this material are those of the authors and do not necessarily reflect the views of the Defense Threat Reduction Agency.

References

[1] M. Racanelli et al., IEDM Tech. Digest, p. 885, 1995.
[2] I. De Wolf et al., Proc. of the Electrochemical Society, Vol. 92-13, p. 307-15, 1992.
[3] M. Mendicino et al., IEEE Int. SOI Conference Proceedings, p. 132, 1997.
[4] J. W. Sleight et al. IEEE EDL, p. 248, 1999.
[5] William G. En et al., IEEE Int. SOI Conference Proceedings, p. 85, 2001.
[6] Y. Jeon et all, IEEE Int. SOI Conference Proceedings, p. 164, 2003.
[7] S.E. Thompson et al., IEEE Transactions on Electron Devices, Vol. 51, No. 11, p. 1790, Nov. 2004.

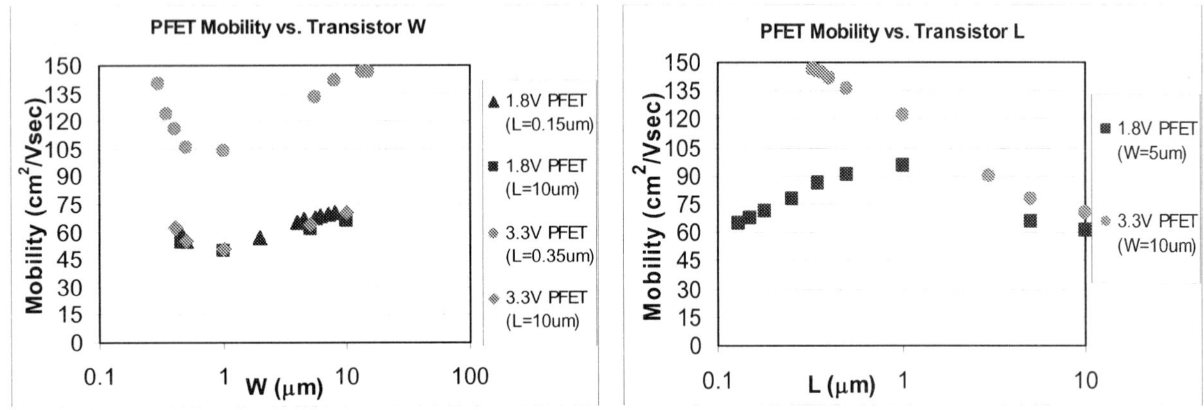

Figure 1. Measured P-channel hole mobility for 3.3V and 1.8V transistors as a function of gate Width (1a left) and Length (1b right).

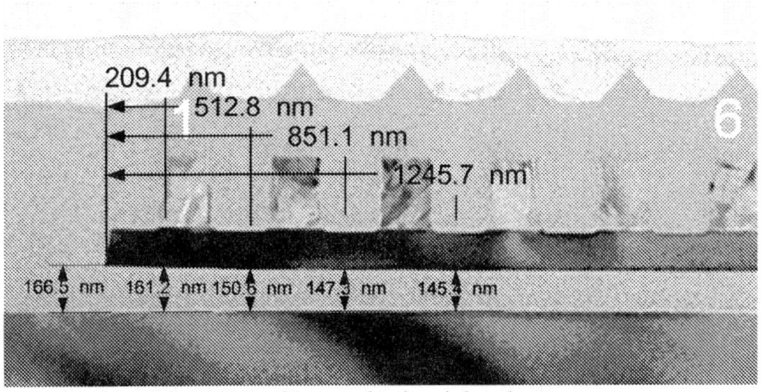

Figure 2a. TEM of large device island.

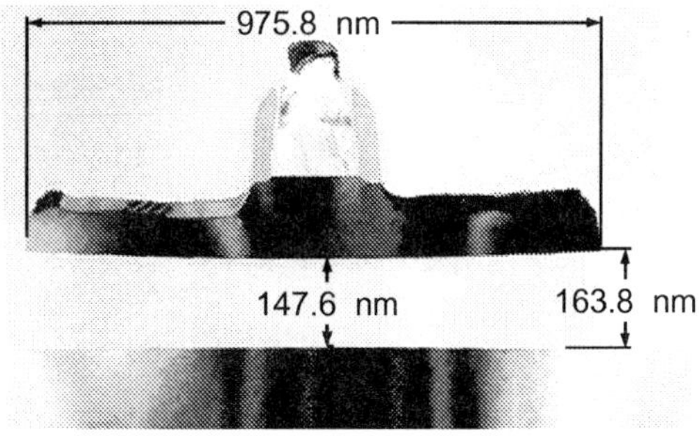

Figure 2b. TEM of 0.15 μm transistor.

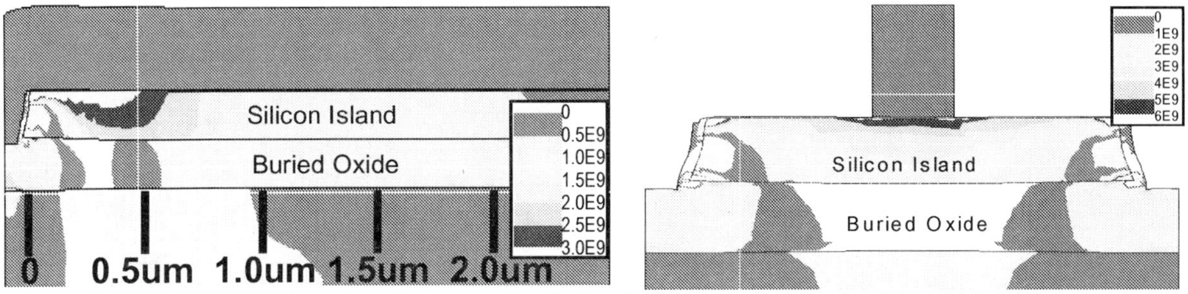

Figure 3a(left). SUPREM4 compressive stress simulations in large device island and 3b(right) 0.15 μm transistor

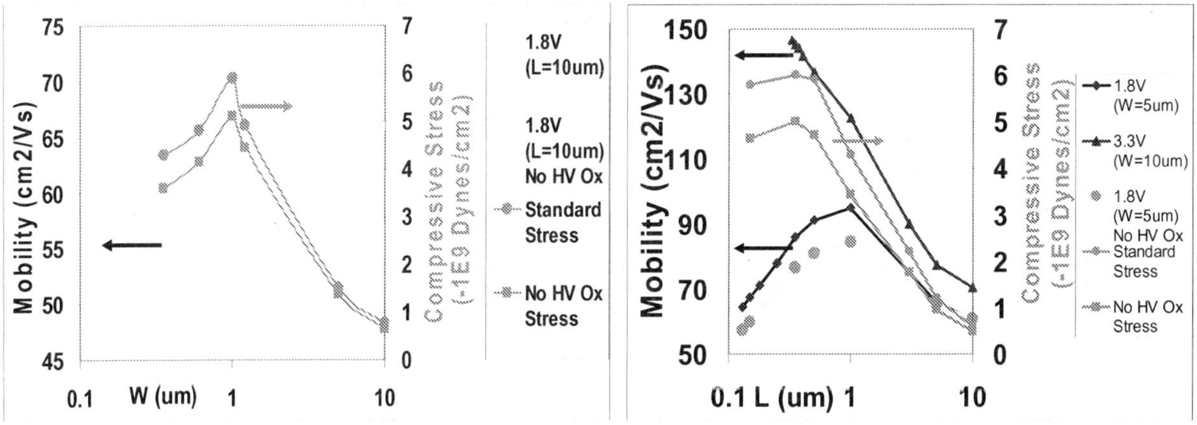

Figure 4. PFET hole mobility and simulated average stress vs. device Width (4a. left) and Length (4b. right)

Session #7

DEVICE PROCESS

CO-CHAIRS

Atsuchi Ogura
Meiji University
Thierry Poiroux
CEA-DRT – LETI/DTS

DEVICE PROCESS — Session 7

2005 IEEE International SOI Conference

A Novel Self-Aligned Substrate-Diode Structure for SOI Technologies

Mario M. Pelella, Gert Burbach[2], Akram Salman, Stephen Beebe, Darin Chan, and Jim Buller[3]
Advanced Micro Devices, Sunnyvale, CA 94088-3453; [2]Dresden, Germany; [3]Austin, Texas
Email: mario.pelella@amd.com

Introduction

The ability to provide a stable suite of nonFET devices, such as diodes, resistors, and decoupling capacitors, with low variability and a broad operating range is paramount in modern-day, high-performance microprocessors [1], [2]. In this work, we describe novel self-aligned diode and resistor structures and their process integration into an advanced 90nm SOI technology. Their superior device characteristics over conventional device structures built within the SOI film is described.

Device Structure and Fabrication

The process integration of the self-aligned substrate-diode structure is shown in Fig. 1. The first step illustrated is a high-energy phosphorus implant that creates the n-type well of the diode and is inserted early in the process flow. Viable implant conditions range from $1x10^{13}$ – $2x10^{14}/cm^2$ for the dose and 200-300KeV for the energy, which is transmitted through the STI and buried SiO_2 (BOX) layers into the substrate (or handle wafer) region. Next, prior to the source and drain processing steps, an oxide-etch step is used to remove the STI and BOX films, which opens the anode and cathode regions of the diode. The standard transistor gate-polysilicon (poly) acts as an oxide-etch stop (or hard mask) that self-aligns the anode and cathode regions and enables the diode length to be defined to the critical dimensions of the technology. Following the MOSFET's source/drain implant and salicide steps, an inter-layer dielectric is deposited and the standard metal contact processing is used to connect to the anode and cathode terminals of the diode. A cross-sectional transmission electron microscope (TEM) micrograph of the final end-of process structure is shown in Fig. 2.

Experimental Results and Discussion

The substrate-diode structure was integrated into a 90nm SOI technology using the process steps noted above with standard bonded wafer material from Soitec (Smart Cut™ process). The measured forward I-V characteristics are shown in the inset of Fig. 3 for the substrate-diode using a $1x10^{14}/cm^2$, 240 KeV phosphorus implant for the deep nWell, which is compared to the conventional SOI-film diode structure that was simultaneously built within the same process flow. For the same layout area, the substrate-diode current at V_F =0.8V is ~7x more than the SOI-film diode, due to the additional junction area (areal component). The ideality factor, defined as

$$\eta = (q/kT)[(V_1-V_2)/\ln(I_1/I_2)], \qquad (1)$$

where q is the electron charge, T is the temperature, and k is Boltzman's constant, I_1 and I_2 are typical sampling currents (~1-100uA), is extracted for both diode structures as a function of current density (per diode width) and is shown in Fig 3. The results show a dramatic improvement in the ideality-factor levels over the entire current range. Moreover, novel transistor improvement features, SOI-film scaling, and self-heating effects will continue to aggravate the characteristics of SOI-film diodes as we migrate to future technology nodes.

Using the same integration as the substrate-diode structure, a substrate-resistor structure can be formed by replacing the P+ anode region with another N+ cathode region. The resistance results and schematic cross-section of the substrate-resistor structure are shown in Fig. 4, as a function of deep well implant energy and dose.

The thin SOI-film region in advanced SOI technologies and the low thermal conductivity of the BOX film, which leads to significant self-heating during an ESD event, severely limits the effectiveness of diode-based ESD protection networks. By creating the diode structure within the substrate region, self-heating effects are greatly mitigated [3] and current is allowed spread throughout the deep-nwell region reducing junction hot spots. The benefits of these attributes are shown by the dramatically improved 100ns transmission-line pulsed (TLP) I-V results in Fig. 5 for the same diode structures of Fig. 3. The It2 point (2nd breakdown of the diode where the device fails) of the substrate-diode structure is increased by >5X over the SOI-film diode and [4]. The on-resistance (slope of the I-V curve) of the diode is also significantly improved providing a much lower clamp voltage for protection networks.

Summary

A new process to form robust diode and resistor structures for SOI technologies has been demonstrated. This novel process offers dramatically improved device characteristics over conventional SOI-film structures for a myriad of analog, I/O, thermal sensing, and ESD applications. A key attribute of the device structures is that they are decoupled from the SOI-film, which provides chip designers stable device characteristics and longevity as high-performance microprocessors migrate to advanced transistor architectures, such as FD/SOI MOSFETs with only 5-20nm SOI films that are unable to support viable conventional nonFET devices.

References

[1] M.M. Pelella, "High performance SOI technology for Advanced Microprocessors," Proc. Of Int. Conf. On Solid State Devices and Materials, 2002, pp. 70-71.

[2] D. Greenlaw, et al., "Taking SOI substrates and low-k dielectrics into high volume microprocessor production," IEDM 2003, pp 277-280.

[3] A. Salman, M.M. Pelella, S. Beebe, and N. Subba, "ESD Protection for SOI Tech. using an Under-The-BOX (Substrate) Diode Structure," IEEE EOS/ESD Sym., Sept. 2004.

[4] C. Putnam, et al., "An Investigation of ESD Protection Diode Options in SOI," IEEE SOI Conference Proc. , Oct. 2005.

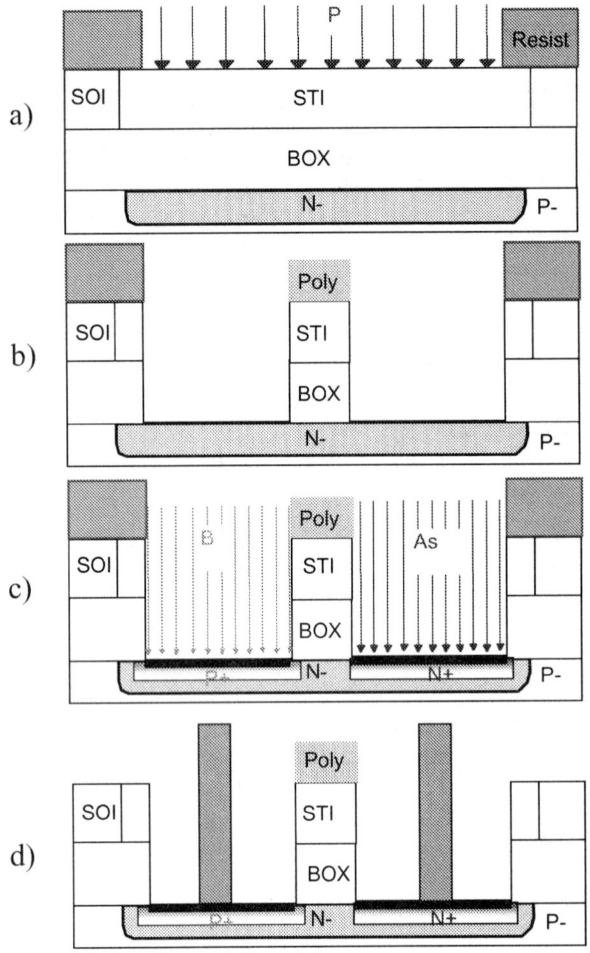

Fig. 1. Schematic cross-sectional views of the process flow for the substrate-diode structure. a) Deep phosphorus implant; b) etch substrate-diode contact window; c) S/D implants and silicidation; d) deposit ILD Std. 1st contact

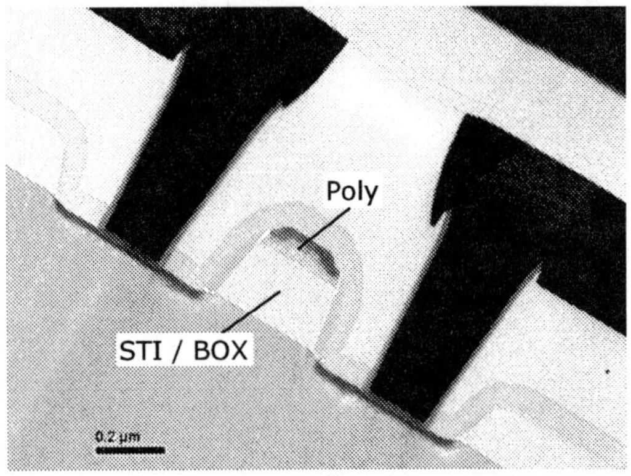

Fig. 2. Cross-sectional transmission electron microscope micrograph of the final end-of process substrate-diode structure.

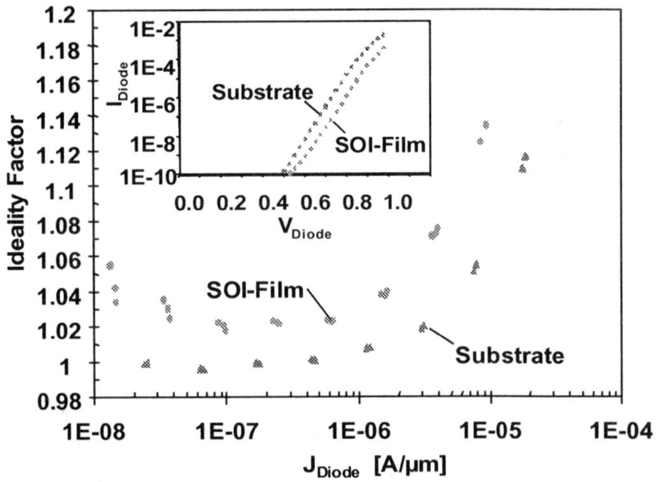

Fig. 3. Extracted ideality factor from the measured I-V characteristics (inset) of the substrate-diode using a $1x10^{14}/cm^2$, 240 KeV phosphorus deep implant for the nWell. 750Å /1450Å STI/BOX film, L_{POLY} = 315nm, W = 5.4um, $T_{Ambient}$ = 25°C

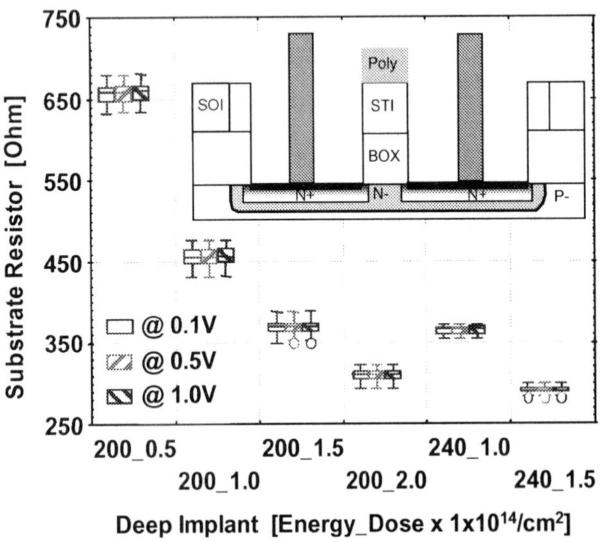

Fig. 4. Measured resistance results for a square (W=27um, L=27um) substrate resistor as a function of implant dose and energy. The schematic cross-section of the substrate-resistor structure is shown in the inset.

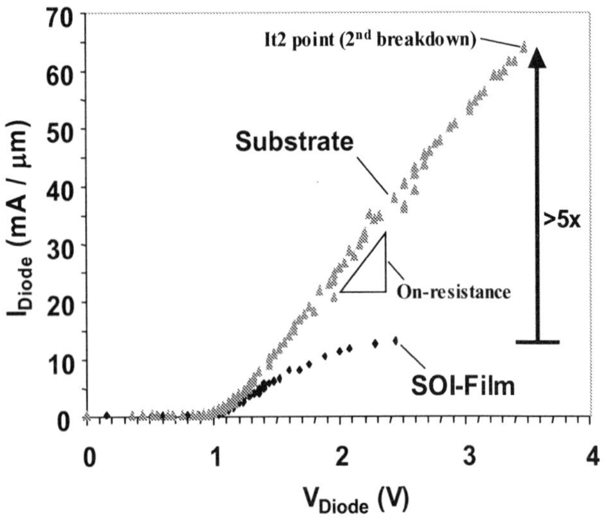

Fig. 5. Measured 100ns transmission line pulsed I-V results for the substrate-diode and SOI-film diode structures of Fig. 3. 750Å /1450Å STI/BOX film, L_{POLY} = 315nm, W = 5.4um, $T_{Ambient}$ = 25°C.

HIGH VOLTAGE DEVICES ADDED TO A 0.13µm HIGH RESISTIVITY THIN SOI CMOS PROCESS FOR MIXED ANALOG-RF CIRCUITS

O. Bon [1,2] , L. Boissonnet [1] , O. Gonnard [1] , S. Chouteau [1] , B. Reynard [1] , A. Perrotin [1] and C. Raynaud [1,3]

(1) STMicroelectronics, Crolles, France - (2) LAAS-CNRS, Toulouse, France

(3) CEA-LETI, Grenoble, France

Email: olivier.bon@st.com, Tel: +33 (0)476383034

Introduction

Thin SOI substrates are mostly used today for high performance applications. However SOI CMOS technologies are also known as good candidates for mixed Analog-RF and digital low power SOC [1]. This paper demonstrates the capability to integrate High Voltage (HV) devices (BV > 15V) in a 0.13µm CMOS process on thin SOI.

Device structure and Process

This SOI HV device is a Lateral Drain Extension MOS (LDEMOS) based on a low doped unsalicided drift region that sustains the drain reverse bias (Fig. 1 – Top layout and AA' cross section). Whereas floating body (FB) is an interesting effect for high speed digital circuits, the body has to be contacted in case of these analog HV devices (Fig. 1 - BB' cross section). The major challenge is the efficiency of this body contact (BC) in order to avoid Kink effect by perfect control of body potential.

The initial SOI layer and the buried oxide are respectively 1600Å and 4000Å thick. High resistivity (>1kOhm-cm) Unibond SOI substrates have been used. Gate oxidation is 20Å for the CMOS Core Process whereas it is 50Å for HV devices. After a Cobalt silicidation process - which shorts source and body - the back-end is processed with six levels of Cu interconnects. The drift region is protected to salicidation by a protection oxide (Fig. 2), that avoids the drift being shorted between drain and body. A NPN bipolar transistor is also implemented in this technology [2].

Compared to the standard CMOS, only four masks are added for complementary N and P HV module. The N(P) drift and P(N) body regions are implanted respectively before and after gate patterning (Fig. 3). The body implant is therefore self-aligned with the gate and allows to target a 0.6V threshold voltage independently of well implantations. 2D simulations have been used to calibrate the dose and energy of the body implantation according to the Core Process thermal budget.

In FB devices the P+ region used to build the BC has been suppressed keeping constant active width.

Results and Discussion

BC transistor BVds(Lext) shows two different breakdown behaviors (Fig. 4). In the Punch-Through breakdown mode, BVds is Lext-dependent, the drift region is fully depleted and the high doped drain contact area limits the transistor voltage capability (Fig. 5). Regarding the theorical expression (Eq. 1a), the breakdown voltage varies with a Lext 2nd order equation. In the "Avalanche" mode, BVds is Lext-independent, only the drift sustains the potential. In this case, the breakdown is only controled by the drift doping level. This behavior is confimed by the theorical equation of BVds vs Lext (Eq. 1b).

Electrical measurements are summarized (Table 1) for the two BC reference NMOS (W/L=2x5µm/0.5µm and Lext=0.6µm). The 15V BVds target is greatly reached with a low S.Ron (8.5mΩ.mm^2) and a low leakage current (<10pA/µm@5V).

Moreover F_T (Fig. 6a) and F_{MAX} (Fig. 6b) reach respectively 19.5GHz and 35GHz for a drain polarisation of 4V [3]. This transistor is therefore a good candidate for power handling in the 2GHz band. Furthermore, by using HR SOI substrate, high quality passives as well as a reduced substrate coupling has been demonstrated [4].

Four devices (W=2x25µm) using 0 to 6 BC have been measured to qualify the BC efficiency. At Vg=1V and for the FB transistor, Kink effect starts from Vd=1V (Fig. 7). In the same test conditions, all BC options are fully efficient all over the Vd range. At higher Vg and for the FB transistor, Kink effect is still present at 4.5V (Fig. 8) but is almost hidden by drain current quasi-saturation and Self-Heating (Fig. 8b). In the same time Single BC transistor starts to present Kink effect at Vd > 6.5V. Adding an extra BC enhances the transistor performances and shifts the Kink effect threshold to 8V. Anyway these two voltages are out of the main transistor operating area. The 6 BC transistor, built with the reference transistor (Table 1) design rules, is not affected by Kink effect. All body contacted transistors have a low leakage current (<10pA/µm@5V) and a high

0-7803-9212-4/05/$20.00 ©2005 IEEE

breakdown voltage (Fig. 6), compared to the FB one which presents a high leakage level.

Conclusion

We have added to a 0.13µm thin SOI CMOS Core Process a high competitive SOI NLDEMOS which presents excellent power switch and analog characteristics. Measurements have demonstrated that both drift and BC design rules allow to obtain HV devices (BV>15V) with a low S.Ron and a low leakage.

References

[1] J-O Plouchart, "SOI nano-technology for high-performance system on-chip applications", Proc. of SOI conference 2003

[2] G. Avenier and al., "A self-aligned vertical HBT for thin SOI SiGeC BiCMOS", submitted to BCTM 2005

[3] S. Matsumoto and al., "RF Performance of a State-of-the-Art 0.5µm-Rule Thin-Film SOI Power MOSFET", in IEEE Electron Devices, vol. 48, pp 1251-1255, n°6, June 2001

[4] C. Raynaud and al., "Is SOI CMOS a promising Technology for SOCs in High Frequency range ?", ECS Conference 2005

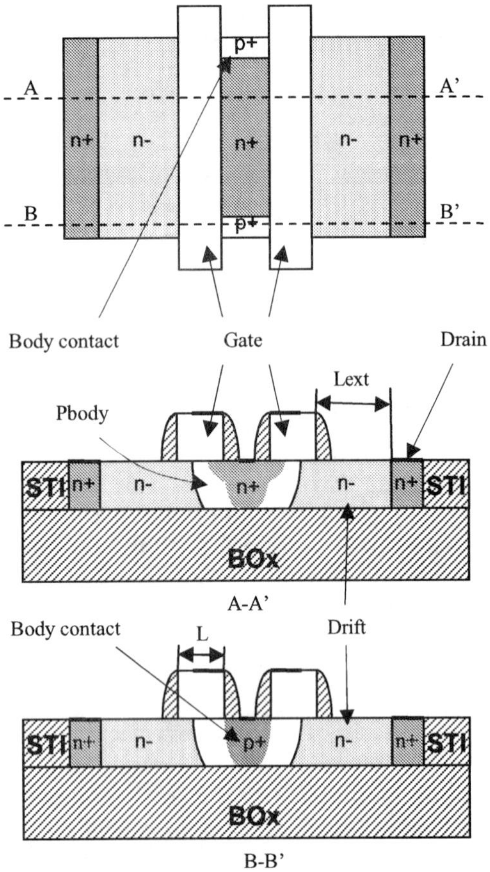

Figure 1: NLDEMOS top layout and cross sections

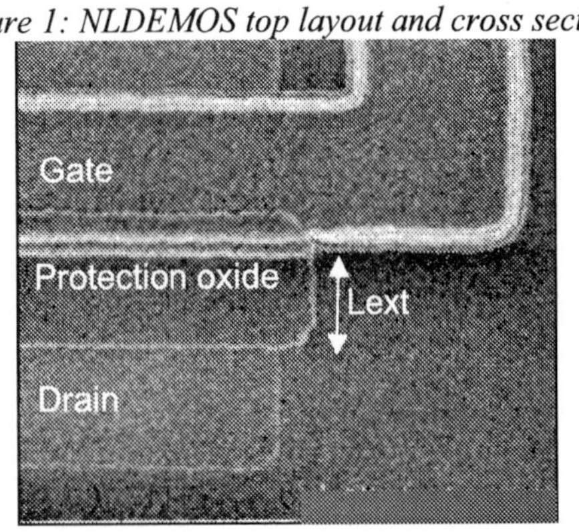

Figure 2: Salicided transistor SEM shots

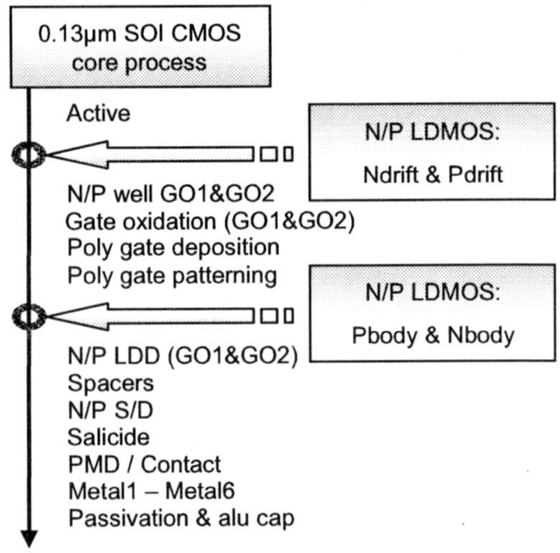

Figure 3: Core Process flow with LDEMOS option

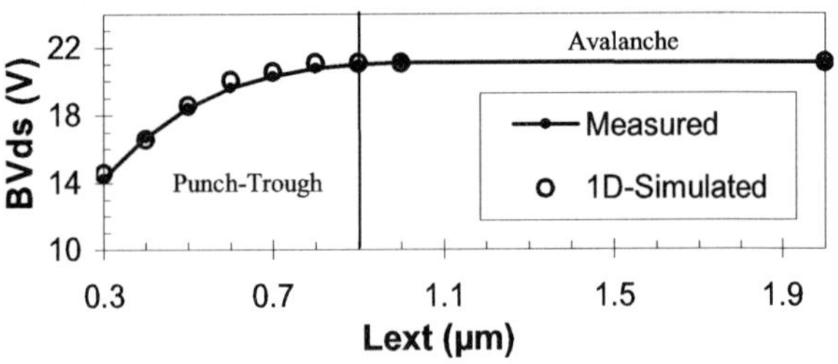

Figure 4: Measured and 1D-simulated BVds vs Lext (Body contacted transistor)

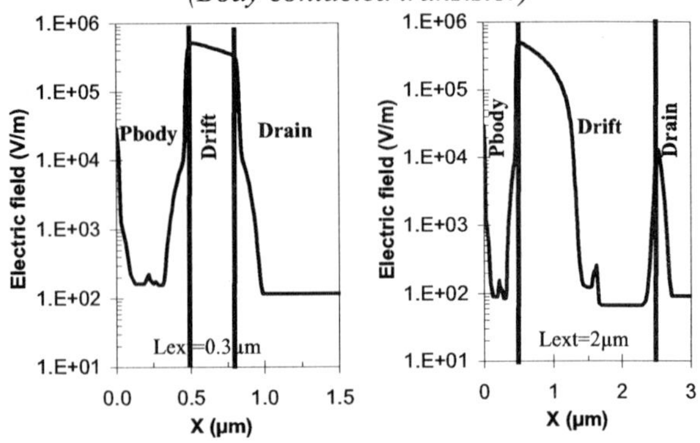

Figure 5: Simulated electric field cross sections for a Punch-Trough (left) and an Avalanche (right) Breakdowns (Body Contacted transistor)

$$V_{bd}^{P-T} = \frac{\varepsilon.E_C^2}{2qN_A} + E_C L_{ext} - \frac{qN_D L_{ext}^2}{2\varepsilon} \quad \text{(a)}$$

$$V_{bd}^{Av} = \frac{\varepsilon.E_C^2}{2qN_A} + \frac{\varepsilon.E_C^2}{2qN_D} \quad \text{(b)}$$

Equations 1: Breakdown voltage vs Lext theorical expression for Punch-Trough (a) and Avalanche (b) modes

Vt (V)	Ion (µA/µm)	Ioff (pA/µm)	S.Ron (mΩ.mm²)	BVds (V)	Ft (GHz)	Fmax (GHz)
	@ Vd=5V		@Vg=2.5V		@Vd=4V	
0.6	375	7.5	8.5	19.5	19.5	35

Table 1: Electrical measurements for a 2 BC transistor with W/L=2x5µm/0.5µm and Lext=0.6µm

(a)

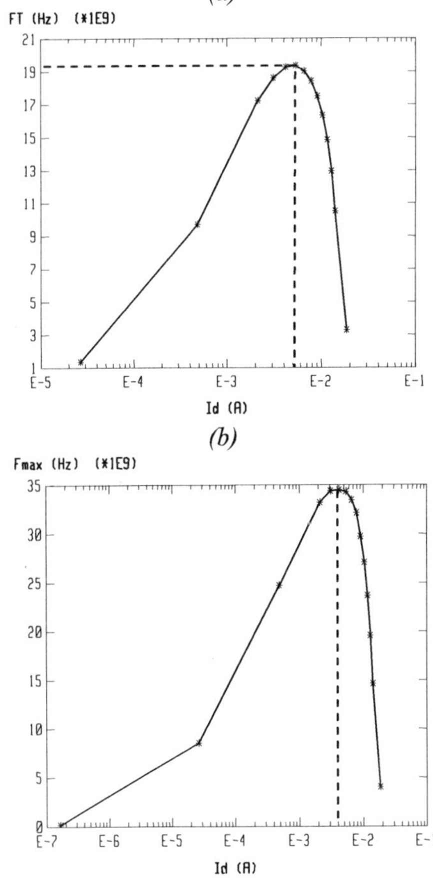

Figure 6: F_T (Fig. 6a) and F_{MAX} (Fig. 6b) vs Id @ Vd=4V for a BC transistor with W/L=2x5µm/0.5µm and Lext=0.6µm

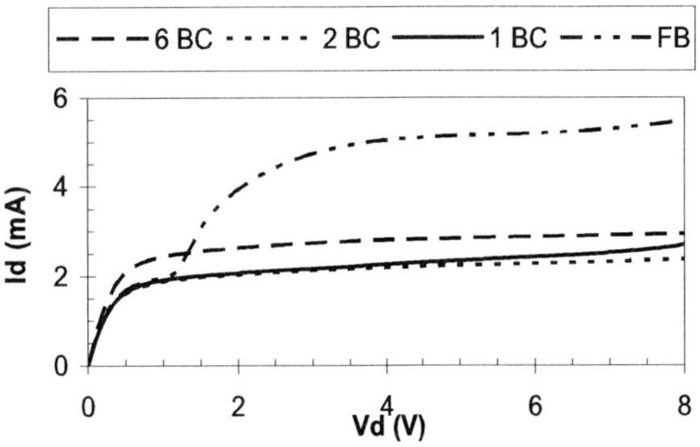

Figure 7: Id vs Vd at Vg=1V for BC (1, 2 or 6 BC) and FB transistors (W/L=2x25µm/0.5µm – Lext=0.6µm)

(a)

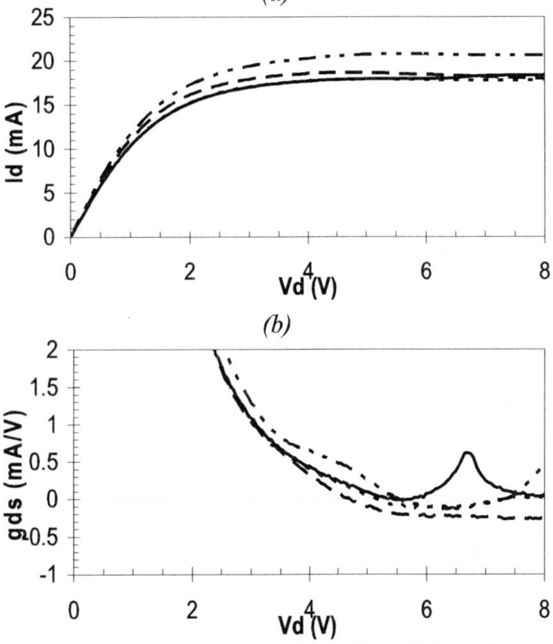

Figures 8: Id (Fig. 5a) and gds (Fig. 5b) vs Vd at Vg=2.5V for BC and FB transistors (W/L=2x25µm/0.5µm – Lext=0.6µm)

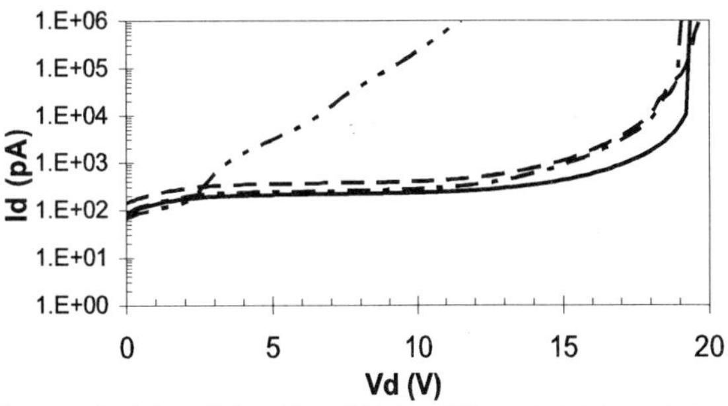

Figure 9: Id vs Vd at Vg=0V for BC and FB transistors (W/L=2x25µm/0.5µm – Lext=0.6µm)

Lateral Integration of Partially Insulated and Bulk MOSFETs Using Partial SOI Process

Sung Hwan Kim, Chang Woo Oh, Kyoung Hwan Yeo, Dong Uk Choi, Min Sang Kim, Sung Min Kim, Jeong Dong Choe, Jeongnam Han*, Young-Pil Kim*, Dong-Won Kim, Donggun Park, and Byung-Il Ryu

*DR and *PD Team, R&D Center, Samsung Electronics Co, San 24, Kiheung-Eup, Yongin-City, Kyungki-Do, KOREA, 449-711* Phone: 82-31-209-4025, Fax: 82-31-209-9861, E-mail: toshk@samsung.com

Introduction

As the transistors are aggressively scaled, their requirements for HP (High Performance), LSTP (Low STand-by Power) and so on are more and more difficult to be achieved on the same wafer with conventional planar structure and multiple IIP (Ion ImPlantation)[1]. In considerations of most IC devices composed of HP and LSTP transistors with different V_{TH}s, I_{ON}s, and I_{OFF}s, complex process such as dual or triple gate oxide process are inevitable to satisfy those requirements, as well as multiple IIP scheme[2]. Thus, if various transistors with different structures and electrical characteristics can be introduced without those complex process, those must be very beneficial to maximize IC performance. Especially, lateral integration of SOI and bulk devices is very attractive because of their complementarity in terms of V_{TH} controllability, body bias effect, and leakage currents[3].

In this paper, we propose and demonstrate the lateral integration of partially insulated and bulk MOSFETs using partial SOI technique even without complex process and SOI wafer.

Device Fabrication

The fabrication process for lateral integration are shown in Fig. 1[4, 5]. At first, the epitaxial growth of SiGe/Si layers is performed before hard mask deposition and its patterning[Fig. 1(a)]. And Si/SiGe layers are selectively etched out to make PiOX (Partially insulating OXide). The mask layers are removed by wet etch, followed by the Si epitaxial growth on the whole wafer[Fig. 1(b)]. Thereafter, pad oxide/SiN mask deposition and trench etching for isolation are carried out. And the SiGe layers are selectively removed by a specially formulated etchant[Fig. 1(c)]. After the selective removal of SiGe, oxidation and nitride/oxide deposition are followed to form the buried insulating layers, namely PiOX. After CMP process, the fabrication of PUSD (PiOX Under Source/Drain), PUC (PiOX Under Channel) PiFETs (Patially insulated MOSFETs) and bulk MOSFETs are completed by applying conventional CMOS process on this template[Fig. 1(d)]. Newly fabricated transistors on various substrates, partial SOI and bulk substrate, are shown in Fig. 2.

Results and Discussion

Electrical characteristics for each structure are shown in Figs. 3~6. As our expectation, three types of transistors show the different V_{TH}s, I_{ON}s, and I_{OFF}s, resulting from their structures. It is clear that those differences are mainly originating from the appearance of PiOX and strongly depending on the overlapping between gate and PiOX layers, the body thickness, and the position of the peak concentration of channel IIP related to PiOXs[4, 5]. Except for the only PUC nMOSFET with L_G=30nm showing the degradation due to fringing field effects and the low I_{ON} current due to mobility degradation and high SDE resistance, their characteristics look quite reasonable even at the small gate length. In terms of short channel effects, patial SOI devices have better DIBL and swing values owing to the self-limited shallow junction and the small depletion capacitance due to PiOXs. In terms of driving current, they show multiple I_{ON}s and I_{OFF}s linearly proportional to their V_{TH}s both for nMOS/pMOS and for different gate length. Driving current levels for transistors with L_G=30nm are also comparable to the best results of unstrained transistors. From the results, it was proven that the various transistors satisfying different driving and stand-by current requirements can be easily implemented on the same wafer with partial SOI process. More detailed characteristics are summarized in Table 1.

Conclusions

We proposed and successfully demonstrated partially insulated and bulk MOSFETs with multiple V_{TH}s, I_{ON}s, and I_{OFF}s by using partial SOI process without complex process and SOI wafer. Both nMOS and pMOS applicable to the HP and LSTP transistors were simultaneously implemented on the same wafer with the same process except partial SOI process. These results must be very useful to implement IC systems requiring various specifications of V_{TH}s, I_{ON}s, and I_{OFF}s.

References

[1] *PIDS and ERD in updated version of ITRS 2003,* 2004.
[2] T. Fukai, *et al., Proc. of VLSI Tech.2003*, p. 83, 2003.
[3] M. Yang, *et al., Proc. of VLSI Tech.2004*, p. 160, 2004.
[4] C. W. Oh, *et al., Proc. of ESSDERC 2004*, p.233, 2004.
[5] K. H. Yeo, *et al., IEEE EDL.*, Vol. 25, p.387, 2004.

0-7803-9212-4/05/$20.00 ©2005 IEEE

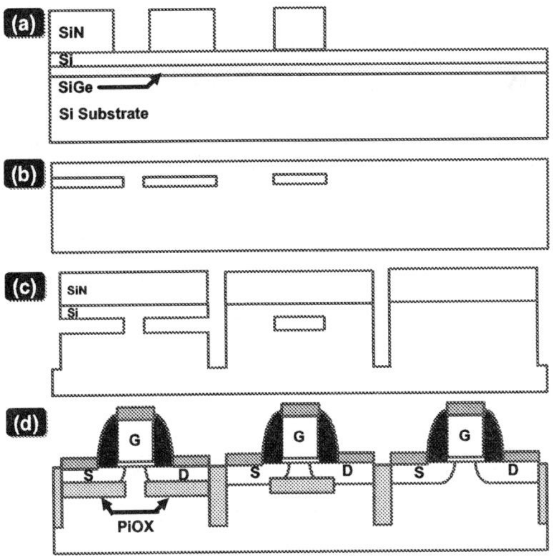

Fig. 1. Process sequence for lateral integraion; (a) SiGe/Si Epi. growth and patterning, (b) SiGe/Si etch and blanket Si Epi. growth, (c) trench etch and SiGe removal, (d) PiOX formation and conventional process.

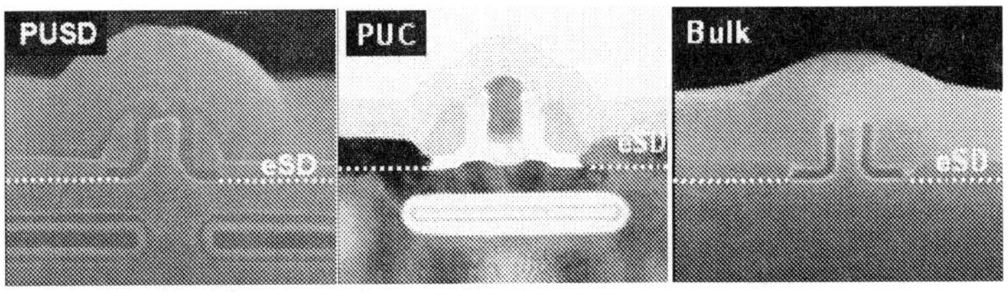

Fig. 2. Cross-sectional microphotographs of partially insulated and bulk transistors fabricated on the bulk silicon wafer using partial SOI process.

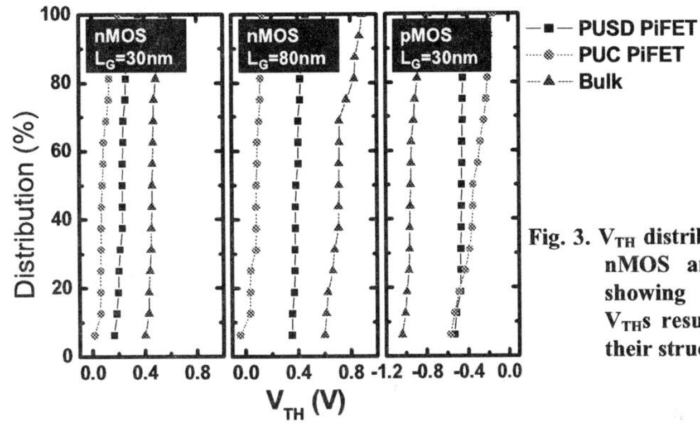

Fig. 3. V_{TH} distributions for nMOS and pMOS showing multiple V_{TH}s resulting from their structures.

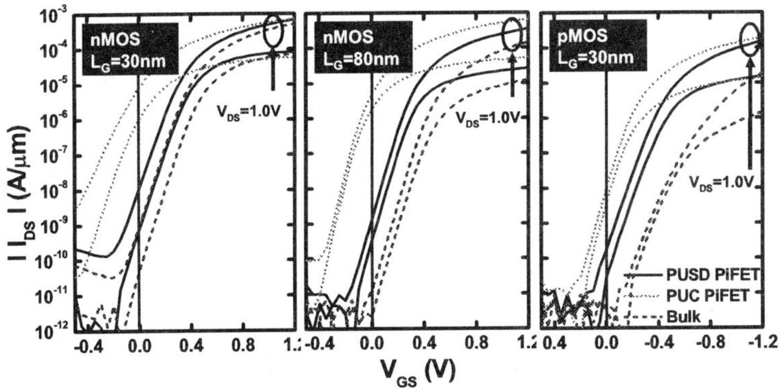

Fig. 4. I_{DS}-V_{GS} characteristics for partial SOI and bulk transistors. Except for the PUC nMOSFET with L_G=30nm, their characteristics are quite reasonable in terms of SCE and I_{OFF} even at the small gate length.

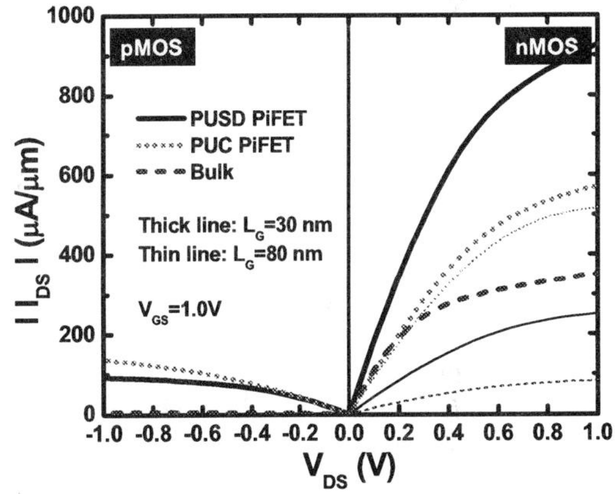

Fig. 5. I_{DS}-V_{DS} characteristics for partial SOI and bulk transistors showing multiple I_{ON}s due to their structures both for nMOS/pMOS and for different gate lengths at V_{GS}=V_{DS}=1.0 V.

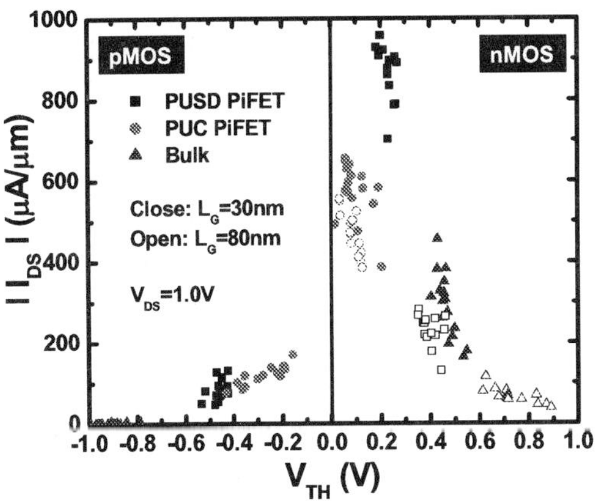

Fig. 6. I_{DS}-V_{TH} characteristics for partial SOI and bulk transistors. I_{ON} currents are linearly proportional to V_{TH} for nMOS/pMOS despite their different structures.

Table 1. Electrical parameters for partial SOI and bulk devices.

	nMOS L_G =30nm			nMOS L_G =80nm			pMOS L_G =30nm		
TYPE	PUSD	PUC	Bulk	PUSD	PUC	Bulk	PUSD	PUC	Bulk
V_{TH} [V]	0.23	0.08	0.46	0.3	0.08	0.71	0.46	0.35	0.95
DIBL [mV/V]	105	150	110	49	20	60	66	37	36
Swing [mV/dec]	100	130	102	78	68	96	109	90	114
Ion [μA/μm]	927	571	351	251	517	84	93	138	6

Undoped Thin Film FD-SOI CMOS with Source/Drain-to-Gate Non-overlapped Structure for Ultra Low Leak Applications

Noriyuki Miura, Yasuhiro Domae, Toyokazu Sakata, Minoru Watanabe,
Tomohiro Okamura, Tadashi Chiba, Koichi Fukuda and Jiro Ida
Semiconductor R&D Division, Oki Electric Industry Co., Ltd., Tokyo, Japan
Email: miura288@oki.com, Tel: +81-426-62-6109

Abstract

In this paper, we present an undoped thin film fully-depleted (FD) silicon-on-insulator (SOI) CMOS with source/drain-to-gate non-overlapped structure for ultra low leak (ULL) transistor. The fabricated device achieved a cutoff frequency f_T of 65GHz with $I_{off}<$ 0.1pA/um (GIDL-free). The proposed inverted-gate-implantation/planar-type SOI is practical and low-cost solution for coin-battery applications.

Introduction

The mobile-electronics market is growing fast and ULL ($I_{off}<0.1$pA/um) device is required especially for coin-battery applications. However with bulk CMOS, gate-induced drain leakage (GIDL) and gate leakage current limit the minimum I_{off} [1]. With conventional FD-SOI, floating-body effects (FBE) and GIDL remain the critical issues (Fig.1) due to the very high SOI body concentration. In this paper, an undoped thin film FD-SOI CMOS with S/D-to-gate non-overlapped structure is firstly fabricated and studied. Although non-overlapped structures have already been reported [1][2][3], they are bulk CMOS devices and for high-speed targets. We have confirmed the "FBE/GIDL-free" features of the device that realize ultimate ULL applications.

Fabrication Process and TCAD simulation

FD-SOI MOSFETs are fabricated with the process flow given in Fig.2. Especially, we focus on inverted-gate-implantation/planar SOI in view of manufacturability (simple and low-cost process). SOI wafers have final silicon layer thickness (T_{si}) of 30nm and buried oxide (T_{box}) of 200 nm. NO-treated gate oxide (T_{ox}= 2.5nm) is followed by undoped poly-silicon deposition, inverted gate implantation, gate activation RTA (P+gate for Nfet and N+gate for Pfet), LP-TEOS deposition for hard-mask and gate patterning. Vt-control, LDD and HALO implantations are not needed in the present process. Then, sidewall spacer and graded S/D are formed. Thin T_{si} specific cobalt salicide process [4] is applied to reduce the S/D sheet resistance. Fig.3 shows the simulated non-overlapped transistor (S/D to gate distance is estimated at 20nm by side). To certify the maintaining drive-current even with the non-overlapped, Fig.4 shows the simulated channel resistance of various SOI types, doped or undoped body, non-overlapped or overlapped S/D. Source-side resistance point of view, non-overlapped structure can be accepted only with low-concentration SOI body. This is confirmed by experimental result (Fig.8) in the next section.

Experimental Results and Discussion

Fig.5 shows the Vt-rolloff of undoped-body and counter doped-body Nfet/Pfet. The threshold voltage can be well controlled with shallow, low-dosage arsenic/BF2 channel (counter) implants for Nfet/Pfet. The I_DV_G characteristics of 140nm-gate transistors are shown in Fig.6. The subthreshold slopes are less than 75mV/ decade. No GIDLs are observed and $I_{off}<0.1$ pA/um targets are achieved. The minimum I_{off} is not limited by GIDL and gate tunneling leakage. Fig.7 shows the I_DV_D characteristics. Since SOI body is "perfectly" full-depleted, no kinks are observed. The intrinsic drive currents (excluding kink-parasitic currents) are more than those of highly-doped SOI (Fig.8). Even with single-drain structure, the S/D breakdown voltages (BV_{ds}) and hot-carrier life time (BV_{hc}: I_{ON} 10% degradation, 10years) are enough ($BV_{ds}>>3V$ in Fig.7 and $BV_{hc}>1.65V$ in Fig.9) for T_{ox}=2.5nm transistor. Since our proposed device has no overlap capacitance component (Fig.11), a cutoff frequency f_T is high enough (f_T=65GHz) for 140nm-gate ULL transistor (Fig.10). In addition, no gate direct in-tunneling occurs due to non-overlap. Thus the proposed structure has adaptability for future ULL device with thin gate oxide and gate length shrinking technologies.

Conclusion

An undoped FD-SOI with source/drain-to-gate non-overlapped structure is introduced. The transistor achieved a cutoff frequency f_T of 65GHz with $I_{off}<0.1$ pA/um. With the proposed FD-SOI structure, the ULL target is fulfilled while maintaining the drive-current. This structure is practical, simple and low-cost solution for FBE/GIDL-free SOI. Our approach strengthens the ability of FD-SOI for the ULL applications.

References

[1] ITRS 2003 Edition, Emerging Research Devices
[2] F. Boeuf et. al., pp.637-640, IEDM Dig. 2001,
[3] H. Lee et. al., pp.219-225, IEEE Trans. Nanotech. , Vol.1, No.4, 2002
[4] T. Ichimori et. al., IEEE SOI Conf. pp.72-73, 2000

0-7803-9212-4/05/$20.00 ©2005 IEEE

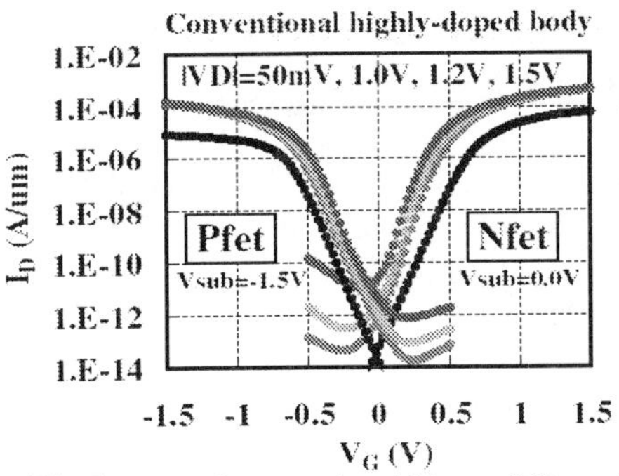

Fig.1 I_DV_G of conventional N+gate/Nfet and P+gate/Pfet with highly-doped SOI, S/D-to-gate overlapped structure (L=140nm). FBE and GIDL are bottlenecks for I_{off} 0.1pA/um.

- SOI Tsi=30nm, Tbox=200nm
- Gate Oxidation SiON 2.5nm
- (Vt control implant: Skip or Counter-doping)
- poly-Si deposition
- Gate implant : B+ for Nfet / Phos+ for Pfet
- Gate Activation RTA
- Hard-Mask LP-TEOS, Gate Patterning
- (LDD/HALO implant: Skip)
- Sidewall spacer
- S/D implant, S/D Activation RTA
- Cobalt Salicide Process [4]

Fig.2 Process flow of non-overlapped, inverted-gate-implantation, planar SOI transistor (P+gate for Nfet, N+gate for Pfet). Vt/LDD/HALO implants are not needed.

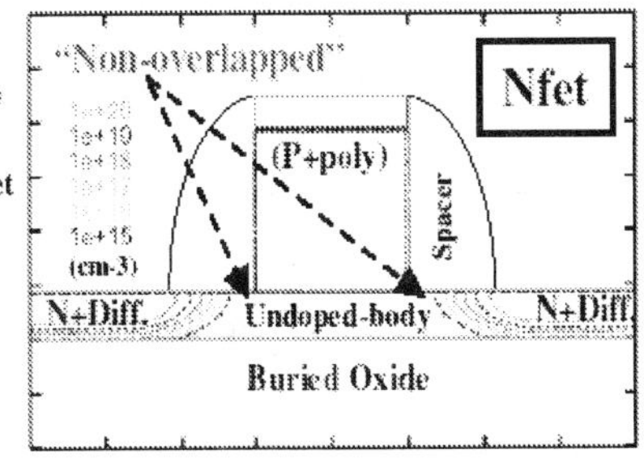

Fig.3 Proposed undoped-body/ non-overlapped SOI structure (TCAD simulation). S/D to gate distance (Lno) is about 20nm.

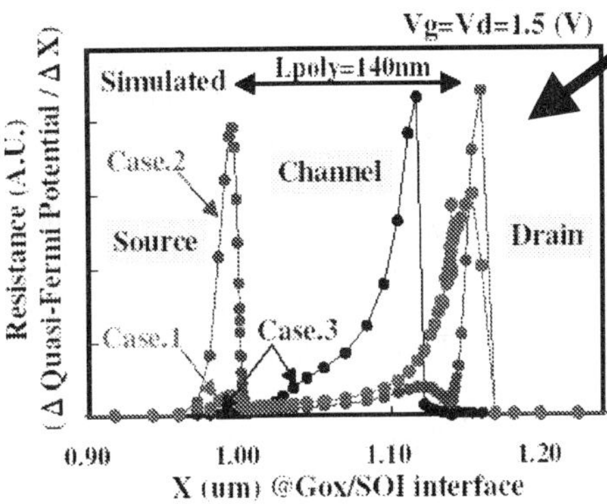

Fig.4 Simulated results of channel resistance (Nfet). Non-overlapped structure can be accepted only with low-concentration body.

Case.1: P+gate, Lno=+20nm, (Non-overlapped) N_{body} 4E15cm^{-3}

Case.2: N+gate, Lno=+20nm, (Non-overlapped) N_{body} 3E18cm^{-3}

Case.3: N+gate, Lno=-20nm, (Overlapped) N_{body} 3E18cm^{-3}

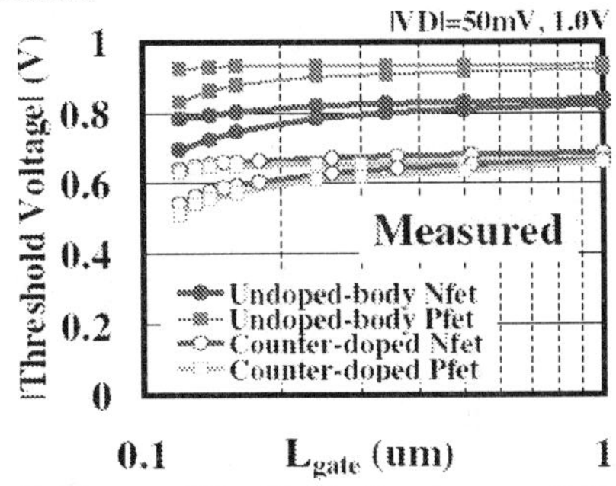

Fig.5 Measured Vt of body-floating type SOI. |Vt| can be well controlled with counter doping.

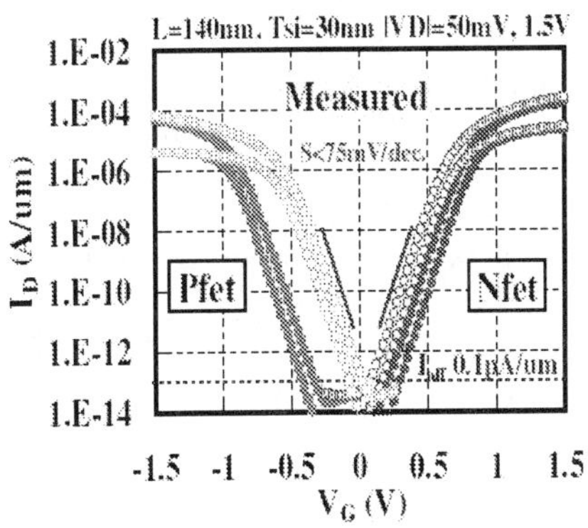

Fig.6 Measured "FBE/GIDL-free" I_DV_G of body-floating type SOI transistors (symbols are same as in Fig.5) .

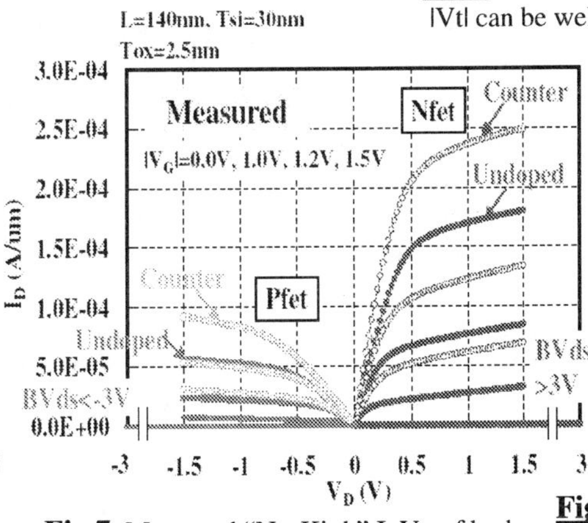

Fig.7 Measured "No-Kink" I_DV_D of body-floating type SOI transistors (symbols are same as in Fig.5) .

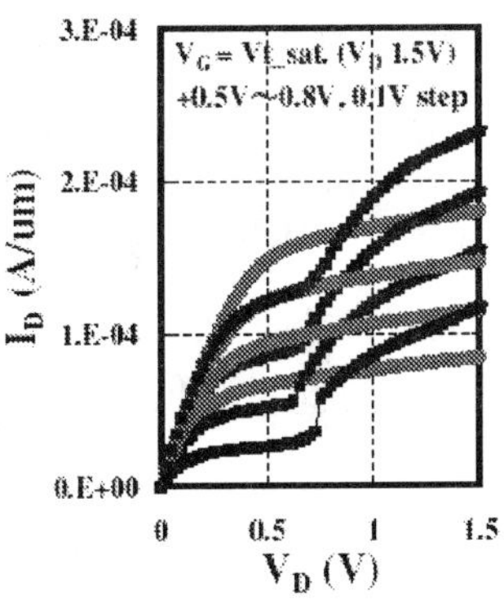

Fig.8 I_DV_D comparison between undoped-body/P+gate (non-overlap) Nfet and highly-doped body/N+gate (overlap) Nfet (L=140nm).

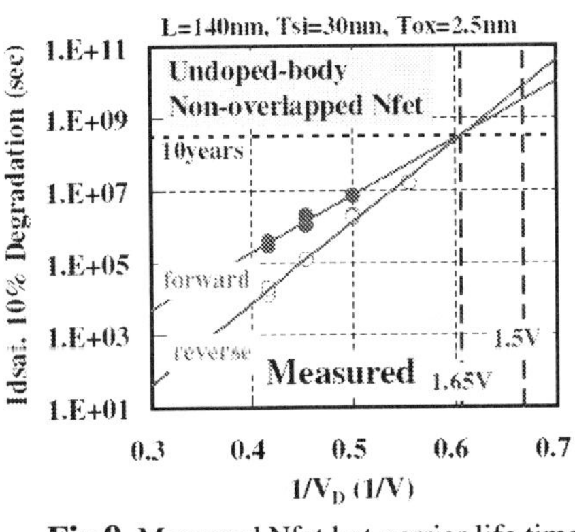

Fig.9 Measured Nfet hot-carrier life time.

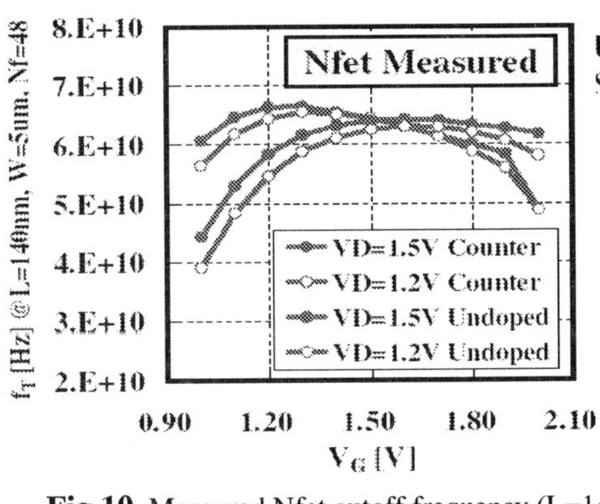

Fig.10 Measured Nfet cutoff frequency (L=140nm: I_{off}<0.1pA/um).

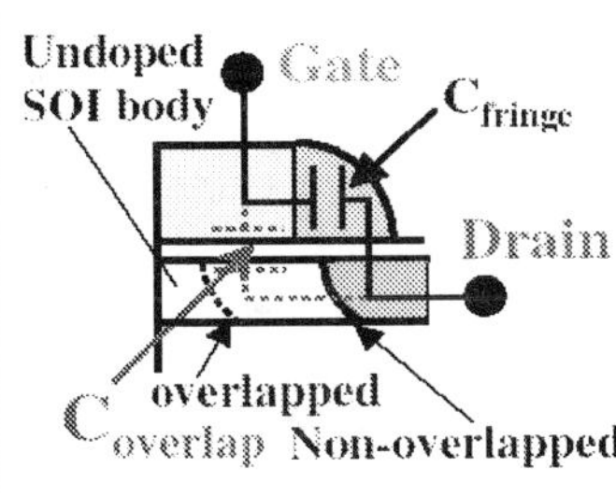

Fig.11 Schematic view of Miller capacitances. Non-overlapped structure has no $C_{overlap}$.

Effect of Contact Liner Stress in High-Performance FDSOI Devices with Ultra-Thin Silicon Channels and 30 nm Gate Lengths

D. V. Singh*, J. M. Hergenrother, J. W. Sleight†, Z. Ren†, H. Nayfeh†, O. Dokumaci†, L. Black**, D. Chidambarrao†, R. Venigalla†, J. Pan**, B. L. Tessier†, A. Nomura**, J. A. Ott, M. Khare†, K. W. Guarini, M. Ieong† and W. Haensch

IBM Semiconductor Research and Development Center, T. J. Watson Research Center, Yorktown Hgts, NY 10598
†*IBM Systems & Technology Group,* ***Advanced Micro Devices, at IBM Semiconductor Research and Development Center, Hopewell Junction, NY 12533*
**Phone: 914-945-3525, Fax: 914-945-2142, email: singhd@us.ibm.com*

Abstract

We have investigated for the first time the effect of stressed contact liners on the performance of fully depleted ultra-thin channel CMOS devices with a raised source/drain. Significant enhancement in mobility and drive current is observed in both nFETs and pFETs. The observed enhancement shows a strong dependence on the Si channel thickness and the height of the raised source/drain, consistent with stress simulations.

Introduction

CMOS technology based on fully depleted (FD) devices provides reduced parasitic capacitance and the potential for improved short-channel control. Poly-silicon gate FDSOI devices have been reported in the past [1-2], however their compatibility with performance enhancing technology elements such as stressed contact liners [3] has yet to be clarified. Simulations indicate that channel stress induced by the contact liner is strongly influenced by both the channel thickness and the source/drain topography. While stress couples more effectively into a thinner channel, the raised source drain region tends to degrade stress transfer. In this work we have investigated the effect of stressed contact liners on high performance ultra-thin SOI (UTSOI) devices with varying channel thickness and raised Source/Drain (RSD) height.

Device Structure

Ultra-thin Si channel MOSFETs with gate lengths of 30nm were fabricated on 300mm <100> bonded SOI wafers. The initial SOI layer was thinned by thermal oxidation to target a final channel thickness, t_{ch}, between 12 and 20nm underneath the gate oxide. Device isolation was achieved using a conventional STI approach. The gate oxide consisted of a nitrided oxide with t_{inv} ~ 18.5Å. Selective epitaxial silicon was grown on the undoped source/drain regions with a SiN disposable spacer encapsulating the gate as shown in Fig. 1(a). A buffer S/D implant and anneal was performed after RSD growth to eliminate resistive pockets between the RSD and the thinner extension regions. The disposable spacer was then selectively removed followed by halo implants, thin offset spacer formation, and extension implants. A final spacer was then formed followed by deep S/D implant, spike RTA and a Ni salicide process. A Si_3N_4 tensile (compressive) stress liner film was deposited over nMOS (pMOS) devices. In the "no stress" case a relaxed liner film was used. A cross-sectional TEM of the final UTSOI device structure is shown in Fig. 2.

Device Characteristics and Discussion

Fig. 3 shows kink-free I_d-V_{ds} curves for nMOS and pMOS devices with t_{ch} = 18nm and L_{poly} = 30nm. $V_{t,sat}$ roll-off and *DIBL vs.* L_{poly} curves are shown in Figs 4 and 5. The devices exhibit excellent short-channel control for L_{poly} down to 25nm. The impact of liner stress on the R_{on} vs. L_{poly} characteristics of nMOS devices with t_{ch} = 18nm and RSD height t_{RSD} = 20nm is shown in Fig. 6. The devices compared are nominally identical (except for the contact liner) and the decrease in dR_{on}/dL_{poly} with liner stress is a measure of mobility enhancement. The corresponding nMOS I_{on}-I_{off} plot is shown in Fig. 7 showing clear enhancement in drive current with tensile stress. Fig. 8 summarizes the enhancement in nMOS I_{on} due to liner stress for various combinations of t_{RSD} and t_{ch}. The measured enhancement is consistent with predictions based on stress simulations using a finite element approach shown in Fig. 9. Figs 10 and 11 shows the impact of compressive liner stress R_{on} *vs.* L_{poly} and I_{on}-I_{off} for pMOS devices. A substantial improvement in drive current is evident. We have also fabricated 101 stage, unloaded FO = 1 ring oscillators (RO) using UTSOI devices. Fig. 12 shows ring oscillator stage delay vs. I_{ddq} plots for the "stress" and "no stress" cases. In the "stress" case the ROs comprised of nFETs with a tensile liner and pFETs with a tensile liner relaxed by Ge implantation.

Conclusions

We have built high-performance 30nm polysilicon-gate UTSOI nFETs and pFETs, and have demonstrated that tensile and compressive stress nitride liners can be used to significantly increase the drive current and mobility in these devices. We have shown that stress transfer to the channel has a strong dependence on the channel thickness and raised S/D height.

[1] B. Doris et al., IEDM Tech. Dig., 2002, p.267
[2] Z. Krivokapic et al., Symp. on VLSI Tech., 2003, p. 13
[3] H. S. Yang et al., IEDM Tech. Dig. , 2004, p. 107

0-7803-9212-4/05/$20.00 ©2005 IEEE

Fig. 1. SEM cross-sections following (a) selective epitaxial growth of RSD, (b) disposable spacer removal and offset spacer formation, and (c) final spacer formation for salicide and pFET deep S/D implants.

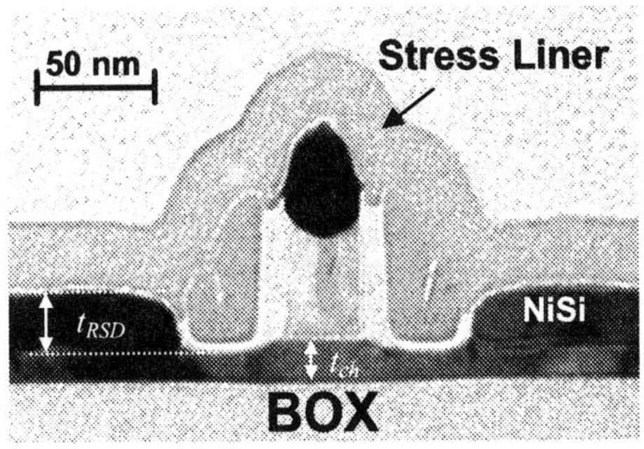

Fig. 2. Cross-sectional TEM image of a completed 30nm device showing 18nm thick body, RSD, NiSi and SiN stress liner.

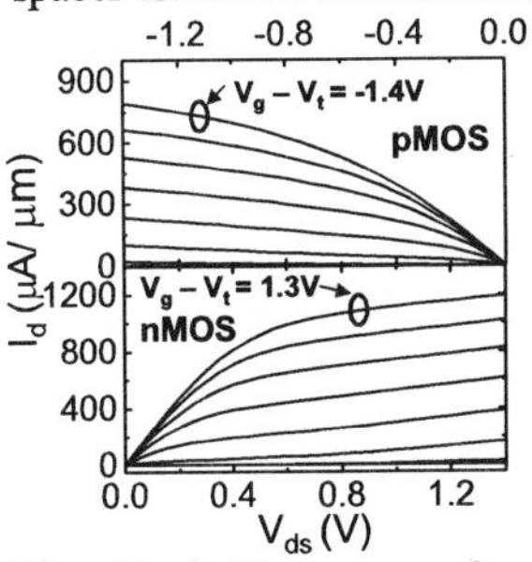

Fig. 3. I_d-V_{ds} curves for nMOS and pMOS devices. ($|\Delta V_{gs}| = 0.2V$)

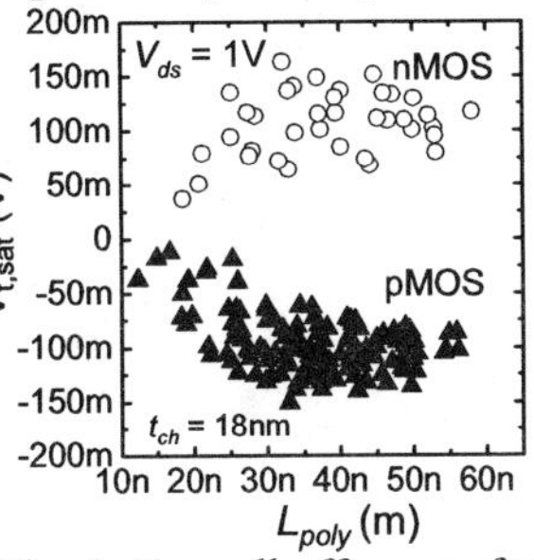

Fig. 4. $V_{t,sat}$ roll-off curves for nMOS and pMOS devices.

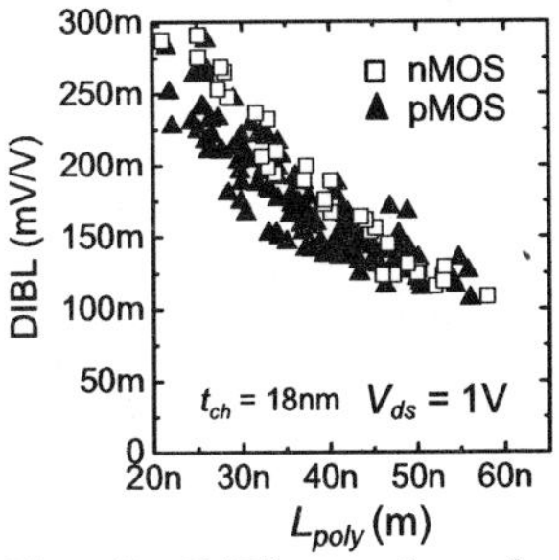

Fig. 5. *DIBL vs.* L_{poly} for nMOS and pMOS devices.

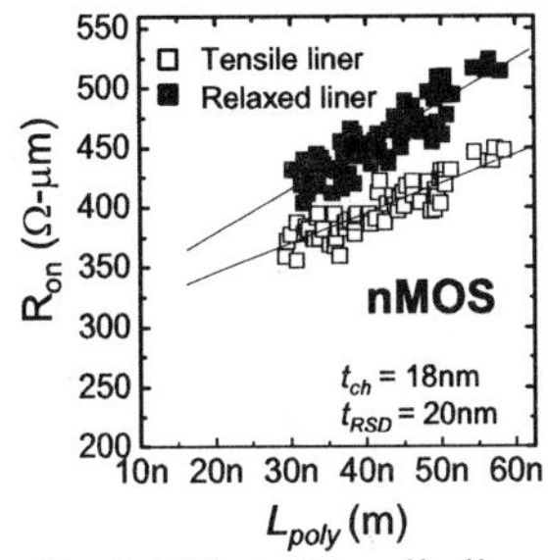

Fig.6. Effect of tensile liner stress on R_{on} vs. L_{poly} (V_{ds} = 50mV, $V_{gs} - V_t$ = 0.7V)

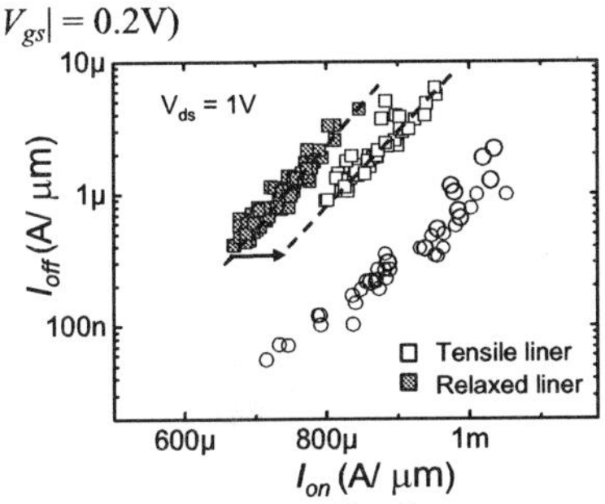

Fig. 7. Impact of tensile liner stress on nMOS I_{on}-I_{off} characteristics (squares) for t_{ch} = 18nm and t_{RSD} = 20nm. Also shown are I_{on}-I_{off} characteristics for more optimal UTSOI devices (circles) fabricated with t_{ch} = 18nm and tensile stress liner.

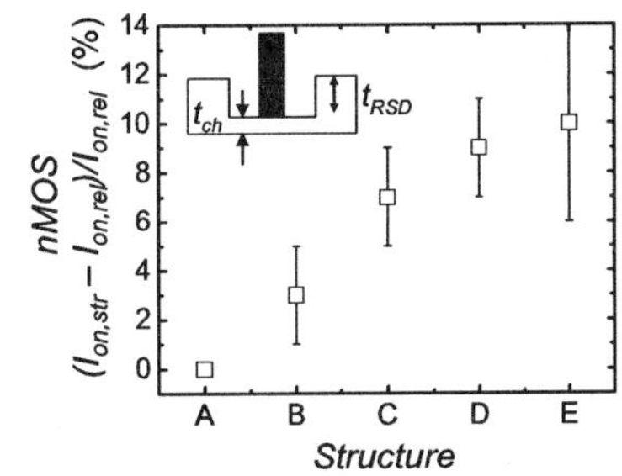

Fig. 8. Summary of improvement in I_{on} due to tensile liner stress for different device structures: A) t_{ch} =14nm, t_{RSD} =36nm, B) t_{ch} =18nm, t_{RSD} =28nm, C) t_{ch} =15nm, t_{RSD} =25nm, D) t_{ch} =18nm, t_{RSD} =20nm and E) t_{ch} =60nm, t_{RSD} = 0

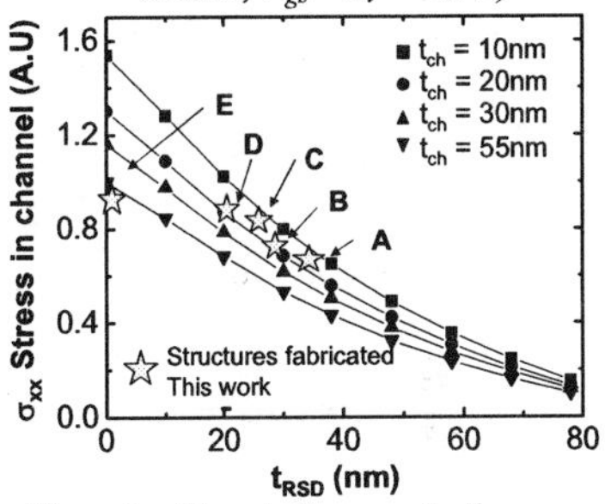

Fig. 9. Simulations of the σ_{xx} component of stress at the mid-point of the channel as a function of t_{ch} and t_{RSD}. The various cases studied in this work are indicated.

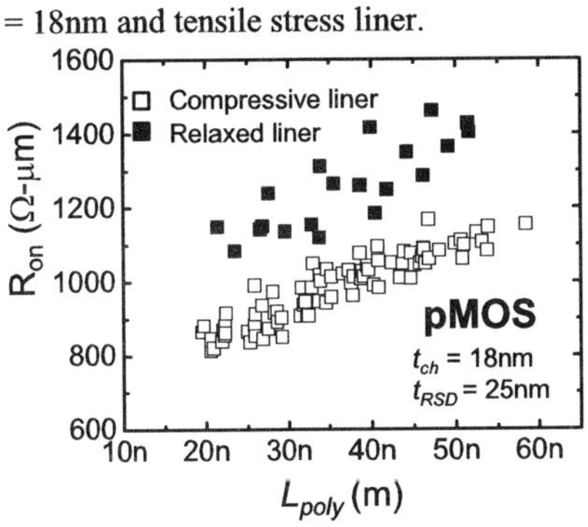

Fig. 10. Effect of compressive stress liner on R_{on} vs. L_{poly} for a pMOS device. (V_{ds} = -50mV, $V_{gs} - V_t$ = -0.7V)

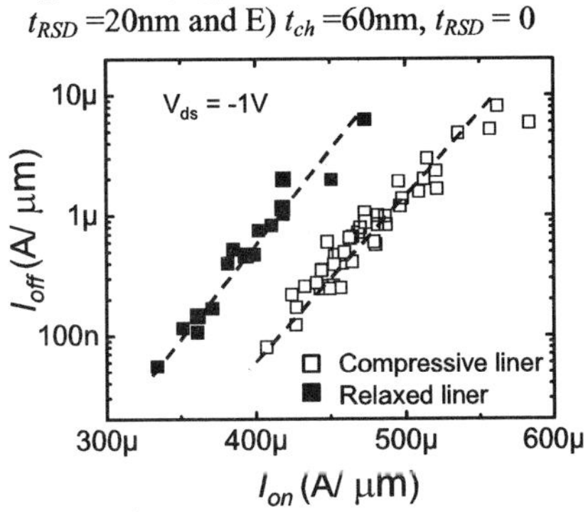

Fig. 11. Impact of compressive liner stress on pMOS I_{on}-I_{off} characteristics for t_{ch} = 18nm and t_{RSD} = 25nm.

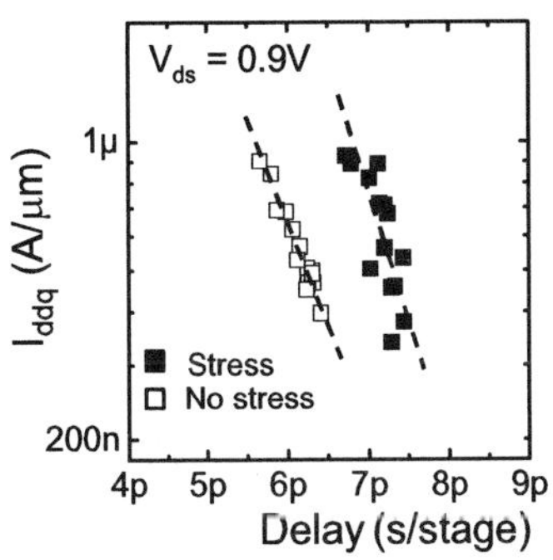

Fig. 12. Impact of liner stress on ring oscillator performance.

Impact of BOX Scaling on 30 nm Gate Length FD SOI MOSFETs

M. Fujiwara, T. Morooka, N. Yasutake, K. Ohuchi, N. Aoki, H. Tanimoto
M. Kondo, K. Miyano*, S. Inaba, K. Ishimaru, and H. Ishiuchi

SoC Research & Development Center, *Process & Manufacturing Engineering Center
Toshiba Corporation Semiconductor Company
8 Shinsugita-cho Isogo-ku Yokohama, Kanagawa 235-8522 Japan
Tel: +81-45-770-3638 Fax: +81-45-770-3571 Email: makoto.fujiwara@toshiba.co.jp

Introduction

Fully-depleted (FD) SOI MOSFETs have been considered as a promising candidate to extend planar CMOS scaling. BOX thickness (T_{BOX}) of FD devices is expected to be scaled with gate length (L_g) in order to suppress short-channel effects [1]. It has been reported that thin BOX is effective in adjusting threshold voltage (V_t) by means of back-gate or substrate bias [2], [3]. However, systematic study of BOX scaling and trade-offs focusing on the short-channel characteristics has not been experimentally carried out yet. This paper presents the first demonstration of ultra-thin BOX FD SOI devices with nominal gate length of 30 nm. The characteristics of FD SOI MOSFETs are investigated in detail as T_{BOX} is varied from 5 nm to 145 nm. In addition, optimum design regions of T_{BOX} for achieving performance requirements are demonstrated.

Device Fabrication

FD SOI CMOS transistors with T_{BOX} of 5, 25, 50 and 145 nm were fabricated. TEM micrograph of a MOSFET with 5 nm thick BOX is shown in Fig. 1. Channel regions were thinned down to about 10 nm by thermal oxidation. After shallow trench isolation formation and channel doping, plasma-nitrided gate oxides (EOT ~ 1 nm) were grown, followed by gate poly-Si/hard mask deposition and patterning. Elevated source/drain (S/D) extension and elevated S/D structures were employed to reduce parasitic resistance (Fig. 2). In order to eliminate agglomeration of ultra-thin Si, process conditions of hydrogen pre-cleaning and selective Si epitaxy were carefully optimized. Halo and S/D extension were implanted after elevated S/D extension formation, which alleviates channel thickness scaling in controlling short-channel effects. The gate hard mask was stripped prior to Ni salicide process. All wafers received the same process sequence to clarify the effects of T_{BOX} on device performance.

Results and Discussion

A. Threshold Voltage Adjustment

As shown in Fig. 3, the V_t roll-off is considerably altered by T_{BOX}. Furthermore, the V_t shift is observed especially at T_{BOX} = 5 nm even when the substrate voltage is fixed at ground. It should be noted that the significant difference in V_t cannot be accounted for by the reduced short-channel effects in thin BOX devices. Fig. 4 shows that 5 nm BOX MOSFETs achieve excellent cutoff characteristics down to L_g = 30 nm. It is evident in Fig. 5 that the substrate doping concentration at the BOX/substrate interface (N_{sub}) increases for short-channel devices with 5 nm and 25 nm BOX, which is attributed to dopant penetration by channel and angled halo implantation through the thin BOX. Fig. 6 compares the simulated V_t shift for different N_{sub}. It is found that 5 nm BOX exhibits the V_t shift of about 0.2 V at high N_{sub}, indicating that the V_t shift in

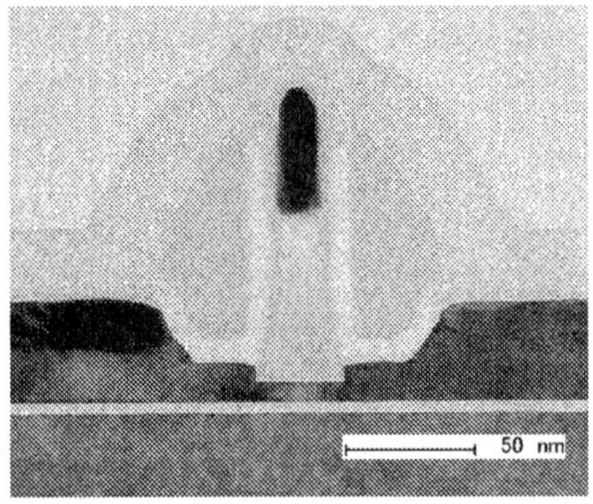

Fig. 1 TEM micrograph of an FD SOI MOSFET with 5 nm thick BOX.

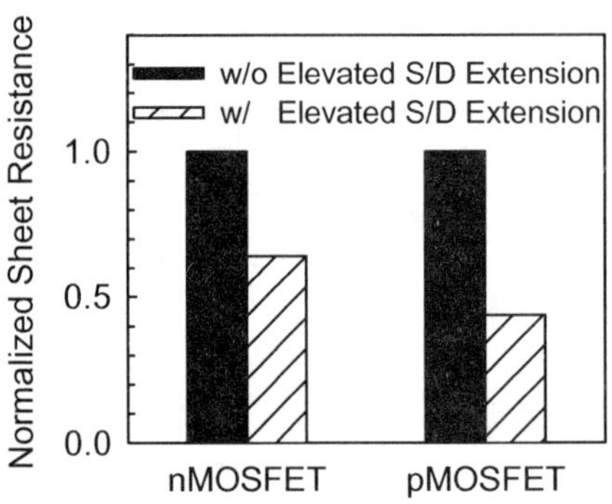

Fig. 2 Comparison of normalized S/D extension sheet resistance.

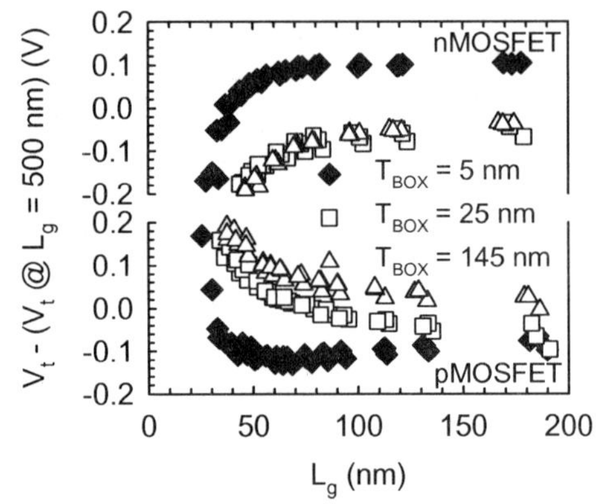

Fig. 3 V_t roll-off for different T_{BOX}. The substrate voltage is 0 V. $|V_d|$ = 0.9 V.

0-7803-9212-4/05/$20.00 ©2005 IEEE

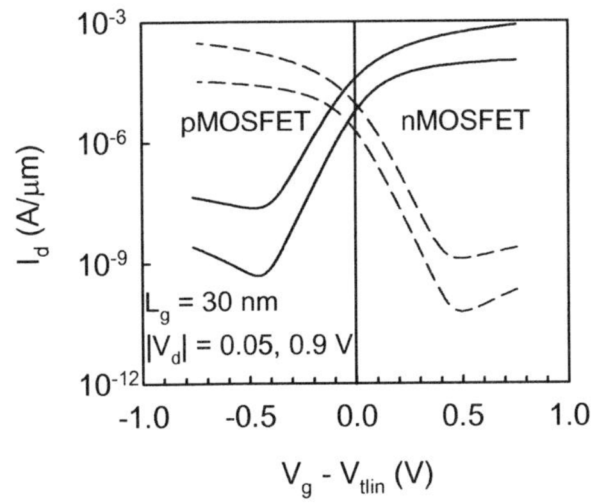

Fig. 4 I_d-V_g characteristics for 30 nm gate length MOSFETs with T_{BOX} = 5 nm. $|V_d|$ = 0.05, 0.9 V. V_{tlin} is V_t at $|V_d|$ = 0.05 V.

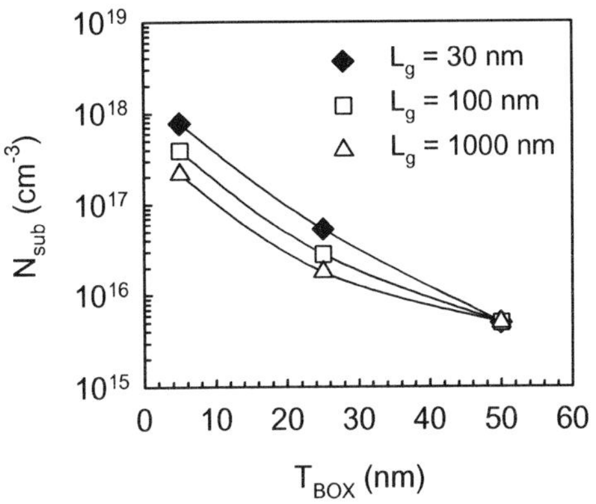

Fig. 5 Simulated substrate doping concentration at the BOX/substrate interface.

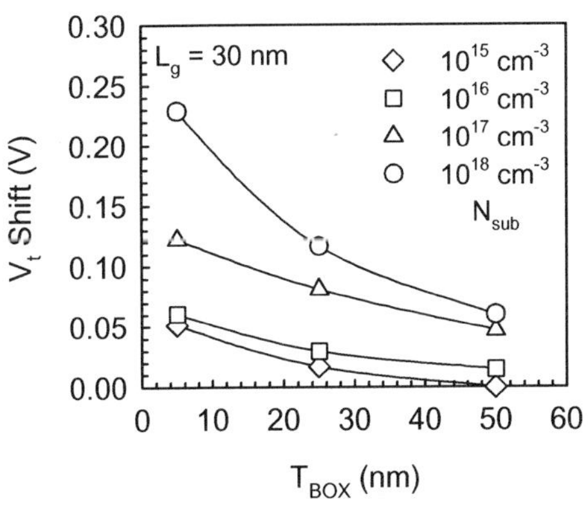

Fig. 6 Simulated V_t shift versus T_{BOX}. L_g = 30 nm. V_d = 0.9 V.

Fig. 3 is caused by the difference in N_{sub}. These results suggest that by tailoring of channel and halo profiles, both V_t and V_t roll-off can be controlled by the appropriate combination of T_{BOX} and N_{sub}, in addition to the workfunction of the gate material.

B. Subthreshold Slope

Tight coupling of the gate-to-channel potential in an FD device provides a steep subthreshold slope (S), which is one of the advantages over a bulk device. Fig. 7 indicates that BOX

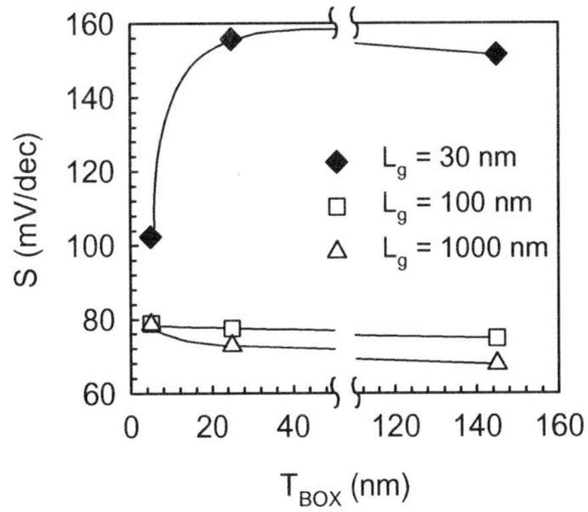

Fig. 7 S versus T_{BOX}. V_d = 0.9 V.

scaling leads to degraded S in the absence of short-channel effects because of an increase in BOX capacitance [4]. At L_g = 30 nm, however, reduced short channel effects at T_{BOX} = 5 nm results in substantially improved S. This is an important advantage in ultra-thin BOX MOSFETs, which allows for a larger gate overdrive for the same off-current and power supply voltage.

C. Drive Current

Fig. 8 shows that in long-channel devices, the drain current (I_d) at a constant gate overdrive decreases as T_{BOX} is reduced. Ultra-thin BOX in conjunction with high N_{sub} results in higher normal electric field in the channel region compared to thick BOX with low N_{sub} (Fig. 9), thereby degrading the transconductance (g_m) at long channel as illustrated in Fig. 10. Meanwhile, Fig. 8 additionally reveals that I_d becomes less sensitive to T_{BOX} in short-channel devices. This comparable I_d at short channel can be interpreted as being due to the effects of S/D parasitic resistance and reduced self-heating effects in thinner BOX.

D. Inverter Delay

Increased junction capacitance resulting from BOX scaling has the potential of reducing the device speed, which should be the restricting parameter to determine the minimum T_{BOX}. Fig. 11 shows the propagation delay (T_{pd}) of unloaded inverters as a metric for device performance. It is found that the delay degradation at T_{BOX} = 5 nm is about 10% at a constant drive current. Considering realistic digital circuits, however, interconnect capacitances contribute to the total capacitive loads, which relaxes the junction capacitance restrictions and allows further scaling of T_{BOX}.

E. Design Space of BOX Thickness

Taking into account the short-channel design considerations, the design space of T_{BOX} is summarized in Fig. 12. The device speed specifications constrain the lower limit of T_{BOX}, where the improvement of the electrostatic integrity should also be considered. Decreasing T_{BOX} significantly improves S. As a result, for the same off-current, V_t can be set lower in ultra-thin BOX, providing more gate overdrive and hence higher on-current. This increases the speed of ultra-thin BOX devices and pushes the minimum limitation of T_{BOX}, leading to wider range of acceptable T_{BOX}. In the mean time, the V_t range from varying N_{sub} places the higher bound for T_{BOX} since the

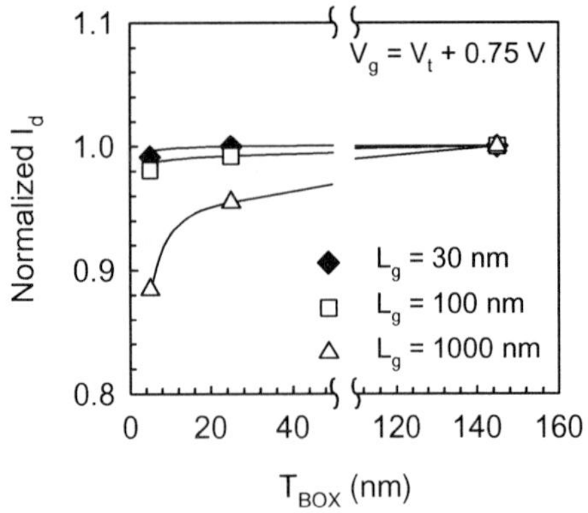

Fig. 8 Normalized I_d at $V_g = V_t + 0.75$ V as a function of T_{BOX}. $V_d = 0.9$ V.

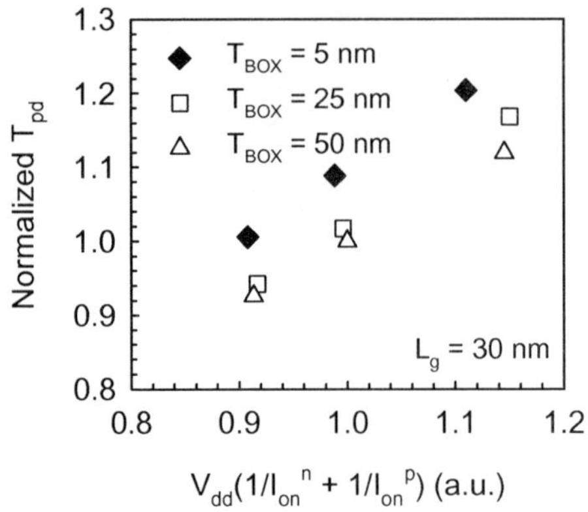

Fig. 11 Normalized T_{pd} of unloaded inverters with fan-out of 1. $L_g = 30$ nm.

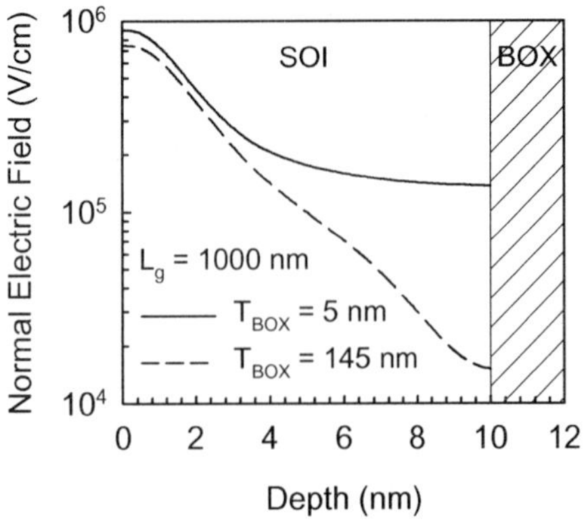

Fig. 9 Simulated normal electric field in the channel region at the same gate overdrive. $L_g = 1000$ nm.

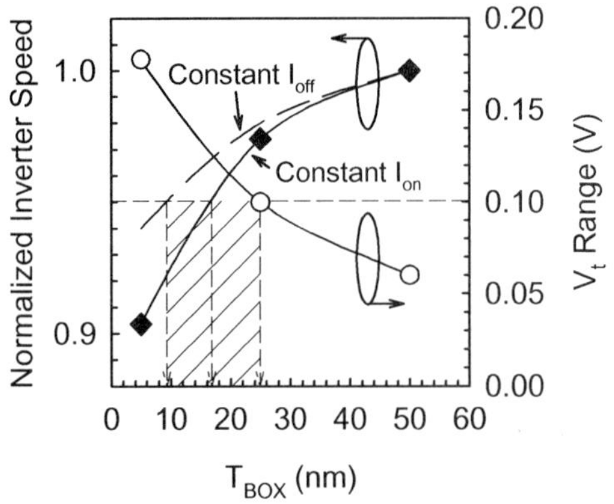

Fig. 12 Design space of T_{BOX}. The shaded region represents T_{BOX} that satisfies criteria of speed degradation ≤ 5% and V_t range ≥ 0.1 V.

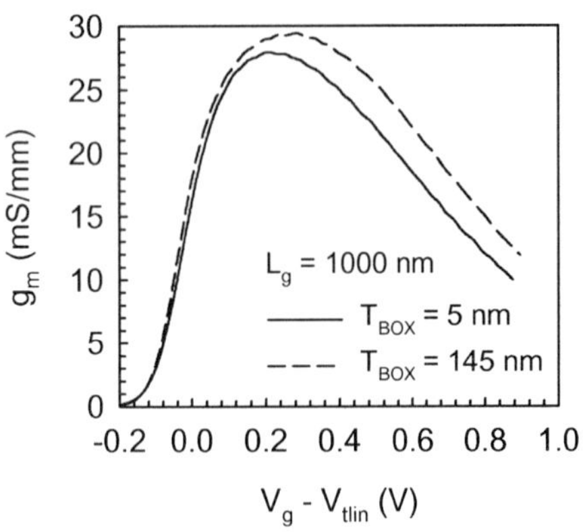

Fig. 10 Linear g_m for $L_g = 1000$ nm. $V_d = 0.05$ V. V_{tlin} is V_t at $V_d = 0.05$ V.

V_t roll-off control and multiple V_t on a chip are also the key factors in device design. At the higher limit of T_{BOX}, the V_t tuning is traded against the device speed, and additional performance enhancement will be achieved by careful tailoring of substrate and halo doping profiles to control the short-channel effect. The optimum T_{BOX} should be determined from performance tradeoffs between the short-channel control and the junction capacitance, depending on specific requirements of the application.

Conclusion

The characteristics of FD SOI devices with varying T_{BOX} are systematically investigated, and the impact of BOX scaling on short-channel device characteristics is identified. It is found that N_{sub} modulates V_t of ultra-thin BOX devices at a constant substrate voltage, indicating that both V_t and V_t roll-off can be adjusted by choosing the appropriate set of T_{BOX} and N_{sub}. BOX scaling coupled with N_{sub} tuning reduces short-channel effects and thus drastically steepens S, leading to higher drive current for the same off-current. This improves the device speed for ultra-thin BOX and allows for lower T_{BOX}. The V_t range and device speed requirements place the boundaries on the T_{BOX} design space. Careful design of substrate doping profiles is essential to extract maximum performance from ultra-thin BOX FD SOI MOSFETs.

References

[1] *ITRS*, 2004.
[2] T. Ohtou, et al., *Ext. Abs. SSDM*, p. 502, 2004.
[3] R. Tsuchiya, et al., *IEDM Tech. Dig.*, p. 631, 2004.
[4] T. Numata, et al., *IEEE Trans. Electron Devices*, p. 2161, 2004.

2005 IEEE INTERNATIONAL
SOI CONFERENCE

Session #8

RADIATIONS AND OPTICS

CO-CHAIRS

Les Palkuti
DTRA/TDNR

Mike Liu
Honeywell SSEC

RADIATIONS AND OPTICS Session 8

Total Dose Performance of Conventional Static and Dynamic Circuits In a Radhard 0.18-μm FDSOI Process*

P. Gouker, B. Tyrrell, P. Wyatt, E. Austin, A. Soares, C.K. Chen and J. Burns
Lincoln Laboratory, Massachusetts Institute of Technology, Lexington, MA 02420-9108
Email: pgouker@LL.mit.edu, Tel 781-981-0460

Abstract

Static and dynamic circuits were fabricated in the MIT-LL 0.18-μm FDSOI CMOS process, and exhibited a high tolerance to total dose radiation up to 1 Mrad (SiO_2). Circuits were designed using conventional design rules and layout techniques, *i.e.*, they are not radhard-by-design. Hardening was done at the circuit fabrication level using process enhancements. These are the first circuit-level hardness results reported to date for these new enhancements.

Introduction

Radiation-induced charge trapping in isolation oxides, *i.e.*, field, trench and buried oxides (BOX) causes large threshold voltage shifts of n and p-channel FETs [1], hence circuit failure typically at total dose below 30 krad (SiO_2). In FDSOI the main effect of total dose irradiation is the threshold shift due to charge trapping in the BOX because the front and back oxides are capacitively coupled. Treating the SOI wafers before CMOS fabrication [2] or removing the substrate below the BOX and thinning the BOX [3] reduce the radiation-induced threshold voltage shift and offstate leakage of n-channel FETs (nFETs). In this work, we show the first total dose results for static multiplier and dynamic shift-register circuits fabricated with these process enhancements on MIT-LL 0.18-μm FDSOI process. We show a direct correlation between total dose effects in the circuits and in nFETs.

Circuit Fabrication and Testing

Multiplier and shift-register circuits were fabricated on standard Unibond™ wafers and on SOI wafers with a treatment in the BOX [2], and on standard wafers after substrate removal below the BOX. The substrate below the BOX was removed on some wafers after circuit fabrication by bonding the wafer to a carrier wafer using a low-temperature oxide bond [3]. Electrical contact to metal bondpads was done through the BOX side. These fabrication splits are summarized in Figure 1. The multiplier is a static arithmetic circuit with 3,500 transistors, two 8-bit inputs, and 50-ohm output drivers. Figure 2 shows a schematic of a one stage shift register, which is a building block of a dynamic finite impulse response filter used in signal processing for radars.

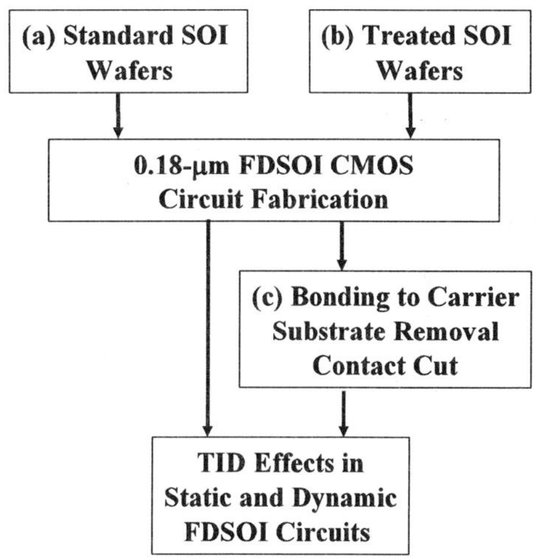

Figure 1: Fabrication processes enabling total dose hardening of MIT-LL 0.18-μm FDSOI circuits: (a) standard SOI wafers (b), versus treated SOI wafers with or without (c) removal of the substrate below the BOX.

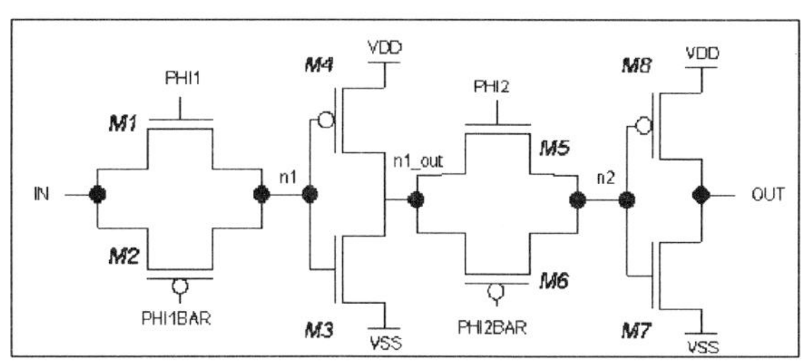

Figure 2: One stage of a dynamic shift register circuit fabricated in MIT-LL 0.18-μm FDSOI process.

*This work was sponsored by the Defense Threat Reduction Agency under Air Force Contract # FA8721-05-C-0002. Opinions, interpretations, conclusions, and recommendations are those of the authors and are not necessarily endorsed by the United States Government.

The circuits were irradiated with a 10-keV X-ray source on an Aracor 4100. During irradiation, the circuits were biased to simulate worst case. Multipliers had their inputs grounded and outputs floating; the main circuit (V_{DD}) and the input/output circuitry (V_{DDIO}) power supplies were at 1.5 V. Shift-registers had all inputs clocking at 1 MHz with a digital square wave input, and V_{DD} at 1.5 V. The circuits were irradiated in incremental doses up to 1 Mrad (SiO_2). Between dose increments, pin continuity and functionality tests were run on an Agilent 93000.

Total Dose Results

All the multiplier circuits remained functional at high-frequency (same as pre-radiation) up to 1 Mrad, however, the standby leakage currents on V_{DD} and V_{DDIO} were noticeably dependent on the circuit fabrication split. Figure 3 shows the radiation-induced standby leakage current on V_{DD} and V_{DDIO} up to 1 Mrad. This current correlates very closely with the leakage of a single nFET shown in Figure 4. This suggests that the only important total dose effect in these static circuits is the negative threshold shift of nFETs due to charge trapping in the BOX [3].

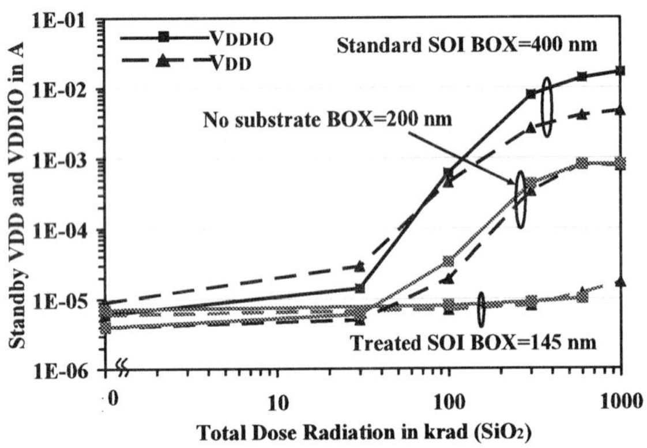

Figure 3: Standby leakage current of 0.18-μm FDSOI multiplier circuits fabricated on standard SOI, on treated SOI wafers or standard wafer after substrate removal.

Dynamic circuits behaved differently, as shown in the shmoo plots of Figure 5. Each plot consists of over 2,500 functionality tests with the clock frequency varying from 1 to 600 MHz (x-axis) and V_{DD} varying from 0.3 to 1.5 V (y-axis). Each point is light gray if the test passes both walking "1" and walking "11", otherwise it is black (failing). If V_{DD} is too low or the clock speed is too high, each stage cannot drive the next stage, and the test fails. If the clock speed is too low, charge cannot be stored long enough on the floating node so circuit failure can be caused by any one leaky nFET.

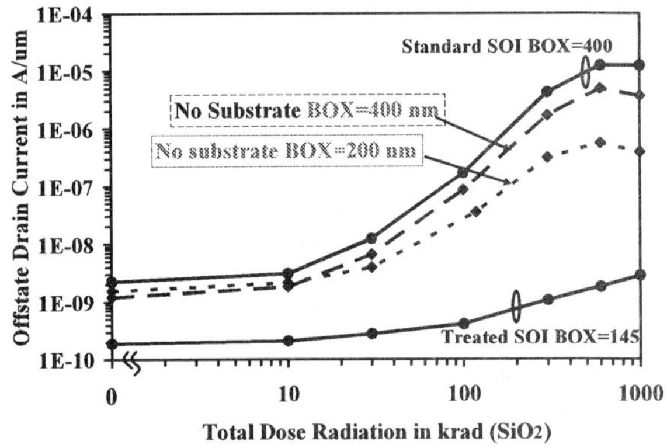

Figure 4: Offstate drain leakage current of W=0.6–μm, L=0.18–μm FDSOI nFETs, the dimensions of most transistors in the multipliers and shift registers. For the circuits with no substrate, the multipliers were on a wafer with a BOX=200 nm; the shift registers were on a wafer with a BOX=400 nm. With no substrate, the radiation-induced offstate leakage of the nFETs increases with the BOX thickness, which is consistent with our earlier results [3].

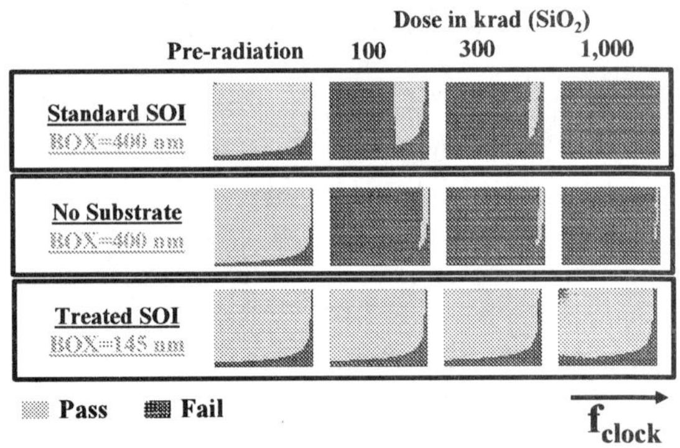

Figure 5: Functionality shmoo plots for 0.18-μm FDSOI dynamic shift-registers on standard wafer, on treated SOI wafers or standard wafer after substrate removal with a clock frequency from 1 to 600 MHz and V_{DD} supply from 0.3 to 1.5V up to 1 Mrad (SiO_2). The circuit passes the test of walking "1" and "11" in the light gray regions. The circuit fails in black regions.

The operating space barely changes up to 1 Mrad (SiO_2) on treated wafers, while it is drastically reduced starting at 100 krad on standard wafers and standard wafers with no substrate due to the rapid increase in the nFET leakage for doses above 30 krad in Figure 4. As discussed earlier [3], nFETs with no substrate are quite sensitive to total dose effects unless the BOX is thinner than 100 nm. The effect of radiation on the p-channel FETs had no noticeable effects on the circuit performance.

Summary

The total dose tolerance of 0.18-μm FDSOI static and dynamic circuits can be enhanced by making simple changes to the fabrication process. Dynamic circuits are more sensitive to total dose effects because only one leaky n-channel transistor can cause the circuit to fail. Static circuits are not as likely to fail at high total dose, though they exhibit severe parametric shifts. Using treated SOI wafers for the circuit fabrication reduces dramatically the radiation-induced standby leakage current of static multipliers, and nearly eliminates the radiation-induced reduction of operating space of dynamic shift registers up to 1 Mrad (SiO_2).

References

[1] J.R. Shwank et al., IEEE Trans. Nucl. Sci., Vol. 50, No. 3, pp. 522–538, June 2003.

[2] M. Alles et al., IEEE SOI Conference, 2001 Proceedings, pp. 131–132, Oct. 2001.

[3] P. Gouker et al., IEEE Trans. Nucl. Sci., Vol. 50, No. 6, pp. 1776–1783, Dec. 2003.

2005 IEEE International SOI Conference

150nm SOI Embedded SRAMs With Very Low SER

David K. Nelson, Harry Liu, Keith Golke, Anuj Kohli
Honeywell DSES Plymouth, Plymouth MN 55441

Abstract

A split word line design technique that improves the soft error rate (SER) of high performance 150nm SOI embedded SRAMs is presented along with SER results.

Introduction

When an energetic particle passes through silicon, it generates excess electron hole pairs which leads to voltage transients on circuit nodes. If the voltage transient disturbs a node which is storing information, it can lead to an upset. This effect is also known as single event upset or SEU. SOI processes significantly improve this situation by cutting off the charge collection along the ion track at the buried oxide [1]. In many space applications, SEU performance is specified as soft error rate in Adam's 90% worst case environment. This is an extremely harsh environment consisting of heavy ions, protons and neutrons. A commercial standard for specifying SER is New York City (NYC) environment which is quite mild in comparison and consists mainly of neutrons. For example, an SRAM that has an SER of 1E-11 errors/bit-day (417 FITS/Mb) in Adam's 90% worst case environment will be virtually SEU immune in NYC environment.

A technique to increase SEU hardness in a memory cell is to insert a resistance which creates a delay element in the feedback loop of the two cross-coupled inverters. However, the delay element increases the write time. To eliminate the write time increase, the resistance in the delay element can be bypassed with a transistor controlled by a word-line. This is called an active delay element.

A previous design with this active delay element used one word-line to control the bit-line pass gates as well as the active delay element during read or write, as shown in Figure 1(a). When the word-line (WL) is low, the memory cell is not accessed through the bit-line pass gates and the active delay element is put into a non-bypass mode resulting in maximum SEU hardness. When WL is high, the memory cell is accessed through the bit-line pass gates and the active delay element is put into a bypass mode resulting in a significant reduction of SEU hardness. This approach is area efficient, and test results have shown that this configuration provides extremely good SEU performance in non-bypass mode [2]. However, this SRAM memory cell's SEU susceptibility is significantly increased during a write or read operation when its active delay element is put into bypass mode. As a result, the dynamic SEU performance of the SRAM is degraded with increasing operating frequency and number of words in a memory cell row.

This paper describes a design approach which addresses this problem and presents SEU test results with corresponding SER calculations.

Design Approach

The design approach involves using separate word-lines to control the active delay element independent from the bit-line pass gates. The separation of the word lines is shown in Figure 1(b). The WL is activated during a read or write while write word-line (WWL) is activated only during a write. Now, during a read operation the active delay element is maintained in a non-bypassed mode, in both accessed and non-accessed memory cells, keeping them in a maximized SEU protected state. However, this is not the case during a write when a physical row in the memory cell array contains memory cells from multiple words. The number of words depends on the size of the column multiplexer (mux) used. For example, Figure 2(a) shows a memory array with a 4:1 column mux. This creates 4 words in a physical row. Figure 2(b) shows the same example but with separate word-lines. During a write, all words are accessed in the row but only 1 of the 4 words is being written. The data state in the memory cells of the the other 3 words must be protected from SEU. However, this is not the case since the active delay element is in a bypass mode. If the SRAM is in a write state for a significant percentage of the time and the bypass mode is significantly more vulnerable to SEU than the non-bypass mode, then the resulting average SER will be dominated by the bypass mode.

Pulsing the write word-line improves the SEU performance of the system when it operates at low frequencies since it reduces the time the memory cell is in the bypass mode. However, when operating at high frequencies and high write duty cycle rates, the bypass mode time increases and the average SER may become dominated by the SRAM memory cells in a bypass mode.

Figure 2(c) shows a concept to further reduce the SER by splitting the common WWL line into multiple WWL lines so that each word is individually accessed within the physical row. Thus, in contrast to a design where a single word line controls every active delay element in the same physical row, as shown in Figure 2(b), the active delay element is in bypass mode only on memory cells being written. When a memory cell is being written, the stored data is being deliberately changed thus the active delay element is not necessary for SEU hardening. This technique is called split word-line write control.

When an SRAM is in a standby state, memory cells are not being accessed. SER contribution occurs solely from memory cells in non-bypass mode. This is defined as the static SER state. When an SRAM is performing a read or a write operation, memory cells are being accessed. SER contribution occurs from memory cells in the non-bypass mode plus memory cells in the bypass mode not being written. The number of bypass mode memory cells is dependent on the word-line configuration per Figure 2 (a-c). This is defined as the dynamic SER state. The split word-line write control has area and associated speed penalties, so this architecture is only used when the static SER (SER_static/bit) is satisfactory, but the dynamic SER (SER_dynamic/bit) is not. The dynamic SER can be approximated with the following equation:

SER_dynamic/bit =
[(SER_static/bit) * Num_bits + (SER_bypass/bit) * Num_bits_word * (Col_mux_size - 1) * (Bypass_factor * WL_pulsewidth/clock_period)] / Num_bits (1)

Equation Variable Definitions:

SER_static	= Memory cell SER non-bypass mode
Num_bits	= Number of memory cells (i.e., bits)
SER_bypass	= Memory cell SER bypass mode
Num_bits_word	= Number of bits in the word
Col_mux_size	= Column mux size
WL_pulsewidth	= Word-line pulse width
Clock_period	= SRAM cycle time
Bypass_factor	= 1.0* for Figure 2(a),
	= 0.5* for Figure 2 (b),
	= 0.0* for Figure 2(c)
	*(Assumes 50/50 read/write operation)

Test Results

A library of 3ns - 5ns access time 2Kb – 512Kb size SRAM macros using this technique was implemented on Honeywell's 150 nm SOI technology.

Heavy Ion and Proton SEU testing was performed on selected SRAM macros. Heavy Ion testing was performed in March 2005 at Lawrence Berkeley National Labs (LNBL) 88 inch cyclotron using a variety of ions. Figure 3(a) shows the worst case (1.95V, 25°C) heavy ion test results in non-bypass mode. The resulting SER is $<6\times10^{-20}$ errors/bit-day in Adam's 90% worst case environment. Figure 3(b) shows the worst case condition (1.95V, 25°C) heavy ion test results in bypass mode. The resulting SER is $<3\times10^{-8}$ errors/bit-day in Adam's 90% worst case environment. Proton testing was performed in April 2005 at Indiana University's proton accelerator with proton energies of 24, 50, 100 and 200 MeV. Proton testing in non-bypass mode produced only one upset after exposing 1Mb worth SRAM memory cells with a proton fluence of $5\text{x}10^{13}$ protons/cm^2. This results in a SER $<8.6\times10^{-12}$ errors/bit-day in an extreme proton environment orbit (inclination = 63deg, appogee = 2058km, perigree = 2058km). Proton testing in bypass mode produced 3442 upsets after exposing 64K worth of SRAM memory cells with a proton fluence of $5\text{x}10^{13}$ protons/cm^2. This results in an SER of $<9.3\times10^{-7}$ errors/bit-day in an extreme proton environment orbit.

The dynamic SER performance of SRAM macros can be calculated using Equation 1. Example calculations are shown in Table I with the following assumptions: 512×32 SRAM macro, 50/50 read/write duty cycle, clock period of 5ns, word-line pulse of 3ns, and 4:1 column mux. The dynamic SERs of Figure 2(b) exhibited marginal improvement over the Figure 2(a) design. The dynamic SERs of Figure 2(c) design are orders of magnitude better than Figure 2(a-b) designs in both Adam's 90% worst case and extreme proton environments.

Summary

An SEU hardened SRAM memory cell with active delay element and split word lines has been presented which significantly reduces dynamic SER in heavy ion and proton environments.

Acknowledgments

This material is based upon work supported by Defense Threat Reduction Agency under Contract No. DTRA01-03-D-0018, Delivery Order 0001.
Any opinions, findings and conclusions or recommendations expressed in this material are those of the author(s) and do not necessarily reflect the views of Defense Threat Reduction Agency.

Reference

[1]K. W. Golke, IEEE Tran. Nuclear Sci., NS-40, 1910 (1993).
[2] H. Y. Liu, D. K. Nelson, K. W. Golke and S. T. Liu, "SEU Hardened 0.15μm SOI SRAMs," 2005 HEART Conference

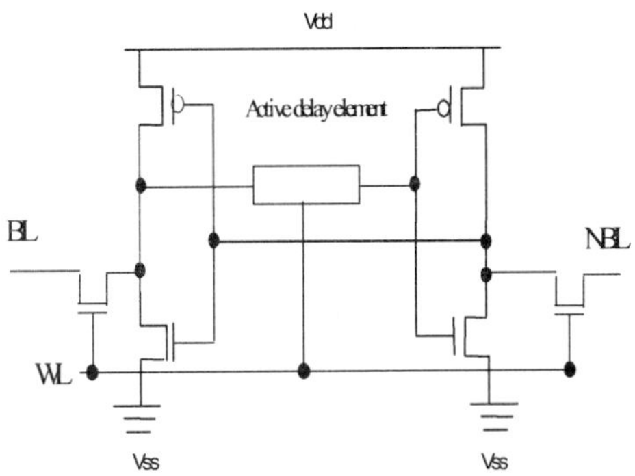

Figure 1 (a)

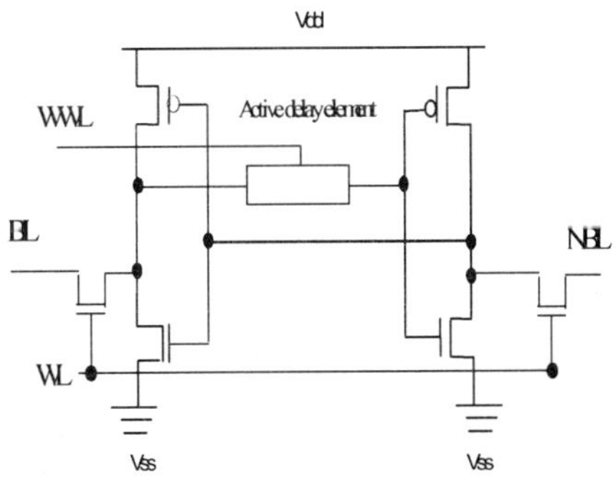

Figure 1 (b)

Figure 1. SEU hardened memory cell with active delay element, using a.) common word-line; b.) separate write word-line

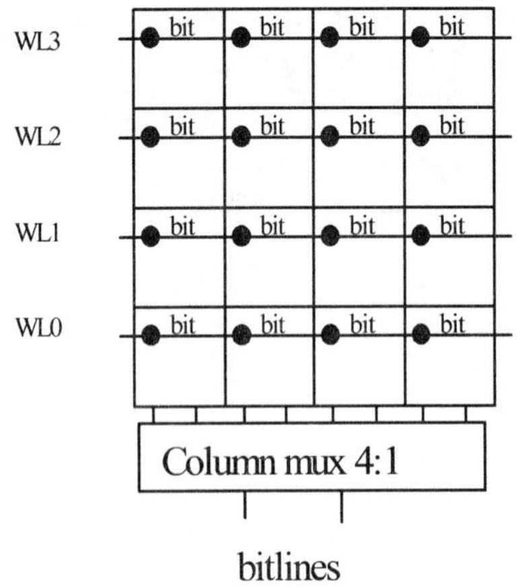

Figure 2(a)

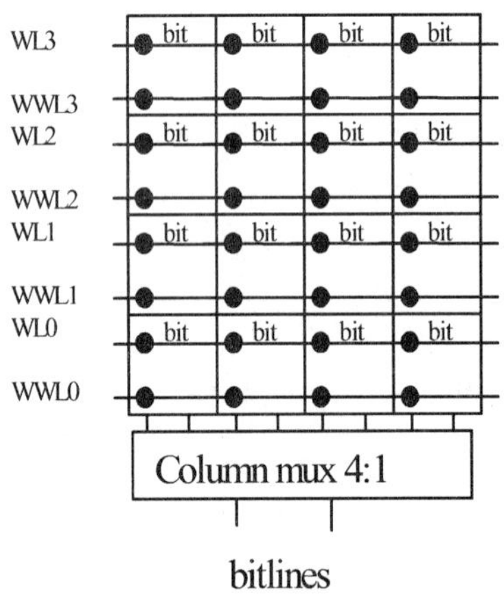

Figure 2 (b)

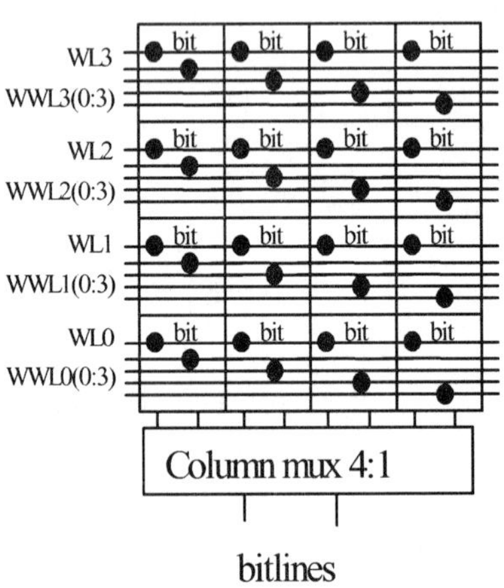

Figure 2 (c)

Figure 2. Three examples of memory arrays. a) memory array with common word-line WLX; b) memory array with separate word-lines WWLX; c) memory array with split word-lines WWLX(0:3)

Table I. SER(errors/bit-day) Calculations from SEU Test Results

	Non-Bypassed Mode	Bypassed Mode	Dynamic(1) per Fig 2a design	Dynamic(1) per Fig 2b design	Dynamic(1) per Fig 2c design
Adam's 90% worst case environment	6.00E-20	3.00E-08	1.05E-10	5.27E-11	6.00E-20
Extreme proton environment(2)	8.60E-12	9.30E-07	3.28E-09	1.64E-09	8.60E-12

1) Assumes 512x32 50/50 Read/Write duty cycle, 5ns clock period, 3ns word-line pulse, 4:1 column mux
2) Solar Quiet, Inside Magnetosphere, 63° inclination, Apogee 2058Km, Perigee 2058Km

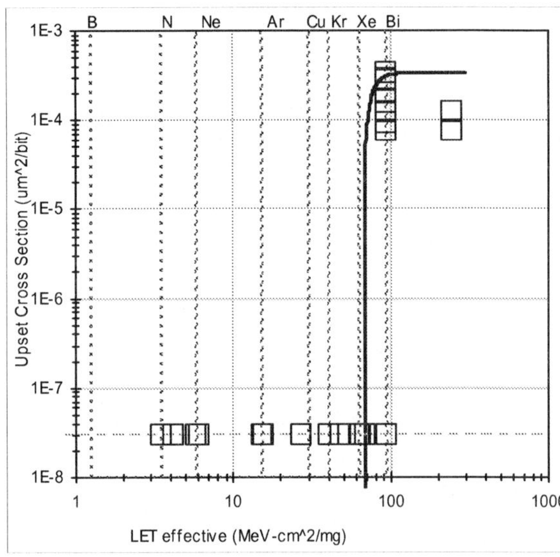

Figure 3(a)

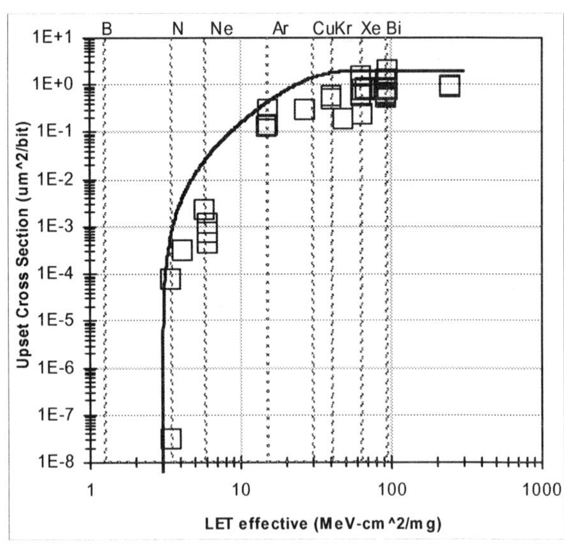

Figure 3(b)

Figure 3. Heavy Ion SEU test results at worst case environment of VDD=1.95V, 25°C from LBNL 88" cyclotron. a) non-bypass mode; b) bypass mode

Considerations for Single Event Effects in Non-Planar Multi-Gate SOI FETs

M. L. Alles, D. R. Ball, L. W. Massengill, R. D. Schrimpf, K. M. Warren, and R. A Weller
Institute for Space and Defense Electronics, Vanderbilt University
Email: Michael.Alles@Vanderbilt.Edu Tel: (615) 343-8829

Reductions in supply voltage and device dimensions, and increased transistor counts in advanced circuits, make single event effects increasingly relevant. Non-planar multiple-gate FETs (MugFETs) are under consideration for $\leq$ 45 nm technology generations. In this work, we examine issues for SEE in advanced MugFETs using energy deposition and device simulations.

Several variations of the MugFET device structure have been reported including the FinFET, Tri-Gate, and OmegaFet [1-7]. The majority of these devices, namely the FinFET and Trigate devices, are built on SOI. While there is some question about which is the preferable embodiment [7], the devices have some common features relevant to SEE: (1) finite active region dimensions are less than carrier diffusion lengths and smaller than typical dimensions of single-event induced ionization charge tracks and (2) dimensions of the source and drain regions are typically larger than those of the channel silicon, including the silicon thickness. The latter is to allow reasonable electrical contact to be made to the source and drain regions.

Three dimensional (3D) technology computer aided design (TCAD) device simulations were conducted to examine the device response to an ionizing single event. The simulated NMOS MugFET is based on that reported in [6]. Figure 1 shows the simulated device structure (gate and spacer materials not shown).This device has a relatively long channel (~60 nm), and the low aspect ratio of the channel (38 nm thick x 32 nm wide) makes it representative of the tri-gate configuration. The raised source and drain are 300 nm x 100 nm x 200 nm. Source and drain dopings are $1x10^{20}$ cm^{-3}, and channel doping is $1x10^{16}$ cm^{-3}. A mid-gap gate work function of 4.6 eV is used to obtain a threshold voltage of 0.2 V.

Single event (SE) strikes were simulated centered at three locations, all normal to the surface of the device: middle of the drain, middle of the channel, and middle of the source. The charge generation was modeled by a Gaussian spatial distribution having a radius of 50 nm and a linear energy transfer (LET) of 10 MeV-cm^2/mg (0.1 pC/μm) [8]. Simulations were performed with 0.1 μm x 0.1 μm drain and source contacts. The resulting currents are shown in Figure 2. Several observations can be made. First, in all cases the current exists over a significantly longer time than the direct collection of the ion-induced charge, indicating that the device is turning on (in a parasitic bipolar mode). Figure 3 shows the drain-body-source potential at various times during the conduction resulting from an ion strike to the channel. Second, although the deposited charge is lowest for the hit to the channel because of the thinner silicon region, the peak current is highest, indicating amplification. Previous work [9-10], in which drain and source contacts were placed on the ends for a FinFET structure, showed that a drain hit produced a higher peak current than a hit to the channel (body). The same results were obtained here; however, with more realistic contact placement on the top of the drain and source regions, the body hit produced the highest peak current. Figure 4 shows the integrated collected charge. The charge collected for the drain hit is <10% smaller than that for the body hit. These results indicate that most of the device (drain, gate, and source) may contribute to the sensitive area.

A third observation is that the dimensions, especially those of the channel silicon, are smaller than typically assumed radii of charge generation tracks. As shown by the charge generation contours in Figure 1, much of the charge track extends outside of the active fin region. Since multiple parallel fins are used to increase effective device width (current drive) of these types of devices, this may affect the response and sensitive area. Here, an ion track may span multiple fingers of the device, depending on fin spacing. Since charge deposition is typically converted from LET to charge/micron assuming some spatial (radial) profile that is smaller than the device dimensions, this has implications for "effective LET". It is not clear how all of the energy is converted to charge in these very small structures; the idea of a conventional charge track is questionable.

To examine this, simulations were repeated for a three fin device, shown in Figure 5 (again shown with the gate and isolation oxide not shown for clarity), for 4 hit locations and with two different fin spacings (s). Results are shown in Figures 6 and 7. For the larger spacing (s=150 nm), hit locations centered in any of the fins produce nearly identical results, while hits located between fins produce a less pronounced response. In the case of smaller spacing (s=75 nm), the hit located between two fins results in the highest peak current, and in an overall higher peak current than for the larger spacing. Figure 8 shows the collected charge for each of the simulated cases. Note that for the larger spacing, the charge varies more with location compared to the smaller spacing, and that the smaller spacing results in a larger maximum collected charge.

These results highlight the dependence of the response, the effective LET and the sensitive area, on the device topology and the assumption for the radius of the charge generation. Historically, assumptions have been

0-7803-9212-4/05/$20.00 ©2005 IEEE

made for energy deposition (i.e., charge generation tracks) based on average energy values and continuous spatial distributions represented by a function (such as Gaussian). For cases where the volumes of interest are much larger than the assumed generation radius, this is reasonable (although neglects possible nuclear reactions). Energy deposition in solids is actually a collection of discrete events, and results in a distribution of energy values and directionality for a given incident ion. For the sub-100 nm dimensions and discrete multiple sensitive volumes of these devices, the quantitative validity of the previously used models for charge generation in TCAD SE simulations is questionable; at the very least there is the need to calibrate such an approach using more advanced models.

Preliminary Monte-Carlo based energy deposition simulations have been performed using the Geant4-based code developed at Vanderbilt [11, 12]. Results shown in Figure 9 for s=150nm illustrate the concept of the reduced effective LET, as well as show the energy deposited in adjacent fins. This statistical simulation approach allows for improved quantitative analysis of single event effects in nano-scale devices. In addition, the impact of the gate materials and over layers can be accounted for using this approach developed at Vanderbilt [13, 14].

This work was supported in part by AFOSR through the MURI program.

References:

1. X Huang et al., *IEDM Tech. Digest*, 67 (1999).
2. R. Chau et al., 2002 International Conference on Solid State Devices and Materials (SSDM 2002), Nagoya, Japan.
3. Linda Geppert, *IEEE Spectrum*, November 1 (2002).
4. Y.-K. Choi et al., *IEDM Tech. Digest*, 259 (2002).
5. J. Kedzierski et al., *IEDM Tech. Digest*, 247 (2002).
6. J Choi et al., *IEDM Tech. Digest*, 647 (2004).
7. J. Fossum et al, *IEDM Tech. Digest*, 613 (2004).
8. Integrated Systems Engineering, Version 9.5 Manual (2004).
9. Silvaco International, *Simulation Standard*, **12**, No. 11 (2002).
10. M. Alles et al., May 2005 ECS Conference, SOI Technology and Devices XII, abstract 539 (2005).
11. A. S. Kobaysahi et al., *IEEE Trans Nucl. Sci.*, **51**, 3312, (2004).
12. C. L. Howe et al., *IEEE Trans Nucl. Sci.*, December 2005, in press.
13 A. S. Kobayashi et. al., *IEEE Trans Nucl. Sci.*, December 2005, in press.
14. K. M. Warren et. al., *IEEE Trans Nucl. Sci.*, December 2005, in press.

Figure 1. Active region of the simulated MugFET. Contours represent the electron-hole pair generation due to the simulated SE strike to the body region. The gate material, spacers, and isolation oxide are not shown to allow the fin to be viewed.

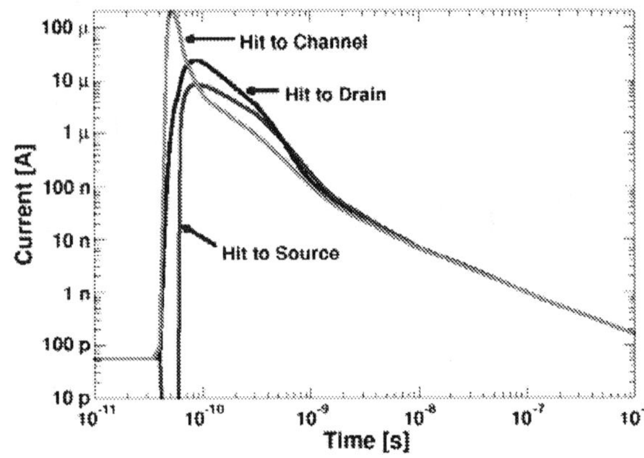

Figure 2. Drain current response to simulated ion strikes to the drain, channel (body) and source.

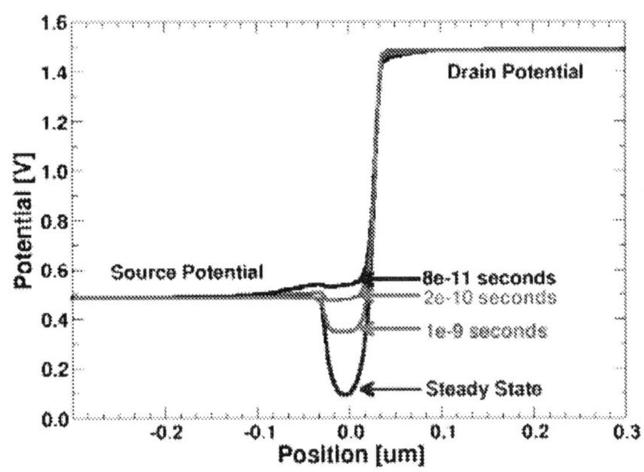

Figure 3. Potential profile prior to, during, and following the simulated ion strike to the channel (body) region.

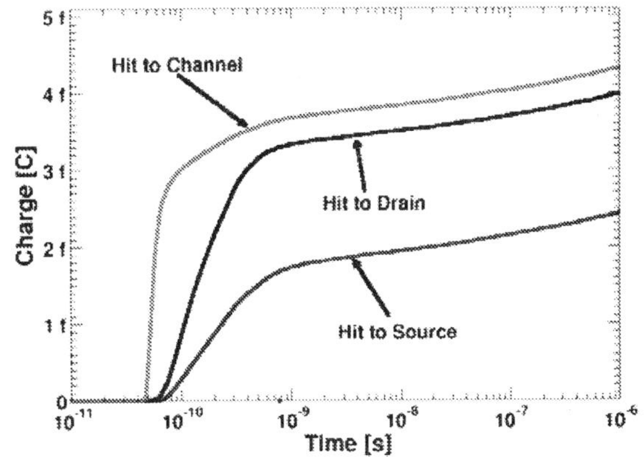

Figure 4. Collected charge for hits to the drain, channel (body) and source. The body and drain hits produce comparable amounts of total charge; the source results in approximately 50% of that of the other regions.

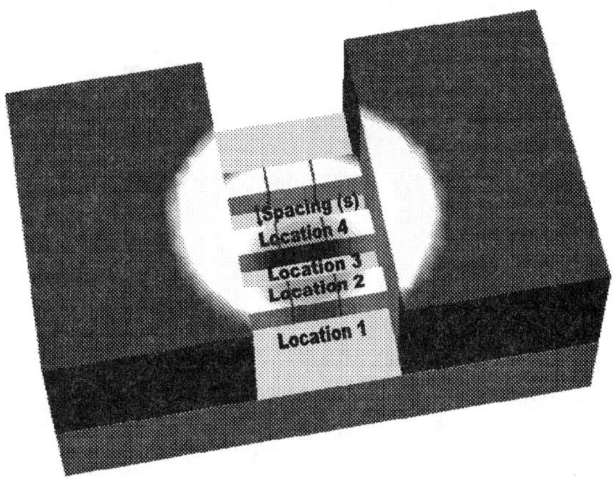

Figure 5. Active region of the 3 fin simulated MugFET. Charge generation contour is shown for a hit to the central fin. The 4 hit location simulated are noted. Simulations were performed for two fin spacings. The gate material, spacers, and isolation oxide are not shown to allow the fins to be viewed.

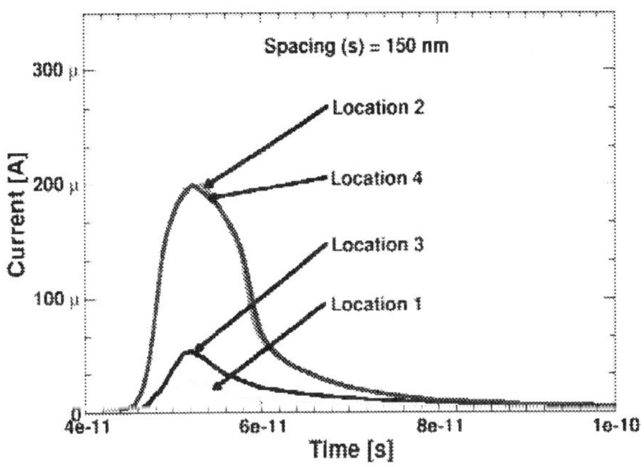

Figure 6. Simulated drain current reponse of the 3 fin MugFET to the hits at the positions noted in Figure 5 for the fin spacing of 150 nm.

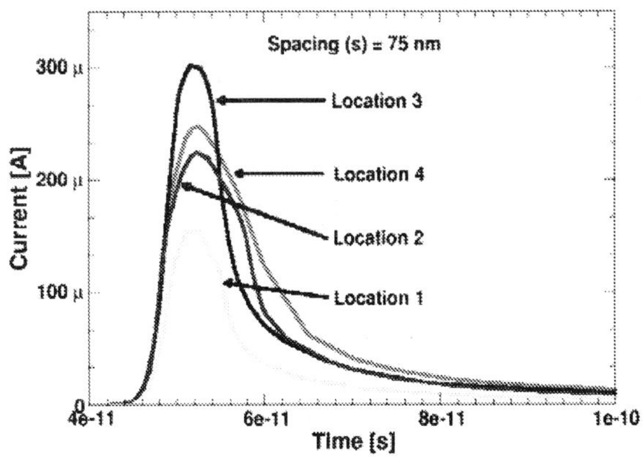

Figure 7. Simulated drain current reponse of the 3 fin MugFET to the hit at the positions noted in Figure 5 for the fin spacing of 75 nm.

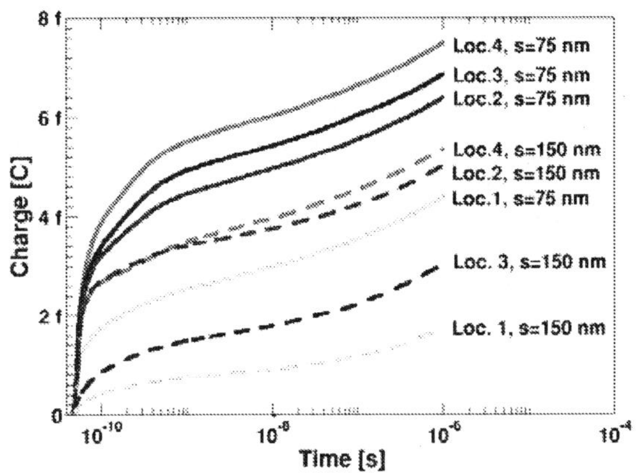

Figure 8. Simulated collected charge of the 3 fin MugFET for hit locations noted in Figure 5 for the fin spacing of 150 nm and 75 nm.

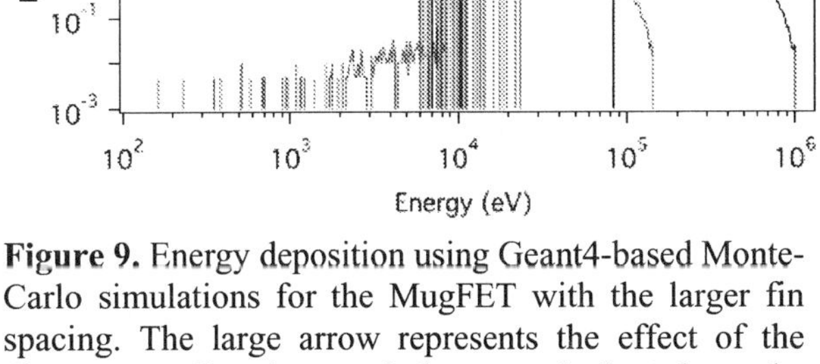

Figure 9. Energy deposition using Geant4-based Monte-Carlo simulations for the MugFET with the larger fin spacing. The large arrow represents the effect of the discrete small volume of the central fin where the simulated strike occurred.

Subterranean Photonics using SIMOX 3-D Sculpting for Optoelectronic Integration in Silicon-On-Insulator

T. Indukuri, P. Koonath and B. Jalali
University of California, Los Angeles, Department of Electrical Engineering, Optoelectronic Circuits and Systems Laboratory, Los Angeles, CA 90095-1594
Email: jalali@ucla.edu

Abstract

Three-dimensional optoelectronic integration can be achieved in SOI wafers using the process of SIMOX 3-D sculpting. Micro-resonators, with unloaded Q of 8000 and extinction ratio >20 dB, were fabricated in a buried silicon layer and MOS transistor structures were fabricated on the surface silicon layer using a patterned SIMOX process.

Introduction

Silicon-On-Insulator (SOI) material system has proven to be an excellent platform for the fabrication of both photonic and electronic devices[1-3], thus offering the prospect of integration of these functionalities on the same substrate. However, to make use of this advantage, it is necessary to develop innovative fabrication technologies that can monolithically integrate optical devices on silicon substrates[4], without compromising the real-estate economics of wafer manufacturing.

We have previously demonstrated a patterned SIMOX process (SIMOX 3-D Sculpting)[5] to realize a variety of optical devices such as microdisk resonators[6] and add-drop filters[7] in vertically-stacked layers of silicon. This process can be extended to obtain efficient optoelectronic integration of electronic and photonic devices on different layers of silicon.

In this paper, we demonstrate the fabrication and characterization of subterranean waveguide-coupled microdisk resonators in a buried silicon layer with unloaded quality factors of 8000 and extinction ratios in excess of 20dB. In addition, we report the realization of MOSFETs in the epi-silicon layer. With optical devices in a buried layer and electronic devices on the surface layer, we demonstrate the feasibility of SIMOX 3-D Sculpting to achieve monolithic optoelectronic integration.

Experiment and Results

The process of SIMOX 3-D Sculpting involves the implantation of oxygen in SOI substrates through a semi-transparent implant mask followed by annealing to realize buried optical structures (see Fig 1). An SOI wafer with 0.6μm of silicon on top of 1μm of buried oxide (made by SOITEC Inc) was thermally oxidized and patterned using reactive ion etching to form an implant mask defining waveguides, microdisk and racetrack resonators. The patterned wafer was then implanted with an oxygen ion dose of 5×10^{17} cm^{-2} and energy of 150 keV. The wafer was subsequently annealed at 1320°C for 7.5 hours in an Ar ambient with 1% O_2.

Fig 2 shows the cross-sectional scanning electron micrograph (SEM) of a rib waveguide in the buried silicon layer that acts as a bus waveguide to the microdisk resonator. It can be seen that our patterned SIMOX process has resulted in a 220nm buried silicon layer separated from a 240nm epi-silicon layer by a continuous 100nm SiO_2 layer. Fig 3 shows an optical micrograph of the top view of a 20μm radius microdisk resonator, coupled laterally to 2μm wide bus waveguides. This resonator is situated in the buried layer, 340nm below the surface.

To characterize the devices, a broadband amplified spontaneous emission (ASE) source was used in conjunction with an in-line fiber polarizer and a polarization controller to control the polarization state of the input light. An optical spectrum analyzer was used to observe the output spectral characteristics. Fig 4 shows the measured transmission spectrum after it has been normalized to correct for the spectral shape of the ASE source. Sharp resonances were observed with a free spectral range of 5nm. The narrowest resonance at ≈1528nm shows an unloaded quality factor of 8000 with an extinction ratio in excess of 20 dB. This is calculated from the measurements of quality factor and extinction ratio from the output spectra.

For the purpose of characterization of the devices in the buried silicon layer, the silicon on the top layer was thermally oxidized in order to prevent the leakage of the optical energy from the buried layer devices to the top silicon layer. However, while integrating electronic devices on the top silicon layer, only a narrow strip of silicon directly above the waveguides needs to be oxidized, as most of the optical energy is confined within the lateral dimensions of the waveguide. Alternatively, the thickness of the buried oxide can be increased to eliminate the coupling.

It must be emphasized that we have developed a process that results in the formation of a surface silicon layer separated from the buried micro-resonators by an isolating dielectric layer. By confining optical circuitry to a buried layer, the silicon real estate on the surface layer is available for the realization of high speed SOI electronics. Fig 5 shows MOS transistor structures that were fabricated on top of buried racetrack resonators and their electrical characterisitics are currently being investigated.

Conclusion

In summary, SIMOX 3-D Sculpting has been used to demonstrate monolithic optoelectronic integration. Specifically, we have demonstrated buried, laterally-coupled microdisk resonators with unloaded quality factors of 8000 and extinction ratios in excess of 20 dB. This approach presents an economically efficient method to monolithically integrate electronic and optical devices on an SOI substrate. Electrical characterization of the MOSFETs is in progress.

Acknowledgements

This work was performed under the CS-WDM program funded by the MTO office of DARPA. The authors would like to thank Dr. Jag Shah of DARPA for his support.

References

[1]. A. Liu, R. Jones, L. Liao, D. Samara-Rubio, D. Rubin, O. Cohen, R. Nicolaescu, M. Pannicia, Nature 427, 615 (2004).

[2]. D. Cristea, F. Craciunoiu, M. Caldararu, Mater. Sci. Eng. B 74, 89 (2000).

[3]. A. V. Krishnamoorthy, K. W. Goossen, IEEE J. Sel. Top. Quantum Electron. 4, 899 (1998).

[4]. L. C. Kimmerling, Appl. Surf. Sci. 159 – 160, 8 (2001).

[5]. P. Koonath, K. Kishima, T. Indukuri, B. Jalali, Appl. Phys. Lett. 83, 4909 (2003).

[6]. P. Koonath, T. Indukuri, B. Jalali, Appl. Phys. Lett. 85, 1018 (2004).

[7]. P. Koonath, T. Indukuri, B. Jalali, IEEE LEOS Annual Meeting 2, 747 (2004).

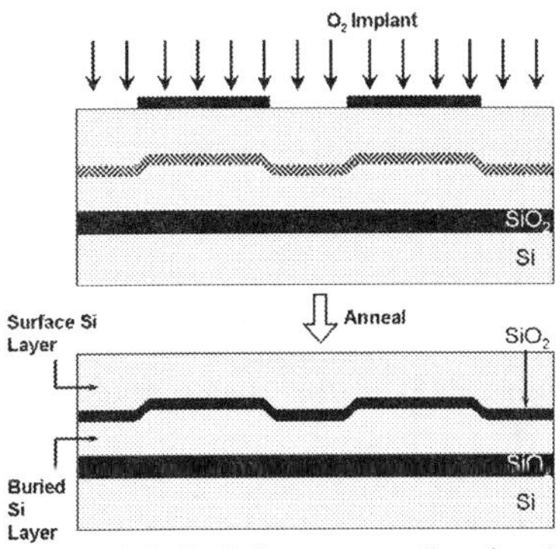

Figure 1 SIMOX 3-D Sculpting process flow involves oxygen implantation through an implant mask followed by high temperature anneal that results in buried waveguiding structures with an isolated surface silicon layer.

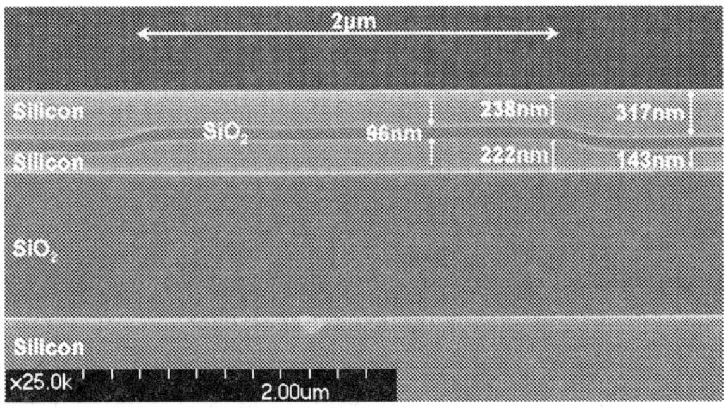

Figure 2 Cross-sectional SEM of input/output port waveguide of the microdisk resonator

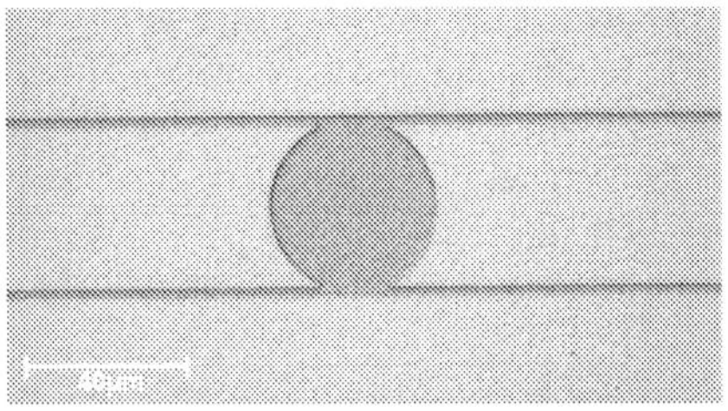

Figure 3: Micrograph of top view of a 20□m radius disk resonator. The entire device is 350nm below the surface of the SOI wafer.

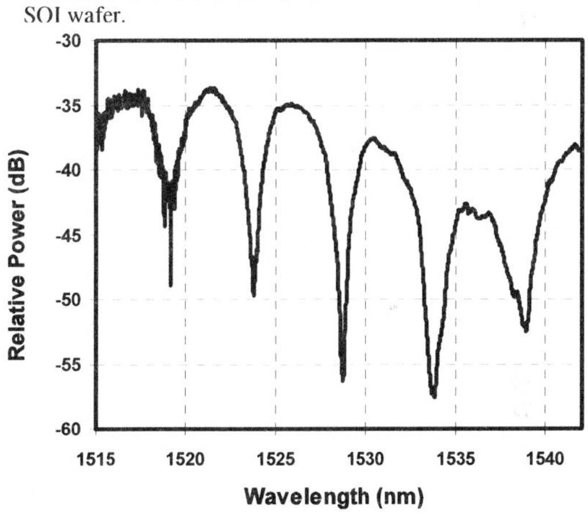

Figure 4: Transmission spectrum of the fabricated disk resonators.

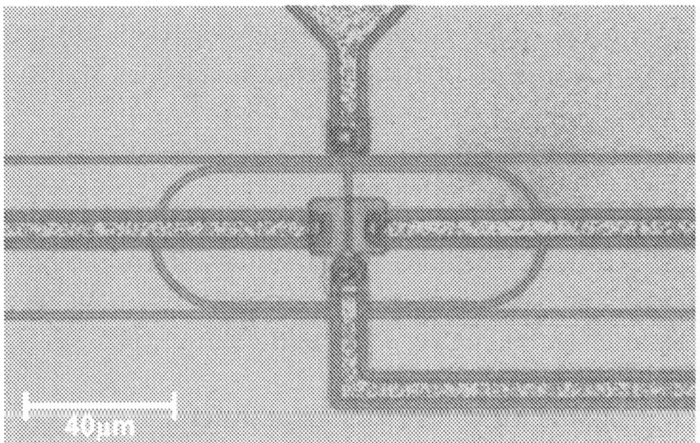

Figure 5: Fabricated MOSFET structures on top of buried racetrack resonators. Electrical characterization is underway.

Raman Amplification and Lasing in SiGe-on-insulator waveguides

V. Raghunathan, R. Claps, O. Boyraz, P. Koonath, D. Dimitropoulos and B. Jalali
Optoelectronic Circuits and systems Laboratory
University of California, Los Angeles
Email: jalali@ucla.edu , Phone: 310-206-4554

Introduction and Outlook

Stimulated Raman scattering in SOI waveguides has received significant attention recently with the demonstration of pulsed [1], cw Raman lasers [2] and high gain Raman amplification [3,4]. However, the limited bandwidth of the Raman signal in Silicon (~105GHz) renders this scheme unsuitable for broad-band WDM amplification unless multi-pumping scheme is employed. The introduction of Germanium into the SOI structure offers new avenues for tailoring the device characteristics, namely:

1. SiGe layers can be grown easily on a SOI subsrate with high uniformity and negligible defects. The presence of buried oxide also ensures that the optical mode does not leak into the bulk Silicon.
2. SiGe grown on SOI substrate gives rise to a doubly clad geometry in the vertical direction which enables efficient fiber-to-waveguide coupling [5].
3. Strain in the SiGe layer grown on Silicon and the compositional effect of SiGe layer can be used to tailor the Stokes shift of the Raman signal.
4. The Raman spectra can also be broadened in the presence of varying Ge composition in the SiGe layer thus offering the possiblity of broadband amplification.
5. The growth of SiGe superlattice can result in zone-folding of the phonon modes in the SiGe layer and this can potentially enhance the spectral capabilities of the Raman/Brillouin nonlinear processes. As an illustration, figure 1 shows the effects of the folded accoustic phonon modes in SiGe superlattice [6]. New Brillouin active modes emerge as a result of the zone-folding. Similar effects are also expected in the Raman active optical phonon branches.
6. Higher carrier mobility of Strained-SiGe can also aid in reducing the carrier lifetime [7] in the waveguide and hence mitigate the effects of losses due to free carriers generated by Two-photon absorption (TPA). However the benefit may be countered by the higher TPA coefficient of Ge.

Waveguide Structure and Experimental results

The SiGe layers used to perform the experiments reported here were grown on a SOI substrate using Chemical vapour deposition (CVD) under non-equilibrim growth conditions [8].

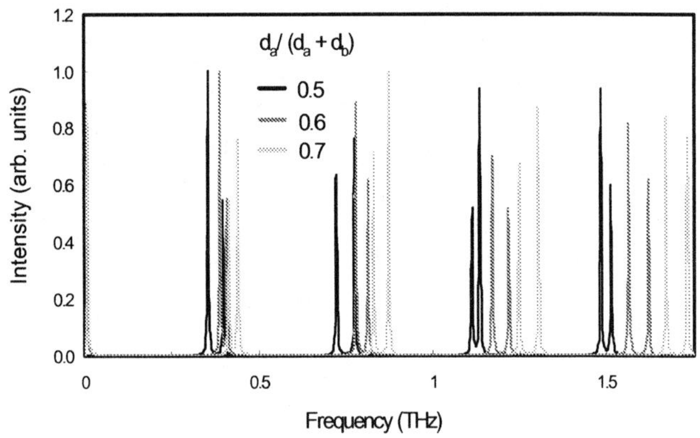

Fig. 1. The zone-folding of phonon modes in SiGe superlattices results in the creation of new Brillouin/ Raman active modes. This figure illustrates the folding of accoustic modes and the new Brillouin active modes for different layer thicknesses: Si (d_a) and Ge (d_b) [6]. For bulk Silicon there is only one Brillouin active mode at 40GHz and one Raman active mode at 15.6 THz.

The structure consisted of a 1μm buried oxide, 2.0 μm Si film, a 0.74 μm layer of $Si_{1-x}Ge_x$ alloy, and a 0.4 μm Si cap layer. The SiGe layer was characterized by SIMS analysis, which indicated a Ge content of 7.5% ±0.5. The rib waveguide formed via photolithography and Reactive Ion Etching. The rib width was 4 μm and the etch depth was 1.4 μm. The modal area of the waveguide was calculated to be 5.5 μm^2 and the propagation losses are estimated to be 1dB/cm.

A series of pulsed-pump measurements of stimulated Raman amplification were performed using the experimental setup outlined elsewhere [4]. Pulsed pumping is employed because of high peak powers that can be attained and to minimize the accumulation of TPA generated free carriers. The pump pulses had 30ps width with 25MHz repetition rate and were centered at 1539nm. The signal laser was CW and was centered at 1673nm. Figure 2 shows the change in the signal power caused by the Raman interaction with the pump. The peak gain of 16dB is obtained at the average pump power of 46mW. The spectrum for Stimulated Raman Scattering (SRS) was measured by tuning the signal wavelength in the presence of a fixed pump wavelength (1539 nm). Figure 3 shows the results for the SGOI waveguide compared to that from an SOI

0-7803-9212-4/05/$20.00 ©2005 IEEE

waveguide. The data indicates a Stokes down shift of 37GHz for SGOI waveguide.

The results for pulsed Raman amplification show that the net gain after the loss in the system (including the coupling losses) was ~5dB. With the objective of creating a lasing device, a cavity was built across the waveguide using a fiber loop, as described elsewhere [1]. The length of the fiber (~8m) was chosen to match the repetition rate of the pump laser. Figure 4 shows the output Stokes power (average) of the Raman laser obtained using a 2% tap coupler from the laser cavity, as a function of the input pump power (average). A linear fit to the data is also shown. The laser threshold is measured to be 20mW

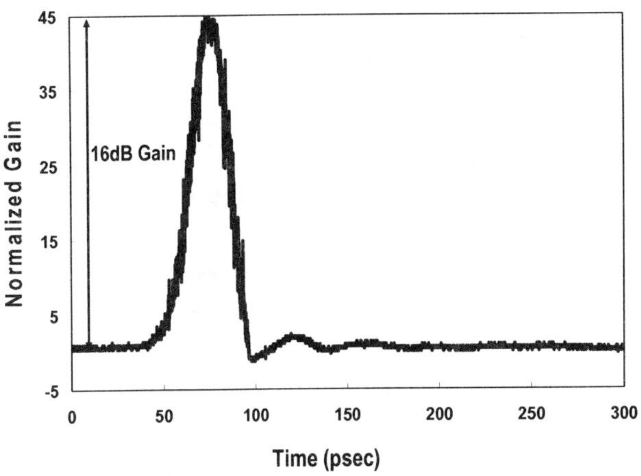

Fig. 2. Time domain trace of the maximum Raman amplification obtained with the SiGe alloy waveguide, using a pulsed pump scheme.

average pump power (25W peak power).The saturation of the Stokes emission is due to the EDFA nonlinearities which broadened the pump spectrum. The pump spectrum begins to broaden in the fiber patchcords and EDFA at about 35 mW average power (45 W peak pulse power). Because of the narrow gain spectrum of Raman in silicon (~ 105GHz) this results in a lowering of the Raman gain coefficient [1]. The possibility that the observed lasing is caused by stimulated Raman amplification in the fiber loop can be readily rejected based on the negligible gain in the short length of fiber [5].

Discussions and conclusions

The SiGe layer in our device was grown under non-equilibrium conditions and hence the 740nm thick layer in the waveguide is found to be coherently strained [8]. In this case the Raman Stokes shift consist of two contributions: the compositional effect and the strain effect as explained in ref [5] and references there in. The compositional and strain effects lead to shifts of -4.87cm^{-1} and +2.17cm^{-1} respectively when compared

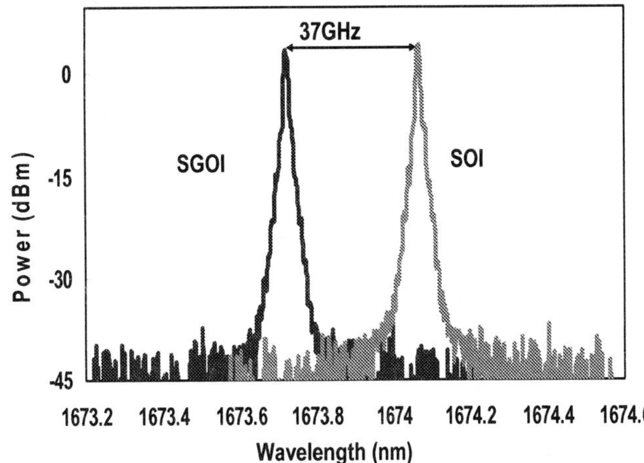

Fig. 3. The signal laser line for the wavelength at which maximum Raman amplification is observed. Two cases are shown: a pure SOI waveguide, and an SGOI waveguide.

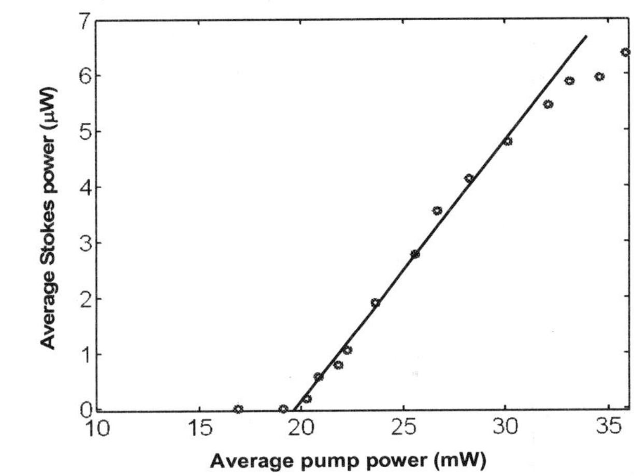

Fig. 4. Variation of the average output Stokes power versus input pump power, for the SiGe laser.

Thus the net theoretical shift (-2.7cm^{-1} or 70GHz) is found to be ~50% larger than that observed in experiments. This difference can be attributed to the fact that not all the optical mode is confined within the SiGe layer owing to the lower Ge concentration [5].

In conclusion, large pulsed gain and lasing have been reported in GeSi waveguides. The SiGe on SOI platform represents a Raman medium with a flexible gain spectrum.

References

[1] O. Boyraz, et. al., Opt. Exp 12, 5269-5273 (2004).
[2] H.S. Rong, et. al., Nature 433, 725-728 (2005).
[3] A.S. Liu, et. al., Opt. Exp 12, 4261-4268 (2005).
[4] V. Raghunathan et. al., CLEO 2005, CMU1
[5] R. Claps, et. al., Opt. Exp 13, 2459-2466 (2005)
[6] R. Claps et. al., submitted to Phys. Rev. B.
[7] R. Claps, et. al., Opt. Exp 12, 2774-2780 (2004)
[8] R. People, et. al., Appl. Phys. Lett, 322-324 (1985).

Retinal Implant Electrode Arrays with 10V SOI CMOS Circuitry

Murat Okandan*, Bruce Draper, Kurt Wessendorf, Sean Pearson, Ralph Young
Sandia National Laboratories, Albuquerque, NM 87185 *mokanda@sandia.gov

The interface between stimulation electrodes (which deliver electrical pulses) and retinal tissue is the most important interface in the retinal prosthesis application. As a member of the DOE Artificial Retina project, we have been developing a micromachined electrode array to address the critical mechanical and electrical coupling at this interface. Our design incorporates mechanical springs at each electrode site to allow controlled mechanical contact between the electrode array and the retinal surface, as well as built-in 10V capable CMOS electronics to handle routing of signals and to monitor integrated sensors. This process is also directed towards addressing other MEMS sensor/actuator systems that require higher voltages (~10V).

Figure 1 shows a schematic of the retinal implant. As can be seen in this figure, the available space for electronics is very limited – and a high level of integration is needed to provide the functionality required.

The requirements for the electrode array can be summarized in two main categories – mechanical and electronic functionality. Delicate nature of the retina is the main driver for our development of spring-supported electrodes (Figure 2). Neurons in the retinal tissue are very sensitive to pressure and can be damaged easily. In addition, the distance between the individual electrodes and the neurons is one of the main parameters that determine currents required for stimulation. In our process, the springs are made in the device layer silicon (1um thick), and released in the final step. Data routing going through the eye-wall becomes the driver for integrating CMOS electronics into the electrode array. In order to get to high electrode densities (100s to 1000s of electrodes), integrated demultiplexing capability on the electrode array is a necessity. Once the CMOS electronics is available in this device format, other functions (such as measuring displacement of electrodes through the change in resistance in the springs, monitoring local potentials and buffering) become much more feasible.

Figure 3 shows some of the mechanical test structures which have been fabricated and tested. Short loop test runs (after several process and device simulation steps) have also been completed with bulk wafers proving the 10V operation of both n and p devices (Figure 4).

We are currently designing the next test chip which will use the SOI process for evaluating the circuitry and mechanical components simultaneously.

Sandia is a multiprogram laboratory operated by Sandia Corporation, a Lockheed Martin Company, for the United States Department of Energy's National Nuclear Security Administration under contract DE-AC04-94AL85000.

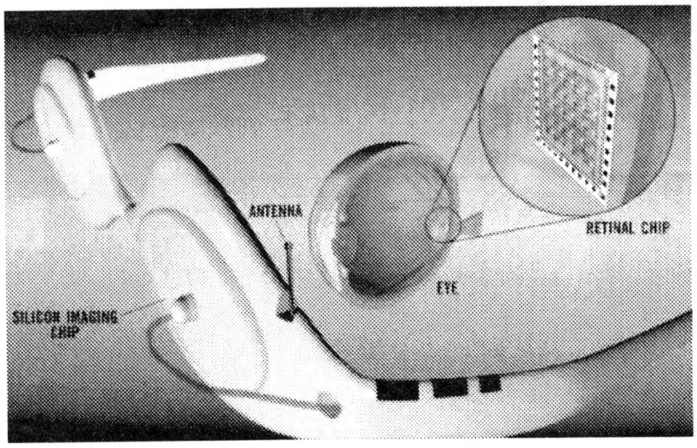

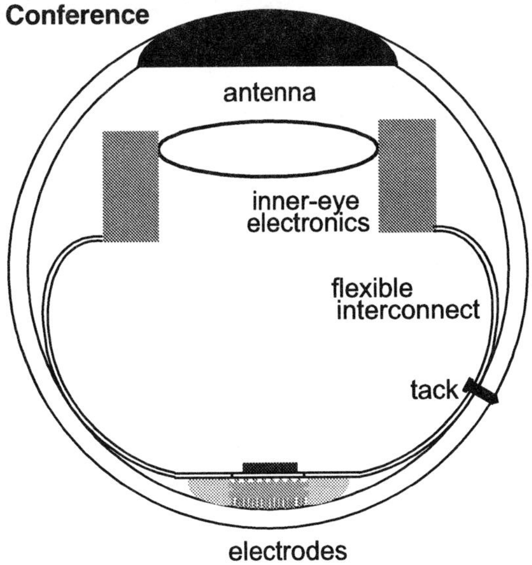

Figure 1. Schematic of the retinal implant (image courtesy of Dr. Mark Humayun's group, Doheny Retina Institute). In order to address both mechanical and electronic functions, the electrode array has built-in CMOS circuitry and spring supported electrodes.

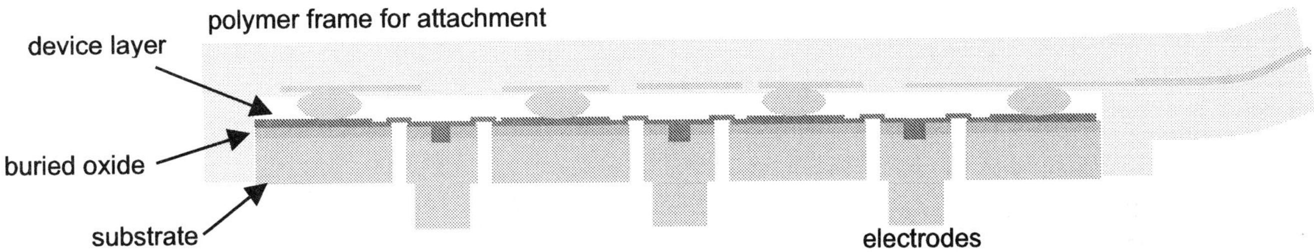

Figure 2. Electrode array with springs in device layer. Devices go through a release step which removes part of the oxide layers to free the electrodes. Polymer frame provides metal lines for routing signals between the electrode array and the rest of the implant.

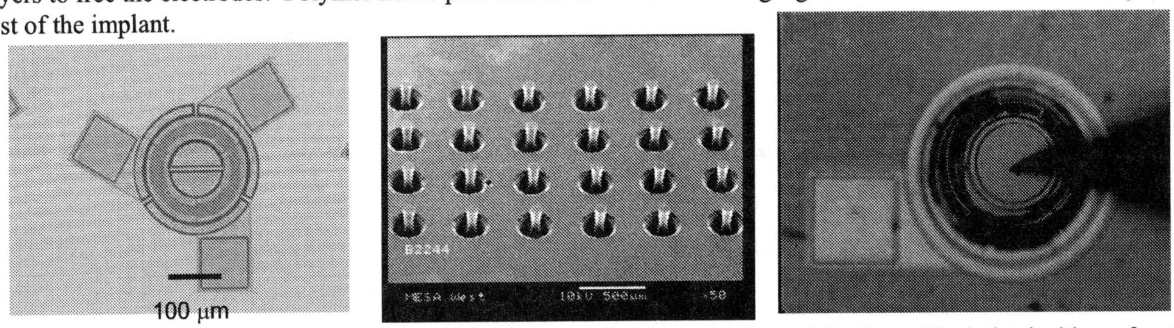

Figure 3. Images of mechanical test structures that have been fabricated and tested (a. front side, b. back side, c. front side). Electrodes are formed by Bosch etching from the back side. The third image shows manipulation of an electrode with a probe tip.

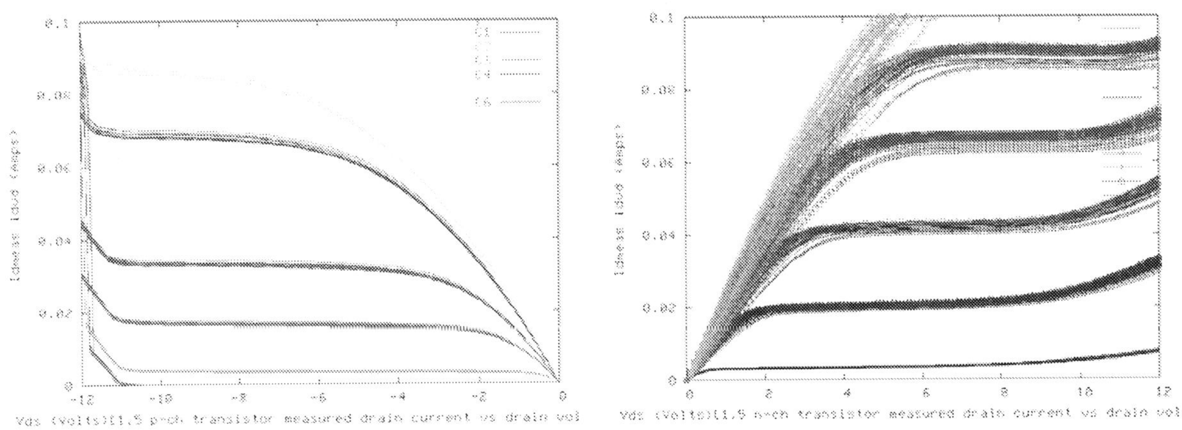

Figure 4. P and N device electrical characteristics obtained in the short loop runs. 10V operation is demonstrated with threshold voltages around 1V. Impact ionization is observed above 11V for devices with channel length of 1.5um.

Session #9

MEMORY DESIGN TECHNIQUES

CO-CHAIRS

Pierre Fazan
Innovative Silicon Inc.
Stephen Parke
Boise State University

Retention Characteristics of Zero-capacitor RAM (Z-RAM) cell based on FinFET and Tri-Gate devices

M. Nagoga[1], S. Okhonin[1], C. Bassin[1], P. Fazan[1], W. Xiong[2], C. R. Cleavelin[2], T. Schulz[3] K. Schruefer[3], M. Gostkowski[4], P. Patruno[5], C. Maleville[5]

[1] Innovative Silicon, PSE-B, CH-1015 Lausanne, Switzerland, ph: +41-21-6939097
[2] Texas Instruments, 13560 North Central Expressway, Dallas TX
[3] Infineon Technologies, Otto-Hahn Ring 6, D-81730 München, Germany
[4]ATDF, Inc. 2706 Montopolis Drive, Austin, TX
[5] SOITEC, Parc Technologique des Fontaines – 38190 Bernin, France
e-mail: mnagoga@z-ram.com

Introduction

Manufacturers of Dynamic Random Access Memory (DRAM) and embedded DRAM (eDRAM) have a tremendous challenge to shrink the memory cell and the main issue in cell area reduction lies with the capacitor integration. In the recently introduced zero-capacitor floating body memory cell (Z-RAM), the conventional storage capacitor is replaced by the body capacitance of a SOI MOSFET [1-6]. The charge stored in the floating body affects the device threshold voltage through the body effect and can be used to distinguish two states. The functionality of a Z-RAM memory array was already demonstrated on silicon using PD or FD SOI devices [7, 8]. To ensure the scalability of this new memory, it is mandatory to demonstrate its operation on advanced devices such as double-gate MOSFETs, FinFETs, etc. Recently TCAD simulations were used to illustrate the memory effect in double-gate FinFET devices [9]. Unfortunately these simulations where not confirmed by measurement results.

In this paper we experimentally study for the first time the retention characteristics of Z-RAM cells based on CMOS FinFET and Tri-Gate devices.

Experimental and results

Standard Unibond® SOI wafers were used as starting material. The silicon film and BOX thicknesses were 90 nm and 200 nm, respectively. The silicon film was doped with ion implant, resulting in channel doping concentrations of $2x10^{17}cm^{-3}$. 50 nm-wide silicon fins were defined with KrF excimer (248nm) lithography and RIE trim etch. The gate dielectric was 2nm thermal oxide. FinFET devices have a thick dielectric hardmask on top of the fins to deactivate the top gate (Fig. 1). For Tri-Gate devices, the top of the fins has the same gate dielectric as the sidewalls of the fins (Fig. 2). Polysilicon gate was deposited and doped by ion implantations. After source and drain formation, NiSi was used to reduce parasitic resistance. Finally an Al/Si metallization was used to complete the process.

Normally the FinFET's body is fully depleted. By applying a negative voltage (V_{bg}) to the back gate, an accumulation layer can be created at the back interface. In this case the transistor shows a partially depleted behavior and can be used as a memory cell (Fig. 3) [10]. DC current-voltage characteristics measured at two back gate biases are shown in Fig. 4. In a Z-RAM cell the state "1" corresponds to an excess of majority carriers in the transistor's body and the state "0" corresponds to their deficiency. The sensing is performed by measuring the channel current. To investigate the memory effect in the FinFET we write "1", read the source current, write "0", and read the source current again. The measured current versus time characteristic is shown in Figure 5. The difference between the currents, measured after write "1" and write "0" characterizes the memory effect. A good programming window is observed in this case.

A critical characteristic of a DRAM is its retention time. Retention times of few tens of milliseconds are observed at room temperature on FinFETs with 300nm channel length (Fig. 6). These values suit well typical embedded DRAM specification. FinFETs provide better retention characteristics compared to Tri-Gate devices (Fig. 7). Retention time decreases at shorter channel length as it is shown in Fig. 8, but even 100 nm devices still show retention time of few *ms*.

Conclusion

The retention characteristics of a Z-RAM memory cell based on CMOS FinFET and Tri-Gate devices are measured for the first time. A retention time of few milliseconds is measured at room temperature on 100 *nm* devices. This FinFET based Z-RAM memory will allow manufacturing of very low cost DRAMs and eDRAMs for 45 and sub 45-nm generations.

References

[1] M.R. Tack et al, *IEEE Trans. Electron Dev.*, 1990, p.1373.
[2] S. Okhonin et al, *IEEE Int. SOI Conf.*, 2001, p. 153.
[3] S. Okhonin et al, *IEEE Electron Dev. Letters*, 2002, p. 85.
[4] T. Ohsawa et al, *IEEE J. of Solid-St. Circ.*, 2002, p. 1510.
[5] P. Fazan et al, *IEEE Int. SOI Conf.*, 2002, p. 10.
[6] C. Kuo et al, *IEEE Trans. Electron Dev.*, 2003, p. 2408.
[7] K. Inoh et al, *Symp. VLSI Technology*, 2003, p. 63.
[8] T. Shino et al, *Symp. VLSI Technology*, 2004, p. 132.
[9] T. Tanaka et al, *IEEE IEDM*, 2004, p. 919.

0-7803-9212-4/05/$20.00 ©2005 IEEE

[10] W. Xiong et al, *IEEE Int. SOI Conf.*, 2004, p.195.

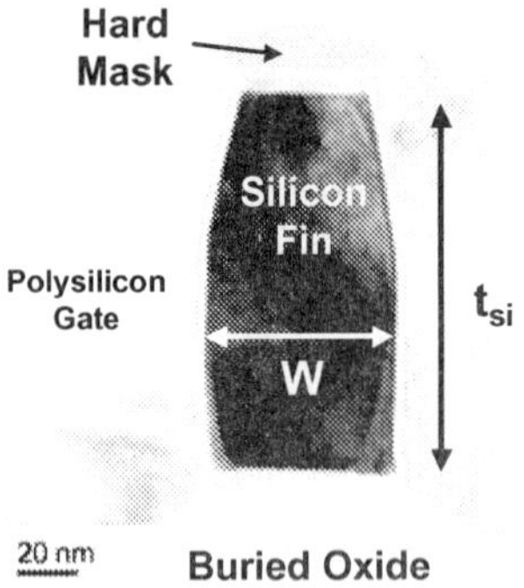

Fig.1 FinFET TEM cross-section.

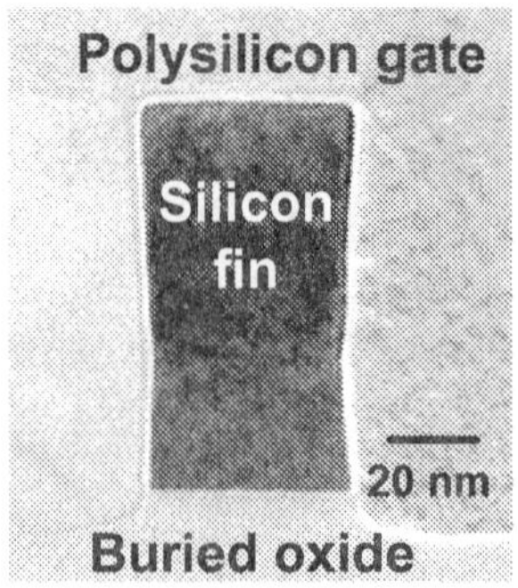

Fig. 2 Tri-Gate TEM cross-section.

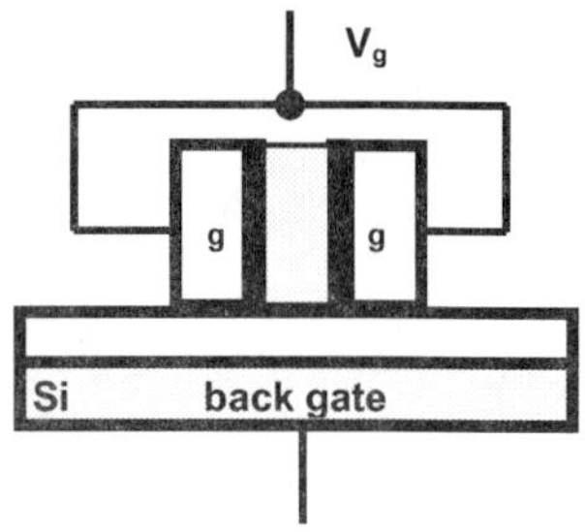

Fig. 3 FinFET operation mode: the back gate is used to form the floating body storage node.

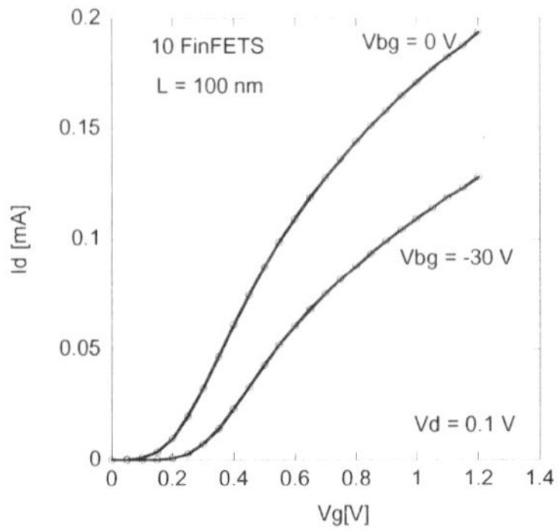

Fig. 4 I-V characteristics measured at Vbg = 0 V and -30 V.

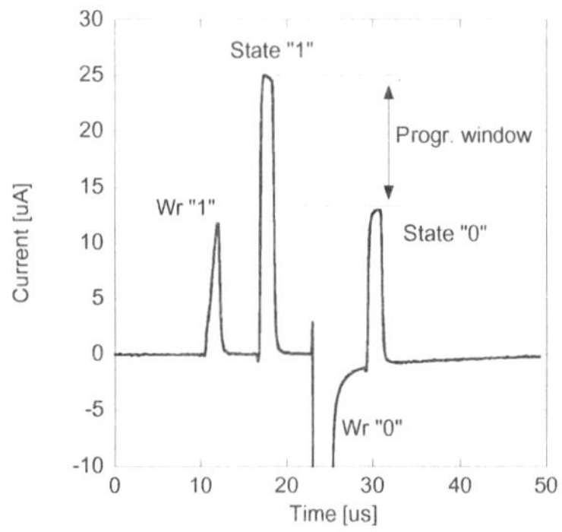

Fig. 5 Source current time dependencies during programming.

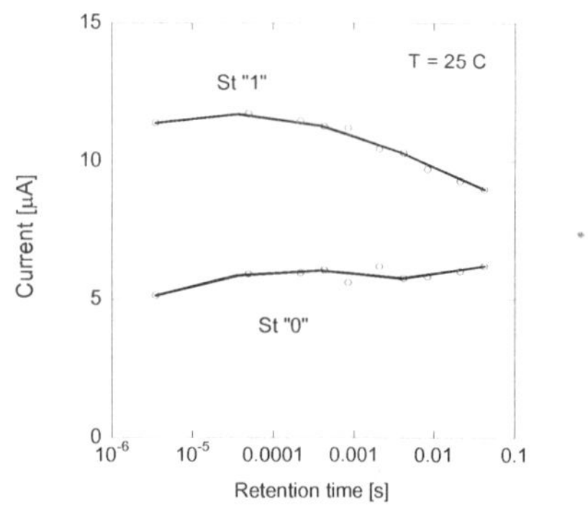

Fig. 6 Retention characteristics of a Z-RAM memory cell on FinFET.

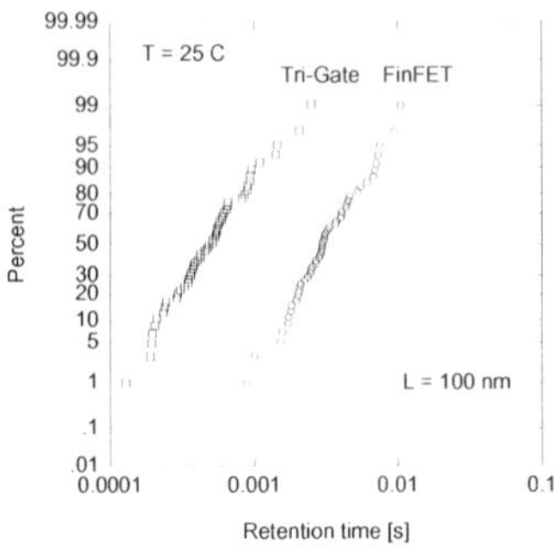

Fig. 7 Retention time measured on FinFET and Tri-Gate devices with L = 100 *nm*.

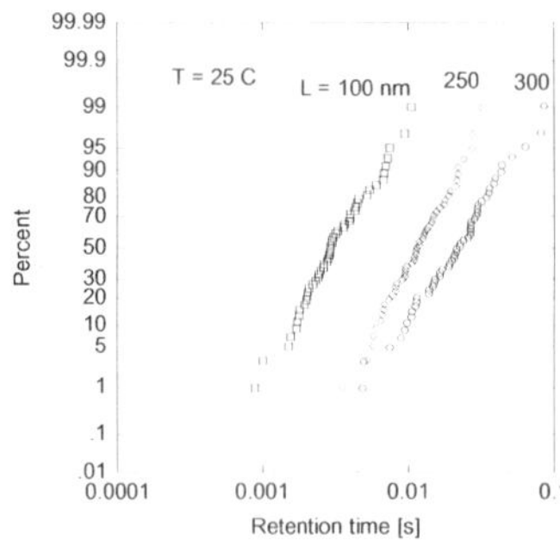

Fig. 8 Retention time for FinFETs with different gate length.

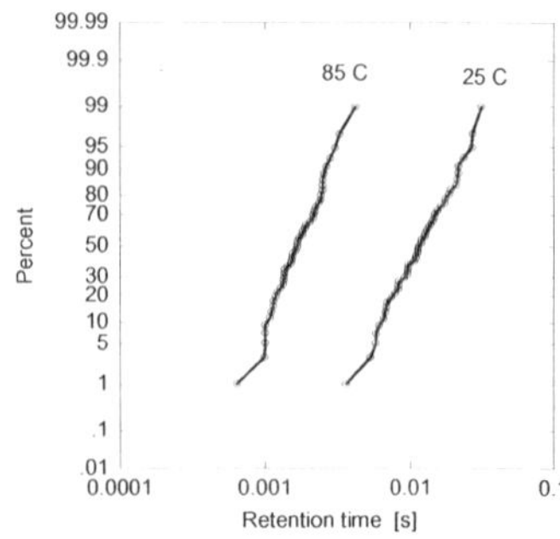

Fig. 9 Retention time for a FinFET with L = 250 *nm* measured at different temperatures.

Ultra-Low-Voltage Current-Sense Read Circuits for CMOS SOI SRAMs

Olivier Thomas, Andrei Vladimirescu and Amara Amara
Institut Superieur d'Electronique de Paris - ISEP, 21 rue d'Assas 75270 Paris Cedex 06, France
University of California at Berkeley, BWRC, 2208 Allston Way, Berkeley, CA 94704, USA
Email: olivier.thomas@isep.fr; andrei@bwrc.eecs.berkeley.edu; amara.amara@isep.fr
Tel: (+33) 1.49.54.52.66

Abstract

For SRAM circuits operated at 0.5V reliable readout of the stored information is challenging due to a voltage swing of tens of mV. Two readout topologies described in this paper for ultra-low-voltage (ULV) SOI-CMOS SRAMs exploit unique features of Partially-Depleted(PD)-SOI transistors to perform current rather than voltage sensing. The analysis presented leads to the dimensioning of the sense amplifier transistors for robust operation with a maximum number of cells per column of the SRAM. Simulations of the presented circuits demonstrate that the information stored in an SRAM with a four-transistor CMOS-SOI cell can be reliably accessed in 3ns with 180nW power consumption.

Introduction

The low-voltage trend for digital circuits has its roots in the lower power consumption of CMOS logic families operating at lower supply voltages and the proliferation of memory real-estate as a dominant part of the entire chip. The stored charge in each memory cell continues to decrease which raises the need for sense amplifiers very sensitive to reduced swings on the bit lines.

Memory readout circuits can sense either a small voltage difference on the bitlines or sense a current difference. Most memories operating under conventional bias use voltage sensing by differential amplifiers or bistable circuits.

Memories operated at ULV generate a very small voltage swing which coupled with the high capacitance of the bit lines lead to a very slow and unreliable readout by traditional voltage-sensing circuits. It is for this reason that current sensing circuits are investigated in this paper for an ULV CMOS-SOI SRAM [1]. A current readout circuit is a current buffer shunting the large capacitive reactance of the bitlines by a small input resitance and producing a current signal at its output which is converted to a voltage difference at the output load. The basic concept of a current sense amplifier is a variable-gain current buffer controlled by Vin on the bit line as shown in Fig. 1.

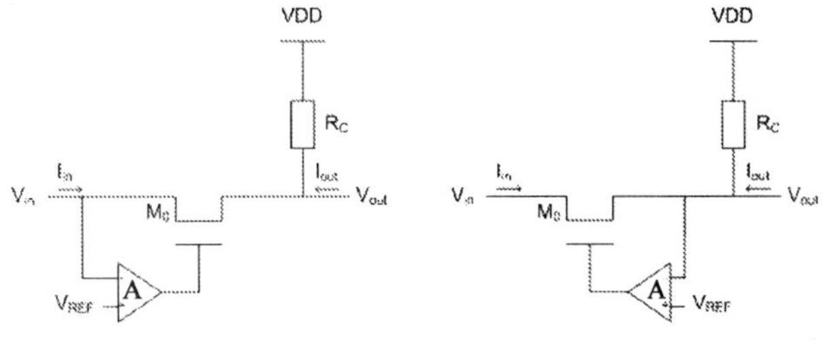

a) b)

Fig. 1 Current sense amp operating principle: a) Input control; b) Output control

The important characteristics of a current sense amp are its input impedance which needs to be low and its transconductance which needs to be high. The simplest circuit which fits this specification is a common-gate amplifier with a low input resistance R_{in}~$1/gm_0$. Unfortunatly MOS transistors present a high R_{in} (higher than a few tens of kΩ) due to a low gm_0. To overcome this problem, a number of different implementations have been reported [2]. The transconductance is improved thanks to an additionnal amplifier (A) connected in either a negative (input control , see Fig 1.a) or positive (output control, see Fig 1.b) feedback loop to control the gate voltage of the common-gate configuration. The effect is an increase of the G_m of the overall circuit, resulting in an important reduction of the input impedance which equal $1/G_m$. The formal analysis presented in [2] shows the output feedback loop to give the best improvement.

All the reported current sense amps are for CMOS bulk operating in strong inversion. The sense amplifiers described in this paper have been developed for an ULV CMOS-SOI SRAM which sets two new challenges: operation in subthreshold and floating-body effects, especially history effects present in PD, SOI.

ULV Current Sense Amplifiers for SOI SRAM

Two sense amplifier topologies shown in Fig. 2 and 3 are proposed. Both are designed to sense the

0-7803-9212-4/05/$20.00 ©2005 IEEE

slight imbalance of the bit-lines, BL, of the SRAM. This is achieved through a positive feedback loop.

The first architecture shown in Fig. 2 with its timing diagram, uses two cross-coupled transistor pairs which evaluate the current imbalance, ME_T, ME_F, and, MC_T and MC_F, which convert the difference to "1" and "0" voltage states, at the complementary outputs DL_T and DL_F. The conversion transistors MC are DTMOS-connected for maximum g_m and the evaluation transistors ME have their bodies connected to the output nodes for maximum overall transconductance.

The outputs DL_T and DL_F are precharged at V_{DD}; after the bit-lines (BL) and the word-line (WL) are activated, the SAE signal comes on enabling the sense amp ME-MC to evaluate and convert the presence of current Icell flowing through the BLs. During the entire read cyle, the precharge signal PRBL is kept high in order to limit the voltage drop on the BLs.

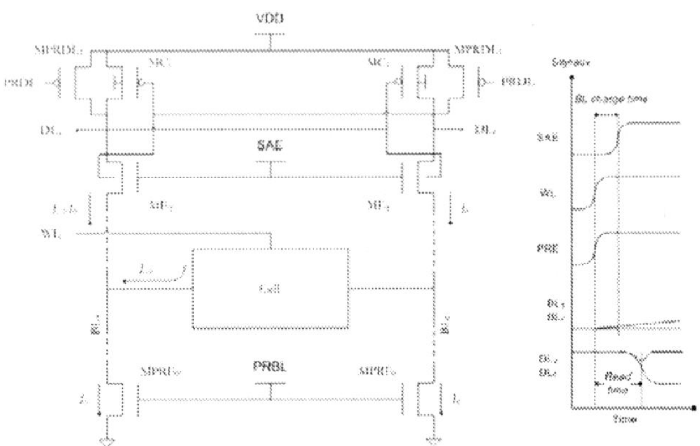

Fig. 2 Current sense amplifier working in ULV

The second sense amp shown in Fig. 3 differs in that it keeps the evaluation transistors ME always on, but disables the entire BL-SA structure until the SAE signal becomes active and the circuit is connected to V_{DD}. The outputs DL are initialized at 0V through the BL during precharge. Once the BL precharge ends, WL is selected and after a short delay SAE goes low starting the readout phase; transistors ME start out in pseudo-linear operation [3], V_{DS}=0, g_m low and g_d high. In subthreshold the constant current region is reached after only 3-4 kT/q or 75-100mV; sufficiently high g_m and g_m/g_d is achieved below this point already, see Fig. 4, enabling the circuit to quickly sense the presence of Icell and set DL_T and DL_F to their correct values.

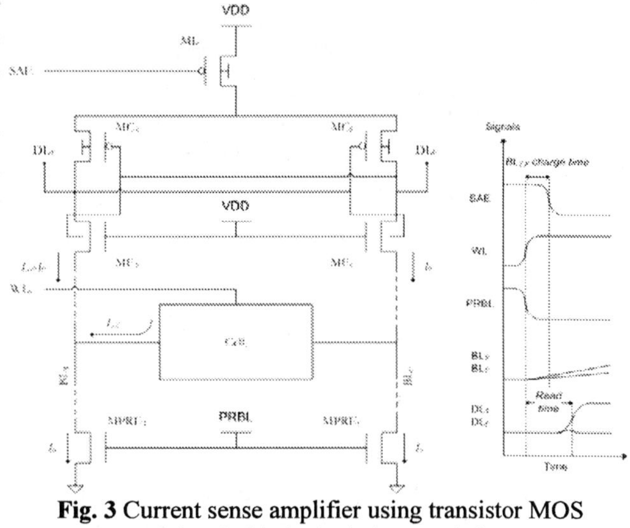

Fig. 3 Current sense amplifier using transistor MOS subthreshold elctrical characteristics

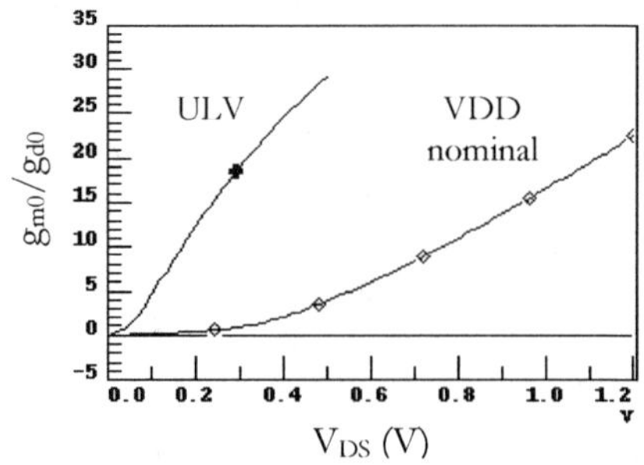

Fig. 4 Ratio gm/gd at VDD nominal and in ULV

This subthreshold behavior can be explained analytically by the subthreshold model reported in [3]. Equations (1) and (2) express the transconductance g_m and the condutance g_d of a transistor in weak inversion derived from the aforementioned model:

$$g_m = \frac{\partial I_{SUB}}{\partial V_{GS}} = \frac{\ln(10)}{S} \cdot K \cdot \left(1 - \exp(-\frac{V_{DS}}{V_{th}})\right) \cdot (a + \lambda \cdot V_{DS}) \tag{1}$$

$$g_d = \frac{\partial I_{SUB}}{\partial V_{DS}} = K \cdot \left(\lambda \cdot \left(1 - \exp\left(-\frac{V_{DS}}{V_{th}}\right)\right) + \frac{1}{V_{th}} \cdot \exp(-\frac{V_{DS}}{V_{th}}) \cdot (a + \lambda \cdot V_{DS})\right) \tag{2}$$

with $K = W \cdot d_0 \cdot 10^{\left(\frac{|V_{GS}| - |V_{Tw}|}{S}\right)}$

For V_{DS} three times higher than the thermal voltage the exponential term becomes negligeable compared to 1. The ratio gm/gd can then be derived:

$$\frac{g_m}{g_d} \equiv \frac{\frac{\ln(10)}{S}}{\frac{\lambda}{(a+\lambda \cdot V_{DS})} + \frac{1}{V_{th}} \cdot \exp(-\frac{V_{DS}}{V_{th}})} >>$$

where $\frac{\ln(10)}{S}$ is a constant gain depending on the subthreshold slope (S) of the technology. The lower S, the higher is the gain. For exemple, if the value of S is 80mV/dec then $\frac{\ln(10)}{S}$ is equal to 28. This value is much higher compared to the denominator which decreases quickly with V_{DS},.

Design Criteria and Simulation Results

For proper operation transistors ME must be biased in pseudo-saturation; values of W and L can be obtained from maximizing the ratio g_m/g_d which leads to the lowest R_{in}. The verification of robust operation of the two designs has been performed for worst-case, at a temperature T=85°C, V_{DD} reduced by 10% and +/- 4% mismatch in ME. The stable operation region for each circuit is plotted in Fig. 5 as a function of mismatch and the number of memory cells per column.

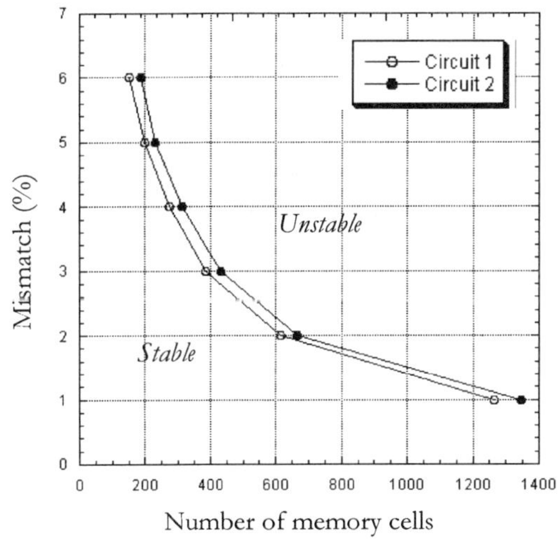

Fig. 5 Stability region of topology 1 and 2

It can be seen that topology 2 is stable for roughly 10% more cells per column than topology 1 when both experience the same mismatch in the ME transistors.

From a power-delay point of view circuit 2 is 6-7% slower but consumes 5-9% less power. Overall circuit 2 leads to less area utilization, important for an SRAM, fewer control signals and a more robust operation.

Conclusion

Two current-mode sense amplifiers have been described for applications in ULV CMOS SRAMs. The two topologies exploit unique features of PD-SOI transistors for achieving typical read access times of 3ns and power consumption of 180 nW. In simulations of a commercial 0.13 μm PD-SOI process with a 4% mismatch between transistors both circuits show stable operation down to V_{DD}=450mV for columns of up to 256 cells. The second topology differs from the first one in that it benefits from the electrical characteristics of a transistor in weak inversion. Sense amplifier 2 stands out by less area utilization (5 transistors), only one control signal and more robust operation.

Aknowledgment – This work was supported in part by the CEA-LETI, Grenoble, FRANCE.

References

[1] O. Thomas, and A. Amara, "An SOI 4 Transistors Self-Refresh Ultra-Low-Voltage memory cell", ISCAS, Bangkok, Thailand, May 2003.

[2] N. Shibata, "Current Sense Amplifiers for Low-Voltage Memories", *IEICE Trans. Electron.*, Vol. E79-C, No. 8, August 1996, pp. 1120-1130.

[3] O. Thomas, A. Valentian, A. Vladimirescu, and A. Amara, "An Accurate Estimation Model for Subthreshold CMOS SOI Logic", *Proc. ESSCIRC*, pp. 275-279, Sep 2002, Florence, Italy.

A Novel Nonvolatile Memory Cell for Programmable Logic

H. Lin and S. Tiwari
Cornell University, Ithaca, NY
Email: lh77@cornell.edu, Tel: (607) 592-1626

Abstract:

A novel, simple and bipolar-injection based nonvolatile field-effect memory cell is demonstrated in silicon-on-insulator CMOS technology. Programming time down to 8 ns are achieved together with erase times of a milli-second. The characteristics, compactness and compatiblity with CMOS processes suggest suitability of the structure for embedded programmable and reprogrammable applications, and extenisble to 3D applications. This extended abstract summarizes the technology and experimental characteristics of the device.

1. Introduction

Several examples of programmable logic approaches (FPGA, CPLD, etc.)[1] utilize floating-gate based storage (EEPROM e.g.) to achieve non-volatile reprogrammability, albeit slowly. Volatile programming, through SRAMs, is the usually preferred approach because of the need of speed. The technology of achieving non-volatile cells necessitates increased process complexity beyond that necessary for implementing CMOS logic, and to date, they are relatively slow in operation. The device structure reported here achieves higher CMOS compatiblity and provides attractive programming capabilities based on a variety of injection processes, including avalanche.

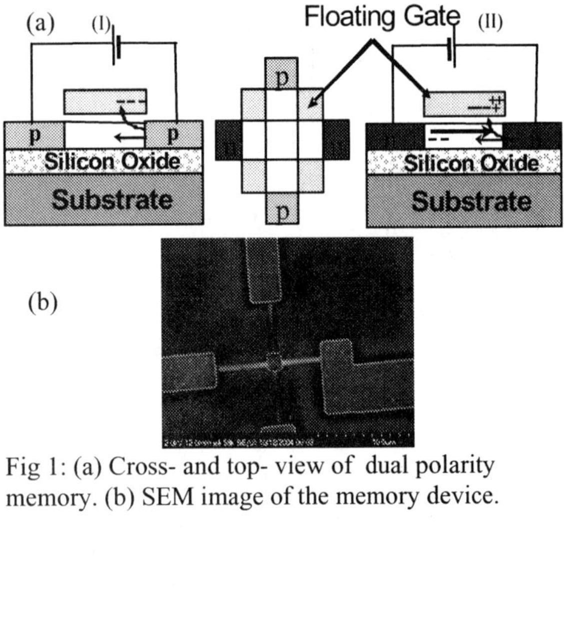

Fig 1: (a) Cross- and top- view of dual polarity memory. (b) SEM image of the memory device.

The highest write speed achieved, 8 ns, is a large improvement over the conventional cells and not too far from the desired characteristics of a universal memory. Use of SOI, isolation of storage node, and self-aligned processes results in smaller parasitc capacitances and superior radiation hardness.

2. Memory Cell Fabrication

The device (Fig. 1) consists of a crossed nFET/pFET structure that has a common channel area and a floating gate on top of a SOI substrate. The coupling of p-channel and n-channel device simultaneously allows the creation of a structure in which electrons and holes can both be injected into the floating gate region. So, programming and erasure is achieved through an injection process where the device is selected through the bias applied to source and drain electrodes rather than an additional control gate. The structure, through bipolar charge injection, also allows the programming and erasure to occur through charge neutralization. There are, however, potential uses where a control gate may also be employed. Like a conventional floating gate cell, charge injected or removed from the floating gate changes the conductance of the device to either a normally-on or normally-off state. This allows the cell to be used as a memory node and as a configuration node for programmability. Devices were fabricated on SOI substrates with a 50 nm thin Si layer and a 377 nm thick buried-oxide layer. Self-aligned source-drain formation was implemented with boron and arsenic implantation. The channel widths are 0.5 μm. Application of a large bias between source and drain creates an avalanche process that provides a large flux of carriers into the floating gate [2]. Both electrons and holes can be injected in this way. To demonstrate the working of the cell, we employ the SOI substrate as a back gate (V_{BG}) to characterize the change of the state (through threshold voltage, e.g.) of the device.

3. Memory Cell Operation and Characteristics

When the normally floating gate is connected[2], the structure shows symmetrical transfer characteristics about 0 V gate voltage. By injecting opposite sign of charges into the floating gate, we can change the conduction state of both sets either from normally-on to normally-off, or the other way, such that we can use them to program an interconnection. For the pFET, when enough negative bias is applied to the drain contact (smaller than –6.5V), avalanche process takes place. The potential of the floating gate is coupled from the drain side. Due to the potential difference between the drain and the floating gate (more positive in the floating gate), electrons can be injected into the floating gate, a so-called avalanche hot electron injection (AHE). On the other hand, avalanche hot hole injection (AHH) can take place in the nFET by applying a positive bias to the contacts. Because of this dual-polarity charge injection capability, we can effectivly accomplish the programming and erasing functions with our novel structure. Charge injection efficiency of AHE injection as shown in Fig. 2, is illustrated as the change of threshold voltage V_t of the pFET with the programming pulse width and pulse voltage. The change of V_t increase with more negative drain voltage and longer pulse time. One significant feature in Fig. 2 is the effective V_t shift by programming at 10's of

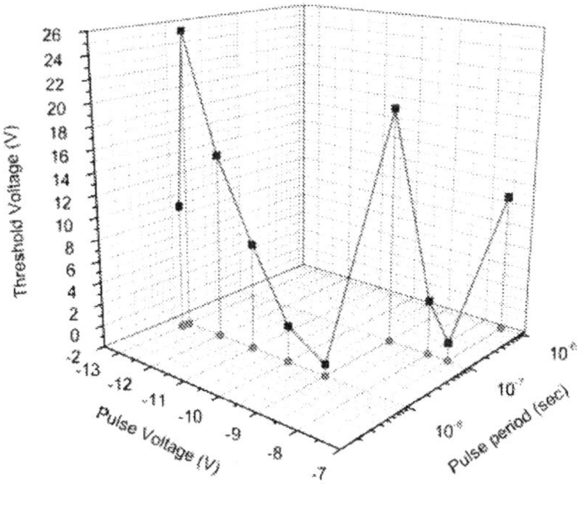

Fig. 2. V_t shift as a function of programming pulse width and pulse voltage applied on pFET.

nanosecond. Similar measurement has been performed on nFET for the AHH injection efficiency. Effective injection of holes requires larger magnitude of programming voltage with much longer pulse width due to larger mass and higher loss of energy through scattering of holes comparted to electrons. This leads to effective programming time of more than hundreds of μs with programming voltage beyond 12 V (Fig. 3).

On the set of nFET, when the applied bias voltage coupled from the drain to the floating gate becomes larger than 20 V, it not only turns on the n-channel but also directs part of the hot electrons close to the drain side into floating gate. This is similar to the channel hot electron (CHE) injection often employed in earlier floating gate devices. Therefore, the feature adds additional capability for the nFET to produce both electrons and holes injection functions. Because of the higher efficiency of the hot electron injection than hot hole injection, channel hot electron (CHE) injection can take place at a much faster time scale than avalanche hot hole (AHH) injection on the nFET. In Fig. 4, such injection (CHE) can be accomplished in several μs with the applied bias increased beyond 20 V. At such a short time scale, we assume the effect of AHH injection can be ignored.

Based on the above characteristics of our device, we have the following two Programming/Erasing schemes:

3.1 P/E with Both Sets (nFET and pFET)

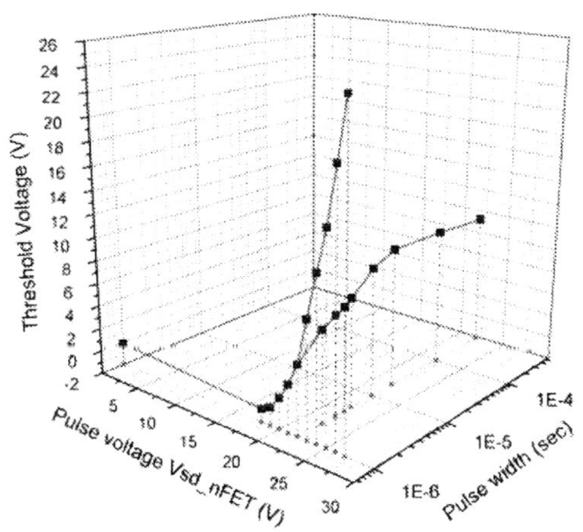

Fig. 4. V_t shift as a function of both programming pulse width and pulse voltage applied on nFET for V_{sd}> 20V.

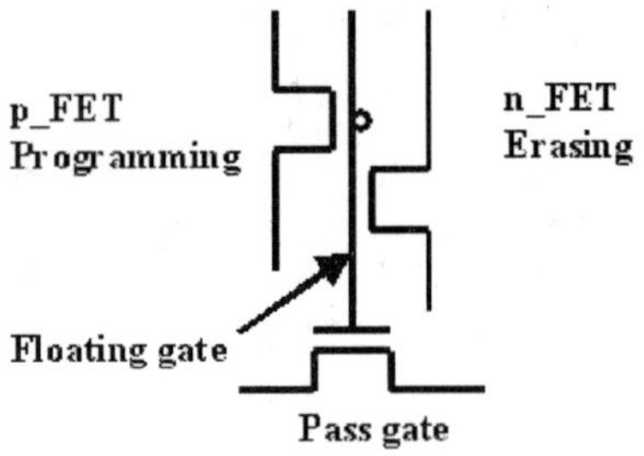

Fig. 5: Schematic of the 1st type of the P/E operation: using pFET for programming (AHE injection) and nFET for erasing (AHH injection). The floating gate is connected to a pass gate.

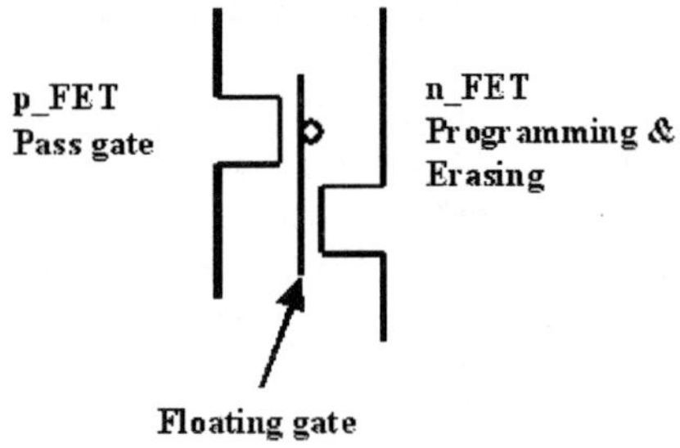

Fig. 6. Schematic of the 2nd type of the operation of the device: nFET for programming (AHE injection) and erasing (AHH injection), while the pFET functions as a pass gate.

TABLE 1
Operation Table of the Memory Cell

	Programming	Erasing
V_{d_nFET}	**"0V"**	**"15V"**
V_{s_nFET}	**"0V"**	**"0V"**
V_{d_pFET}	**"-12V"**	**"0V"**
V_{s_pFET}	**"0V"**	**"0V"**
Time	**10ns**	**1ms**

With an additional pass gate transistor connected, we can effectively employ the program/erase functions provided by pFET/nFET to program an interconnection (Fig. 5). On the pFET, when negative bias smaller than –6.5V is applied to the source drain contact, avalanche hot electron (AHE) injection takes place. As shown in Fig.2, the smallest programming time can be achieved at 8ns with -13V applied bias. On the other hand, avalanche hot hole (AHH) injection can be made to occur in the nFET at a time scale close to ms. This is 3 devices memory node configuration. When one set of the device is employed for charge injection, the other is left grounded. An operation scheme is shown in Table 1.

3.2 P/E with One Set (nFET)

Because the nFET can produce both functions of electrons injection and holes injection, by selectively choosing the magnitude and time of the biased voltage (Table 2), we can choose to inject either electrons or holes into the floating gate and effectively accomplish P/E operation. The set of pFET can then be used as a pass transistor for interconnection in programmable logic circuits (Fig. 6), thus simplifying the wiring at the expense of higher programming voltage.

4. Conclusions

Fast programming capabilities with relatively low voltage has been accomplished in a novel nonvolatile memory cell structure that employs avalanche hot electron (AHE) injection, avalanche hot hole (AHH) injection and channel hot electron (CHE) injection for programming/erasing. Two programming/erasing schemes are proposed for their potential application for programmable logic circuits.

TABLE 2
Operation Table of the Memory Cell

	Programming	Erasing
V_{d_nFET}	**"25V"**	**"15V"**
V_{s_nFET}	**"0V"**	**"0V"**
Time	**1us**	**1ms**

References:

[1] J. Rose et. al., *Proceedings of the IEEE*, Vol.81, No.7 (1993)

[2] H. Lin and S. Tiwari, "Novel Dual Polarity Non-Volatile Memory," unpublished

[3] D, Frohman-Bentchkowsky, *IEEE Journal of Solid State Circuits*, Vol. sc-6, No.5, (1971)

Back-Gate Controlled READ SRAM with Improved Stability

Jae-Joon Kim, Keunwoo Kim, and Ching-Te Chuang
IBM T. J. Watson Research Center, Yorktown Heights, NY 10598
Email: {jjkim2,kkim,ctchuang}@us.ibm.com

I. Introduction

As device size shrinks aggressively in advanced VLSI technology, increased process variation causes significant threshold voltage fluctuation. As a result, stability of SRAM deteriorates due to large threshold voltage mismatch between two neighbor transistors in a cell. The conventional 6-transistor (6T) SRAM (Fig. 1) has its worst stability during READ operation because the voltage at the storage node which has "0" logic value goes up during READ cycle. The stability of SRAM can be improved by decoupling the storage nodes from the bit lines in READ mode as in the 10-transistor (10T) SRAM cell (Fig. 2) which is often used in multi-port register file [1]. In this case, voltage at the storage node is not increased by READ current so degradation of stability in READ mode does not occur. The 10T SRAM cell, however, needs much larger area than 6T cell due to the increased number of transistors in a cell (from 6 to 10). In this paper, we propose a new 8T SRAM cell structure using back-gate controlled device in the READ path, which maintains stability advantages of the 10T SRAM cell structure while reducing the number of transistors.

II. Back-Gate Controlled READ SRAM

Device Structure: The proposed scheme is particularly suited for back-gated Ultra-Thin SOI (UTSOI) or asymmetrical double-gate (DG) devices. For the verification of the proposed scheme, we use UTSOI with back-gate control [2] having 25nm effective channel length, thin silicon film (10nm), scaled front-gate oxide (1nm) and back-gate oxide (10nm). MEDICI is used for mixed –mode device/circuit simulations [3].

Operating Principle: Let us discuss the operating principle of the proposed SRAM cell shown in Fig. 3 and Fig. 4. In the proposed scheme, the cell storage nodes (Qb and Q) are connected to the back gate of READ access transistors TR1 and TR2. Let's assume the cell node Q has "Low" logic value and the node Qb has "High" logic value in STAND-BY mode without loss of generality. In the case, the back gate voltage of TR1 is "High" and that of TR2 is "Low". Due to the difference of back-gate voltages, TR1 and TR2 have different threshold voltages. Since the voltage at "Qb" is higher than the voltage at "Q", TR1 has lower threshold voltage than TR2, hence the on-current (I_{on}) of TR1 is higher than that of TR2. In the READ mode, the RWL is energized and READ transistors are turned-on. Consequently, currents from the precharged READ bit lines RBLb and RBL start to flow through TR1 and TR2. Since I_{on}(TR1) is larger than I_{on}(TR2), the voltage drop at RBLb is higher than the voltage drop at RBL. A sense-amplifier senses the difference between the two bit lines and completes the READ operation.

Worst Case READ Condition: Let us discuss the more rigorous condition for worst case READ in terms of functionality. Fig. 5 shows the worst case for READ operation. We assume the number of cells per bit line is n and the cell(1) is accessed in the READ mode without loss of generality. In the READ mode, on-current flows through access transistors of cell(1) and the on-current (I_{on}) from RBLb to cell(1) is higher than that from RBL as explained above. When the cell(1) is accessed, the read word line (RWL) of the other cells (cell(2) – cell(n)) has zero voltage so that only leakage current flows through READ access transistors of cell(2)-(n). For correct READ functionality in Fig. 5, the sum of each current from RBLb must be always larger than the current from RBL. In the worst case, every cell except cell(1) has logic value different from cell(1) as shown in Fig. 5 because the off-current(I_{off}) of a transistor with high back-gate voltage is much larger than the off-current of a transistor with low back-gate voltage. Hence, devices must be designed to satisfy the following condition.

$$I_{on}@(V_{bg}=VDD)+(n-1)xI_{off}@(V_{bg}=0)$$
$$> I_{on}@(V_{bg}=0)+(n-1)xI_{off}@(V_{bg}=VDD)$$

Design Technique for Leakage Reduction: An important design consideration in the proposed scheme is the large off-current of the device with high back-gate voltage. The large off-current of the device with high back-gate voltage not only causes potential problem in functionality but also increases the stand-by power. Hence, we apply the negative voltage to the un-selected read word lines (RWL) to reduce the off-current through the access transistors [4]. Since the sub-threshold current exponentially depends on gate-voltage, the leakage power can be substantially reduced by applying the negative voltage to the gate of access transistors in stand-by mode. Fig. 6 shows the MEDICI-predicted leakage reduction for negative WL voltage with 0.8V back-gate bias. The sub-threshold leakage with 0V back-gate bias ($I_{off}@V_{fg}=V_{bg}=0$) is also shown for the reference. In this case, it is observed that -0.21V WL voltage with 0.8V back-gate bias gives the same sub-threshold current as 0V WL voltage with 0V back-gate bias.

Static Noise Margin (SNM): The main advantage of the proposed scheme is the improved stability. Fig. 7 shows comparison of SNM between a conventional 6T SRAM and the proposed scheme. The SNM of the proposed scheme is 136mV higher than the conventional 6T SRAM cell at VDD=0.8V. Decoupling the storage node from the READ current path is the main reason behind the improved SNM in the proposed scheme. More importantly, the proposed scheme maintains high SNM for low supply voltages as shown in Fig. 8, hence can be used for scaled supply voltage.

Read Delay Analysis: To compare the performance between the conventional 6T SRAM and the proposed scheme, we measure the delay from the input of the word line (WL) driver to the time when 100mV difference is developed between BL and BLb in the 128x64b SRAM array structure. Fig. 9 shows the delay versus supply voltage. The delay in the proposed scheme is 7.5% larger than that for the 6T array at VDD=0.8V but 2.3% smaller at VDD=0.7V. As the supply voltage goes down, read delay degradation in the proposed scheme decreases because the ratio of $I_{on}@V_{bg}$=VDD over $I_{on}@V_{bg}$=GND becomes larger [5]. The delay in the proposed scheme increases due to the negative WL stand-by voltage and

0-7803-9212-4/05/$20.00 ©2005 IEEE

the use of current difference between two READ transistors. Meanwhile, the delay in the proposed scheme decreases due to the fact that the READ path of the proposed scheme uses only one transistor compared with two serially stacked transistors in the READ path of 6T cell.

III. Summary

We have presented a novel back-gate controlled UTSOI SRAM cell structure and associated design considerations. The proposed scheme offers improved stability compared with conventional 6T cell, and has less number of transistors than the conventional 10T cell. Due to over 3x improvement in SNM even at low VDD (Fig. 8) and reduced area penalty compared with 10T cell, the proposed 8T cell could be aggressively scaled down. This scheme can also be applied to SRAM cells with asymmetrical DG devices.

Acknowledgement

K. Kim and C. T. Chuang were supported by DARPA contract NBCH30390004 for this work.

References

[1] A. Chandrakasan et. al, Design of High-Performance Mircroprocessor Circuits, *IEEE Press,* pp. 285-308
[2] M. Yamaoka et. al, VLSI Symp, pp. 288-291, 2004
[3] MEDICI, Synopsis, Inc., 2003.
[4] K. Itoh et. al, VLSI Symp, pp. 132-133, 1996
[5] J. Fossum et al., *IEEE TED,* pp.167-170. 2001

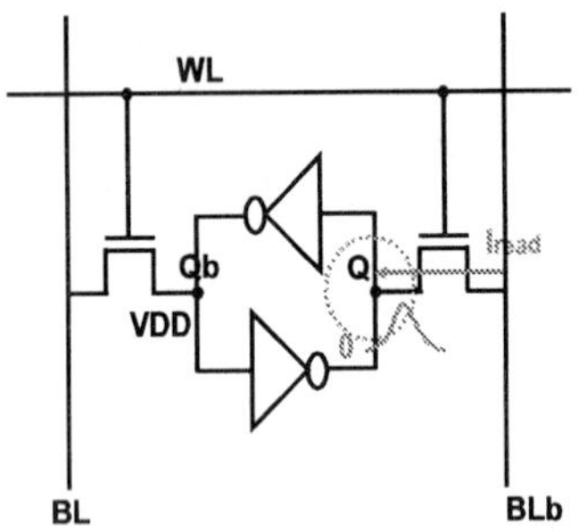

Fig. 1. Read disturb in the conventional 6T SRAM. VQ goes up due to the read current Iread.

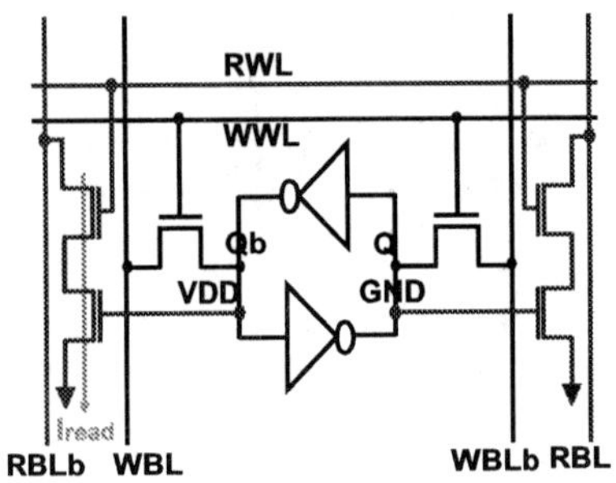

Fig. 2. Conventional 10T dual-port SRAM [1]

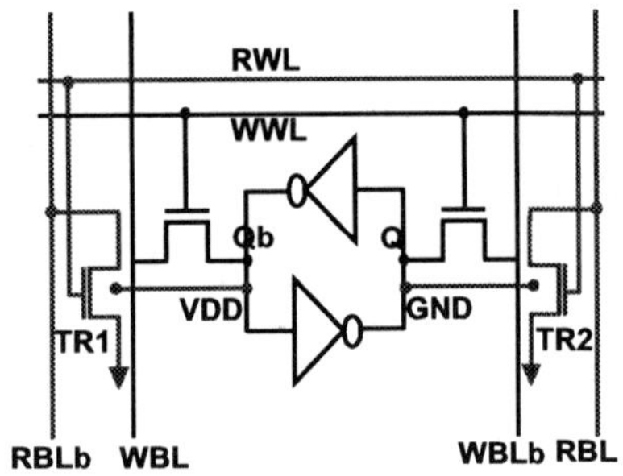

Fig. 3. The proposed 8T back-gate controlled dual-port SRAM

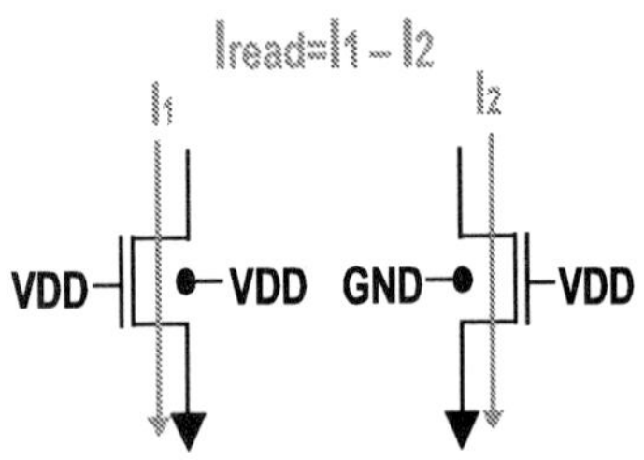

Fig. 4. Effective read current (Iread) in the proposed scheme is the difference between currents through two READ transistors.

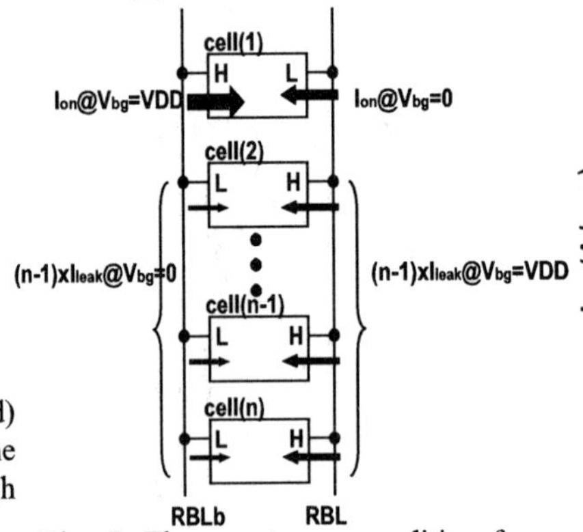

Fig. 5. The worst case condition for READ in the proposed scheme

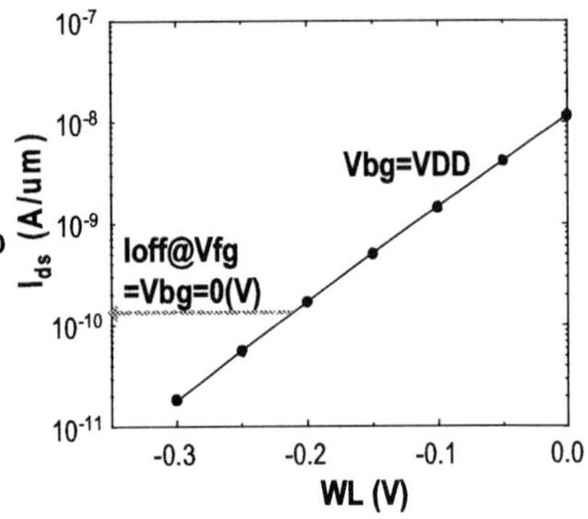

Fig. 6. Leakage reduction with negative WL voltage

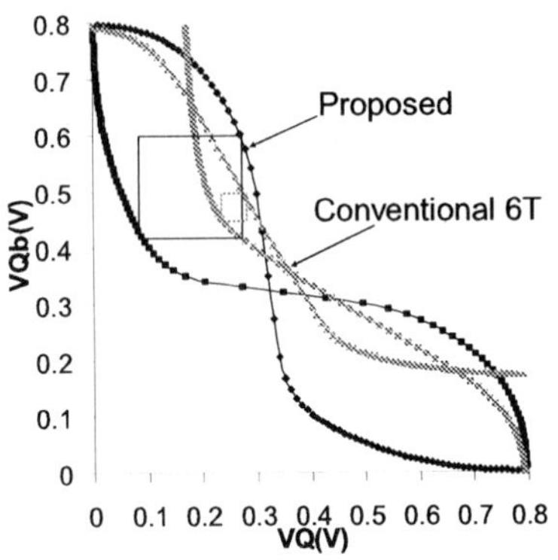

Fig. 7. SNM Butterfly Curve Comparison @Vdd=0.8V

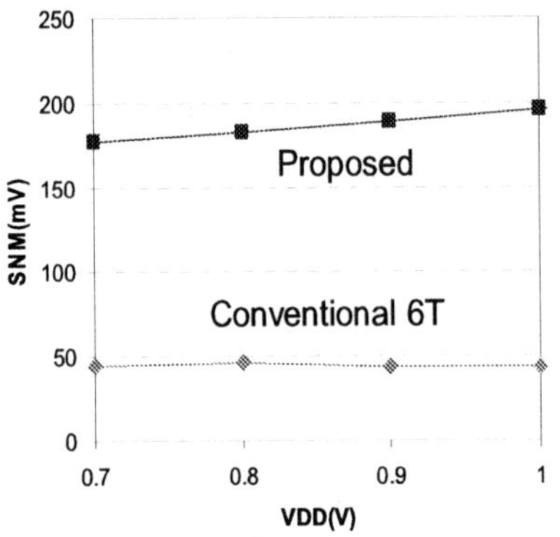

Fig. 8. SNM results over various supply voltages

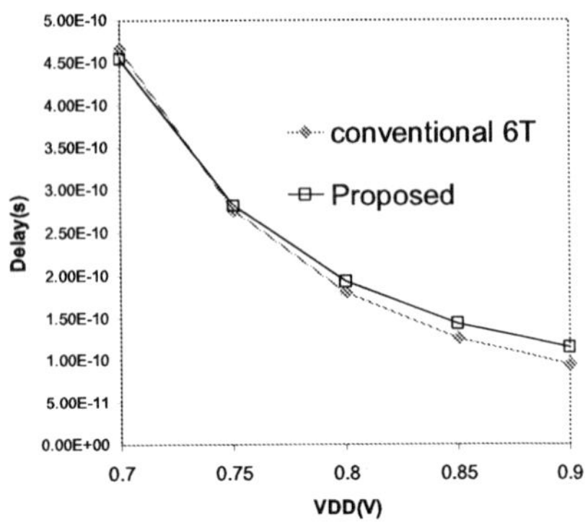

Fig. 9. Read delay comparison between the conventional 6T and the proposed 8T SRAM array

Impact of new transistor scaling methods on SOI SRAM cell stability

G. Burbach*, T. Feudel*, M. Horstmann*, D. Greenlaw*, R. Seltmann*, M. Craig[#]
S. Krishnan[+], M. Leary[+], N. Kepler[+], M. Raab*
Advanced Micro Devices
*AMD Saxony LLC, Wilschdorfer Landstrasse 101, Dresden 01109, Germany
[+]AMD, 1 AMD Way, Sunnyvale CA, 94088, USA
[#]AMD, 5900 E. Ben White Blvd, Austin TX, 78741, USA
Email: Gert.Burbach@amd.com, Tel: +49 (351) 277-4535

Abstract

The size of the embedded SRAM is steadily increasing in high-end microprocessors like Athlon64™ and Opteron™. So the demand for small cell footprints and improved stability has become more challenging. We describe how AMD has recognized and addressed the competing aspects of technology scaling and improved stability in the 90nm technology.

Introduction

Continuously improving SRAM stability requires collaboration between technology, design, and product engineering. We discuss how minimizing transistor Vt variations due to doping and gate CD control improves the static noise margin of the cell in our 90nm technology [1].

New methods to improve transistor performance like dual stress liner [2] introduces a new source for device mismatch. The proximity effects of locally introduced strain on the stability of a 6T SRAM cell will be discussed.

Technology Variations: Vt matching and CD control

Matched transistor Vts are critical to good bitcell noise margins, and with scaling transistor Vt's are becoming increasingly sensitive to non-uniformities in the implantation and activation of dopants. Fig.1 shows a probability plot for simple p+ Rs structures. Comparing activation by conventional radiative anneal to conductive heating, we see that the conductive heating split shows about half the variation of the control [3]. This uniformity improvement is due to the more rapid cool-down rates available on these tools and can improve transistor matching enough to reduce acrossfield ring-oscillator speed variation by one-third, as shown in Fig. 2.

Transistor Vts are also increasingly sensitive to gate CD matching as technology scales. Systematic gate CD matching data shown in Fig. 3 compares a group of isolated poly lines and poly lines in a constant pitch neighborhood for two steppers. "Comatic" aberrations cause a systematic mismatch of more than 5% for the "bad" stepper (lens). Caches with poly lines of uniform orientation and pitch throughout the chip avoid this effect and provide a more manufacturable SRAM.

Random components of gate CD variation are also detrimental, and line-edge roughness along a single transistor is an important one. Fig. 4 shows an AFM trace along the side of one resist line with a 10nm min-max range over a 2um trace length. We have optimized the gate patterning process to mimimize this line-edge roughness.

In Fig. 5 is shown how this line edge roughness impacts the matching of transistor drive currents at different widths. The 6 sigma range increases from ±15% upto ±25%.

These transistor matching concerns have been translated into cell stability margin through SPICE simulation. We see over a fairly wide range of interest that each additional 2mV of matched pair mismatch eliminated translates into 1mV of bitcell static noise margin gained.

Proximity effects of local stress

To overcome the technology barriers of conventional transistor scaling new methods like locally induced strain are developed [4], [5]. But while n-channel devices benefit most from tensile the p-channel devices prefer compressive strain. The proximity of the border between the different regions impacts the device performance as shown in Fig. 6. So overlay and CD tolerances in definition process of the different stressed regions are becoming an additional source of transistor mismatch.

Conclusion

We have shown that SRAM stability in 90nm SOI microprocessors must be managed through an optimization of technology and design choices. Beyond the 90nm node it will be critical to find new process tools and best layout practices to more uniformly activate dopants, continue reducing gate CD variation, and managing proximity effects for locally induced strain under the most demanding product requirements.

0-7803-9212-4/05/$20.00 ©2005 IEEE

References

[1] Greenlaw et al., IEDM 2003, p.277.

[2] Yang et al., IEDM 2004, p. 1075

[3] T.Feudel et al., EMRS 2004 Symp. B, B-I2

[4] M. Horstmann, ICICDT 2004, p. 65-71

[5] M. Horstmann, ECS Int. Symp. On SOI technology and devices, Montreal 2005

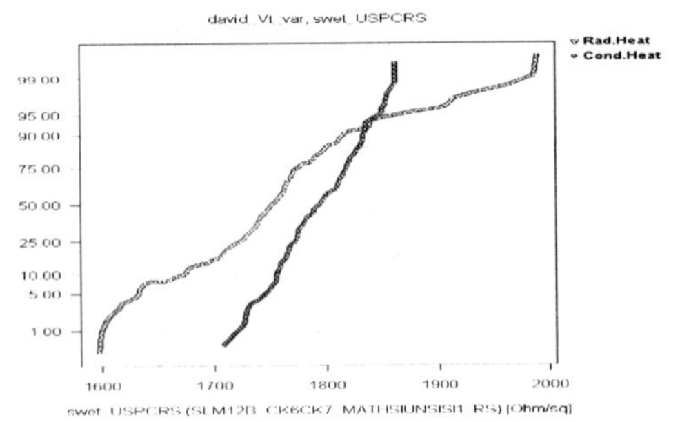

Fig. 1. Dopant activation by conductive heating shows a tighter spread in simple Rs structures compared to activation by conventional radiative heating. Variation is cut approximately in half.

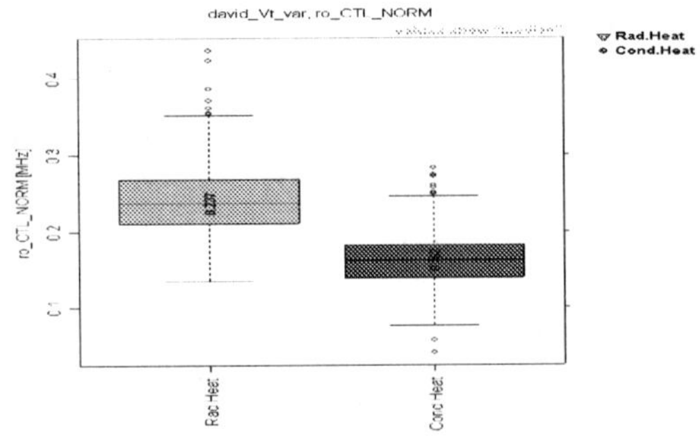

Fig. 2. Dopant activation by conductive heating shows a tighter spread in percent across-chip Ring-Oscillator speed variation. The total normalized variation is reduced by ~30%.

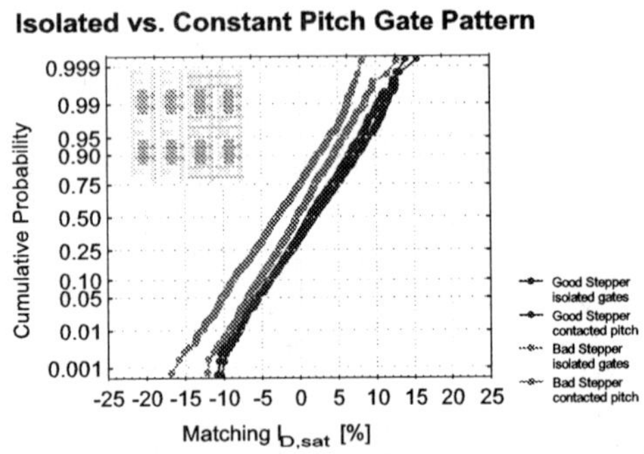

Fig. 3 Comatic Aberrations of stepper lenses causes systematic mismatch of isolated gates. Poly lines in a constant pitch neighborhood are less prone to this effect.

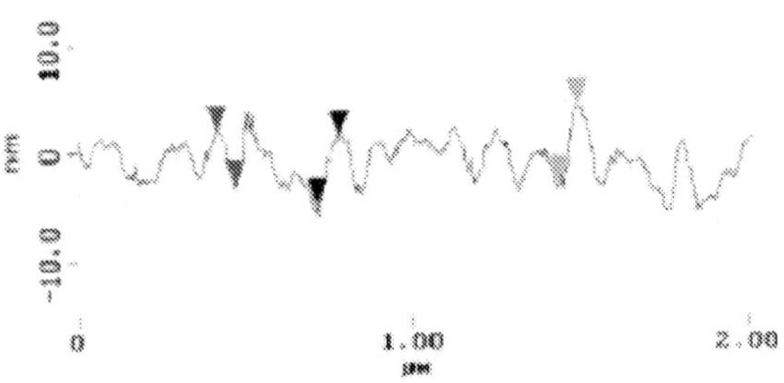

Fig. 4 Gate resist line edge roughness as measured by AFM. This major random component of gate CD variation shows ~10nm min-max range along a 2µm length of gate.

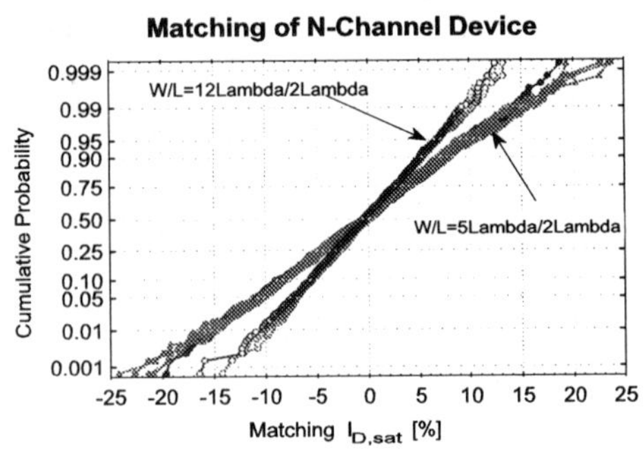

Fig. 5 shows how this line edge roughness impacts the matching of transistor drive currents of different widths. The 6 sigma range increases from ± 15% upto ±25%.

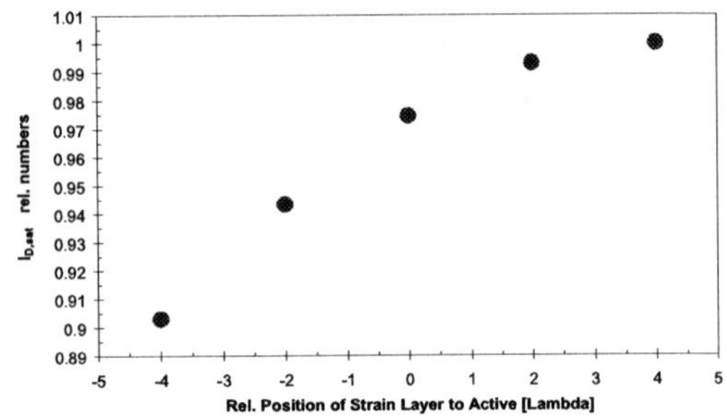

Fig. 6 The proximity of the border between different locally induced strain modulates the drive strength of the transistor.

Optimization of Substrate Doping for Back-Gate Control in SOI T-RAM Memory Technology

M. Ershov, F. Nemati, R. Gupta, V. Gopalakrishnan, R. Gooty, M. Tarabbia, K.J. Yang, S. Banna, D. Hayes, H.J. Cho, and S. Robins

T-RAM Semiconductor, 100 Headquarters Dr., San Jose, CA 95134, USA
E-mail: maxim@t-ram.com, Tel: (408) 597-3633

Abstract
This paper presents various considerations for substrate doping optimization in SOI T-RAM technology. Back gate (substrate voltage) control is used in an SOI T-RAM technology for optimizing cell characteristics. However, it is reported for the first time that typical low-doped substrates used in SOI logic technologies can create unusually slow transient effects in T-RAM cell. It is also demonstrated that the optimization of substrate doping resolves this slow transient problem and improves back gate control of SOI T-RAM memory arrays.

Introduction
T-RAM (Thyristor-based Random Access Memory) is a novel memory cell technology that breaks through the performance-density limitations imposed by conventional memory cell technologies such as 6T-SRAM and DRAM, by providing SRAM-like performance at 1/4th the cell area. T-RAM cell consists of a thin capacitively coupled thyristor (TCCT) and an NMOS access transistor (Figs.1,2). The memory effect in a T-RAM cell is based on the bistable S-type I-V characteristic of a thyristor (Fig.3), allowing a wide read margin, high read current, and a non-destructive read. The high-speed switching of TCCT, enabled by pulsing WL2 gate, provides a fast write speed. These features enable functionality and speed equal to 6T-SRAM. A T-RAM 9Mb memory array (Fig.4) has been previously demonstrated in a 130nm PD-SOI CMOS logic technology, simply integrated by adding photo-masks and implantation steps. XTEM image of T-RAM cell is shown in Fig.5. The cell size is 0.562 μm^2 (0.36 μm x 1.56 μm). A detailed description of this 130nm SOI T-RAM technology can be found in Ref. [1].

Back-gate control of T-RAM cell
The optimization of the gains of NPN and PNP bipolar devices constituting a TCCT can significantly improve cell stability. Controlling the voltage applied to the back substrate is an effective method for controlling and optimizing the NPN bipolar gain in a T-RAM cell without any adverse impact on MOSFETs. Voltage applied to the substrate induces an electric field in the BOX, which modulates the hole concentration in the p-base, the base Gummel number, and the NPN gain (Fig.6). Fig. 7 shows the effect of back gate bias on 9Mb test array measurements.

Parasitic slow transient effects
It was found that the number of array fails is sensitive to the delay between applying Vsub and starting array measurements (Fig. 8). These slow transients should be suppressed for T-RAM array to be functional immediately after power-up. Possible root causes of slow transients include RC delays in the substrate and substrate depletion and inversion. An important point is that commercially available SOI wafers typically have very low substrate doping (~1-2x10^{15} cm^{-3}, p-type), which makes them vulnerable to depletion. A low doped substrate may not supply the voltage to the region of substrate beneath the TCCT. The substrate may be even inverted if the voltage difference between the top Si layer and substrate exceeds threshold voltage of BOX MOS capacitor (where Si layer plays the role of gate). Since there is not good electrical contact to the inversion layer, its generation may take a very long time (especially at low temperatures) when voltages are changed.

To verify the mechanism of slow transients, we performed measurements on the test structures comprising large area BOX MOS capacitors. Figure 9 shows measured transient current in response to step-like voltage signal. With positive Vsub (substrate in accumulation), no slow transients were observed. For negative Vsub, very slow transients were measured with time constant on the scale 10-100 s. Time constant was greatly reduced at high temperatures or under illumination. This confirms that the root cause of slow transients is due to slow generation of the electron inversion layer, which is sped up by increased generation in the depletion region by high temperature or light.

Substrate doping for suppressing slow transients
To suppress parasitic slow transients, substrate inversion should be avoided. This can be accomplished by high doping of the substrate near its interface with the BOX. Impurity concentration should be high enough to make BOX MOS threshold voltage much higher than applied voltage. High doping can be achieved either by using SOI wafers with heavily doped substrate, or by ion implantation of impurities into substrate through the top layers (Fig.10). Figure 11 shows the results of TCAD simulations of MOS structure with low (10^{15} cm^{-3}) and high (2x10^{17} cm^{-3}) substrate doping. Vsub=-3V was applied to the substrate at time t=0, and transient current was calculated. At Vsub=-3V, the low doped substrate is inverted and there is a significant voltage drop in the substrate due to depletion. It takes a very long time (>10 s) to establish a steady-state condition. None of these effects are present in a high doped substrate.

T-RAM arrays were fabricated on highly doped substrates (using blanket ion implantation). Implantation conditions were optimized to achieve high doping near the BOX, while avoiding implantation tail in the top Si layer. Array results shown in Fig.12 clearly confirm that the slow transient problem has been eliminated.

References
[1] F.Nemati et al., IEDM Tech. Digest, p.273, 2004.

0-7803-9212-4/05/$20.00 ©2005 IEEE

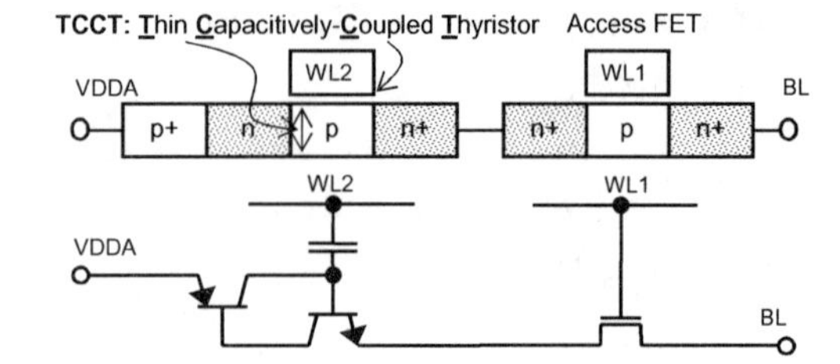

Fig. 1: Structure and equivalent circuit model of T-RAM cell.

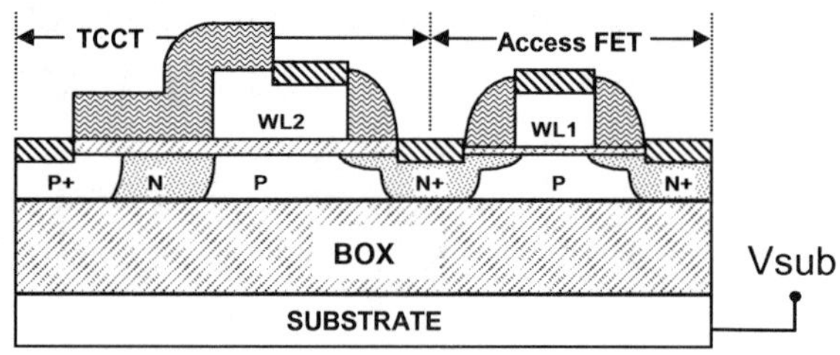

Fig. 2: Cross-sectional diagram of planar SOI T-RAM cell.

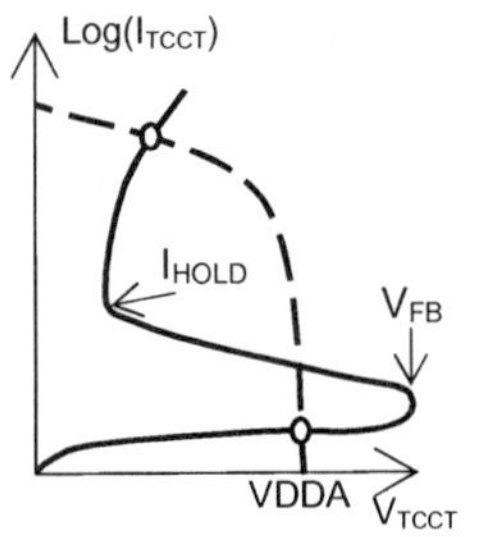

Fig.3. Load-line diagram of T-RAM cell.

Fig. 4: Image of 9Mbit T-RAM test chip with full SRAM functionality.

Fig. 5: Cross-sectional TEM image of T-RAM cell implemented in standard 130nm SOI technology.

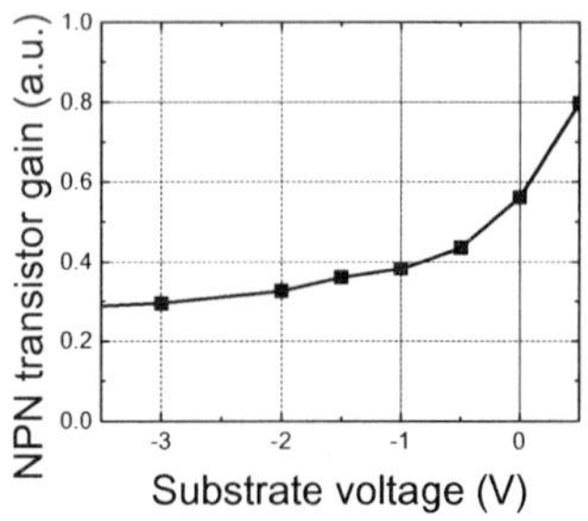

Fig. 6. Vsub dependence of NPN gain.

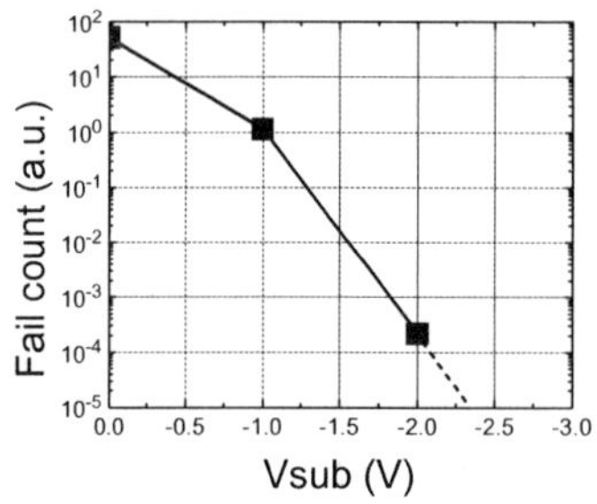

Fig.7. Array fail count vs back-gate bias Vsub.

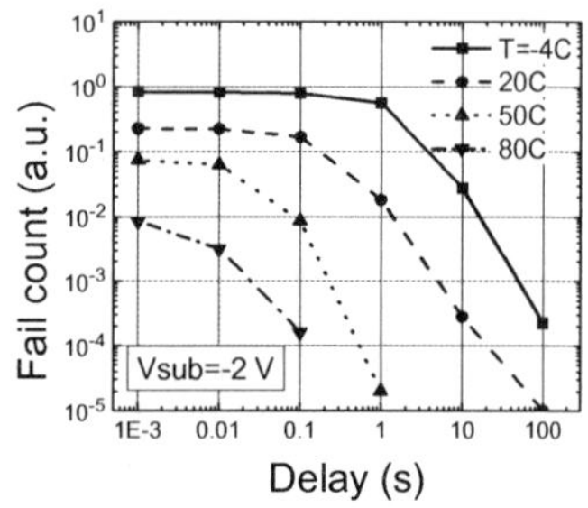

Fig.8. 9 Mbit array fail count vs delay between Vsub and measurements.

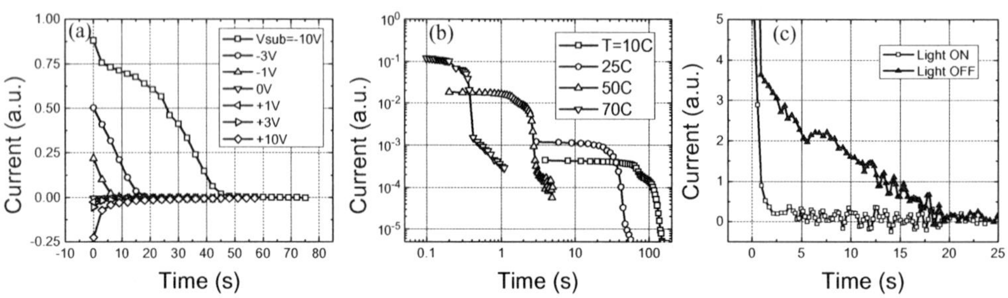

Fig.9. Transient current in MOS capacitor in response to a step-like voltage signal (SOI Si layer – gate, BOX – oxide, substrate – semiconductor) for different conditions: (a) applied voltage, (b) temperature, and (c) illumination.

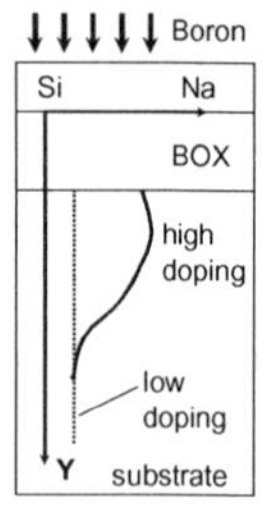

Fig.10. SOI substrate doping diagram.

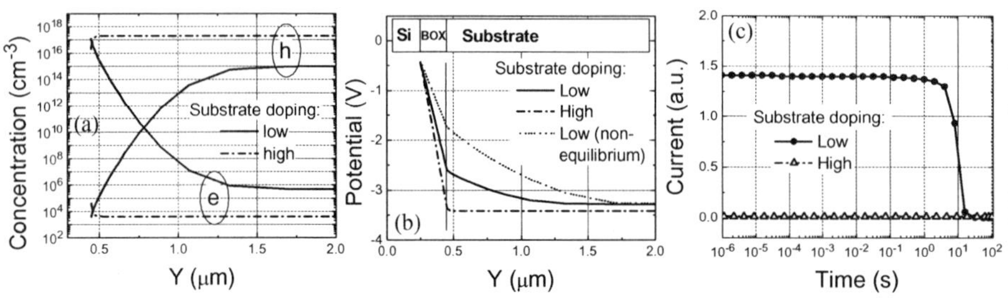

Fig.11. TCAD simulation results at Vsub=-3V: (a) electron and hole concentration, (b) potential (dotted line - right after applying Vsub), and (b) transient current - for low (1e15 cm-3) and high (2e17 cm-3) doping.

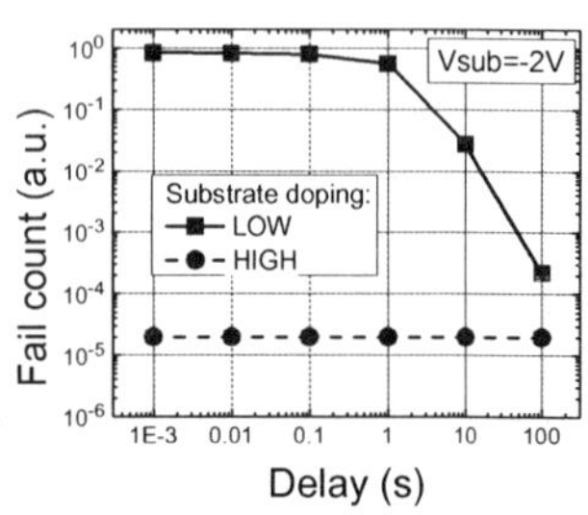

Fig. 12. 9 MBit array fail count for low and high doped substrate.

2005 IEEE INTERNATIONAL
SOI CONFERENCE

Session #10

LATE NEWS SESSION

CHAIR

James Burns

MIT Lincoln Labs

Advanced FinFET Technology: TiN Metal-gate CMOS and 3T/4T Device Integration

Y. X. Liu, K. Endo, M. Masahara, E. Sugimata, T. Matsukawa, K. Ishii, H. Yamauchi, T. Shimizu, K. Sakamoto, S. O'uchi, T. Sekigawa and E. Suzuki
National Institute of Advanced Industrial Science and Technology (AIST)
1-1-1 Umezono, Tsukuba, Ibaraki 305-8568, Japan, Tel: +81-29-861-2417, Email: yx-liu@aist.go.jp

Introduction

Recently, fin-type double-gate (DG) [1] MOSFETs (FinFETs [2]) have received considerable attention as the most promising candidate for future Si VLSI circuits, thanks to their high short-channel effects immunity, high current drivability, and flexible threshold voltage (V_{th}) controllability by the independent double gates [3-8]. To meet both the inevitable demands of high-performance and low power consumption for future VLSIs, further development of FinFET technology is continuously required. Namely, a suitable metal gate is necessary to set a proper V_{th} for high-performance connected-double-gate 3-terminal (3T) FinFETs without channel doping. Co-integration of the 3T- and the V_{th} controllable independent-double-gate 4-terminal (4T) FinFETs is also required, which has been difficult by the conventional CMP process. This paper presents the newly developed fabrication technique for co-integration of the TiN gated 3T- and 4T-FinFETs by reactive sputtering and resist etch-back processes for the first time. The well symmetrical V_{th} of the fabricated N- and P-channel FinFETs and high V_{th} controllability in the 4T-FinFETs are experimentally demonstrated.

Experimental results and discussion

A three-dimensional schemetic diagram of the proposed TiN gated 3T- and 4T-FinFETs is shown in Fig. 1. Prior to device fabrication, the purity of sputted TiN and the interface trap density (D_{it}) were examined as shown in Figs. 2 and 3. It is found that no impurities such as carbon and oxygen are detected and less D_{it} is introduced in the sputtering process. To realize the uniform TiN deposition on the sidewalls of the Si-fins, the sputtering pressure was optimized to be 1 Pa [9].

In the device fabrication, we used a lightly-doped p-type (110)-oriented SOI wafer as a staring material, and Si-fin channels were fabricated by using the orientation-dependent wet etching.

To evaluate the carrier mobility and V_{th}, we fabricated the large TiN gated N- and P-channel multi-FinFETs by using wet etching with an APM (NH_4OH : H_2O_2 : H_2O = 1: 2: 5) solution at 60 °C as shown in Fig. 4. The cross-sectional SEM images of the fabricated TiN gated N- and P-channel multi-FinFETs are shown in Fig. 5. Note that the TiN is uniformly deposited on the sidewalls of the upright Si-fin channels. This indicates that the pressure-optimized reactive sputtering can be used in the TiN gate FinFET fabrication. It is noteworthy from Fig. 6(a) that the well symmetrical I_d-V_g characteristics are realized by adopting the TiN gate. It is also remarked from Fig. 6(b) that the Si-fin height (H_{fin}) control for the N- and P-channel FinFETs is effective to adjust the CMOS drive current. The peak effective carrier mobility in the (111) channel surface was estimated to be 168 and 112 cm^2/Vs for electrons and holes.

The fabrication flow for the co-integration of the TiN gated N-channel 3T- and 4T-FinFETs is shown in Fig. 7. At first, we select the area for 4T-FinFETs, and remain a thick photoresist film on the regions for 3T-FinFETs. Then, a thin EB-resist film is coated on the wafer as shown in Fig. 7(b). Note that the resist thickness on top of the Si-fin is much thinner than that on the saidewall area of the Si-fin. Finally, the thin EB-resist etch-back and RIE are carried out to separate the continuous TiN film on the top of Si-fin as shown in Fig. 7(d). The cross-sectional STEM image of the fabricated TiN gate 4T-FinFET with an 8.6-nm-thick Si-fin channel is shown in Fig. 8.

Figure 9(a) shows the I_d-V_{g1} characteristics with V_{g2} as a parameter for the fabricated TiN gate 4T-FinFET. It is noted that the V_{th} control is more flexible than that for the previously reported n^+-poly-Si gate one [3, 4], thanks to the desirable V_{th} of 0.4 V in the case of 3T-FinFET as shown in Fig. 9(a). However, S-slopes of the 4T-FinFET are worse than that of 3T-FinFET due to the symmetric thin gate oxide and ultra-thin Si-fin channel in the demostrated device. The S-slope should be designed by introducing asymmetric gate oxide thicknesses, $t_{ox2} > t_{ox1}$, and/or optimizing the fin thickness through the relationship of $S = 60\,(3t_{ox1} + 3t_{ox2} + T_{Si})/(3t_{ox2} + T_{Si})$ [6]. On the contrary, in the synchronously driving double gates operation with an offset voltage (V_{OS}), the I_d-V_{g1} characteristics can be controlled with no degradation in the S-slope as shown in Fig. 9(b).

Conclusion

As advanced FinFET technologies, we have developed the co-integration techniques of the TiN gated high-performance 3T- and flexible V_{th} 4T-FinFETs. By using the conventional reactive sputtering of TiN, the well symmetrical V_{th} N- and P-channel 3T-FinFETs and the high V_{th}-controllable 4T-FinFETs using the resist etch-back process have been demonstrated. The developed technologies are attractive to materialize the high-performance and power-managed FinFET CMOS circuits.

References

[1] T. Sekigawa et al., Solid-State Electronics, 27, 827 (1984). [2] D. Hisamoto et al., IEEE Trans. ED. No. 47, 2320 (2000). [3] Liu et al., IEDM. Tech. Dig. 986 (2003). [4] Liu et al., IEEE EDL No. 25, 510 (2004). [5] D. M. Fried et al., IEEE EDL No. 24, 592 (2003). [6] M. Masahara et al., Proc. IEEE Int. SOI Conf., 100 (2004). [7] M. Masahara et al., ESSDERC. Tech. Dig. 73 (2004). [8] L. Mathew et al., VLSI Symp. Tech. Dig. 200 (2005). [9] Liu et al., IEEE Silicon Nanoelectronics Workshop, 22 (2005).

0-7803-9212-4/05/$20.00 ©2005 IEEE

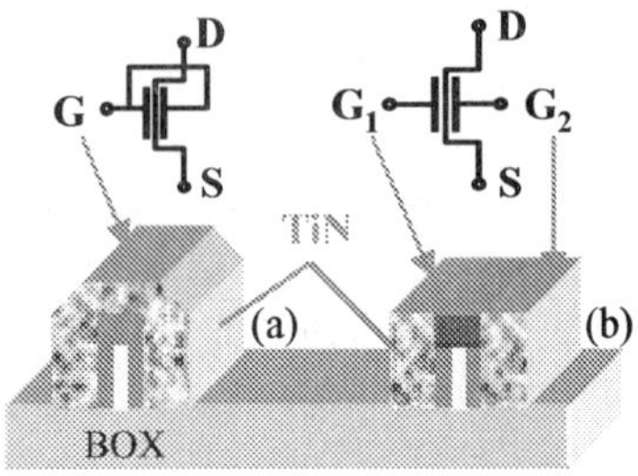

Fig. 1. Three-dimensional schematic diagram of the proposed TiN gated (a) 3-terminal and (b) 4-terminal FinFETs integration.

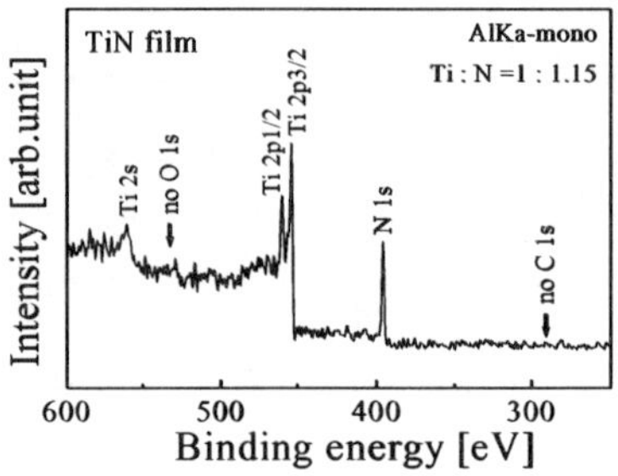

Fig.2. XPS spectrum of the TiN film deposited at $R_N = N_2/(N_2{+}Ar)$ 83 %. No carbon and oxygen are observed, indicating the high purity of TiN.

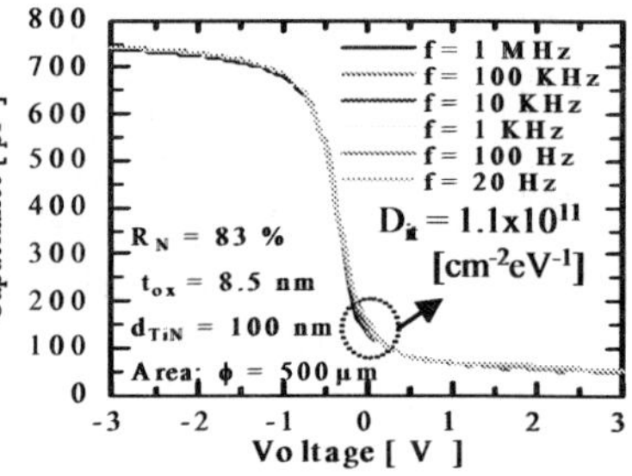

Fig. 3. Frequency dependence of C-V curves. Almost the same C-V curves indicate that less D_{it} is introduced in the sputtering.

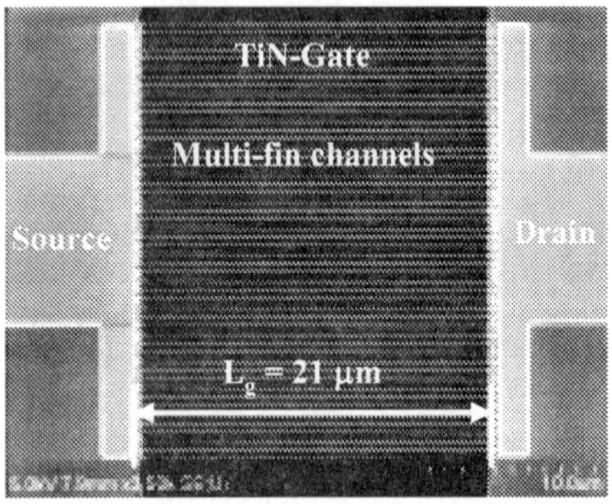

Fig. 4. Plane-SEM image of the fabricated multi-FinFET after TiN gate wet etching which is very effective for large gate formation.

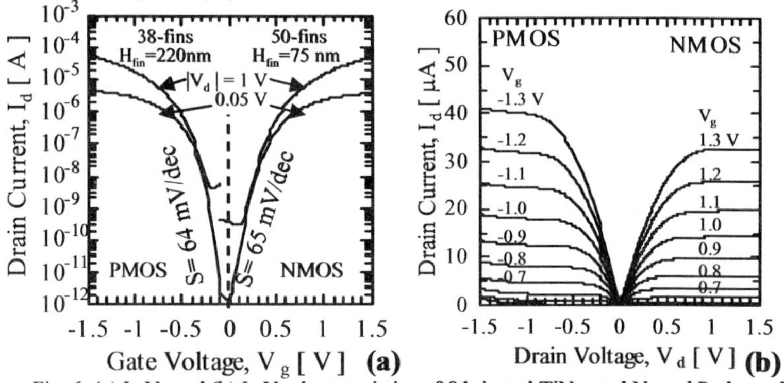

Fig. 6. (a) I_d-V_g and (b) I_d-V_d characteristics of fabricated TiN gated N- and P-channel Multi-FinFETs with the same $L_g = 21$ μm, $t_{ox} = 2.2$ nm and $T_{Si} = 75$ nm.

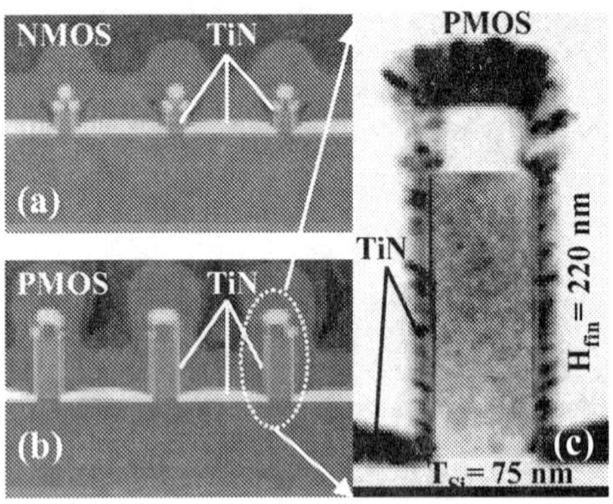

Fig. 5. Cross sectional STEM images of the fabricated (a) N-, and (b) & (c) P-channel multi-FinFETs. The uniform TiN on the sidewalls of Si-fins is confirmed.

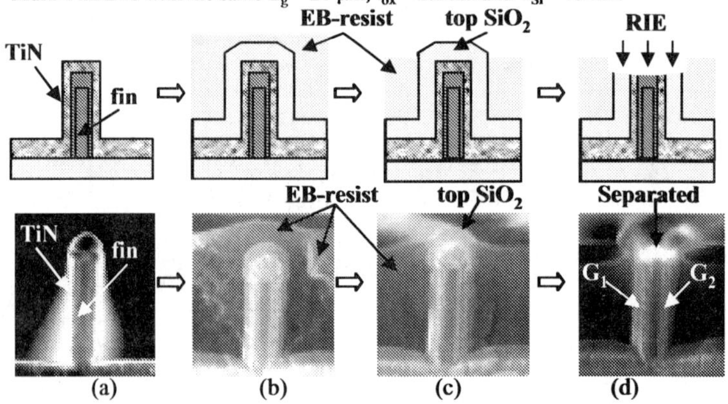

Fig.7. Fabrication process flow for TiN gate 3T- and 4T-FinFETs co-integration and cross-sectional SEM images of the Si-fin at each step. (a) TiN deposition, (b) opening 4T-FinFET regions by lithography and thin EB-resist coating, (c) EB-resist etch-back to the top SiO_2, (d) SiO_2 & TiN RIE for separating connected TiN gate.

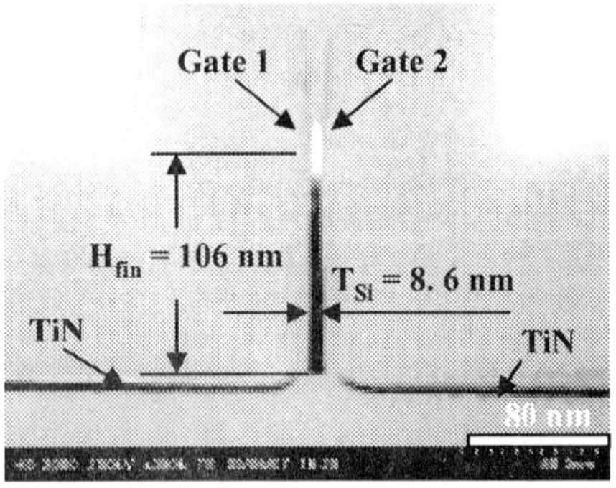

Fig. 8. Cross-sectional STEM image of the fabricated TiN gate 4T-FinFET with a 8.6-nm-thick Si-fin channel. The highest aspect ratio of $H_{fin}/T_{Si} = 12$ is obtained thanks to the wet etching.

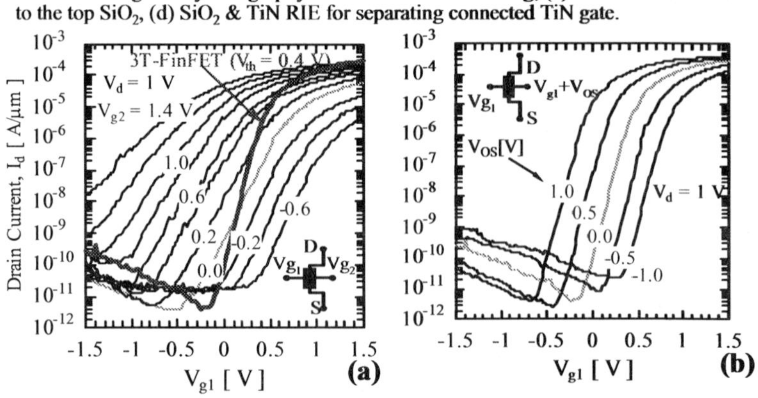

Fig. 9. (a) Single-drive (SD) mode and (b) synchronized double gates drive (DD) mode operation of the fabricated TiN gated 3T- and 4T-FinFET with $T_{Si} = 8.6$ nm, $H_{fin} = 106$ nm, $t_{ox} = 2.2$ nm and $L_g = 350$ nm.

$0.525\mu m^2$ 6T-SRAM Bit Cell using 45nm Fully-Depleted SOI CMOS Technology with Metal gate, High K Dielectric and Elevated Source/Drain on 300mm wafers.

A. Vandooren [1], C. Hobbs[1], O. Faynot [4], P. Perreau[4], S. Denorme[3], C. Fenouillet-Beranger[4], C. Gallon[3], C. Morin[4], A. Zauner[2], G. Imbert[3], H. Bernard[3], P. Garnier[2], L. Gabette[2], M. Broekaart[2], M. Aminpur[1], S. Barnola[4], N. Loubet[3], D. Dutartre[3], T. Korman[2], G. Chabanne[1], F. Martin[4], Y. Le Tiec[4], N. Gierczynski[2], S. Smith[2], C. Laviron[4], M. Bidaud[2], I. Pouilloux[2], D. Bensahel[3], T. Skotnicki[3], H. Mingam[3], A. Wild[1].

(1) Freescale Semiconductors., (2) Philips, (3) ST Mircoelectronics, 850 Rue Jean Monnet, 38926 Crolles, France.
(4) LETI-CEA, 17 Av. des Martyrs, 38054 Grenoble CEDEX 9, France

INTRODUCTION: This paper describes the fabrication and performance of a 45nm fully-depleted SOI technology using a direct metal gate (TiN) and high k dielectric (HfSiON) integrated for the first time on 300mm SOI wafers for low power applications. Performance at the device and at the circuit level are demonstrated. As CMOS technology scales down, thin film devices become more and more attractive in order to control short channel effects [1-3]. In these devices, short channel effects are controlled by the use of thin ~13nm) silicon film in the channel while selective epitaxial growth is performed in the Source/Drain extension regions to reduce the series resistance. While two metal gates are required to adjust the threshold voltage of n and pMOSFETs for bulk technology [4], the use of a single midgap metal gate on SOI offers the advantage of adjusting the threshold voltage of n-channel and p-channel devices simultaneously while keeping the channel undoped. Furthermore, the use of a metal gate offers the additional advantages of suppression of polysilicon depletion effects, elimination of boron penetration and reduction of the gate sheet resistance. Hafnium silicate is used as gate dielectric to reduce gate leakage.

FABRICATION: The devices were fabricated on 300mm UNIBOND® SOI wafers featuring 700Å initial silicon film and 1450Å buried oxide using conventional CMOS integration. The superficial silicon was thinned down to 200Å by oxidation. A 30 Å $HfSi_yO_x$ (60% Hf) gate dielectric was deposited by MOCVD followed by NH_3 nitridation. 100Å PVD TiN was reactively sputtered to form the gate electrode. After offset spacer formation and Hf silicate removal from the S/D regions, a 250Å-thick selective epitaxial silicon was grown in the extensions regions at 750°C. The channel was left essentially undoped with a resulting resistivity of 10 Ω-cm, corresponding to a doping concentration of $\sim 10^{15} cm^{-3}$. RTA at 950°C was performed after SD implantation. Nickel silicidation and copper backend processing were used. No process-induced or substrate-induced strain was used to boost the device performance. A cross-section of the final 45nm device is illustrated in Fig.1.

ELECTRICAL RESULTS: Long channel device I_{DS} vs. V_{GS} curves (Fig. 2) show symmetrical n and p threshold voltage due a TiN gate workfunction close to midgap. The subthreshold slope is near its ideal value (62mV/dec) demonstrating a low interface trap density at the high k/Si interface. I_{DS} vs. V_{GS} and I_{DS} vs. V_{DS} electrical characteristics of 45nm long n and p-channel devices are illustrated in Figs. 3a and 3b, respectively. The short channel effects are well controlled, as illustrated in Figs. 4a, 4b and 5, for subthreshold slope, DIBL and threshold voltage, respectively. As the SD RTA time decreases, the control of short channel effects improves. The I_{on} vs I_{off} curves (Figs. 6a&b) show the best pMOSFET with I_{off}=1nA/µm and I_{on}=300µA/µm and best nMOSFET with I_{off}=1nA/µm and I_{on}=400µA/µm @V_{DD}=1V. These include self-heating effect, which can be important on SOI substrates. The overlap capacitance versus front gate voltage curves (Fig. 7) exhibit no polysilicon depletion effect. The extracted CET_{inv} is 20Å. The effective mobility at the front interface was extracted using the split C-V technique on long channel devices [5] and is illustrated versus effective electric field in Fig. 8. In comparison to the universal mobility, this corresponds to a mobility degradation of ~40% and ~25% for electrons and holes, respectively at operating voltage. Part of this mobility degradation is due to the use of high k dielectric as already observed on bulk substrate. $0.92um^2$ and $0.525um^2$ SRAM bit cells are fully functional. The performance is summarized in Table 1. The butterfly curves and signal to noise margin are illustrated in Fig. 9.

FDSOI, $0.92um^2$	V_{DD}=0.8V	V_{DD}=1.2V
SNM (mV)	230	275
Icell (uA)	2.2	10
Isb (pA)	170	543
Ileak (pA)	0.07	0.15
WM (mV)	230	460

Table 1: SRAM bit cell data

CONCLUSION: A Low power 45nm Fully-depleted SOI technology is demonstrated for the first time on 300mm SOI wafers, using direct metal gate on high k dielectric and selective silicon epitaxy. Short p-channel devices exhibit very good performance. SRAM bit cells are fully functional down to $0.525um^2$ with good SNM and low leakage.

ACKNOWLEDGEMENTS: The authors wish to thank the physical analysis lab in Austin and Crolles and the electrical characterization lab in Crolles for their support. A part of this work was supported by the European NanoCMOS project.

REFERENCES: [1] R. Chau et al. , IEDM 2001, p. 621 [2] B. Doris et al. IEDM 2002 p. 267 [3] Z. Krivokapic et al., 2003 Symp. VLSI p. 131. [4] S. Samavedam et al., 2002 Symp. VLSI Tech. Dig., p.24 [5] C.G. Sodini et al., Solid-State Electronics, vol. 25, pp. 833-841,1982.

0-7803-9212-4/05/$20.00 ©2005 IEEE

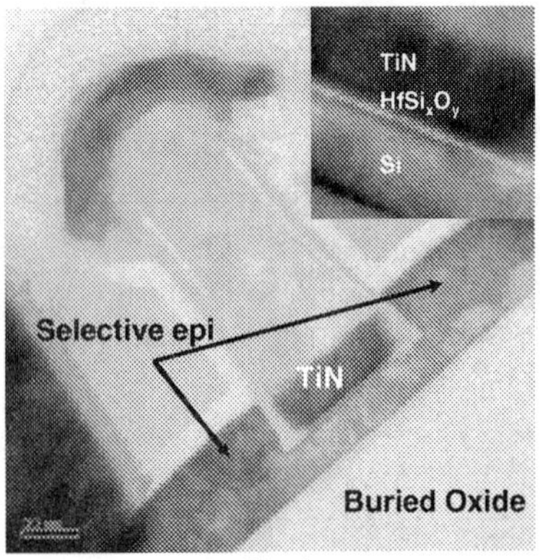

Fig. 1.

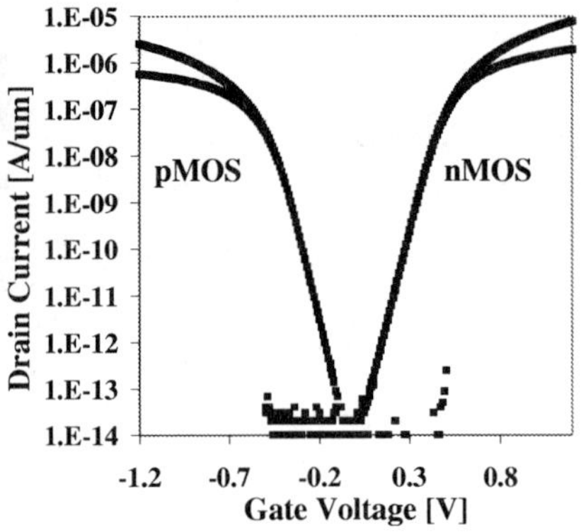

Fig.2

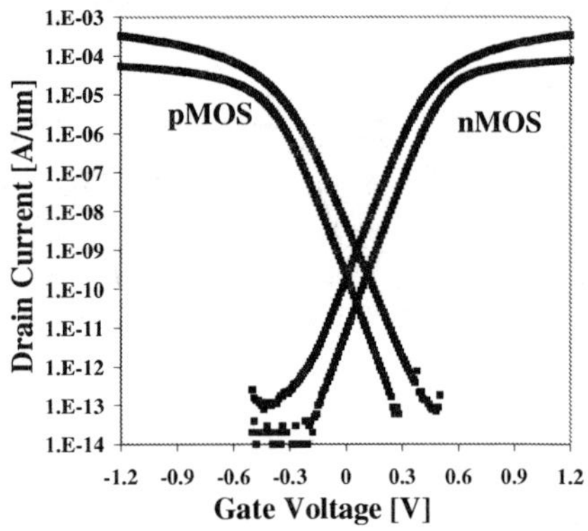

Fig.3a

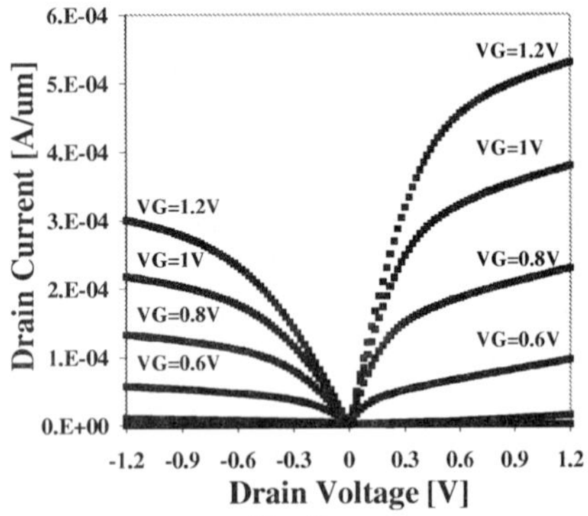

Fig. 3b

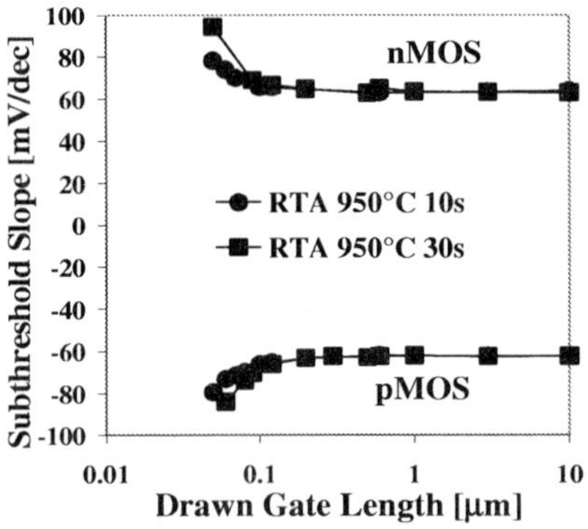

Fig. 4a

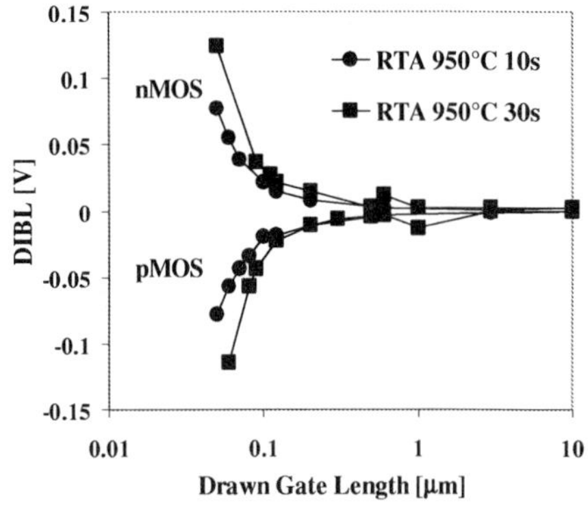

Fig.4b

Fig. 1: HR-TEM micrograph of the 45nm FDSOI device. **Fig. 2**: I_{DS} vs. V_{G1} at V_{DS}=+/-0.1V and +/-1.2V and V_{G2}=0V in 10μm n and p-channel FDSOI MOSFETs. **Fig. 3**: (a) I_{DS} vs. V_{G1} at V_{DS}=+/-0.1V and +/-1.2V (b) I_{DS} vs. V_{DS} in 45nm n and p-channel FDSOI MOSFETs. **Fig. 4**: (a) Subthreshold slope (b) DIBL in n and p-channel FDSOI MOSFETs.

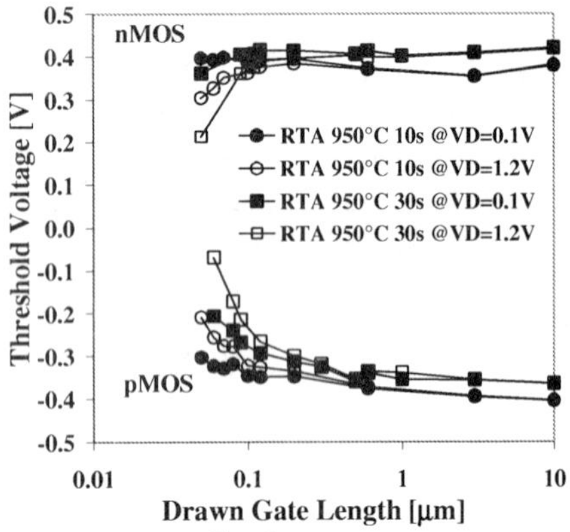

Fig. 5

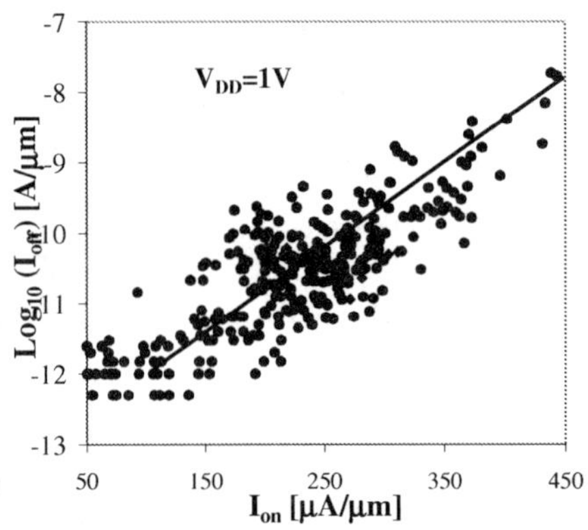

Fig. 6a

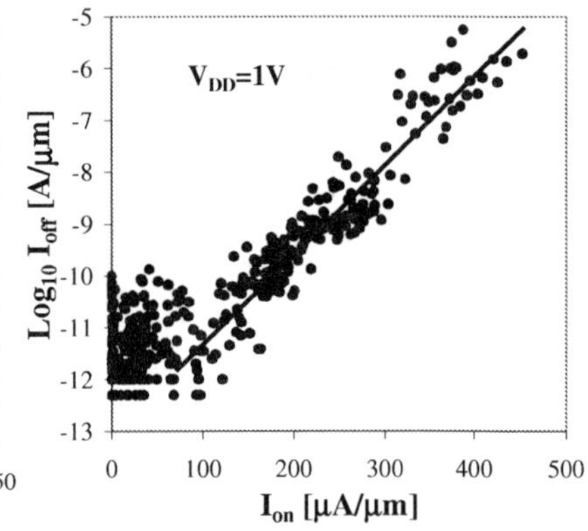

Fig.6b

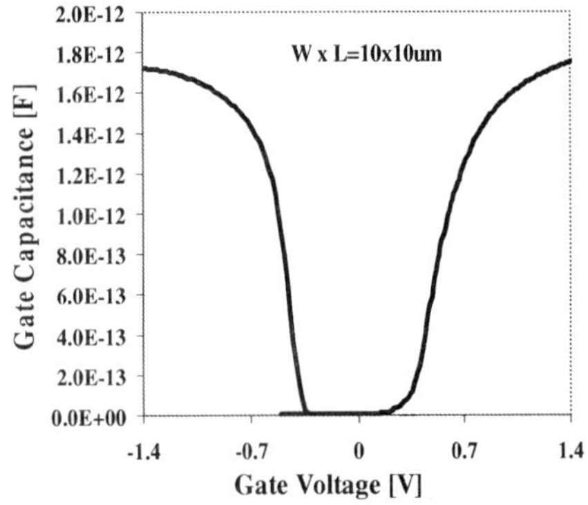

Fig. 7

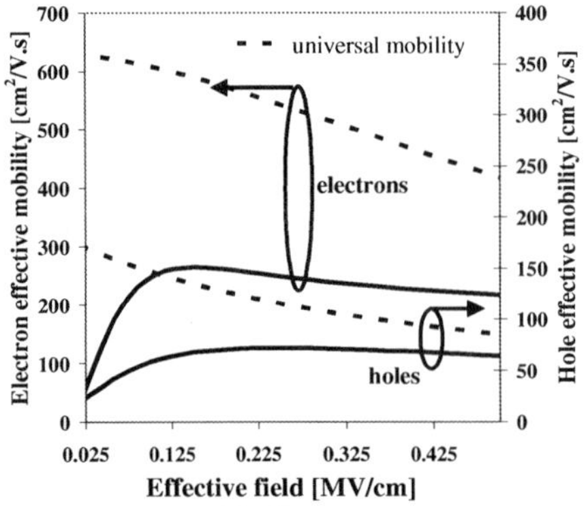

Fig.8

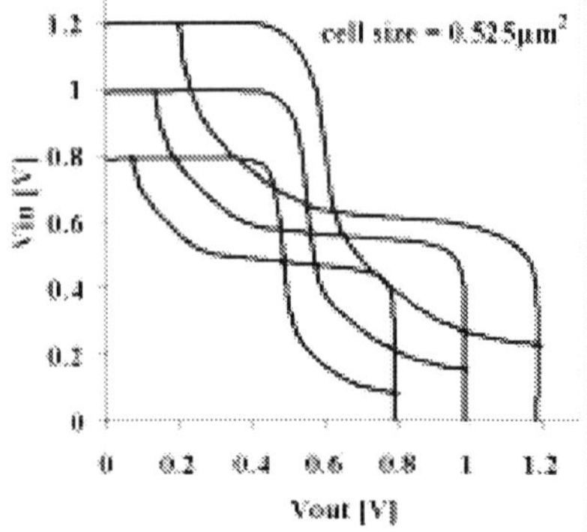

Fig.9

Fig. 5: Threshold voltage vs gate length in n and p-channel FDSOI MOSFETs **Fig. 6**: I_{on} vs I_{off} curves at V_{DD}=1V in (a) n-channel and (b) p-channel FDSOI MOSFETs **Fig. 7**: Capacitance vs. front gate voltage in p and n-channel devices. CET_{inv}=20Å **Fig.8**: Front interface electron and hole effective mobility vs. effective field in 10μm n- and p-channel FDSOI MOSFET **Fig. 9**: Buttefly curves in 0.525um² 6T-SRAM bit cell.

Co-integrated Dual Strained Channels on Fully Depleted sSDOI CMOSFETs with HfO_2/TiN Gate Stack down to 15nm Gate Length

F. Andrieu[1,2], T. Ernst[1], O. Faynot[1], Y. Bogumilowicz[3], J.-M. Hartmann[1], J. Eymery[1], D. Lafond[1], Y.-M. Levaillant[4], C. Dupré [1], R. Powers [5], F. Fournel [1], C. Fenouillet-Beranger[1], A. Vandooren[5], B. Ghyselen [4], C. Mazure [4], N. Kernevez [1], G. Ghibaudo[2] and S. Deleonibus [1]

(1) CEA/LETI and DRFMC, 17 avenue des Martyrs, 38054 Grenoble, France, (2) IMEP, BP 257, 38016 Grenoble, France
(3) STMicroelectronics, 850 rue J. Monnet, 38926 Crolles Cedex, France (4) SOITEC, , 38190 Bernin, France,
(5) Freescale Semiconductor, 3501, Ed. Bluestein Blvd, Austin, TX 78721, USA

Abstract

We report an original Dual Channel Fully Depleted CMOSFET architecture on insulator (DCOI) co-integrating strained-Si (nMOS) and strained-$Si_{0.6}Ge_{0.4}$ (pMOS) with HfO_2/TiN gate stacks down to 15nm gate length. We demonstrate for the first time an I_{ON} improvement for short channel SOI of 10% at 35nm gate length (25% at 75nm, 100% on long channels) for both n- and p-MOSFETs and a more than 3 decades gate leakage reduction compared to a SiO_2 dielectric. Meanwhile, thanks to the dual channel engineering, a threshold voltage adjustment is performed with a mid gap single metal gate suitable for high performance (HP) CMOS.

Introduction

A 125% long channel mobility enhancement has already been demonstrated for nFETs with Fully Depleted strained Silicon Directly On Insulator (FD sSDOI) ([1] and Table 1). However, a strained $Si_{1-x}Ge_x$ or Ge channel provides higher performance enhancements for pFETs (see μ_h in Table 1). For this reason, we integrated a dual strained channel on sSDOI with compressive $Si_{0.6}Ge_{0.4}$ for pMOS. This Dual Channel On Insulator architecture is prefered to a single SiGe/sSi stack for nMOS and pMOS [2,4] because of a better pMOS scalability due to our surface channel operation mode.

Ref	Substr.	Gate stack	EOT (Å)	Long channel μ_h gain	Long channel μ_e gain	T_{Si} (nm)	Short channel I_{ON} gain (%)
[3]	Co.int FD sSi/GeOI	SiO_2/poly-Si	>140	**110%** 500%	**65%** 0%	172-32 10	-
[4]	FD sSi/sSiGe	SiO_2/poly-Si	>40	**107%**	90%	25	-
[5]	PD sSDOI	SiON/poly-Si		50%	63%	30	n: 8% @ L_G=40nm p: 12% @ L_G=150nm
[1]	FD sSDOI	SiON/poly-Si		21%	**125%**	**19**	-
This work	**Co int. FD DCOI**	**HfO_2/TiN**	20	**100%**	**100%**	**14-30**	**n&p**: 25% @ L_G=75nm **n&p**: 10% @ L_G=35nm

Table 1: Comparison between this work and the results of advanced strained devices On Insulator.

In this paper, the sSDOI material stability (by Grazing Incidence X-Ray Diffraction and Raman analyses), as well as the gate stack properties, the long/short channel transport mechanisms (by low-T measurements) and the short channel DCOI performance will be assessed.

sSDOI Material and Device Process

sSDOI substrates were fabricated starting from tensily strained (001) Si/relaxed $Si_{0.8}Ge_{0.2}$ epitaxial layers, transferred using the Smart Cut™ technology [6-7]. After a photo-lithography step (see Fig.2 for more details on the process), a 16nm thick, compressively strained $Si_{0.6}Ge_{0.4}$ layer and a sacrificial 1nm Si cap were selectively grown on the pMOS sSDOI active areas.

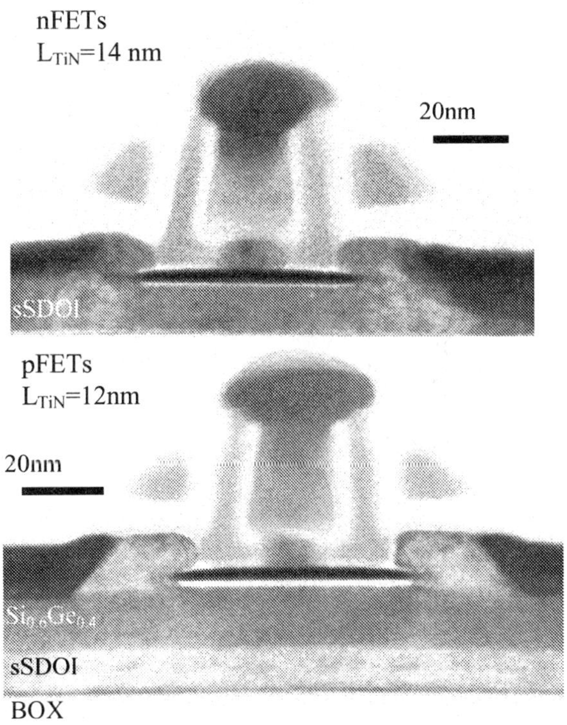

Fig.1: Cross sectional TEM pictures of the co-integrated dual channels MOSFETs with a HfO_2/TiN/Poly/NiSi gate stack.

sSDOI pMOS devices have also been processed for comparison purposes. The 1nm strained silicon cap is necessary to optimise the interface quality [8]. A thin offset spacer was formed to protect the TiN gates during the selective epitaxy of 15nm thick Si raised sources and drains. Tilted implants were then used to dope the

0-7803-9212-4/05/$20.00 ©2005 IEEE

extensions. Fig.1 shows the final dual channel, Ni-salicidated n&pMOSFETs with 15 nm gate length on 14nm thick sSDOI. Their electrical characteristics are presented on Fig. 3.

sSDOI Material Stability

The tensile strain measured by Raman spectroscopy in the starting sSDOI is equal to 1.6 GPa (Fig.4). Different experiments were carried out to evaluate the sSDOI material stability after (i) a 1050°C Rapid Thermal Anneal, (ii) 300nm wide MESA processing and (iii) extensions implantations. First, 2.10^{15} Ge atoms/cm^2 were implanted at 3keV in SOI and sSDOI. Ge was chosen because it is not a doping atom for silicon and its atomic mass is similar to As. Fig.4 proves that this implantation followed by a 600°C re-crystallisation anneal does not induce any strain relaxation in 15 to 33nm thick sSDOI. Second, we proved that even after a 1050°C RTA, the strain reduction is only 15% (Fig.5). Moreover, if the sSDOI is etched before the anneal to form 300nm-wide multi-MESA (insert of Fig. 6), the strain is reduced by the same magnitude in the narrower direction only (Fig.5). The anisotropy is clearly shown by the shape of (220) and (2-20) Bragg peaks in the reciprocal space (Fig.6). These results suggest that, even in narrow active areas, the sSDOI strain could be maintained in the <110> transport direction, boosting nMOSFETs performances.

sSDOI CMOS potential and limitations

SOI and sSDOI transistors with a SiO_2/TiN or a HfO_2/TiN gate stack were processed and compared. On one hand, the high-k dielectric allows a more than 3 decades J_G reduction (Fig.7) for the same EOT=2nm, for both SOI and sSDOI. On the other hand, the electron mobility in HfO_2 devices is reduced, this mainly for low inversion charge densities (N_{inv}) (Fig.8). It evidences the well-known HfO_2-induced extra charges. At $N_{inv}=10^{13}cm^{-2}$, sSDOI/HfO_2 exhibits a 100% electron mobility enhancement compared to SOI/HfO_2. Both ellipsometry measurements and coupling experiments proved that the thicknesses of the sSDOI and the SOI films are very close. Thus, only the strain can explain the 100% electron mobility enhancement. As expected, the hole mobility is not improved in sSDOI at high N_{inv} (Fig.9). This could be explained by quantum confinement effects [1] at high electric fields and for ultra-thin films. To solve this issue, we have successfully co-integrated compressive $Si_{0.6}Ge_{0.4}$ p-channels in the DCOI architecture.

Dual Channel (DCOI) with HfO_2/TiN gate stack

DCOI leads to a 100% hole mobility enhancement at high electric field, for $N_{inv}=7.10^{12}cm^{-2}$ (Fig.9).

The offset of both the conduction band in strained Si and the valence band in strained $Si_{0.6}Ge_{0.4}$ reduces the $|V_T|$ of both n&p MOSFETs on DCOI (Figs.10 and 11). This demonstrates the possible V_T matching for FDSOI undoped channel MOSFETs with a single TiN metal gate for HP applications. To highlight the intrinsic short channel DCOI mobility enhancement, the drain current must be V_T-normalized.

The DCOI exhibits about 50% Id_{lin} enhancement at L_G=75nm and 30% at 35nm (Fig. 12). Even at high V_D, 25% and 10% I_{ON} enhancement are demonstrated for such gate lengths (Fig.13). This is the first comparison ever been carried out between short channel nMOSFETs on sSDOI and SOI. Thanks to the $sSi_{0.6}Ge_{0.4}$ channel, the I_{ON} enhancement for pMOS is larger than previously reported gains coming from a substrate induced strain at such short gate lengths.

However, the Id_{lin} gain reduces with the gate down-scaling (Fig.12). The assumption of a strain relaxation for the shortest gate lengths is not coherent (i) with the parallelism of $V_{Tsat}(L_G)$ (that suggests a constant band offset with L_G) and (ii) with our experiments on sSDOI material stability. The Id_{lin} reduction is rather caused both by serial access resitances and differences in mobility limiting scattering mechanisms between long and short channels (Fig.14). Indeed, on bulk, the strained Si induced short channel I_{ON} enhancement was proved to be partly reduced by the scattering on pockets impurities and neutral defects [9]. In undoped short DCOI channels, low field mobility (μ_0, extracted as in [9]) saturates at low temperatures (Fig.14). This reveals an additional scattering mechanism which is dominant for the shortest gate lengths. This scattering could be due to long range interactions with S/D dopants [10] or to neutral crystalline defects in the shortest channels.

Conclusion

We have reported the succesful co-integration of dual strained-channels on fully depleted sSDOI CMOSFETs with HfO_2/TiN gate stack, down to 15nm gate length. 100% electron and hole mobility improvement has been achieved with 10% I_{ON} for 35 nm gate length. The V_T adjustement provided by the strain enables this device to be suitable for the SOI HP 32nm technology node.

Acknowledgements: This work has been carried out in the frame of CEA-Leti / ALLIANCE collaboration (Freescale, ST Microelectronics and Philips). A part of this work was supported by the RMNT Project "SMARTSTRAIN". The GIXRD measurements were carried out in the European Synchrotron Radiation Facilities. The authors would like to thank J. S. Micha, F. Triolet and C. Villeneuve for GIXRD and Raman technical support and Dr F. Ducroquet, C. Gallon, Dr S. Barraud, Dr. M. Kennard, Dr D. Bensahel and Dr T. Skotnicki for stimulating discussions and support.

References

[1] K. Rim et al., IEDM Tech. Dig., p. 49, 2003
[2] J. Jung et al., IEEE EDL, vol 25 n°8, p. 562, 2004
[3] T. Tezuka et al., Symp. VLSI Tech., p80, 2005.
[4] I. Åberg et al., IEDM Tech. Dig., p. 173, 2004.
[5] A. V. Y. Thean et al., VLSI Tech., p. 134, 2005.
[6] B. Ghyselen et al., Solid State Elec. 48, p. 1285, 2004.
[7] I. Cayrefourcq et al., SOI Tech.-XII, ECS Proc., 3, p. 191, 2005.
[8] O. Weber et al., in proc. of Symp. of VLSI Tech., p. 42, 2004.
[9] F. Andrieu et al., VLSI Tech., p.176, 2005.
[10] M. V Fischetti et al. , Ap. Phys. Let., vol 76, n°16, p. 2277, 2000.

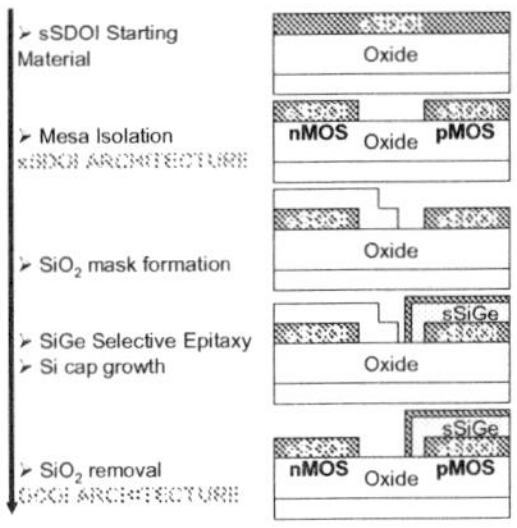

Fig.2: Scheme of the DCOI integration.

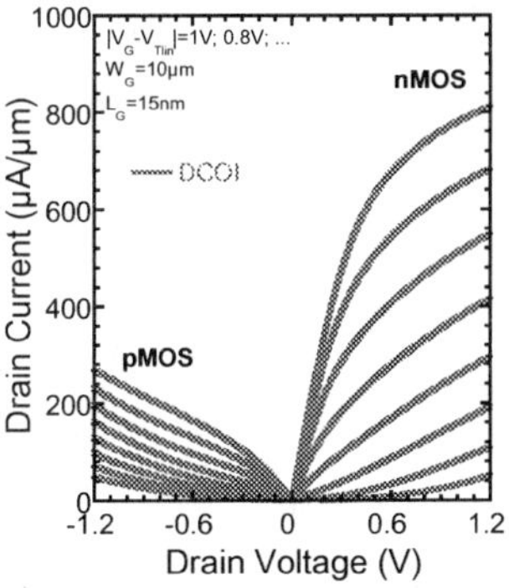

Fig.3: DCOI $I_D(V_D)$ at L_G=15nm.

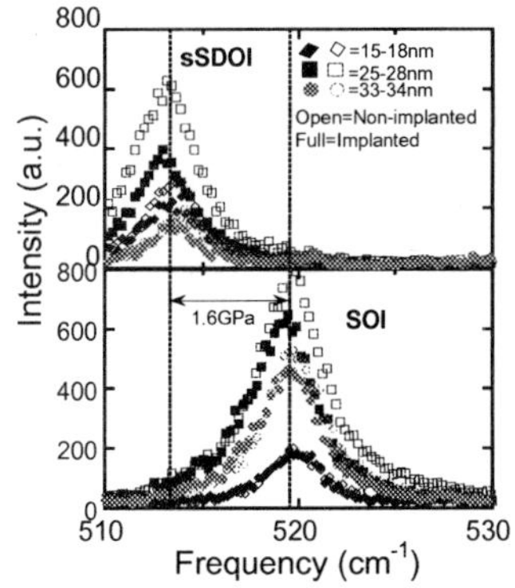

Fig.4: Raman spectra of implanted and not impl. sSDOI (up) and SOI (down).

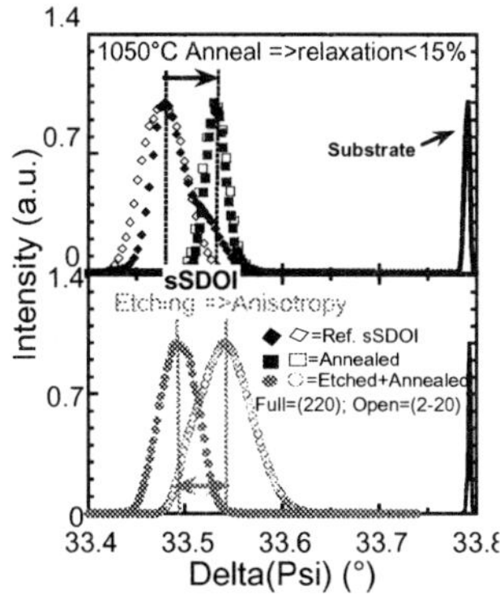

Fig.5: GIXRD of the (220) & (2-20) after anneal (up) & etching (down).

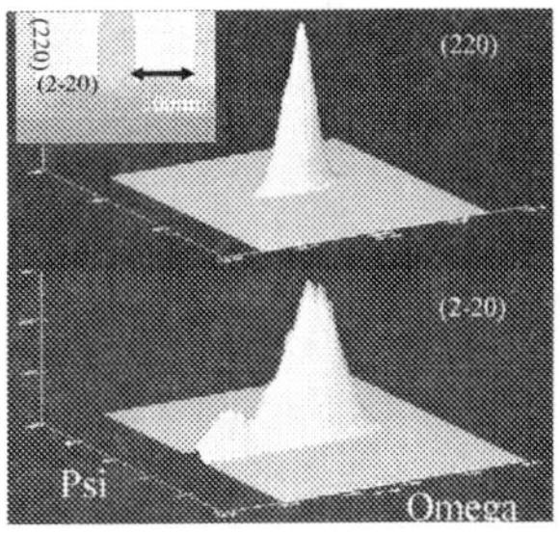

Fig.6: sSDOI (220) and (2-20) diffraction Bragg peaks of the <110> etched sample.

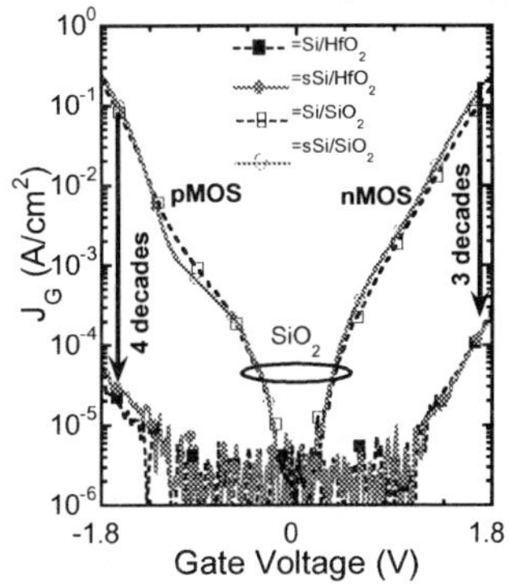

Fig.7: J_G vs. V_G for SiO_2 and HfO_2 devices.

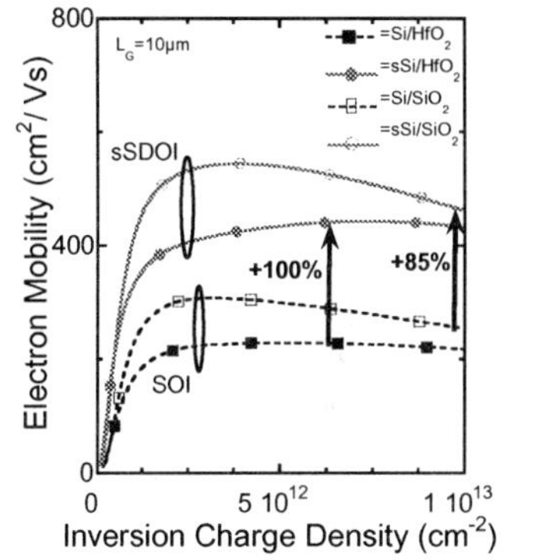

Fig.8: 100% electron mobility enhancement with sSDOI/HfO_2 for nMOSFETs.

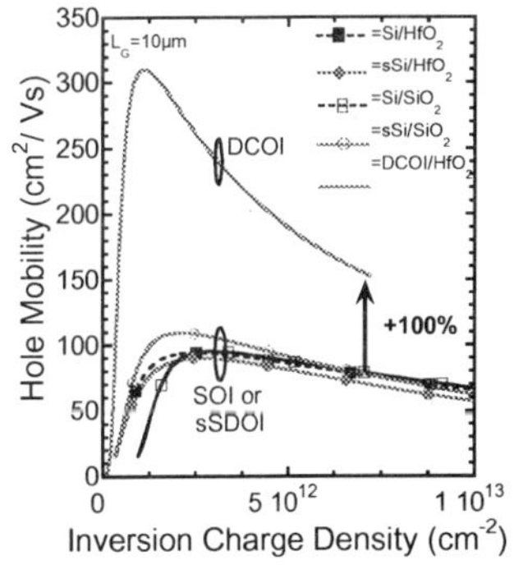

Fig.9: Hole mobility with DCOI, sSDOI and SOI.

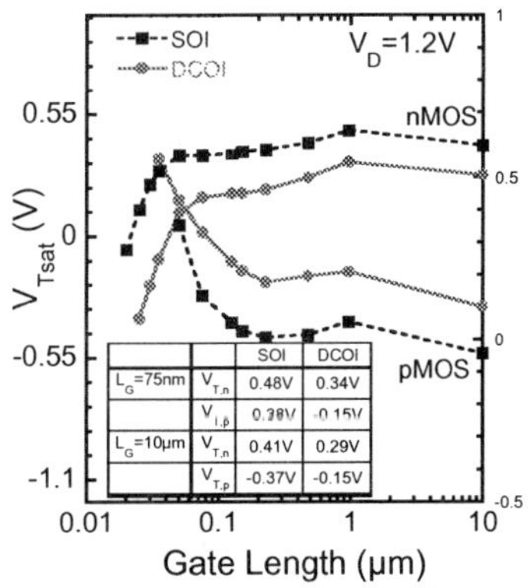

Fig.10: Saturation V_T values for n&pMOS DCOI and SOI. Insert: V_T @ V_D=50mV

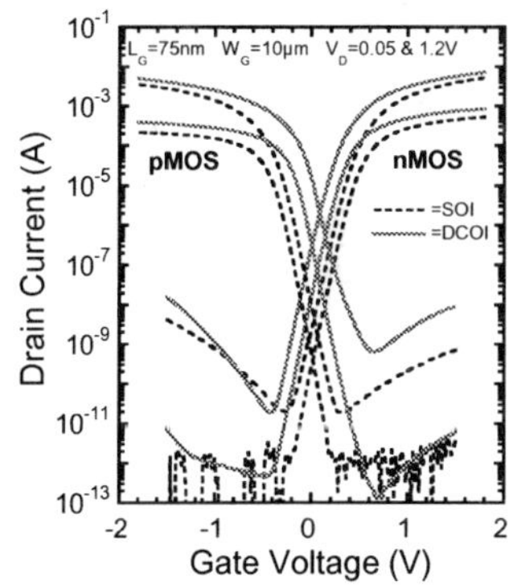

Fig.11: 75nm sub-threshold characteristics of DCOI.

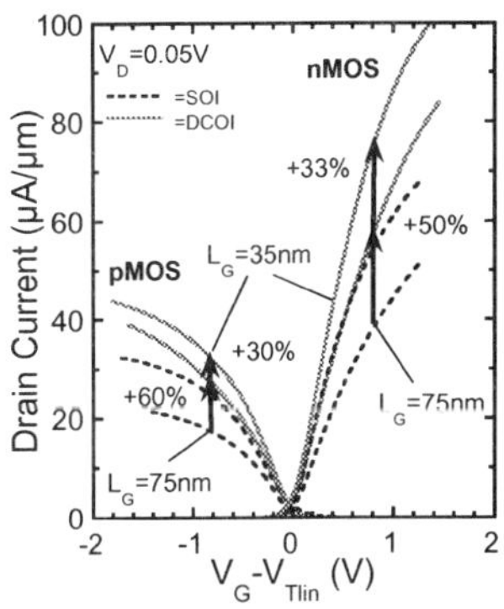

Fig.12: 75nm and 35nm $I_D(V_G-V_{Tlin})$ characteristics of DCOI at V_D=50mV..

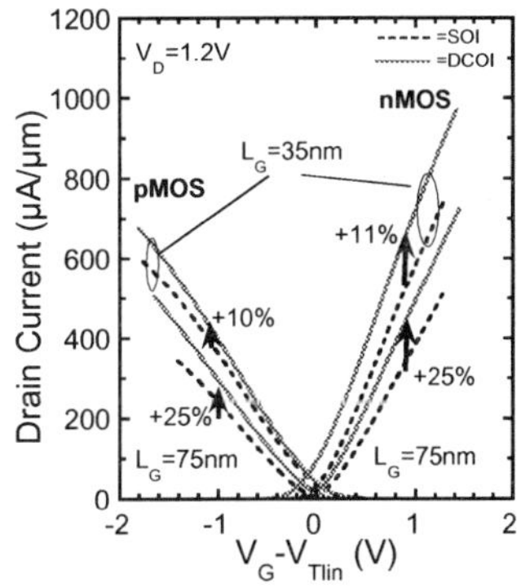

Fig.13: 75nm and 35nm $I_D(V_G-V_Tlin)$ characteristics of DCOI at V_D=1.2V.

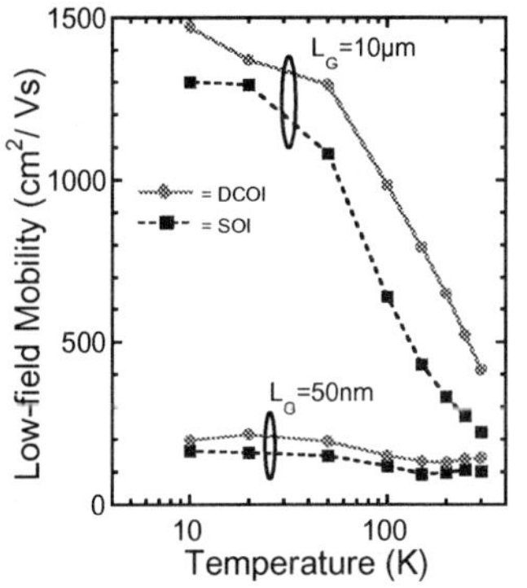

Fig.14: $\mu_0(T)$ for SOI and DCOI nMOSFETs at 10µm and 50nm gate lengths.

2005 IEEE International SOI Conference

Parasitic Source/Drain Resistance Reduction in N-channel SOI MuGFETs with 15nm Wide Fins

A. Dixit [(1,2)], K. G. Anil [(1)], N. Collaert [(1)], R. Rooyackers [(1)], F. Leys [(1)], I. Ferain [(1,2)], A. De Keersgieter [(1)], T. Y. Hoffmann [(1)], R. Loo [(1)], M. Goodwin [(3)], P. Zimmerman [(4)], M. Caymax [(1)], K. De Meyer [(1,2)], M. Jurczak [(1)], and S. Biesemans [(1)]

(1) IMEC, Kapeldreef 75, B-3001 Heverlee, Belgium (2) K. U. Leuven, ESAT-INSYS, Kasteelpark Arenberg 10, B-3001, Heverlee, Belgium
(3) Assignee of Texas Instruments Inc., Dallas USA at IMEC Belgium (4) Assignee of Intel Inc. USA at IMEC Belgium.
Email: Abhisek.Dixit@imec.be, Tel: +32 16 28 82 77

Abstract

We report on a set of process improvements leading to parasitic S/D resistance reduction in n-channel MuGFETs with 60 nm tall and 15 nm wide fins. We experimentally determine the contributions to the parasitic S/D resistance arising from the specific regions of the S/D geometry. A 50% reduction in the S/D resistance (from 1235 to 600 Ω-μm) is achieved by introducing various scaled processes, including EOT reduction, elevated S/D, undoped fins and removal of halos, reduction in S/D spacer width, and smaller gate lengths. The combination of these process enhancements leads to a 2X increase in I_{DSAT}, measured at constant I_{OFF} (=1nA/μm) and V_{DD}=1.3V.

Introduction

Multiple Gate Field Effect Transistors (MuGFETs) are considered as promising device architectures for the sub-45 nm CMOS applications [1]-[2]. Utilization of tall and narrow fins has been reported to help with achieving the target performance [3]. Mobility improvement and reduction of parasitic S/D resistance (R_{SD}) are reported to be the key issues for narrow fin MuGFET technologies [4]-[6]. Due to higher drain current densities than the PFETs, more degradation in performance of the NFETs can be expected in presence of a high R_{SD}. Various components of the parasitic S/D resistance in MuGFETs have been determined by combining the experimental data with device simulations. Key technological challenges for reduction of parasitic S/D resistance are outlined.

Simulations and Device Fabrication

Starting with a stack of 75 nm c-Si on 150 nm buried oxide (BOX), n-channel MuGFETs down to 50 nm physical gate lengths and 15 nm fin widths have been fabricated using 193 nm optical lithography. A summary of the process flow used for device fabrication is shown in Table 1. The cross-sectional TEM and SEM images of the fabricated devices are shown in Fig. 1 and 2 respectively. Results of electrical characterization on the NFETs with 15 nm wide, 60 nm tall, and 30 fins are reported in the following sections. Reported currents are normalized to the effective width of the devices, calculated as W_{EFF}=number of fins × (W_{FIN} + $2H_{FIN}$). Three-dimensional device simulations [7] have been performed on the device structures shown in Fig. 3. The total parasitic S/D resistance (R_{SD}) can be divided into components, as shown in Fig. 4 with the help of device simulations. Minimization of the various components of R_{SD} is discussed in the following three sections.

The S/D-Gate Overlapped Region

In the on state, existence of an accumulation layer in the gate overlapped extension regions interrupts the flow of majority carriers into the bulk of extensions. This gives rise to a gate dielectric thickness- and gate overlapped extension length-dependent component of R_{SD}, namely the accumulation resistance (R_{ac}) [6]. The NFET I_S-V_{GS} curves are shown in Fig. 5 for SiON-poly Si (SiON-Poly) and HfO_2-TiN (HK-MG) gate stacks, where the HK-MG gate stack has been used on undoped fins with respect to the reference SiON-Poly devices (see Table 1). Also, the S/D spacer bottom width has been scaled down by 25% for the HK-MG devices as compared to the reference SiON-Poly devices. Gate leakage current density for the long channel devices, as shown in Fig. 6, helps with verification of Equivalent Oxide Thickness (EOT) of the gate dielectrics for these narrow fin devices. A comparison between the two SiON-Poly stacks shows more $g_{m,MAX}$ improvement for the short-channel devices with 1.4 nm EOT, as shown in Fig. 7. This implies an R_{SD} improvement with EOT scaling, as R_{SD} is known to be the dominating parasitic for degradation in measured $g_{m,MAX}$ for the short channel devices [8]. Such a dependence of R_{SD} on EOT could be attributed to R_{ac}, as discussed earlier in this section. A comparison between HK-MG and 1.4 nm EOT SiON-Poly gate stacks shows an improvement in $g_{m,MAX}$ for HK-MG devices in Fig. 7. This gate length independent improvement can be attributed to lower equivalent oxide thickness of the gate dielectric in inversion (t_{inv}) for HK-MG devices due to the absence of poly-depletion in metal gates. This t_{inv} lowering leads to a better short channel control, as depicted by the HK-MG devices in V_T roll- off curves shown in Fig. 8. The total output resistance plotted as a function of the gate length in Fig. 9 shows a significant improvement (≈20%) in R_{SD} for 1.4 nm EOT SiON-Poly devices over those with 1.8 nm EOT, as is expected from Fig. 7. As compared to devices with SiON-Poly gate stack, the length of the accumulation layer in the S/D extensions is expected to increase for the HK-MG devices due to the absence of halos. However, the S/D spacer width scaling (reduction in sheet resistance, R_{sh} in Fig. 4) and lower t_{inv} for the HK-MG devices compensates for this increase in R_{ac}. Consequently, HK-MG devices achieve low R_{SD} values similar to those for 1.4 nm EOT SiON-Poly devices even in the absence of halos.

The S/D-Extension Region

The spreading and sheet-resistance (R_{sp} and R_{sh} in Fig. 4) exist in the extension region of the S/D. The former depends primarily on the shape of the extension doping profile, whereas the latter depends on the surface doping concentration of the extensions. The as-implanted Arsenic concentration profiles are shown in Fig. 10 for an extension cross-section. It can be seen from Fig. 10 that the use of 45^0 tilt in a 2-quadrant extension implant results in profiles with uniform concentration across the fin width as compared to the 10^0 tilt case. This increase in the conformality of extensions with 45^0 tilt can be seen to result in a lower R_{SD} (≈30%) in Fig. 11 and a corresponding increase in the V_T compensated gate overdrive-current (I_{od}) for the short channel devices in Fig. 12. However, for the narrow fin pitches, corresponding to high-density 45 nm CMOS applications, use of the large tilt implants results in dose loss on the fin sidewalls due to implant shadowing [2]. Thus, integration of a conformal doping process, such as the plasma doping (PLAD), to form S/D extensions could be foreseen to bring significant performance improvement for high-density MuGFET applications [9].

The S/D-Contact Region

The contact resistance in MuGFET S/D comprises of the S/D silicide thickness dependent contact resistance R_{conA} and the S/D-silicide contact area dependent resistance R_{conB} (see Fig. 4). In order to minimize R_{conB}, an SEG of Si on the S/D-HDD regions can be used to increase the S/D-silicide contact area [10]. A 40 nm SEG of Si, performed on the S/D-HDD regions, is shown to achieve a lower R_{SD} (≈15%) in Fig. 13. Approximately 25% improvement in I_{od} at a constant I_{OFF} (=1nA/μm) can be seen for the devices with SEG of Si on the S/D in Fig. 14. However, an increase in the as-deposited thickness of Nickel from 6 to 10 nm, prior to S/D silicidation, shows an increased R_{SD} (≈25%) in Fig. 15. Consequently, a decrease in I_{od} (<10%) at a constant I_{OFF} (=1nA/μm) can be seen in Fig. 16 for the devices with 10 nm as-deposited Ni. This degradation in performance for devices with 10 nm as-deposited Ni can be related to full silicidation of the S/D, which isolates S/D-HDD regions from rest of the device, as can be seen from Fig. 3. Avoiding full silicidation of the S/D regions in narrow fins could be a challenging task and would hinder the minimization of R_{conB}. As an alternative, R_{conA} can be minimized in devices with fully silicided S/D (see Fig. 3a) by introducing novel silicides with a lower contact resistivity to S/D-extensions [12]-[13], as can be seen from Fig. 17, where 3-D device simulations are reported. It can also be seen from Fig. 17 that the benefit of SEG is reduced as the silicide contact resistivity decreases. This implies that integration of novel silicides in MuGFETs is beneficial. The relative improvement in I_{ON} resulting from the contact resistivity reduction is quantified in Fig. 18. Fig. 19 summarizes the relative impact of the various scaled processes on the I_{od} and R_{SD}.

Conclusion

A 22% downscaling of the gate dielectric thickness is shown to result in 20% reduction in R_{SD} due to lower accumulation resistance. Removal of halos without degrading the accumulation resistance has been achieved by a 25% reduction in the S/D spacer bottom width and use of a lower t_{inv} through integration of HfO_2-TiN gate stack. More than 30% decrease in the extension resistance is achieved by increasing the implantation tilt angle from 10^0 to 45^0. A raised S/D-SEG process is shown to reduce R_{SD} by 15%. Highest reduction in the parasitic S/D resistance is shown to result from low contact resistivity-novel silicides and conformal S/D-extensions, which can be the main process enhancements, required in order to meet the ITRS target parasitic S/D resistance (R_{SDN}=110 Ω-μm) for the 45 nm technology node.

References

[1] Watanabe et al, IEEE Tran. Elec. Dev., 50(10), pp. 2073-80, 2003.
[2] Nackaerts et al, IEDM 2004, pp. 269-72.
[3] Yang et al, IEDM 2002, pp. 255-58.
[4] Yang et al, VLSI Symp. 2003, pp. 137-38.
[5] Verheyen et al, VLSI Symp. 2005, pp. 194-95.
[6] Dixit et al, IEEE Trans. Elec. Dev., 52(6), pp. 1132-40, 2005.
[7] ISE-TCAD, Release 9.5, 2005.
[8] Chung et al, IEEE Tran. Elec. Dev., 36(1), 1989.
[9] Goodwin et al, submitted to IEDM 2005.
[10] Kedzierski et al, IEEE Trans. Elec. Dev., 50(4), pp. 952-58, 2003.
[11] Dixit et al, to be published in ESSDERC, Groneble France, September, 2005.
[12] Zhu et al, IEEE Elec. Dev. Letters, 25(8), pp. 565-67, 2004.
[13] Varahramyan et al, Solid-State Electronics, 39(11), pp. 1601-07, 1996.

Process flow
Fin Patterning
Corner rounding in H_2 ambient
Fin implants
Gate stack deposition and patterning
Extensions and *Halos*
Source/Drain Spacer formation
SEG of Si on the S/D regions
HDD implantation
Spike anneal at 1050 °C
Silicidation using Ni

Table 1: Process flow used for the device fabrication. The HfO_2-TiN devices have been fabricated with a reduced S/D spacer width and without the processes listed in italics.

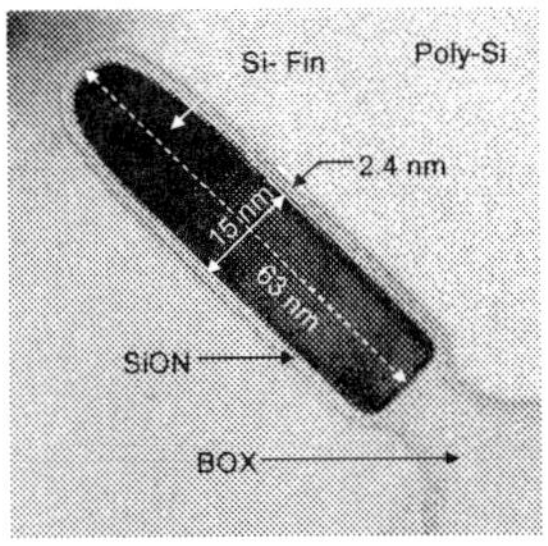

Figure 1: XTEM image of a fin and SiON-Poly gate stack, showing rounded fin corners, vertically straight fin sidewalls, and a controlled BOX under-etch.

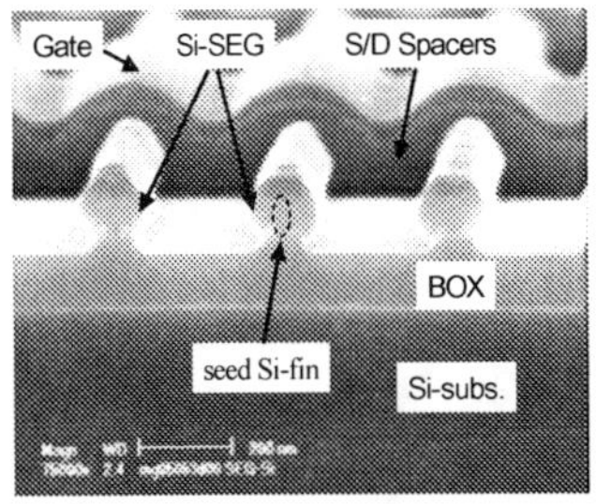

Figure 2: XSEM image of fins after the S/D SEG process. Due to crystal orientation dependence of the Si growth rate (faceting), the resulting S/D regions have a diamond like shape.

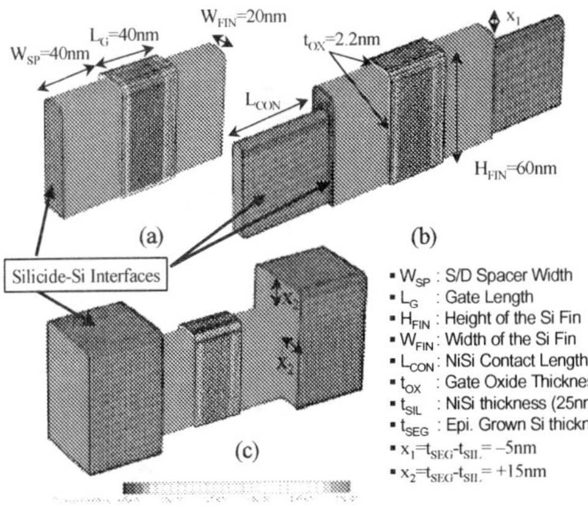

Figure 3: Simulated device structures having silicide on the S/D (a) without an SEG layer (b) with a thin SEG layer (c) with a thick SEG layer. The S/D contacts have been simulated as distributed resistance. Abrupt, box-shaped doping profiles have been used. The S/D spacers are not shown.

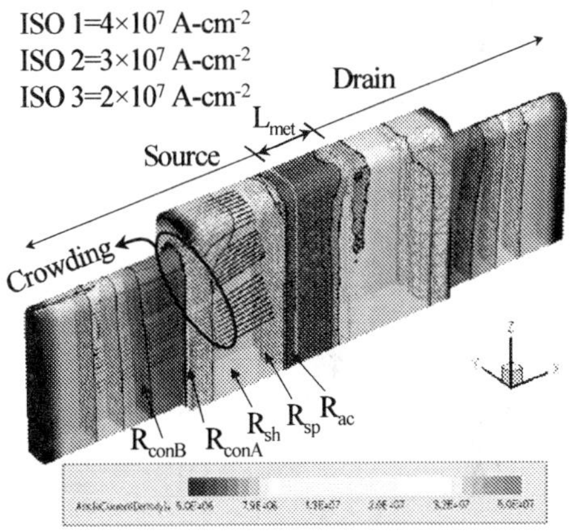

Figure 4: Simulated electron current density for the device (Fig. 3b) in saturation. Equi-current density surfaces (ISO) are shown to help with locating components of R_{SD}. Current stream traces are shown in the current crowded region.

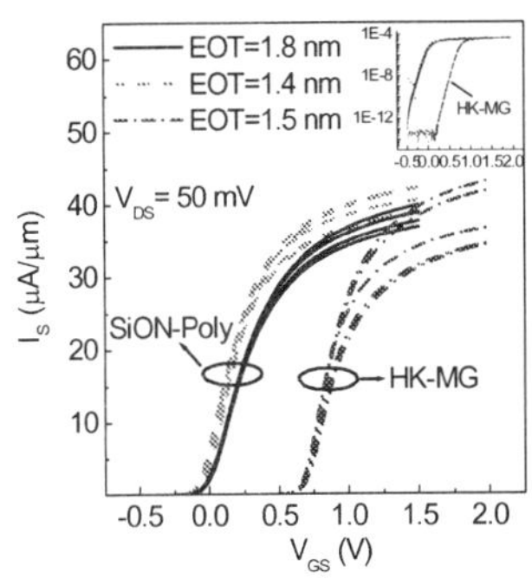

Figure 5: Measured I_S-V_{GS} curves for 90 nm gate length devices. Inset shows the plot in semi-log scale. A V_T-shift of 600 mV and an identical sub-threshold swing for HK-MG devices is shown.

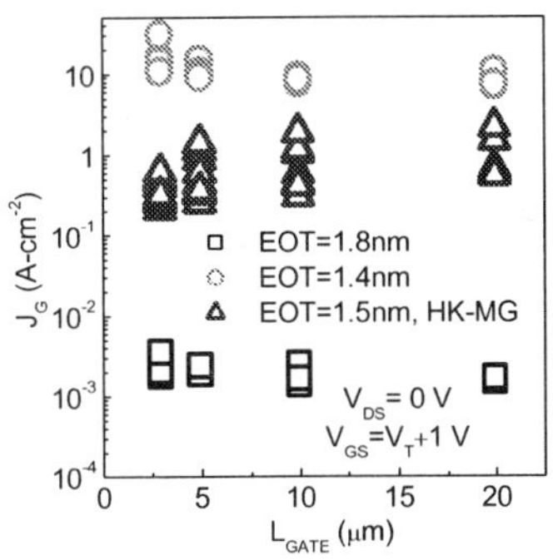

Figure 6: Measured gate leakage density for the long channel devices. J_G trends correlate well with the EOT values for these gate stacks, as calibrated on the planar devices.

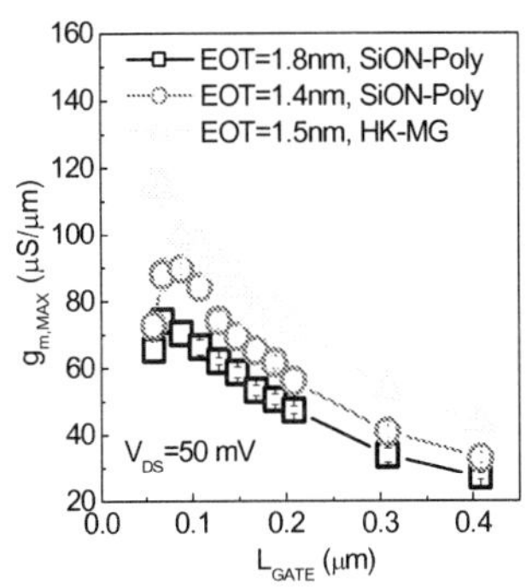

Figure 7: Extracted peak trans-conductance as a function of gate length for the three gate stacks.

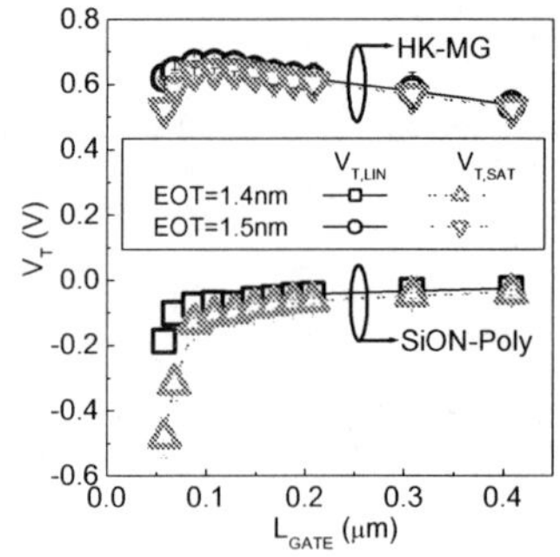

Figure 8: Measured threshold voltage as a function of the gate length. No degradation in the short-channel performance is seen for the HK-MG devices.

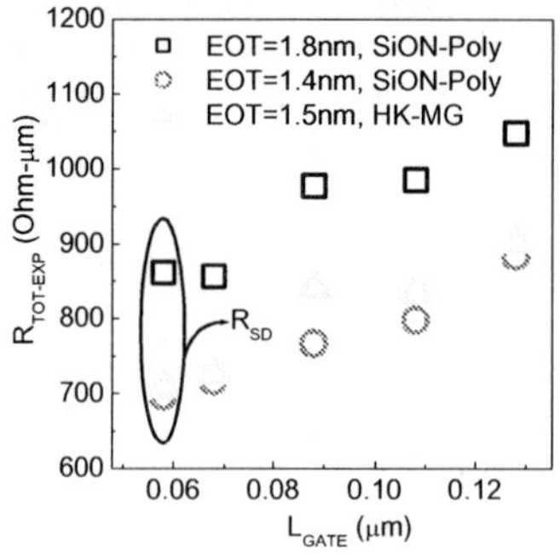

Figure 9: Extracted total output resistance ($=V_{DS}/I_D$), extrapolated to V_{GS}=5V. Resulting values at short gate lengths converge to the S/D resistance (R_{SD}).

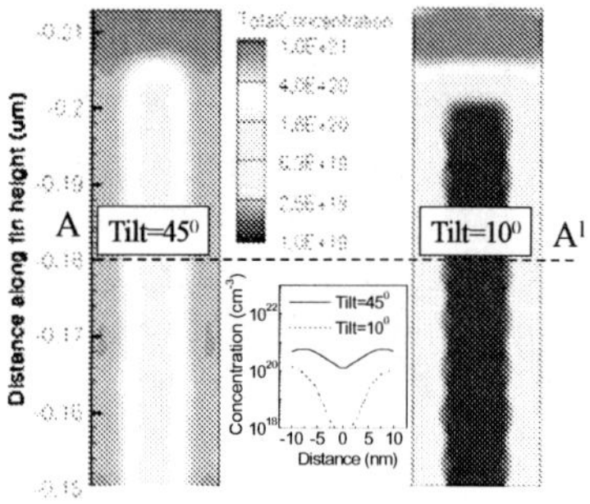

Figure 10: 2-D process simulation of as-implanted Arsenic concentration, comparing implantation tilts for isolated 20 nm wide, 63 nm high fins. Inset shows dopant profiles along the line A-A'.

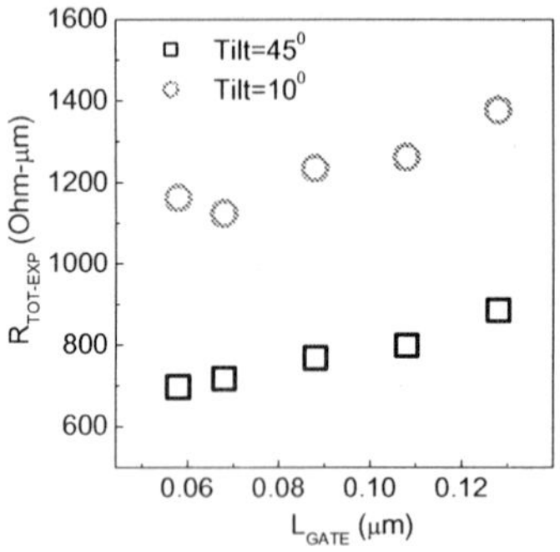

Figure 11: Extracted $R_{TOT-EXP}$ vs. L_{GATE}. A comparison of extension implantation tilt angles is shown for NFETs with SiON-Poly gate stack.

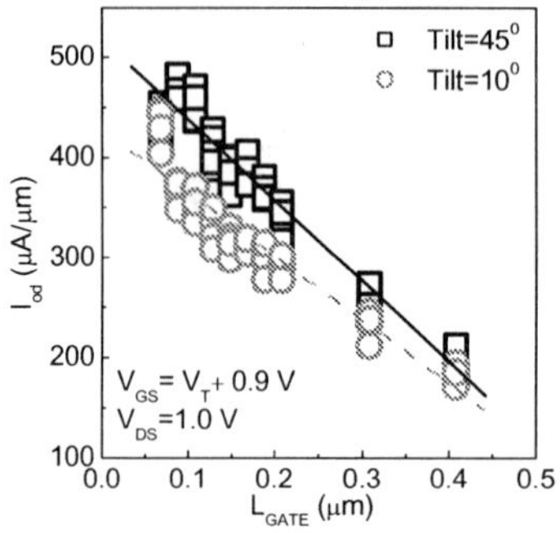

Figure 12: Measured I_{od} vs. L_G for NFETs with SiON-Poly gate stack, showing the necessity for conformal S/D extensions.

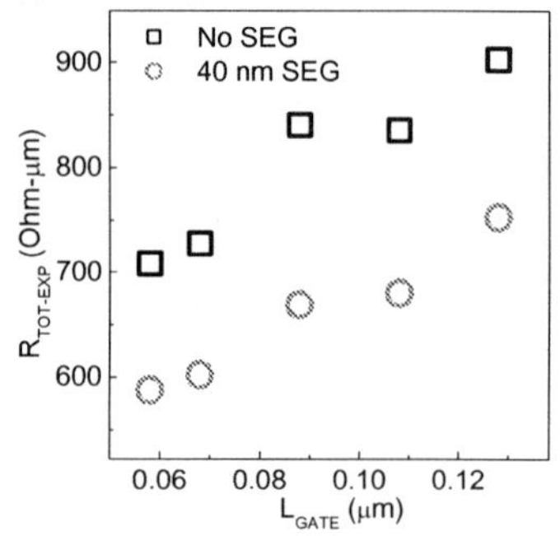

Figure 13: Extracted $R_{TOT-EXP}$ vs. L_{GATE} of NFETs with HK-MG gate stack. A 6 nm Ni thickness was used for S/D silicidation.

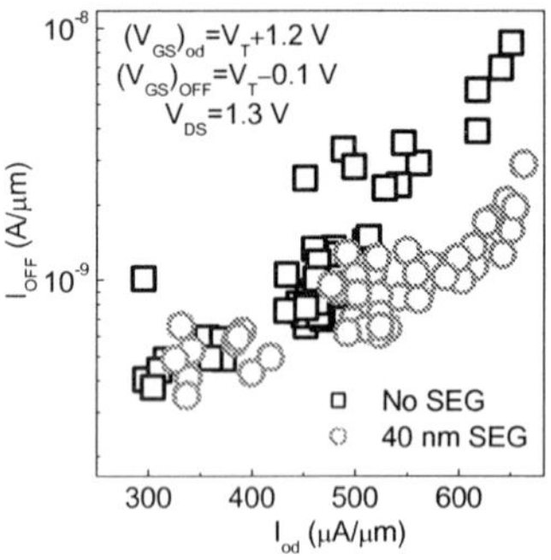

Figure 14: Measured I_{OFF} vs. I_{od} for 40 nm SEG and the no-SEG reference for NFETs with HK-MG gate stack. A 6 nm Ni thickness was used for S/D silicidation.

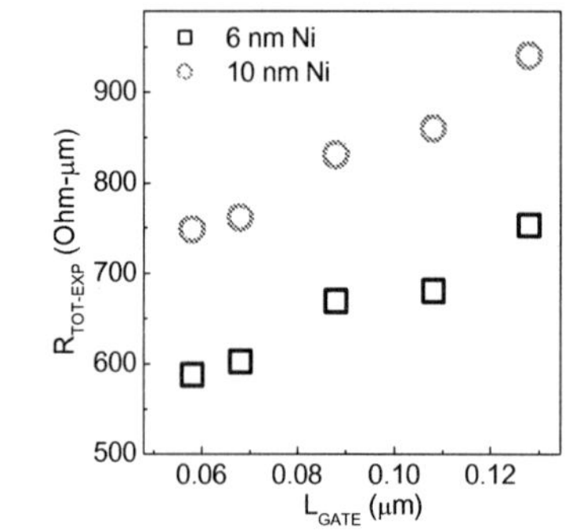

Figure 15: Extracted $R_{TOT-EXP}$ vs. L_{GATE}. A 40 nm SEG of Si was performed on these NFETs with HK-MG gate stack, prior to the S/D silicidation.

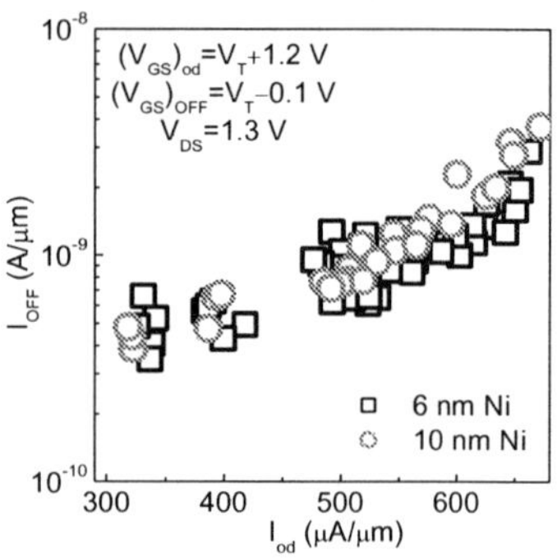

Figure 16: Measured I_{OFF} vs. I_{od}. A 40 nm SEG of Si was performed on these NFETs with HK-MG gate stack, prior to the S/D silicidation.

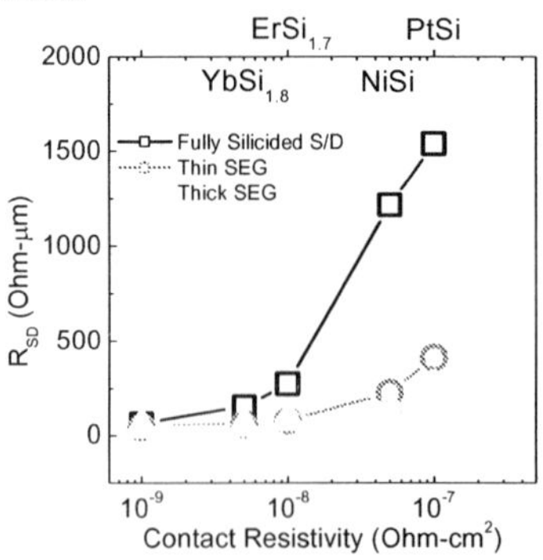

Figure 17: 3-D device simulation of R_{SD} as a function of silicide contact resistivity for the device structures shown in Fig. 3. Abrupt, box-shaped extensions with 3E20 cm^{-3} uniform doping have been used.

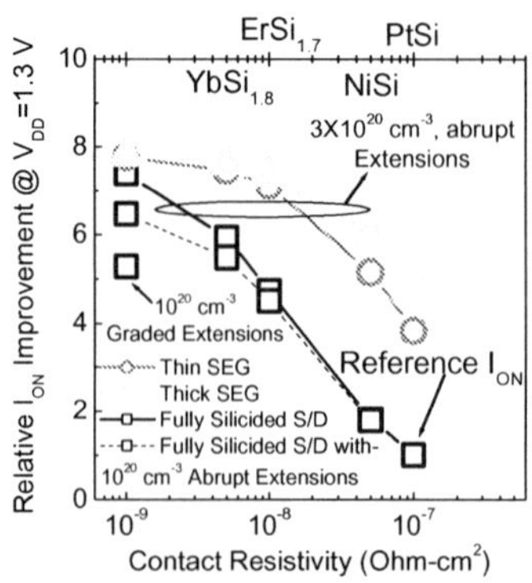

Figure 18: 3-D device simulation of Relative I_{ON} improvement vs. contact resistivity. Novel silicides are listed on the top X-axis for the corresponding contact resistivities. Both abrupt and graded extensions have been used.

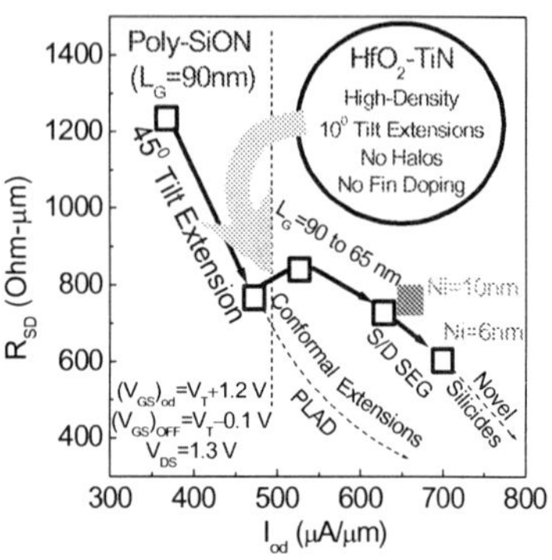

Figure 19: Extracted R_{SD} vs. measured I_{od} at a constant I_{OFF}=1 nA/μm for the bias conditions shown in the figure.

L

M

N

O

P

Q

R

S

T

U

V

W

X

Y

Z